ATLAS DETAILS IN LANDSCAPE DESIGN

景观细部设计图集 I

◎主编　刘少冲　王博　卢良

水景 | 景墙 | 栏杆、围墙 | 桥

中国林业出版社

图书在版编目（CIP）数据

景观细部设计图集 . Ⅰ / 刘少冲，王博，卢良主编 . -- 北京：中国林业出版社，2016.5
ISBN 978-7-5038-8493-1

Ⅰ . ①景… Ⅱ . ①刘… ②王… ③卢… Ⅲ . ①景观设计—细部设计—图集 Ⅳ . ① TU-856

中国版本图书馆 CIP 数据核字 (2016) 第 082900 号

中国林业出版社 · 建筑家居出版分社
责任编辑：李 顺 唐 杨
出版咨询：（010）83143569

出 版：中国林业出版社（100009 北京西城区德内大街刘海胡同 7 号）
网 站：http://lycb.forestry.gov.cn/
印 刷：北京卡乐富印刷有限公司
发 行：中国林业出版社
电 话：（010）83143500
版 次：2016 年 6 月第 1 版
印 次：2016 年 6 月第 1 次
开 本：889mm × 1194mm 1 / 16
印 张：17.625
字 数：200 千字
定 价：168.00 元

本书编委会

主　　编：刘少冲　王　博　卢　良

副 主 编：张大海　尹立娟　郭　超　杨仁钰　廖　炜

编委人员：郭　金　王　亮　文　侠　王秋红　苏秋艳　孙小勇　王月中　周艳晶　黄　希　朱想玲　谢自新　谭冬容　邱　婷　欧纯云　郑兰萍　林仪平　杜明珠　陈美金　韩　君　李伟华　欧建国　潘　毅

支持单位：北京筑邦园林景观工程有限公司

北京久道景观设计有限责任公司

原朴建筑园林设计工程有限公司

三河市草木生园林养护有限公司

《世界园林》杂志

《新楼盘》杂志

CONTENT 目录

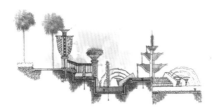

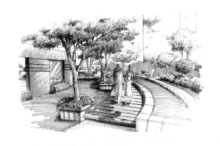

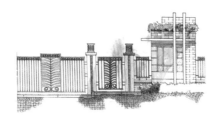

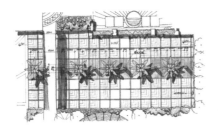

1 水景

一、水景概述

水体景观越来越多的应用于园林景观设计中。通过大量的水景设计观察，从多个角度论述了水景设计形态和理念的多样性和丰富性，总结出当代城市设计中的水景设计的三大原则，并着重强调了水景设计在城市设计中的重要地位。在我国城市应用水景，应重点注意节水问题。为此，各个城市要根据本地区的缺水现状和气候条件进行水景观设计，节约用水，做合理到位的水景。

二、水景在园林设计中的作用

在园林水景规划设计中，水景已占据了很重要的地位，它具有水的固有的特性，表现形式多样，易与周围景物形成各种关系。它具有灵活、巧于因借等特点、能起到组织空间、协调水景变化的作用；更能明确游览路线、给人明确的方向感。所以说，分析水景的特性，明确水景的作用，了解水景的设计形式，利用水景和各种景观元素的关系，以表达设计的意图，是值得我们耗费精力、气力去追求、探讨的课题。

三、水的形态

1. 静水

宁静、轻松而平和，常见于湖泊、水池和水塘中。园林中常用宁静水面形成具凝聚力的空间，映射天空和周围景物。

2. 动水

激越，如流动的溪流、水道、水涧、瀑布、水帘、壁泉以及喷涌的喷泉等。水的一般流动规律是由动到静，即从喷涌的源头到流动的溪流河段，最后汇集起来而趋于平静。

四、水景的种类

现代景观设计中，水景按建造方式的不同可分为天然水景和人工水景。天然水景是以江河湖海等自然水资源为背景的人文环境，人工水景则是以水为主体的人工构筑物。天然水景讲求借景，以观赏为主。构思中借助自然地形，顺应环境设置景观。人工水景按照其设计手法又可分为几何形式和仿天然式人工湖。仿天然式人工湖所表现出的景观特征应该是纯天然的风貌，人造水景忌讳显露人造痕迹。其意图是以人工手段仿造天然湖的效果。平面形式以天然曲线为主，结合周边地形进行设计。设计时应考虑多设置大小不一的天然沿河石块。河岸为自然式的斜坡。沿河多以水生植物为主。而水景常见的形式有四种：静水、流水、落水和喷水。

◆**静水**

静水是现代水型设计中最简单、最常用又最宜取得效果的一种水景设计形式。室外筑池蓄水，或以水面为镜，倒影为图，作影射景；可赤鱼戏水，水生植物飘香满地；可池内筑山、设瀑布及喷泉等各种不同意境的水式，使人浮想联翩，心旷神怡。水池的设计主要讲究平面形式的变化，可方，可圆，或曲折成自然形。静水的设计类型可分为规则式水池和自然式水池（湖或塘）。

◆**流水**

流水主要包括自然溪流、河水和人工水渠、水道等。在水景设计中主要通过控制水量、水深、水宽的大小及流水的形状和在流水中设置主景石来设计流水的效果及引导景观的变化。

◆**落水**

常见的有瀑布、水帘、叠水等。人工瀑布设计时要估计水量的大小，如果瀑身的水量不同，就会营造出不同的气势。瀑布本是一种自然景观，是河床陡坎造成的，水从陡坎处滚落下跌形成气势恢宏的景观。瀑布可分为面形和线形。不同的形式表达不同的感情。人在瀑布前，不仅希望欣赏到优美的落水形象，而且还喜欢听落水的声音。

◆**喷水**

喷泉是喷水的主要形式之一，也是城市动态水景的重要组成部分，常与声效、光效配合使用，形式多种多样。喷泉的设计千变万化，分布极广，从城市广场到街道小区，从公共场所到私家花园，都可以发现喷泉的存在，喷泉越来越成为人们所喜爱的水景构造。

五、水景设计的基本准则

1. 宜“活”不宜“死”的原则

城市有了水，就有了生机，而可以流动的活水可以带给城市灵气与活力。如果将城市水系比喻为城市的血脉，那么流动的城市水系就是保证城市血液流动的基本条件，城市血脉流动和更新又是保证城市肌体健康的前提。

2. 宜“弯”不宜“直”的原则

河流的自然性、多样性弯曲是河流的本性，所以设计水体时，要随弯就弯，不要裁弯取直。河流纵向的蜿蜒性，形成了急流与缓流相间；深潭与浅滩交错；天然河道没有一条是笔直的，如果修建一条笔直、而且等宽的河道，它势必等速，等速的河道里水生动植物难以生长。只有蜿蜒曲折的水流才有生气、灵气。尽量避免直线段太长，能弯则弯，用蜿蜒、蛇形、折线等代替直线；在河道转弯时，也不要用一个半径去完成转弯，尽量多一些变化，甚至弧线、折线共用，这样做不但有其美学价值，而且在水文学和生态学方面有其独特的功能。

3. 虚实结合的原则

“仁者乐山，智者乐水”，“上善若水。水善利万物而不争，处众人之所恶，故几于道。”就是说，最高的善像水一样，水善利用万物而不与之相争。它甘心处在人不愿呆的低洼之地，很相似于道。“浊而静之徐清，安以重之徐生。”浑水静下来慢慢就会变清，安静的东西积累深厚会动起来而产生变化。水中有哲理，水中有道意，水中有禅味。

六、水景设计的手法

1. 形－静态水面划分

水的形态因水体的形状而定，风景园林中的静态湖面，多设置堤、岛、桥、洲等，目的是划分水面，增加水面的层次与景深，扩大空间感；或是为了增添园林的景致与趣味。城市中的大小园林多有划分水面的手法，且多运用自然式，只有在极小的园林中才采用规则几何式。

2. 光影因借

（1）倒影成双

四周景物反映水中形成倒影，使景物变一为二，上下交映，增加了景深，扩大了空间感，一座半圆洞的拱桥，可起到了功半景倍的作用。水中倒影是由岸边景物生成的，故园林水面旁，一定要精心布置各种景物，以获得双倍的光影效果，取得虚实结合，相得益彰的艺术效果。

（2）借景虚幻

由于视角的不同，岸边景物与水面的距离、角度和周围环境也不同。岸边景物设计，要与水面的方位、大小及其周围的环境同时考虑，才能取得理想的效果，这种借虚景的方法，可以增加人们的寻幽乐趣。

（3）动静相随

风平浪静时，湖面清澈如镜，微风送拂，送来细细的涟漪，为湖光倒影增添动感，产生一种朦胧美。若遇大风，水面掀起激波，倒影顿时消失。雨点则会使倒影支离破碎，形成另一种画面。

七、水景在园林中的作用

1. 背景作用

平静的水面，无论是规则式的，还是自然式的，都可以像草坪铺装一样，作为其他园林要素的背景和前景。同时，平静水面还能映照出天空和主要景物的倒影，如建筑、树木、雕塑和人。如果需要加强水面的反射功能，可以从以下几方面来考虑：首先，水面应布置在需要映射的景物之前，观景者于景物之间；而水面大小则取决于景物的尺度和所需映照的面积多少而定。另一应考虑的因素是水体的深度和水体表面色调。水面越暗越能增强倒影。要使水色深沉，可以增加水的深度，另外，也可在池壁和池底漆上深蓝色或黑色。再有，对于希望保持水面反射功能的水体，应保持水体清澈。

2. 纽带作用

在园林中，水体可以作为联系全园景物的纽带。例如，扬州瘦西湖的带状水面延绵数千米，众多景物或临水而建，或三面环水，水体使全园景物逐渐展开，相互联系，形成有机整体。而苏州拙政园中的许多单体建筑或建筑组群都与水有不可分割的联系，水面将不同的建筑组合成为一个整体，起到统一的作用。

3. 焦点作用

流动的水通常令人神往，如瀑布和喷泉激越的水流和声响引人注目，会成为某一区域的焦点。世界上有许多天然的著名瀑布，如尼亚加拉瀑布，约瑟米提公园的布里达尔维瀑布，中国贵州的黄果树瀑布等。天然喷泉比较罕见，因此黄石公园的老忠实喷泉非常著名。如果在园林中塑造瀑布飞流直下的景观，需用水泵将水打上瀑布顶部，然后水流再沿着人工堆叠的山石顺流而下。另外设和于城市环境的变形瀑布叫做水墙瀑布。如曼哈顿的帕利公园中的水墙瀑布。在城市环境中的人工喷泉是利用压力，使水自喷嘴向空中，达到一定高度后再回落到地面。人工喷泉的喷水形式非常丰富，有些喷泉还配以音乐和灯光，因此成为布局中的焦点。另外，水帘、壁泉也常常作为焦点水景来布置。

在水资源日益缺乏的今天，如何去营造宜人的水景，如何去满足人们亲水的这种需求，成为摆在我们设计师面前一个很重要的问题。

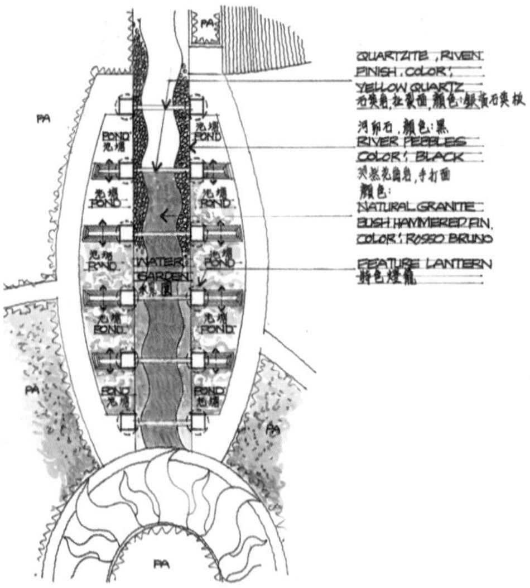
QUARTZITE, RIVEN FINISH, COLOR: YELLOW QUARTZ
河卵石,颜色:黑
RIVER PEBBLES COLOR: BLACK
天然花岗岩,手打面 颜色:
NATURAL GRANITE BUSH HAMMERED FIN. COLOR: ROSSO BRUNO
FEATURE LANTERN
特色灯笼
WATER GARDEN
POND

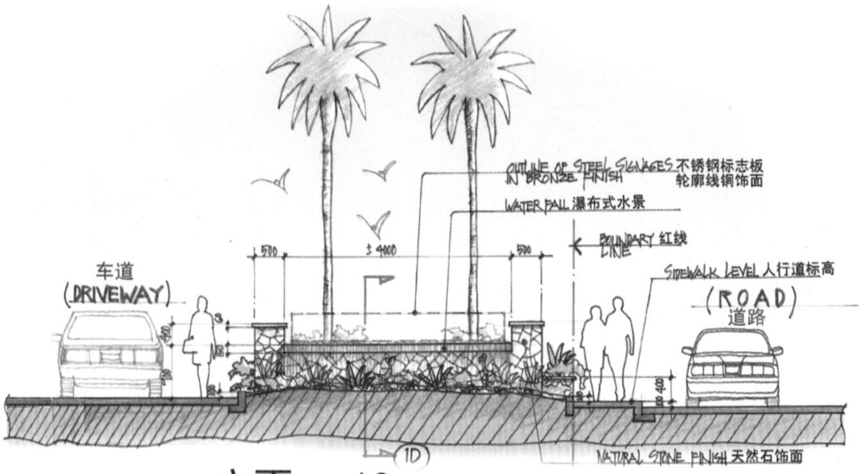
OUTLINE OF STEEL SIGNAGES IN BRONZE FINISH 不锈钢标志板 轮廓线铜饰面
WATER FALL 瀑布式水景
BOUNDARY LINE 红线
SIDEWALK LEVEL 人行道标高
车道 (DRIVEWAY)
(ROAD) 道路
500
4000
500
NATURAL STONE FINISH 天然石饰面
1D

立面 - 1C (标志/水景)

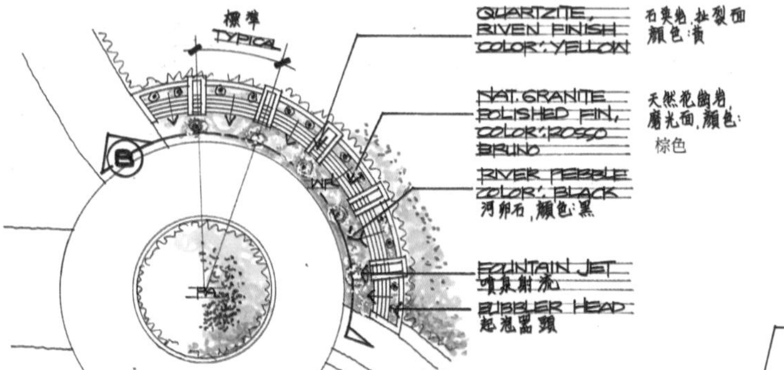
標準 TYPICAL
QUARTZITE, RIVEN FINISH COLOR: YELLOW
石英岩 拉裂面 颜色:黄
NAT. GRANITE POLISHED FIN. COLOR: ROSSO BRUNO
天然花岗岩 磨光面,颜色: 棕色
RIVER PEBBLE COLOR: BLACK
河卵石,颜色:黑
FOUNTAIN JET
喷泉射流
BUBBLER HEAD
起泡喷头

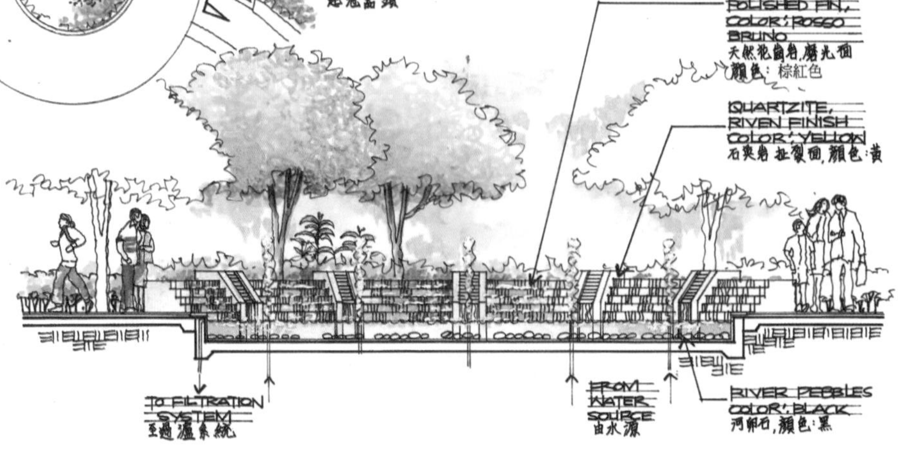
NATURAL GRANITE POLISHED FIN. COLOR: ROSSO BRUNO
天然花岗岩,磨光面 颜色:棕红色
QUARTZITE, RIVEN FINISH COLOR: YELLOW
石英岩 拉裂面,颜色:黄
TO FILTRATION SYSTEM
至过滤系统
FROM WATER SOURCE
由水源
RIVER PEBBLES COLOR: BLACK
河卵石,颜色:黑

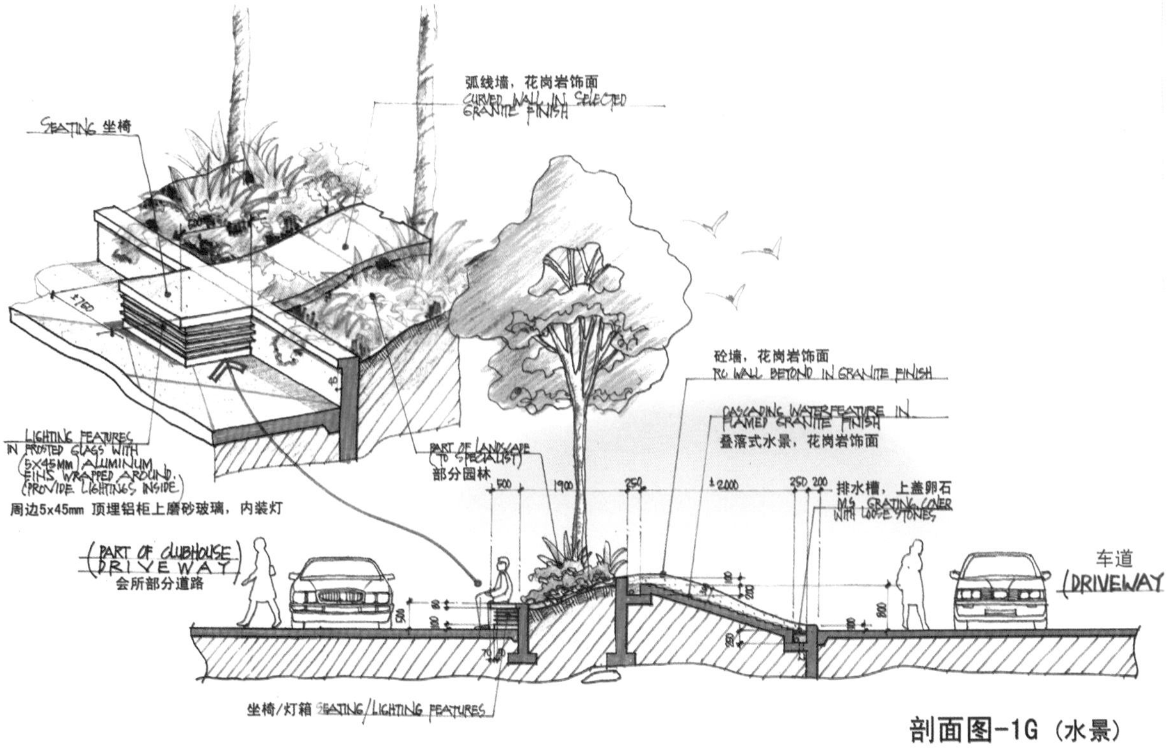
弧线墙，花岗岩饰面
CURVED WALL IN SELECTED GRANITE FINISH
SEATING 坐椅
砼墙，花岗岩饰面
RC WALL BEYOND IN GRANITE FINISH
CASCADING WATERFEATURE IN FLAMED GRANITE FINISH
叠落式水景，花岗岩饰面
LIGHTING FEATURES IN FROSTED GLASS WITH (5X45MM) ALUMINUM FINS WRAPPED AROUND. (PROVIDE LIGHTINGS INSIDE.)
周边5x45mm 顶埋铝柜上磨砂玻璃，内装灯
PART OF LANDSCAPE (TO SPECIALIST)
部分园林
排水槽，上盖卵石
MS GRATING COVER WITH LOOSE STONES
PART OF CLUBHOUSE DRIVEWAY
会所部分道路
车道
DRIVEWAY
坐椅/灯箱 SEATING/LIGHTING FEATURES

剖面图-1G（水景）

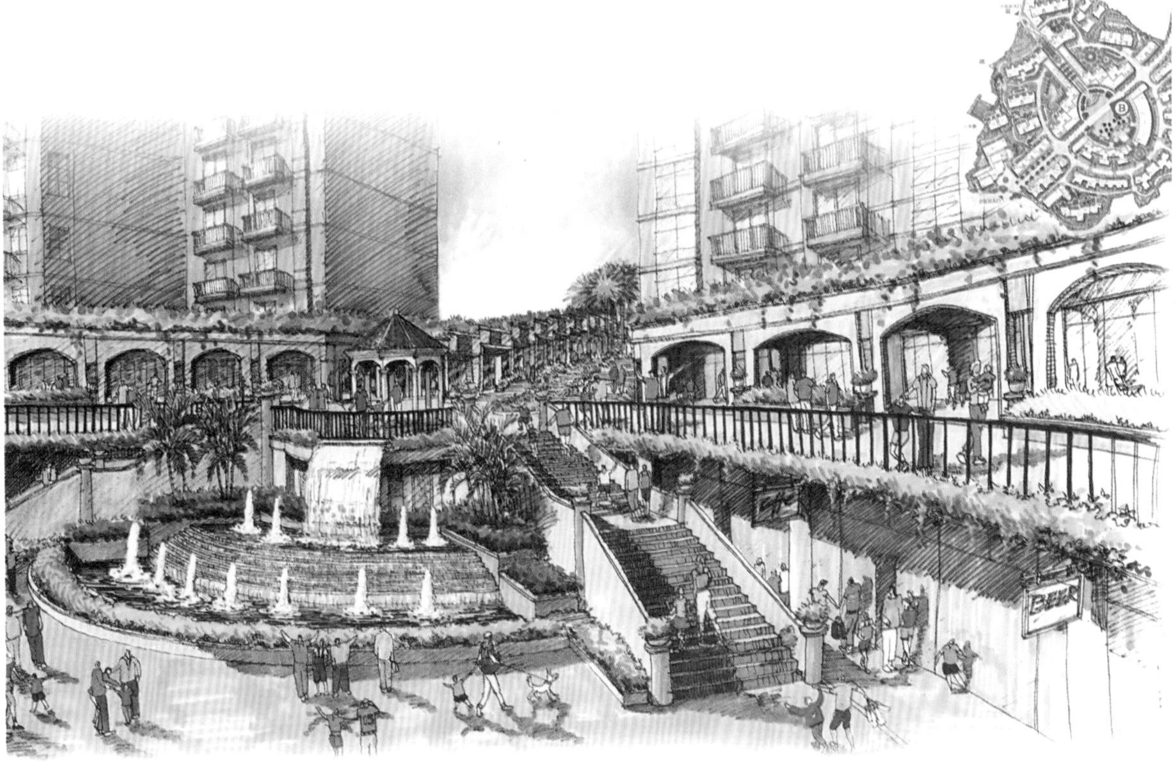

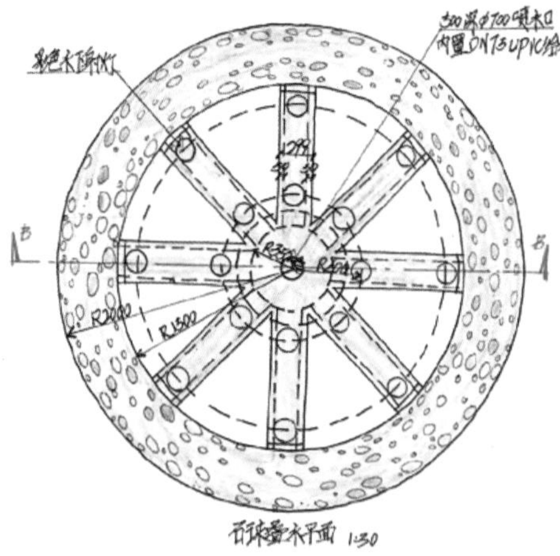

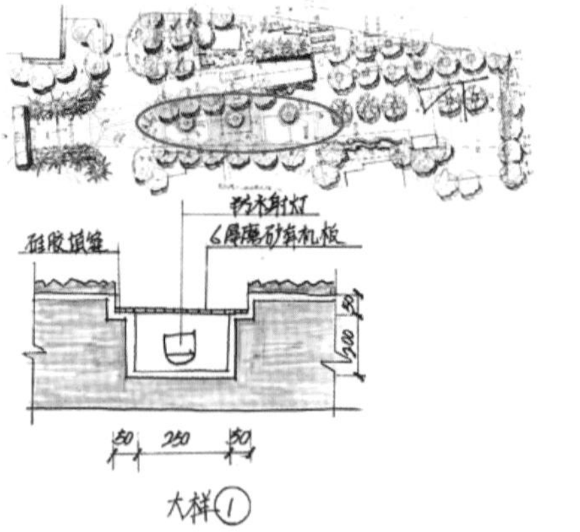
大样①

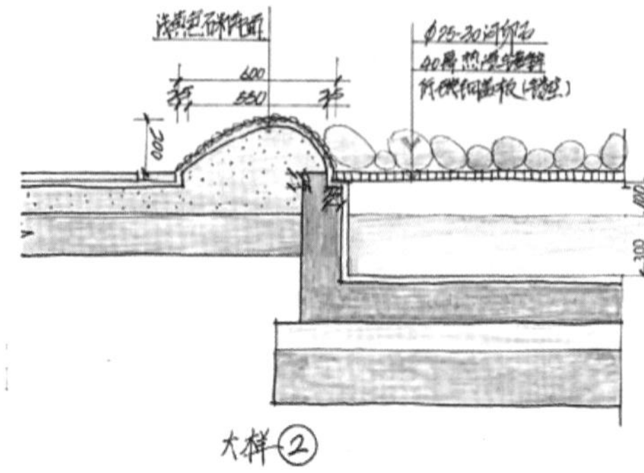
大样②

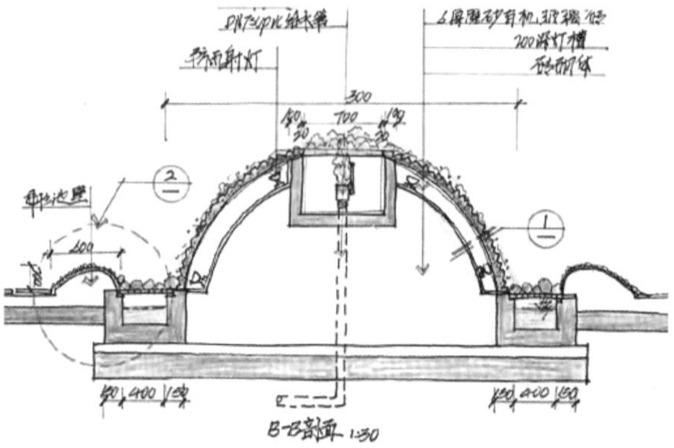
B-B剖面 1:30

会所街跌水喷泉透视图
锦绣华城

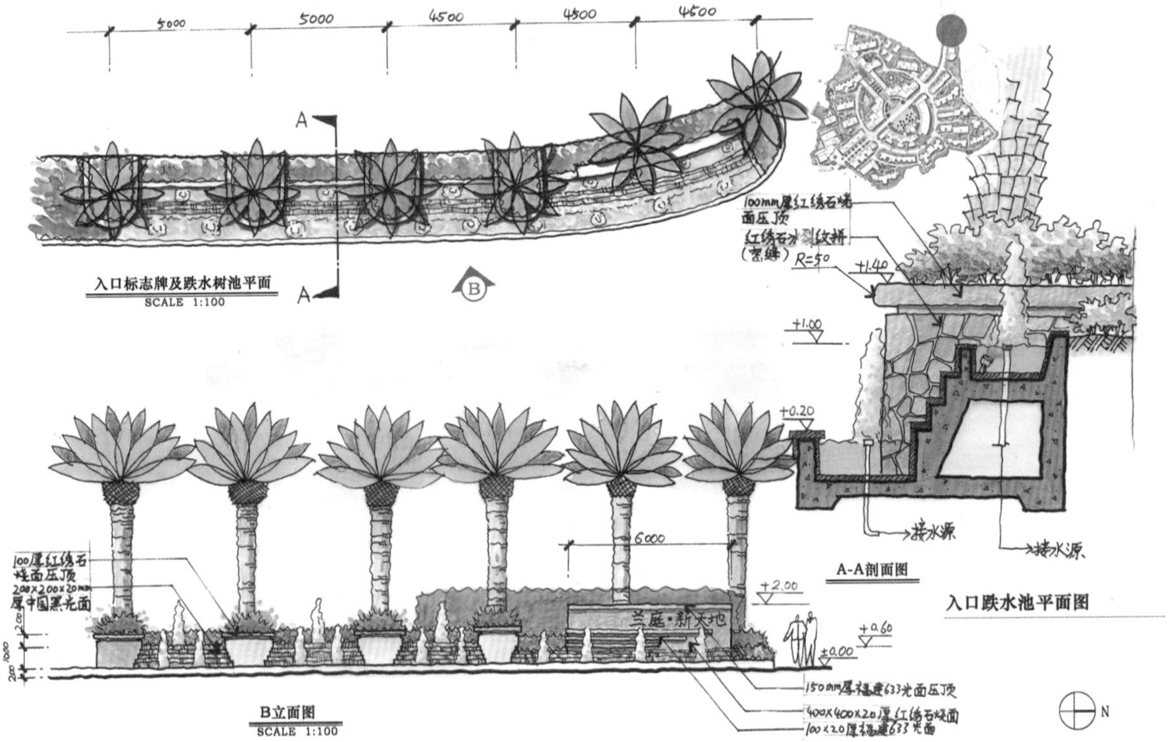
5000
5000
4500
4500
4500
A
A
入口标志牌及跌水树池平面
SCALE 1:100
B
100mm厚红绣石烧面压顶
+1.40
+1.00
+0.20
接水源
A-A剖面图
入口跌水池平面图
6000
+2.00
兰庭·新天地
+0.60
±0.00
150mm厚福建633光面压顶
B立面图
SCALE 1:100
N

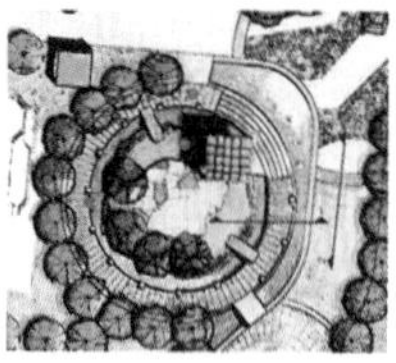

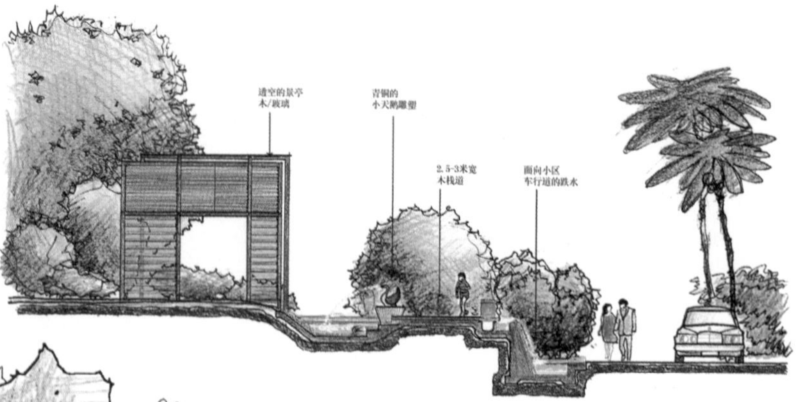
通空的景亭
木/玻璃
青铜的
小天鹅雕塑
2.5-3米宽
木栈道
面向小区
车行道的跌水

剖面图-2

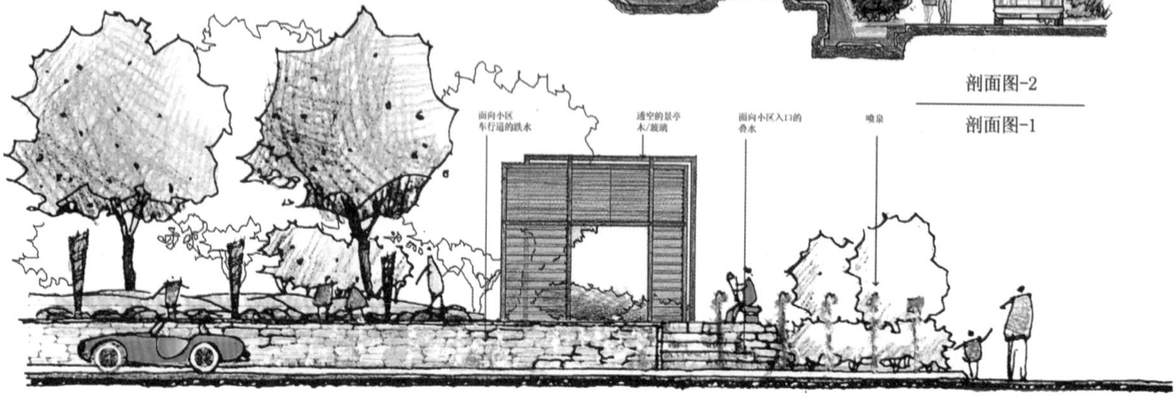
面向小区
车行道的跌水
通空的景亭
木/玻璃
面向小区入口的
叠水
喷泉

剖面图-1

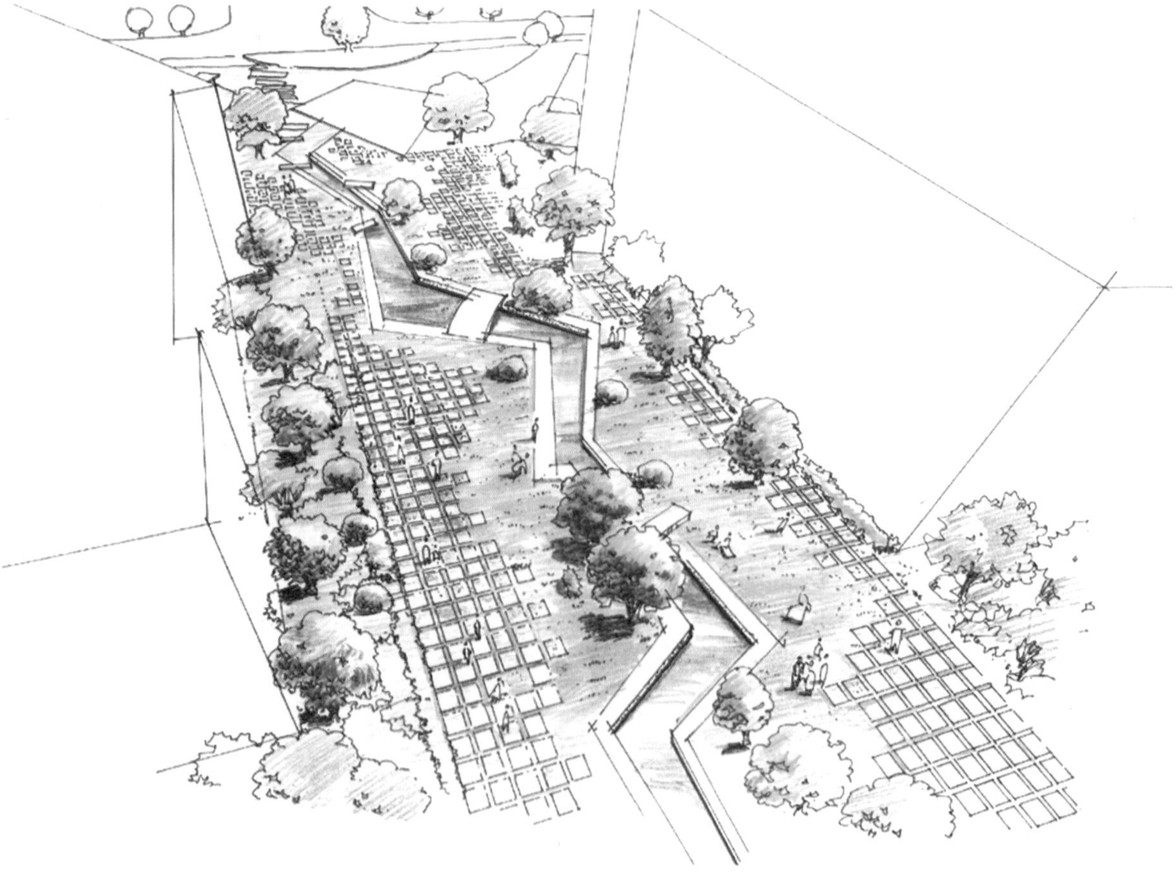

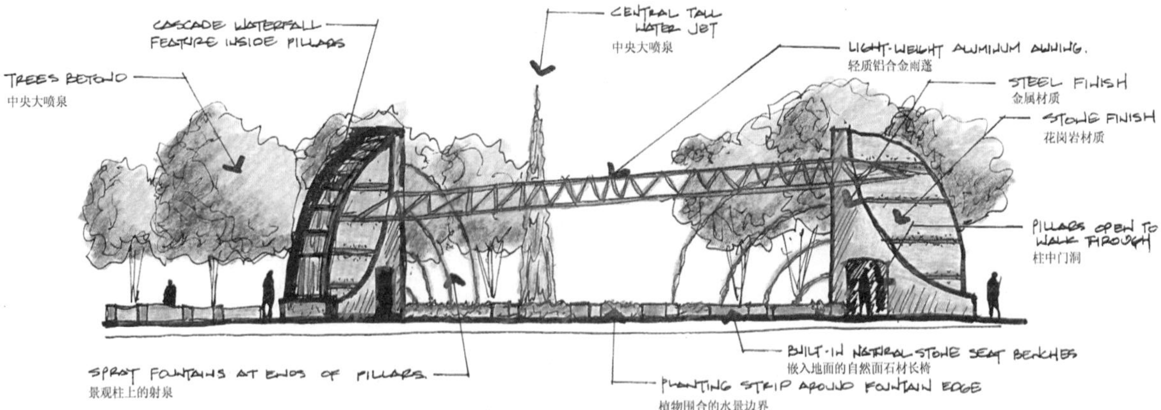
CASCADE WATERFALL FEATURE INSIDE PILLARS
CENTRAL TALL WATER JET
中央大喷泉
LIGHT-WEIGHT ALUMINUM AWNING.
轻质铝合金雨蓬
TREES BEYOND
中央大喷泉
STEEL FINISH
金属材质
STONE FINISH
花岗岩材质
PILLARS OPEN TO WALK THROUGH
柱中门洞
BUILT-IN NATURAL STONE SEAT BENCHES
嵌入地面的自然面石材长椅
SPRAY FOUNTAINS AT ENDS OF PILLARS
景观柱上的射泉
PLANTING STRIP AROUND FOUNTAIN EDGE
植物围合的水景边界

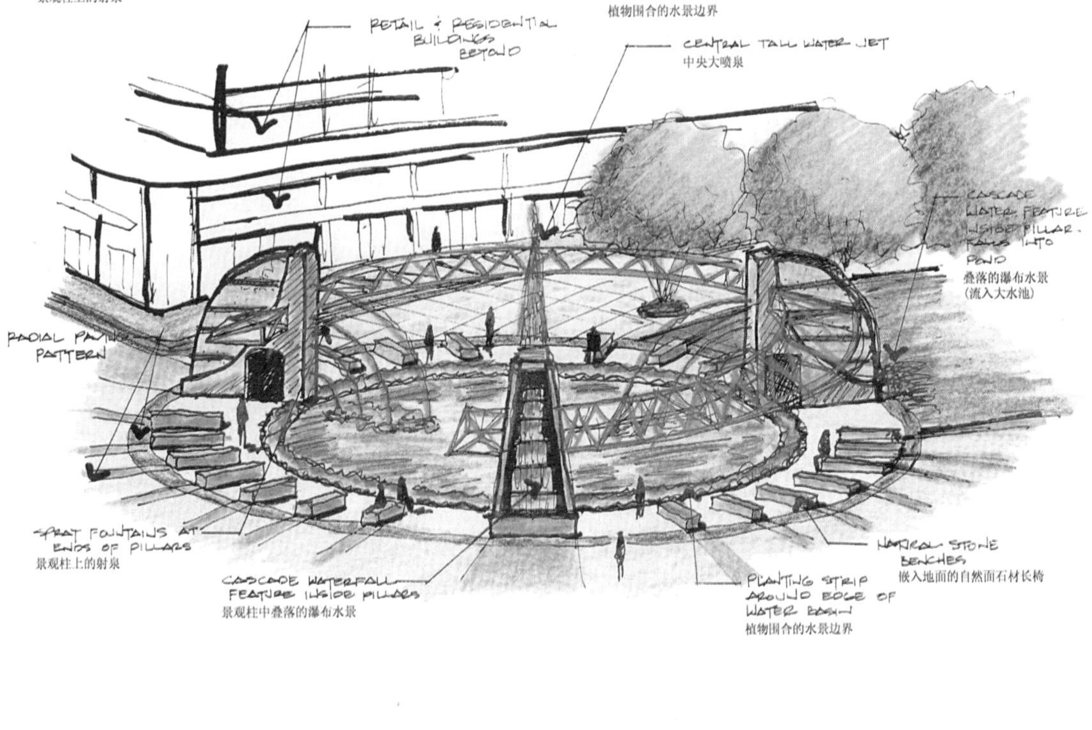
RETAIL & RESIDENTIAL BUILDINGS BEYOND
CENTRAL TALL WATER JET
中央大喷泉
CASCADE WATER FEATURE INSIDE PILLAR, FALLS INTO POND
叠落的瀑布水景
(流入大水池)
RADIAL PAVING PATTERN
SPRAY FOUNTAINS AT ENDS OF PILLARS
景观柱上的射泉
NATURAL STONE BENCHES
嵌入地面的自然面石材长椅
CASCADE WATERFALL FEATURE INSIDE PILLARS
景观柱中叠落的瀑布水景
PLANTING STRIP AROUND EDGE OF WATER BASIN
植物围合的水景边界

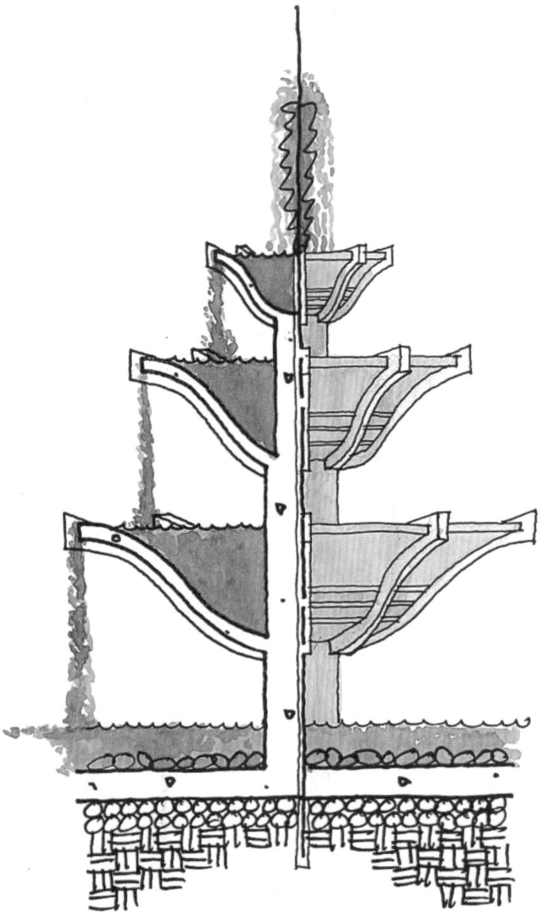

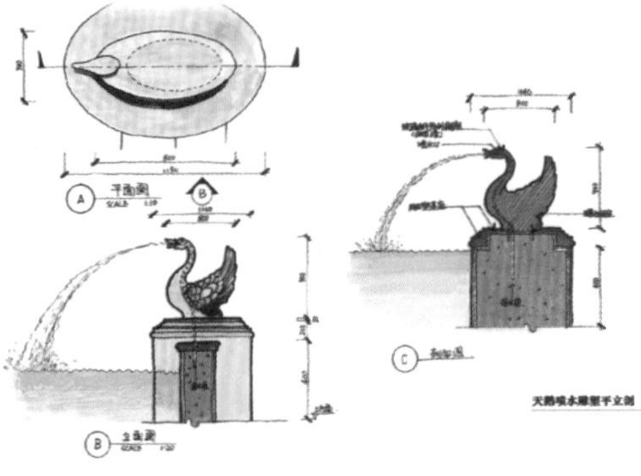

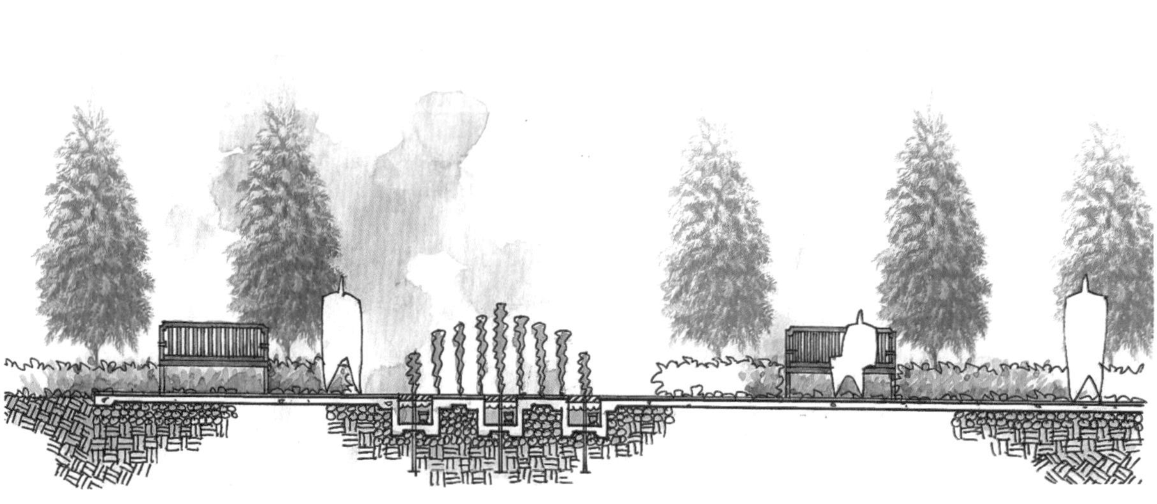

水
景

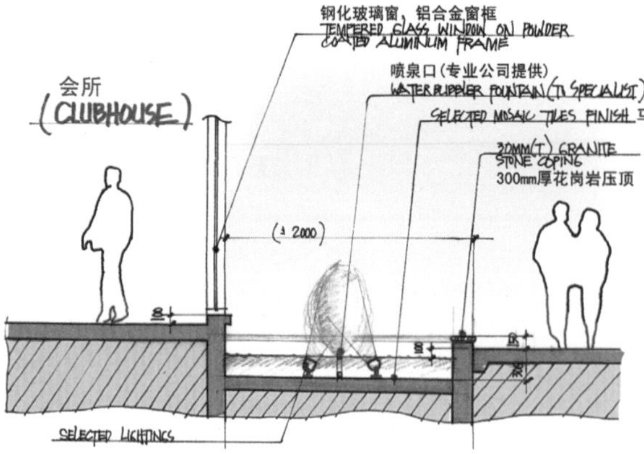
钢化玻璃窗、铝合金窗框
TEMPERED GLASS WINDOW ON POWDER COATED ALUMINUM FRAME
会所
(CLUBHOUSE)
喷泉口(专业公司提供)
WATER BUBBLER FOUNTAIN (TI SPECIALIST)
SELECTED MOSAIC TILES FINISH 马赛克饰面
30MM(T) GRANITE STONE COPING
300mm厚花岗岩压顶
(±2000)
SELECTED LIGHTINGS

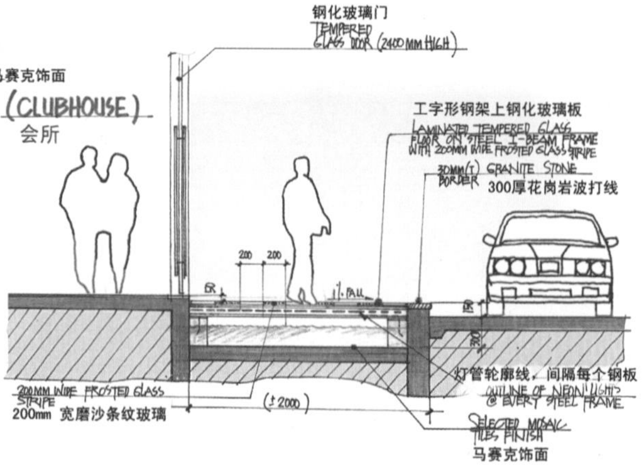
钢化玻璃门
TEMPERED GLASS DOOR (2400 MM HIGH)
(CLUBHOUSE)
会所
工字形钢架上钢化玻璃板
LAMINATED TEMPERED GLASS FLOOR ON STEEL I-BEAM FRAME WITH 200MM WIDE FROSTED GLASS STRIPE
30MM(T) GRANITE STONE BORDER
300厚花岗岩波打线
200 200
灯管轮廓线，间隔每个钢板
OUTLINE OF NEON LIGHT @ EVERY STEEL FRAME
200MM WIDE FROSTED GLASS STRIPE
200mm 宽磨沙条纹玻璃
(±2000)
SELECTED MOSAIC TILES FINISH
马赛克饰面

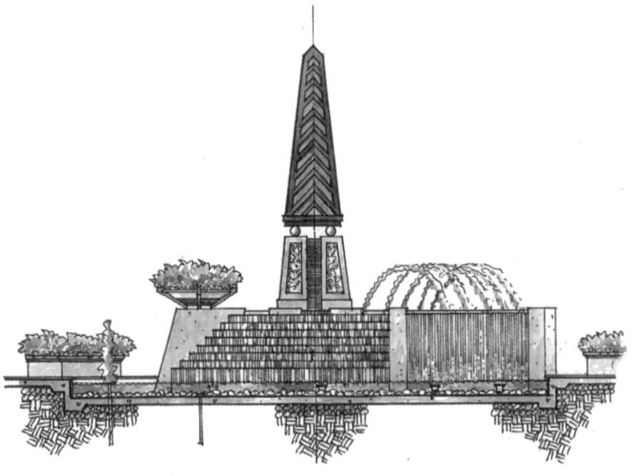

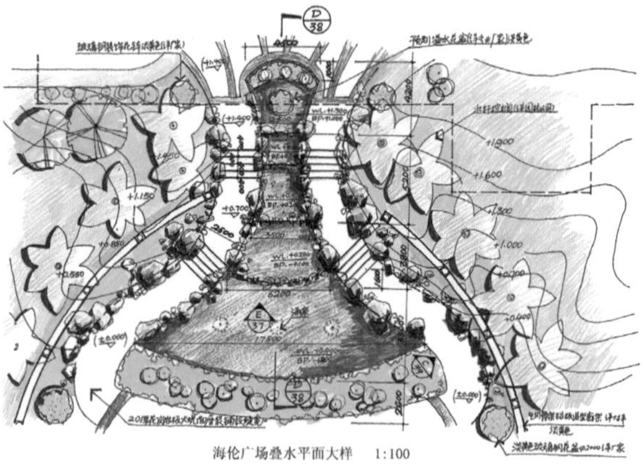

海伦广场叠水平面大样 1:100

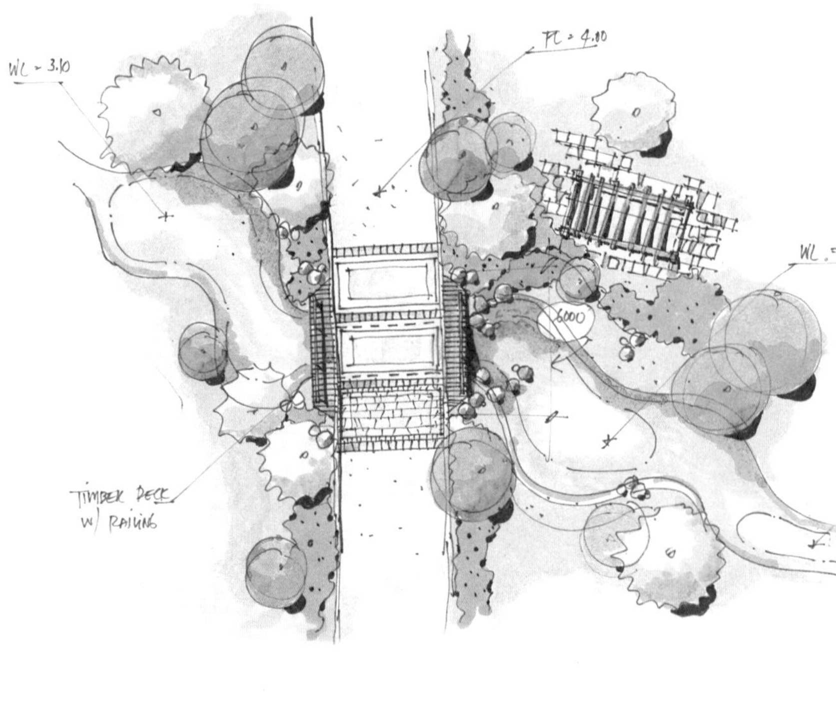
WL = 3.10
FL = 4.80
WL = 3.10
6000
TIMBER DECK W/ RAILING

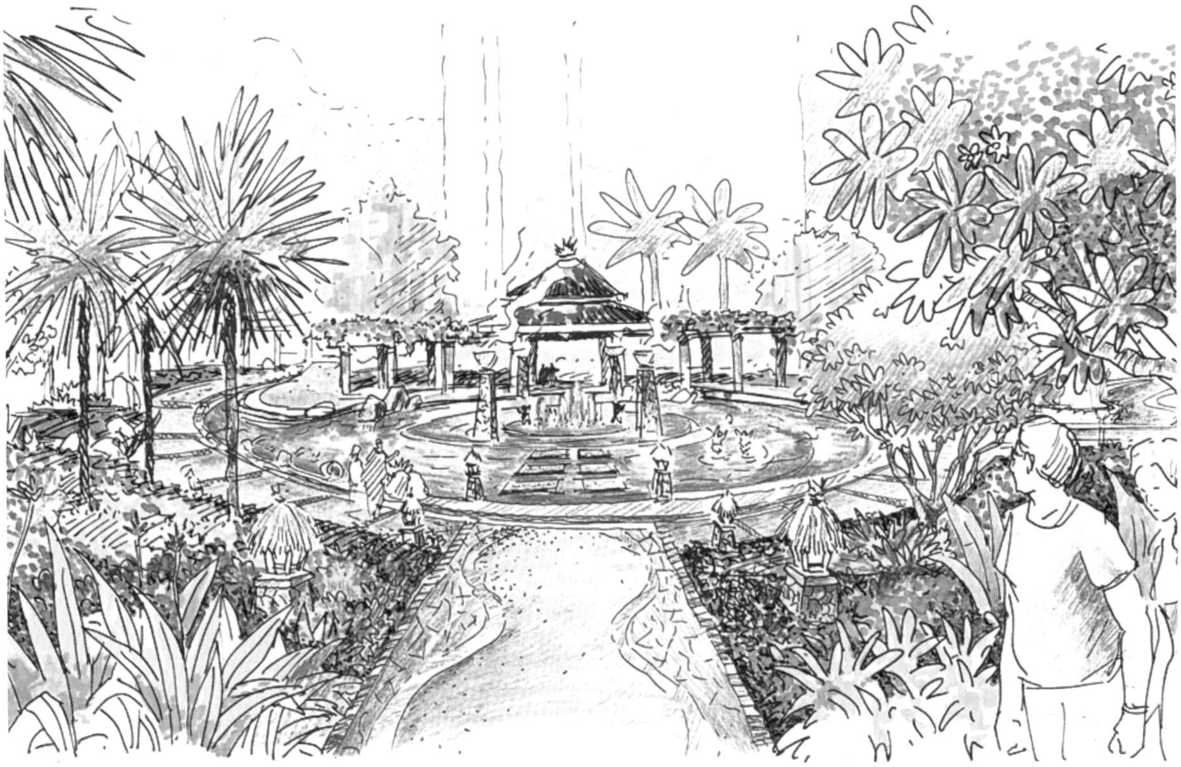

海伦广场叠水立面 1:30

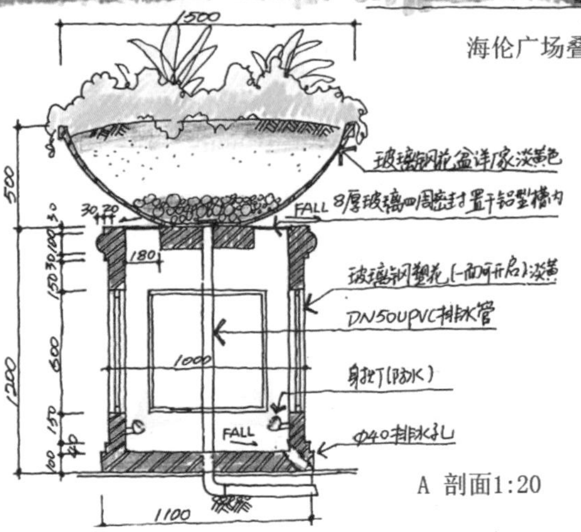

A 剖面1:20

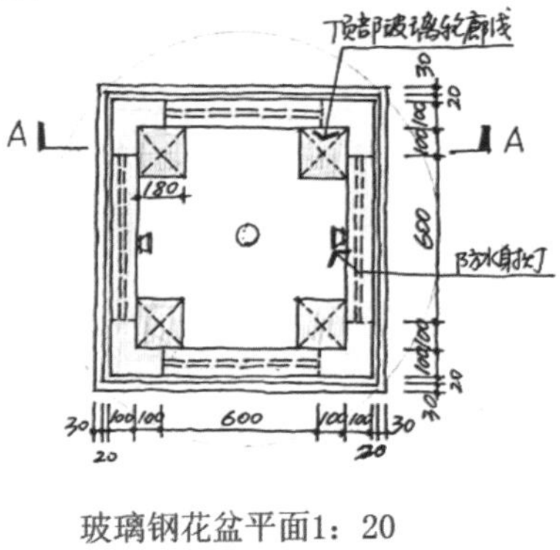

玻璃钢花盆平面1：20

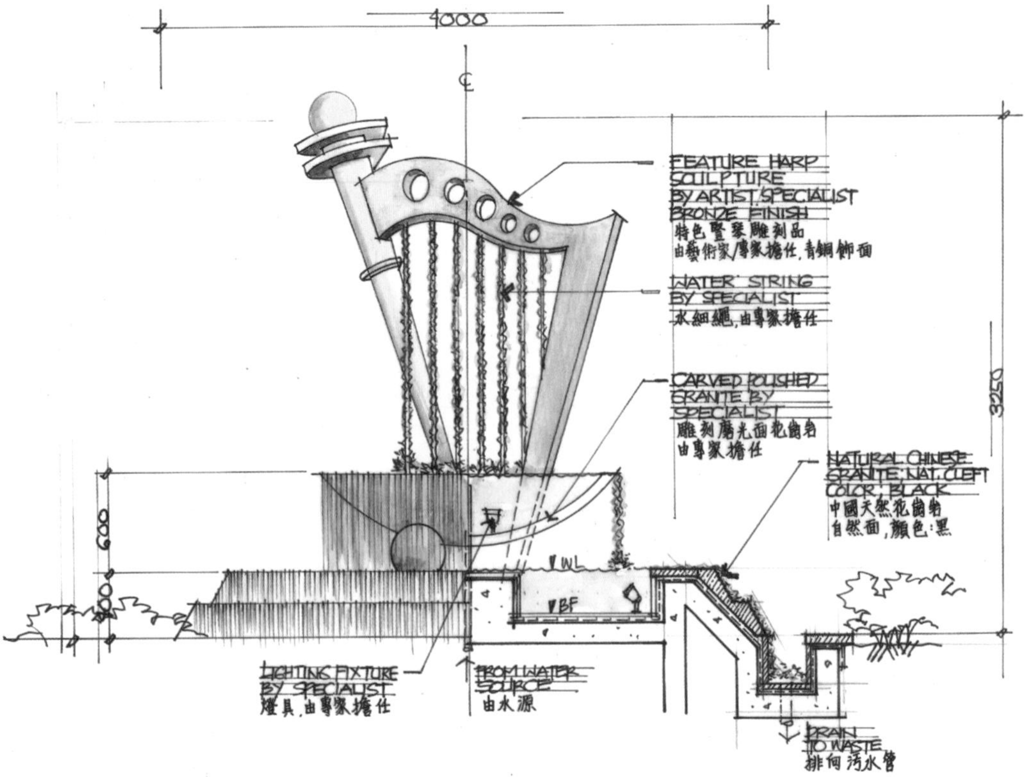
4000
FEATURE HARP SCULPTURE BY ARTIST, SPECIALIST BRONZE FINISH
特色豎琴雕刻品 由藝術家/專家擔任,青銅飾面
WATER STRING BY SPECIALIST
水細繩,由專家擔任
CARVED POLISHED GRANITE BY SPECIALIST
雕刻磨光面花崗岩 由專家擔任
NATURAL CHINESE GRANITE; NAT. CLEFT COLOR: BLACK
中國天然花崗岩 自然面,顏色:黑
3250
600
400
WL
BF
LIGHTING FIXTURE BY SPECIALIST
燈具,由專家擔任
FROM WATER SOURCE
由水源
DRAIN TO WASTE
排向污水管

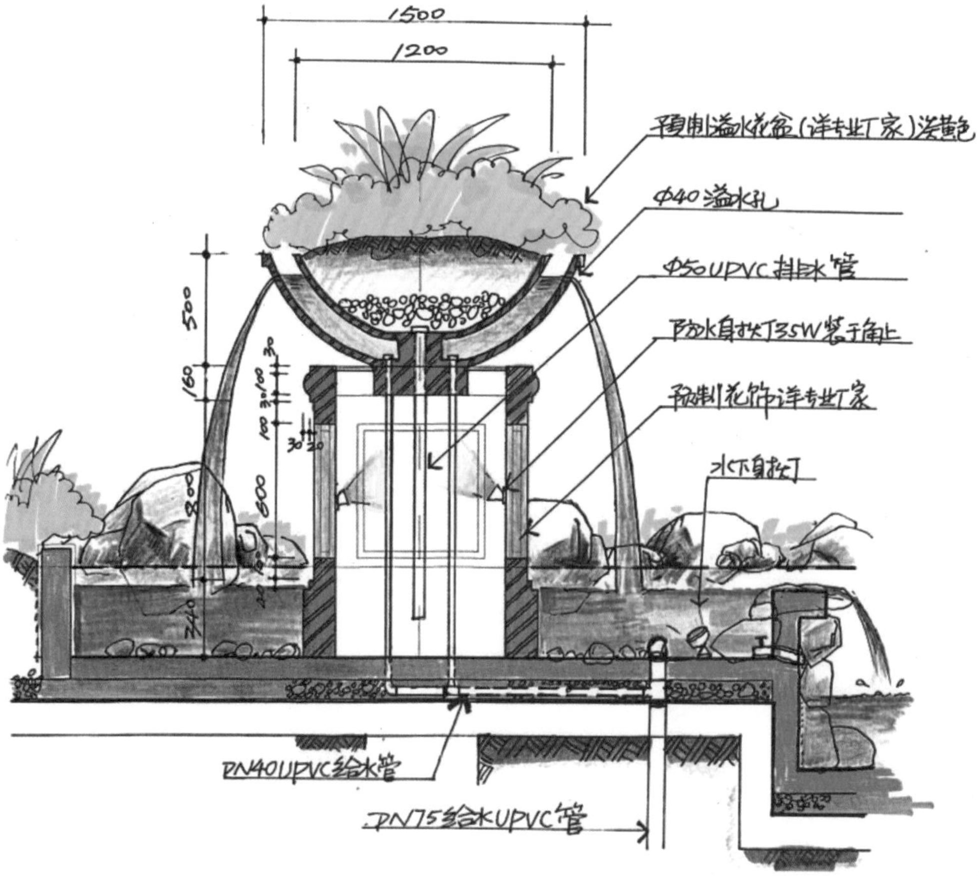
1500
1200
预制溢水花盆(详专业厂家)淡黄色
Φ40溢水孔
Φ50UPVC排水管
防水射灯35W装于角上
预制花饰详专业厂家
水下射灯
500
150
800
600
240
DN40UPVC给水管
DN75给水UPVC管

海伦广场小品大样 1:20

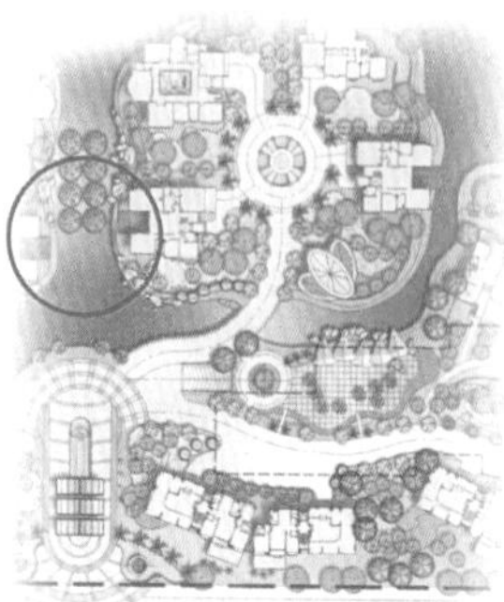

HIGH BRANCHING TREE (GINKGO TREE)
高大喬木(銀杏)
SHRUB AND ARBOR
喬灌木混合種植
WATER SURFACE ON A ARTIFICIAL LAKE
人工湖水面
CURVATURE WATERBANK
曲綫水岸綫
PLANTINGHOLE WITH HIGH BRANCHING TREE SHRUB AROUND
TREE CAVE IN WATER WITH SYMETRICAL ARRANGEMENT
對稱排列的水中樹穴

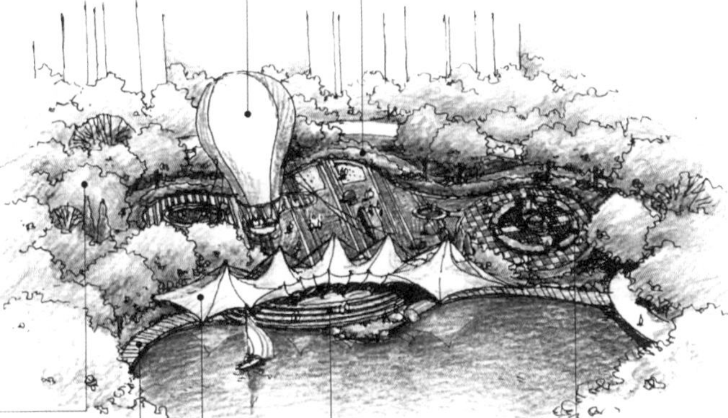
FORESTATION LINK AREA AROUND THE SQUARE
COLOR BALLOONS FOR PRETTIFYING
THE HOLIDAY ATMOSPHERE
PUBLIC SPACE
SEPERATED WITH PRIVATE
SPACE BY PLANTING
PATH LEADING TO WAVE BANK
通向波浪水岸的小径
PLASTIC ARTS ASSEMBLE CURTAIN
组合张拉膜造型
SIDESTEP AND STONE WALL ON THE RIVER EDGE
PLACES FOR LEISURES
COVERED WITH BORDER

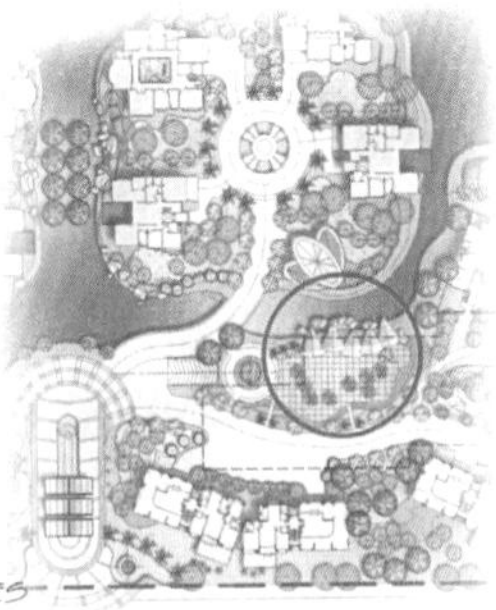

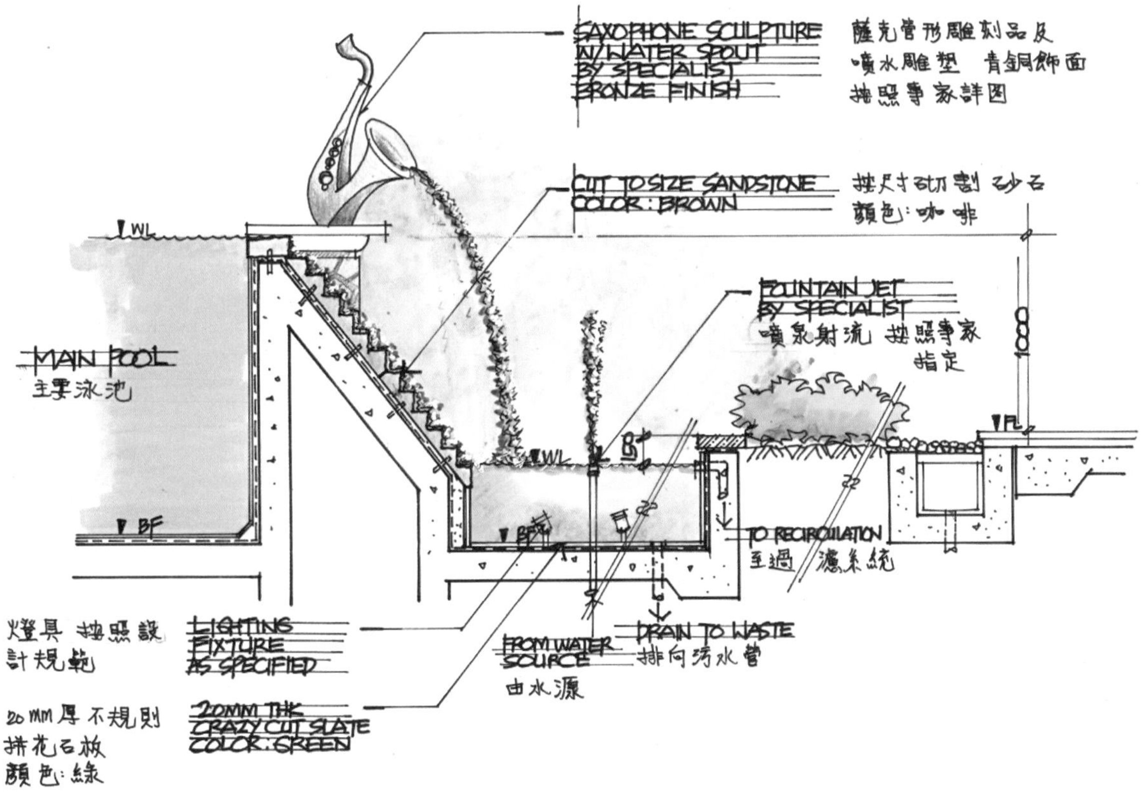
SAXOPHONE SCULPTURE
W/ WATER SPOUT
BY SPECIALIST
BRONZE FINISH
薩克管形雕刻品及
噴水雕塑 青銅飾面
按照專家詳图
CUT TO SIZE SANDSTONE
COLOR: BROWN
按尺寸切割砂石
顏色: 咖啡
WL
FOUNTAIN JET
BY SPECIALIST
噴泉射流 按照專家
指定
1000
MAIN POOL
主要泳池
FL
BF
WL
TO RECIRCULATION
至過濾系統
燈具 按照設
計規範
LIGHTING
FIXTURE
AS SPECIFIED
FROM WATER
SOURCE
由水源
DRAIN TO WASTE
排向污水管
20MM 厚不規則
拼花石板
顏色: 綠
20MM THK
CRAZY CUT SLATE
COLOR: GREEN

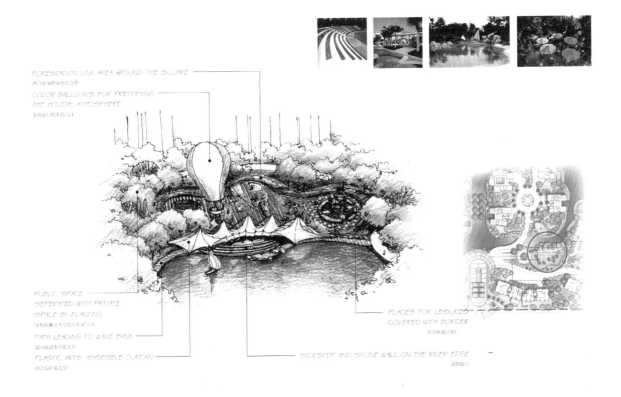

ARBOR AND LAWN
ARC PATH
弧形的小径
STAGE AND LEISURE SEATING
LAWN LIGHTING
草坪灯
TERRACING SIDESTEP

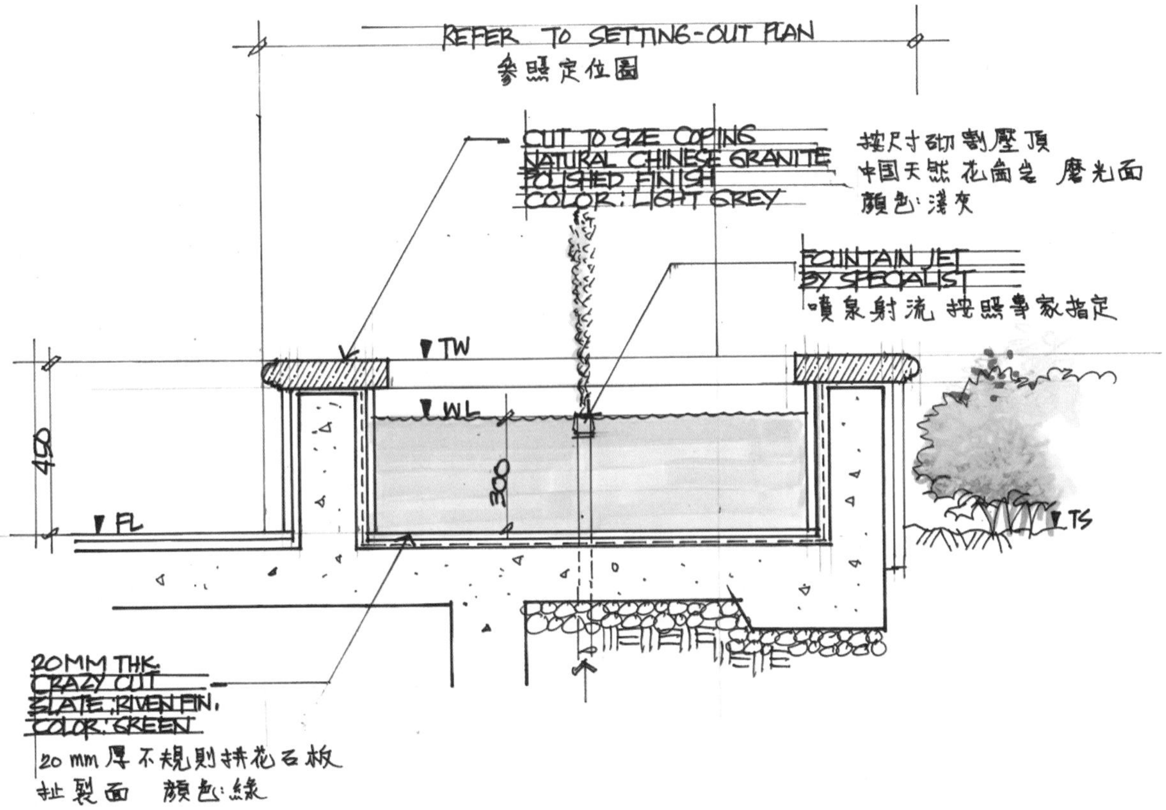
REFER TO SETTING-OUT PLAN
参照定位图
CUT TO SIZE COPING
NATURAL CHINESE GRANITE
POLISHED FINISH
COLOR: LIGHT GREY
按尺寸切割压顶
中国天然花岗岩 磨光面
颜色:浅灰
FOUNTAIN JET
BY SPECIALIST
喷泉射流 按照专家指定
TW
WL
FL
TS
450
300
20MM THK.
CRAZY CUT
SLATE; RIVEN FIN.
COLOR: GREEN
20 mm 厚不规则拼花石板
扯裂面 颜色:绿

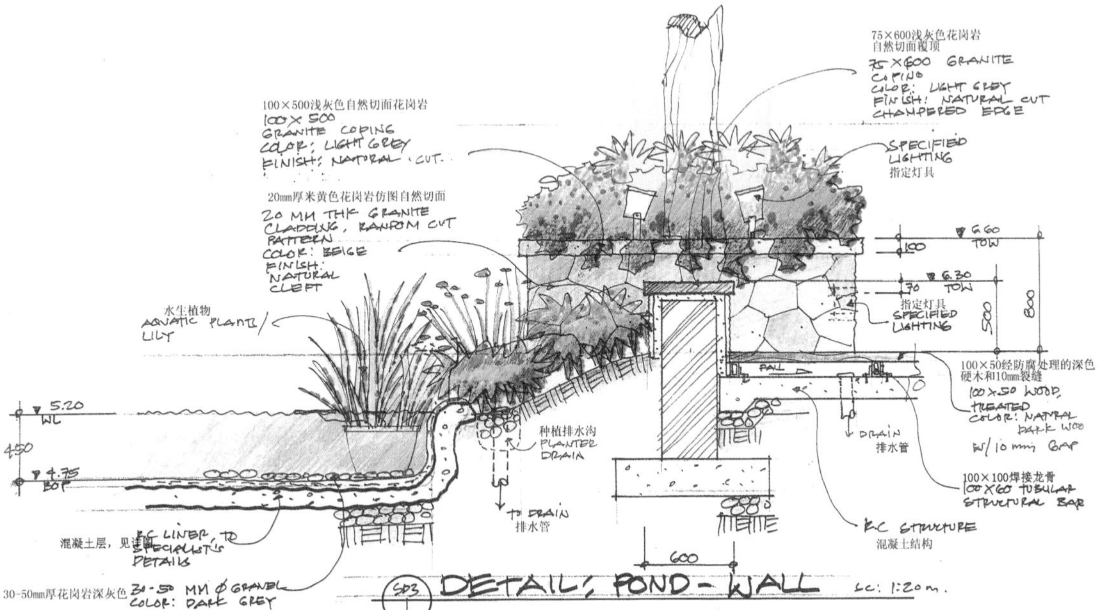
100×500浅灰色自然切面花岗岩
100×500
GRANITE COPING
COLOR: LIGHT GREY
FINISH: NATURAL CUT.
20mm厚米黄色花岗岩仿图自然切面
20 MM THK GRANITE
CLADDING, RANDOM CUT
PATTERN
COLOR: BEIGE
FINISH:
NATURAL
CLEFT
水生植物
AQUATIC PLANTS/
LILY
75×600浅灰色花岗岩
自然切面覆顶
75×600 GRANITE
COPING
COLOR: LIGHT GREY
FINISH: NATURAL CUT
CHAMFERED EDGE
SPECIFIED
LIGHTING
指定灯具
6.60
TOW
100
6.30
TOW
70
指定灯具
SPECIFIED
LIGHTING
500
800
FALL
100×50经防腐处理的深色
硬木和10mm裂缝
100×50 WOOD,
TREATED
COLOR: NATURAL
DARK WOO
W/10 mm GAP
5.20
WL
450
4.75
BOP
种植排水沟
PLANTER
DRAIN
DRAIN
排水管
TO DRAIN
排水管
100×100焊接龙骨
100×60 TUBULAR
STRUCTURAL BAR
混凝土层，见详图
RC LINER, TO
SPECIALIST'S
DETAILS
RC STRUCTURE
混凝土结构
600
30-50mm厚花岗岩深灰色
30-50 MM Ø GRAVEL
COLOR: DARK GREY
SD3
DETAIL: POND-WALL
SC: 1:20m.

池塘/挡墙剖面大样

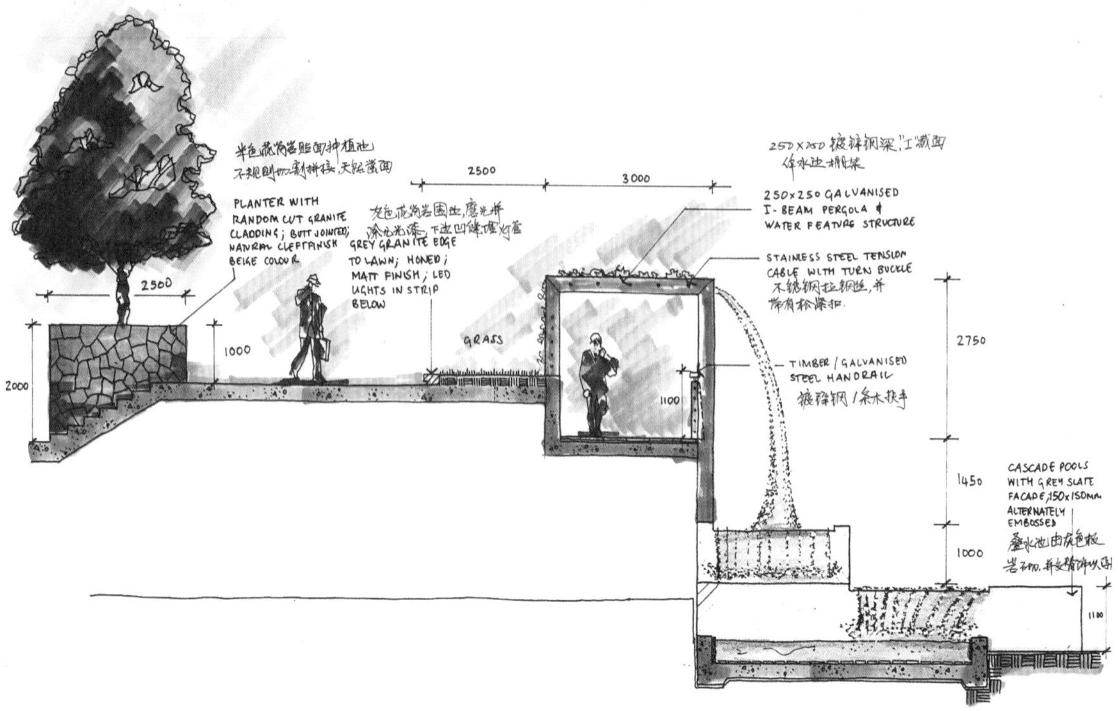
PLANTER WITH RANDOM CUT GRANITE CLADDING; BUTT JOINTED; NATURAL CLEFT FINISH BEIGE COLOUR
GREY GRANITE EDGE TO LAWN; HONED; MATT FINISH; LED LIGHTS IN STRIP BELOW
GRASS
250×250 GALVANISED I-BEAM PERGOLA & WATER FEATURE STRUCTURE
STAINLESS STEEL TENSION CABLE WITH TURN BUCKLE
TIMBER / GALVANISED STEEL HANDRAIL
镀锌钢 / 条木扶手
CASCADE POOLS WITH GREY SLATE FACADE, 150×150mm ALTERNATELY EMBOSSED
2500
2500
3000
2000
1000
1100
2750
1450
1000
1100

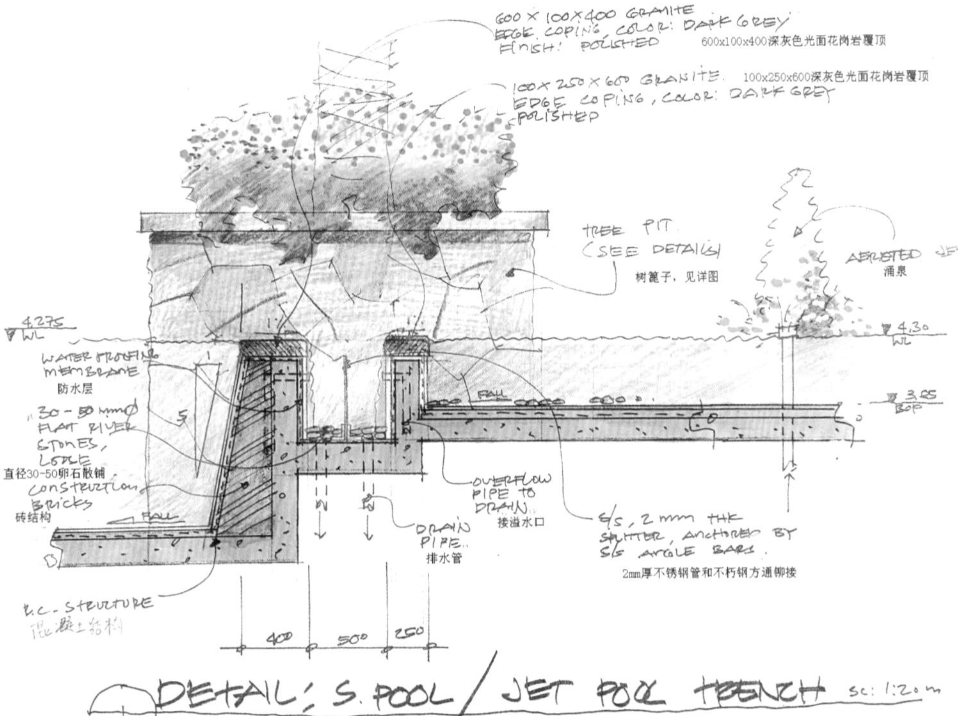
600 X 100 X 400 GRANITE
EDGE COPING COLOR: DARK GREY
FINISH: POLISHED
600x100x400深灰色光面花岗岩覆顶
100 X 250 X 600 GRANITE
EDGE COPING, COLOR: DARK GREY
POLISHED
100x250x600深灰色光面花岗岩覆顶
TREE PIT (SEE DETAILS)
树篦子，见详图
涌泉
WATERPROOFING MEMBRANE
防水层
30-50 MM FLAT RIVER STONES, LOOSE
直径30-50卵石散铺
CONSTRUCTION BRICKS
砖结构
OVERFLOW PIPE TO DRAIN
接溢水口
DRAIN PIPE
排水管
2mm厚不锈钢管和不朽钢方通铆接
R.C. STRUCTURE
400
500
250
DETAIL: S.POOL / JET POOL TRENCH

泳池/入口水景剖面祥图 1:20

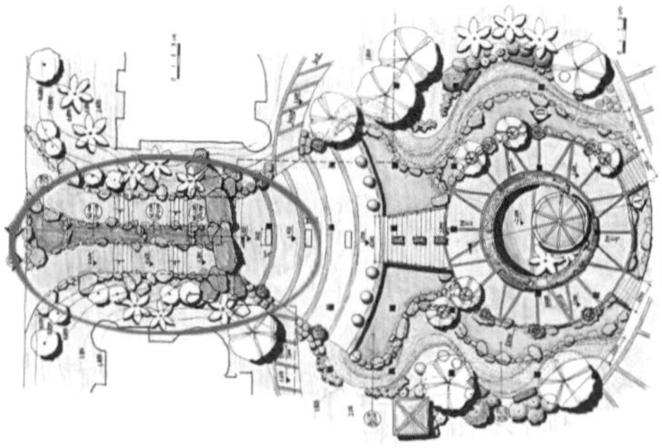

索引平面

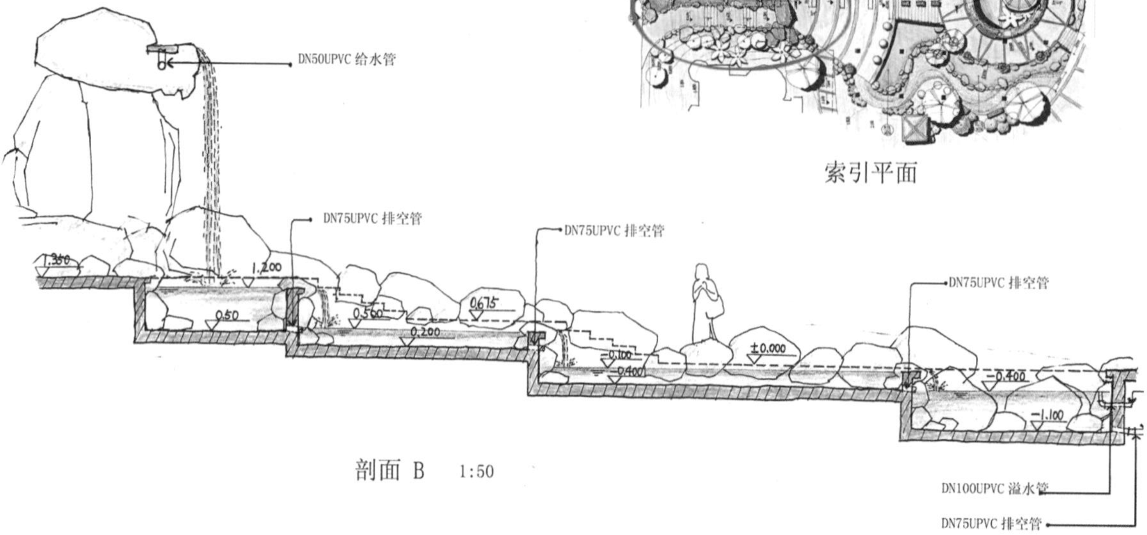
DN50UPVC 给水管
DN75UPVC 排空管
DN75UPVC 排空管
DN75UPVC 排空管
1.350
1.200
0.50
0.560
0.675
0.200
-0.100
-0.400
±0.000
-0.400
-1.100
DN100UPVC 溢水管
DN75UPVC 排空管

剖面 B　1:50

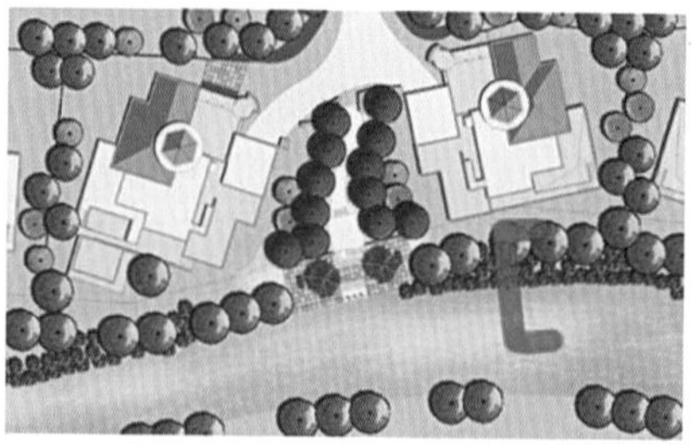

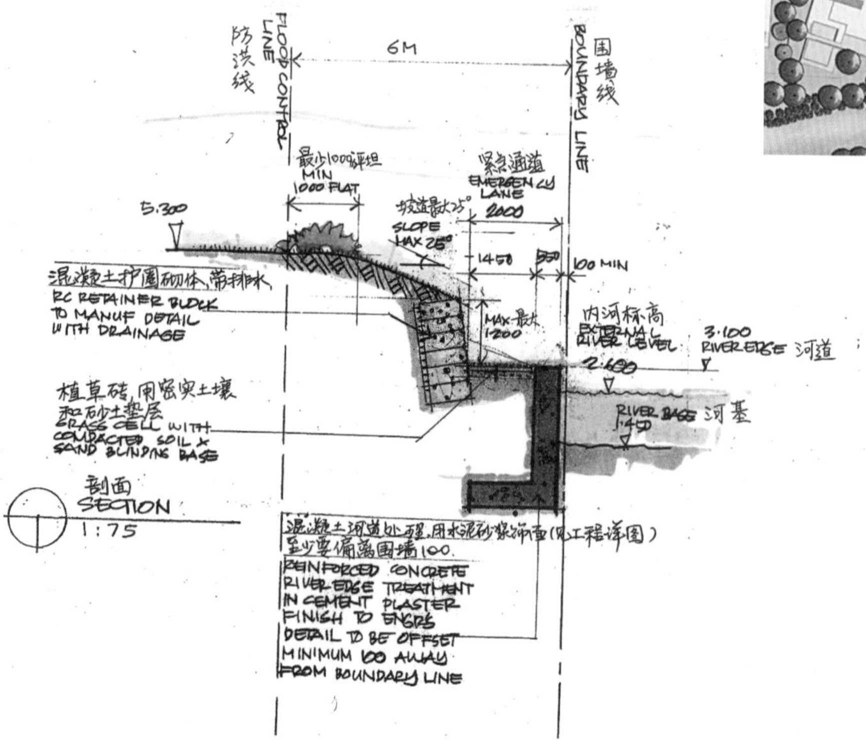
防洪线
FLOOD CONTROL LINE
6M
围墙线
BOUNDARY LINE
最少1000平坦
MIN 1000 FLAT
紧急通道
EMERGENCY LANE
2000
5.300
坡道最大25°
SLOPE MAX 25°
1450
100 MIN
混凝土护围砌体，带排水
RC RETAINER BLOCK TO MANUF DETAIL WITH DRAINAGE
MAX 最大 1200
内河标高
EXTERNAL RIVER LEVEL
2.400
3.100
RIVER EDGE 河道
植草砖，用密实土壤和砂垫层
GRASS CELL WITH COMPACTED SOIL & SAND BLINDING BASE
RIVER BASE 河基
1.450
剖面
SECTION
1:75
混凝土河道处理，用水泥砂浆饰面(见工程详图)
至少要偏离围墙100.
REINFORCED CONCRETE RIVER EDGE TREATMENT IN CEMENT PLASTER FINISH TO ENGRS DETAIL TO BE OFFSET MINIMUM 100 AWAY FROM BOUNDARY LINE

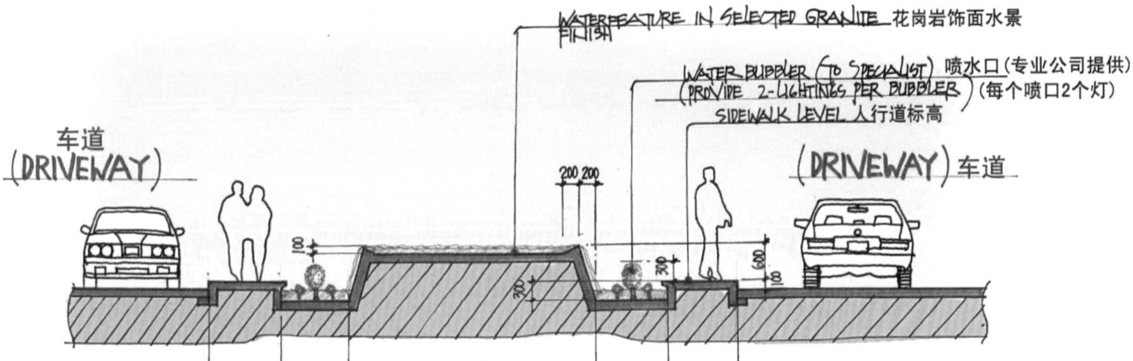
WATERFEATURE IN SELECTED GRANITE FINISH 花岗岩饰面水景
WATER BUBBLER (TO SPECIALIST) 喷水口(专业公司提供)
(PROVIDE 2-LIGHTINGS PER BUBBLER) (每个喷口2个灯)
SIDEWALK LEVEL 人行道标高
车道
(DRIVEWAY)
(DRIVEWAY) 车道
200 200
1200
1200
±4000
1200
1200

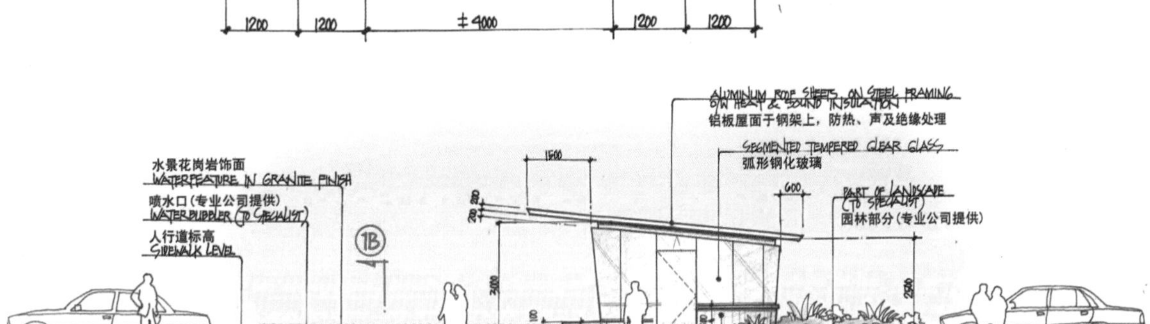
ALUMINUM ROOF SHEETS ON STEEL FRAMING C/W HEAT & SOUND INSULATION
铝板屋面于钢架上，防热、声及绝缘处理
SEGMENTED TEMPERED CLEAR GLASS
弧形钢化玻璃
水景花岗岩饰面
WATERFEATURE IN GRANITE FINISH
喷水口(专业公司提供)
WATERBUBBLER (TO SPECIALIST)
人行道标高
SIDEWALK LEVEL
1500
600
PART OF LANDSCAPE (TO SPECIALIST)
园林部分(专业公司提供)
1B
SELECTED GRANITE FINISH
花岗岩饰面

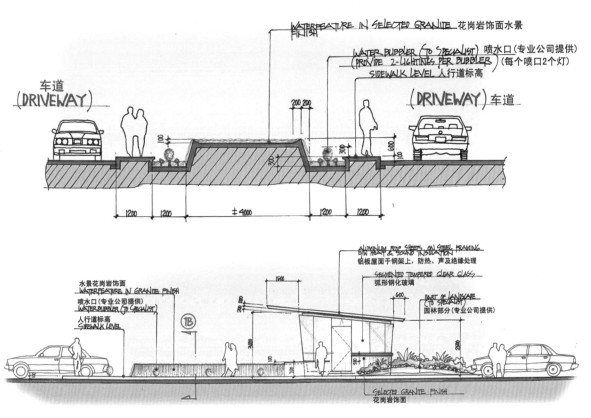

铝制水槽内供水管（专业公司提供）
WATER SUPPLY PIPE (TO SPECIALIST)
ON ALUMINUM GUTTER
OUTLINE OF SIGNAGE IN
STEEL BRONZE FINISH
不锈钢标志板，轮廓线铜饰面
园林部分（专业公司提供）
PART OF LANDSCAPE
(TO SPECIALIST)
400
400
300
道路标高
ROAD LEVEL
2300
300
1200
SELECTED LIGHTINGS 灯饰

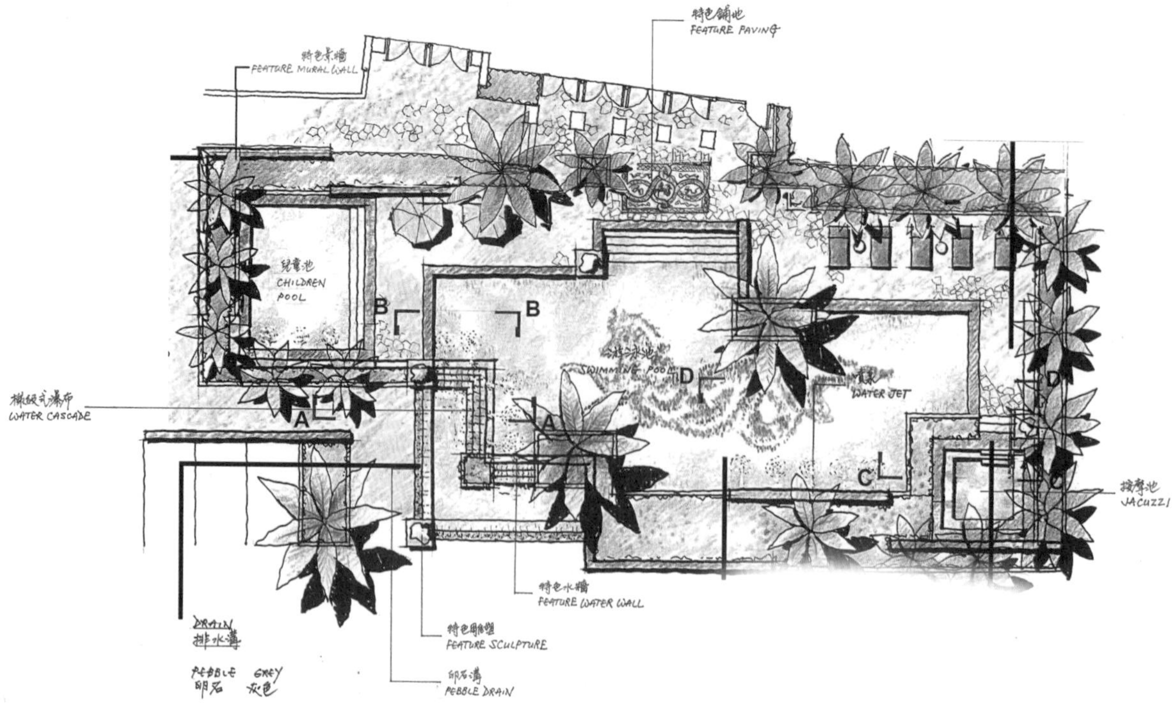
FEATURE PAVING
FEATURE MURAL WALL
CHILDREN POOL
B
B
WATER CASCADE
A
A
SWIMMING POOL
D
D
WATER JET
C
JACUZZI
FEATURE WATER WALL
FEATURE SCULPTURE
PEBBLE DRAIN
DRAIN
排水沟
PEBBLE
GREY
灰色

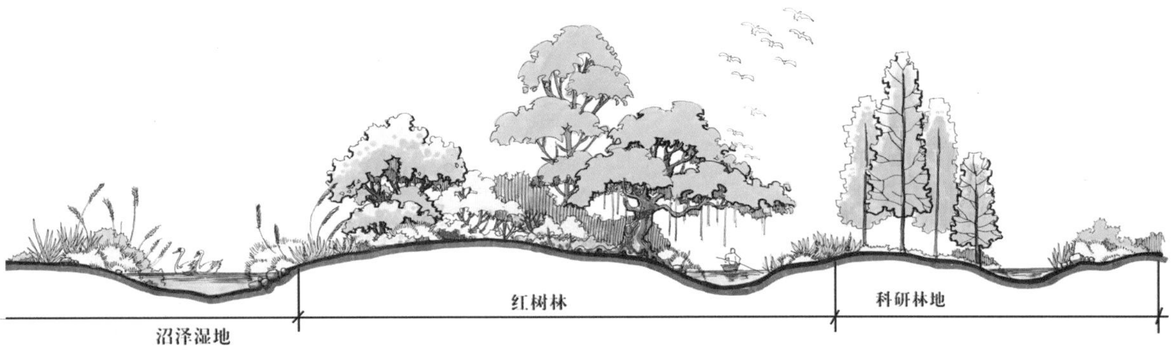
红树林
科研林地
沼泽湿地

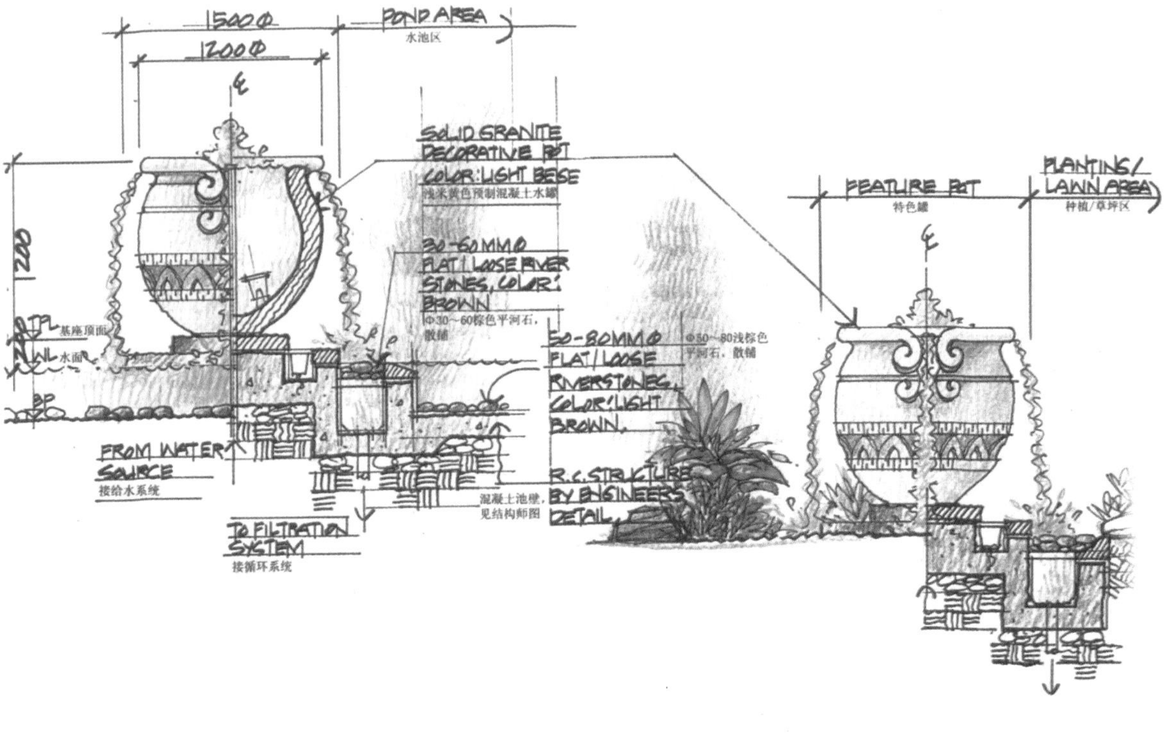
1500Φ
1200Φ
POND AREA
水池区
SOLID GRANITE
DECORATIVE POT
COLOR: LIGHT BEIGE
浅米黄色预制混凝土水罐
30-60MMΦ
FLAT/LOOSE RIVER
STONES, COLOR:
BROWN
Φ30～60棕色平河石，
散铺
TPL 基座顶面
WL 水面
BP
1200
50-80MMΦ
FLAT/LOOSE
RIVERSTONES,
COLOR: LIGHT
BROWN.
Φ50～80浅棕色
平河石，散铺
FROM WATER
SOURCE
接给水系统
R.C. STRUCTURE
BY ENGINEERS
DETAIL.
混凝土池壁，
见结构师图
TO FILTRATION
SYSTEM
接循环系统
FEATURE POT
特色罐
PLANTING/
LAWN AREA
种植/草坪区

4m.

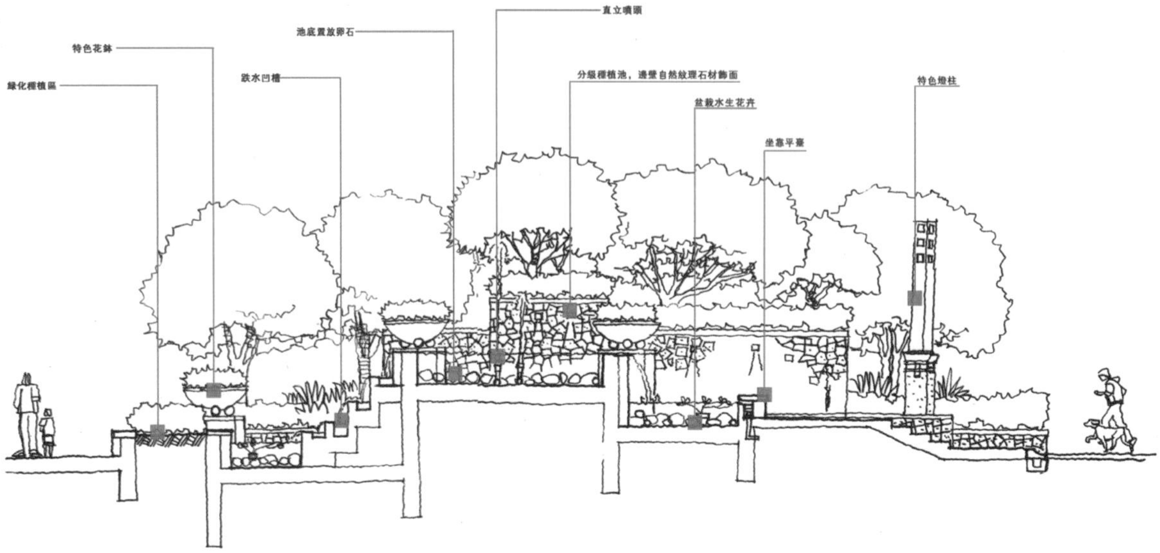
直立噴頭
池底置放卵石
特色花鉢
綠化種植區
跌水凹槽
分級種植池，邊壁自然紋理石材飾面
特色燈柱
盆栽水生花卉
坐靠平臺

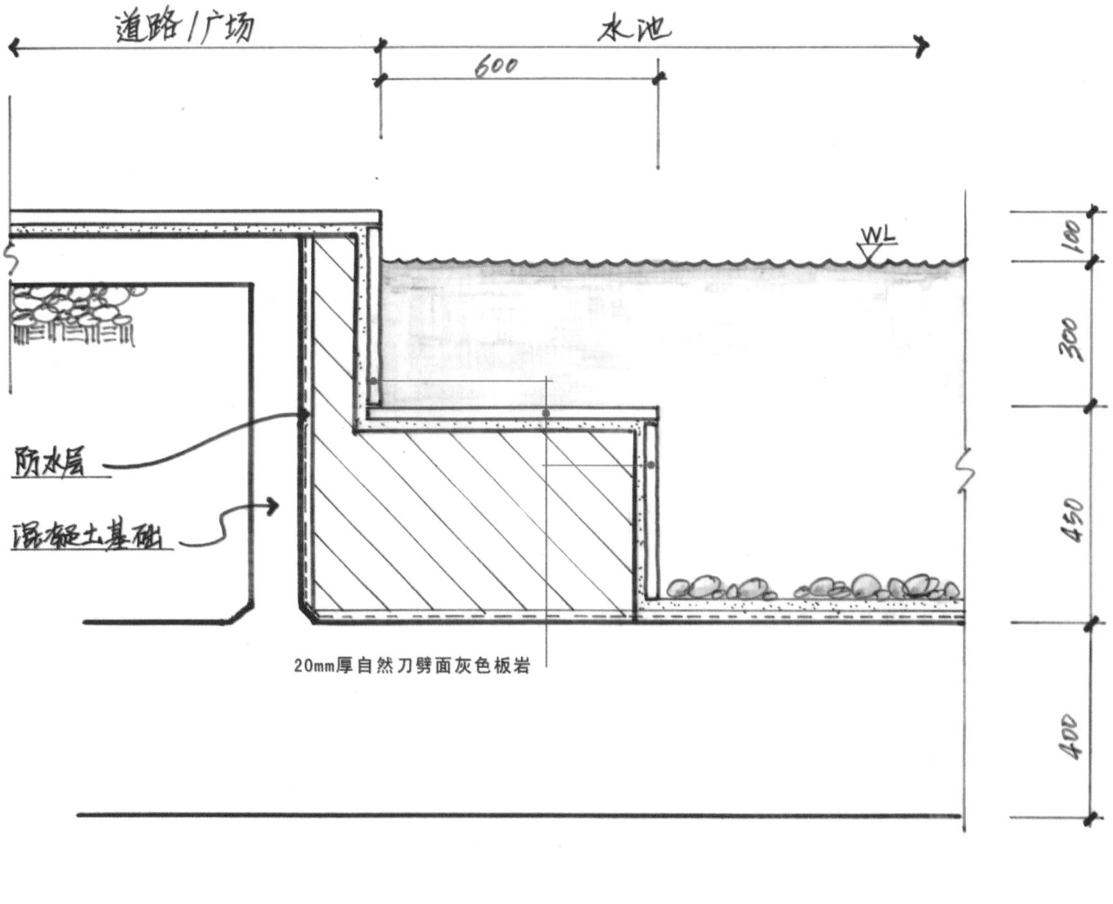
道路/广场
水池
600
WL
100
300
450
400
防水层
混凝土基础
20mm厚自然刀劈面灰色板岩

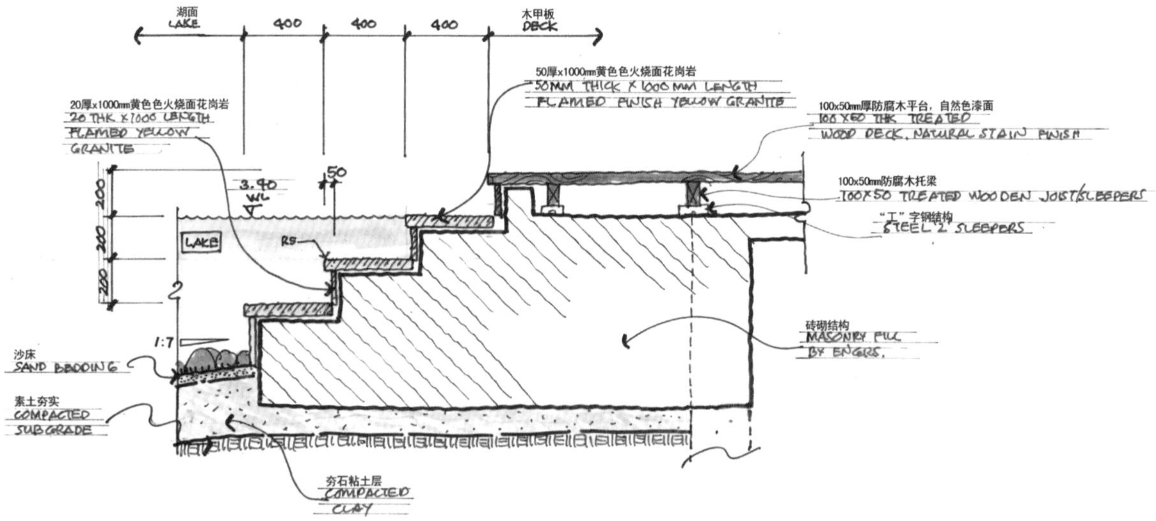

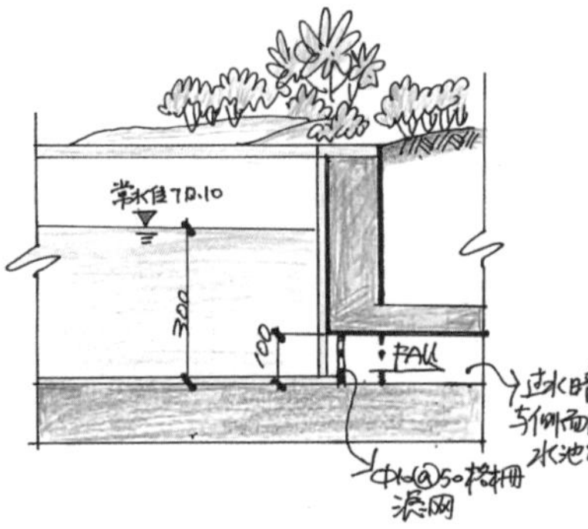

① 过水涵洞示意
SCALE 1:10

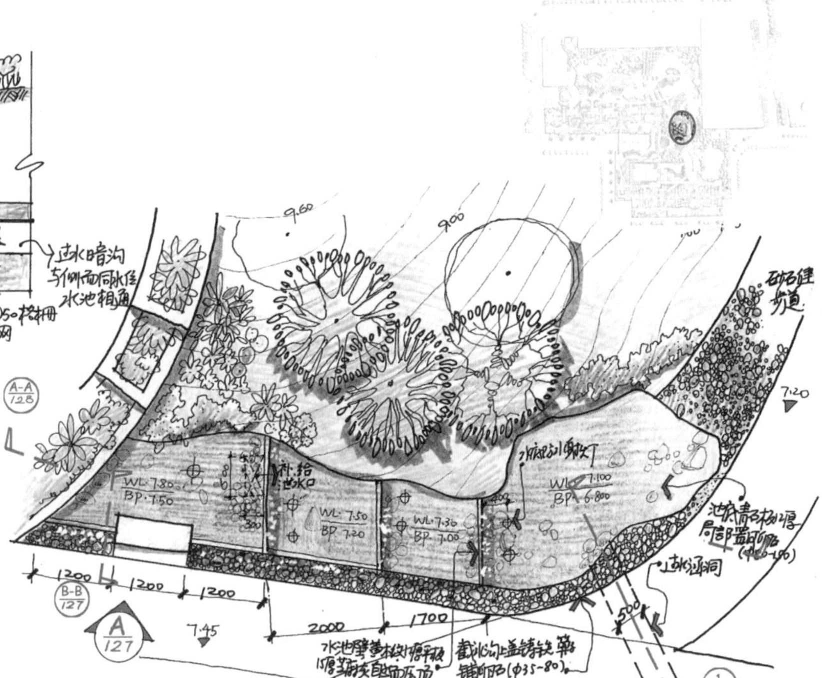

凸起式跌水景平面大样
SCALE 1:50

湖面
LAKE
400
400
400
木甲板
DECK
50厚x1000mm黄色色火烧面花岗岩
50MM THICK X 1000 MM LENGTH
FLAMED FINISH YELLOW GRANITE
20厚x1000mm黄色色火烧面花岗岩
20 THK X 1000 LENGTH
FLAMED YELLOW
GRANITE
100x50mm厚防腐木平台，自然色漆面
100X50 THK TREATED
WOOD DECK, NATURAL STAIN FINISH
100x50mm防腐木托梁
100X50 TREATED WOODEN JOIST/SLEEPERS
"工"字钢结构
3.40
WL
200
200
200
LAKE
R5
1:7
砖砌结构
MASONRY FILL
BY ENGRS.
沙床
SAND BEDDING
素土夯实
COMPACTED
SUBGRADE
夯石粘土层
COMPACTED
CLAY
300
FALL
过水暗沟
与侧面同水位
水池相通
A-A
128
9.60
9.00
7.20
WL 7.80
BP 7.50
WL 7.50
BP 7.20
WL 7.30
BP 7.00
BP 6.800
1300
1200
1200
2000
1700
500
B-B
127
127
7.45

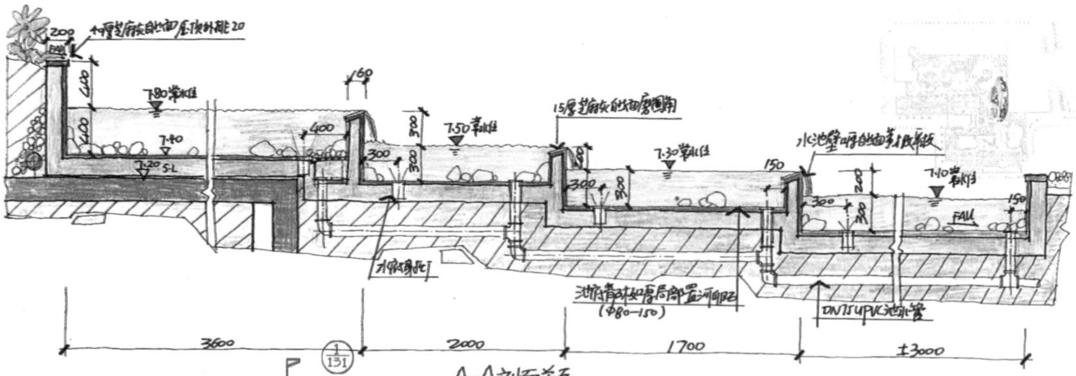

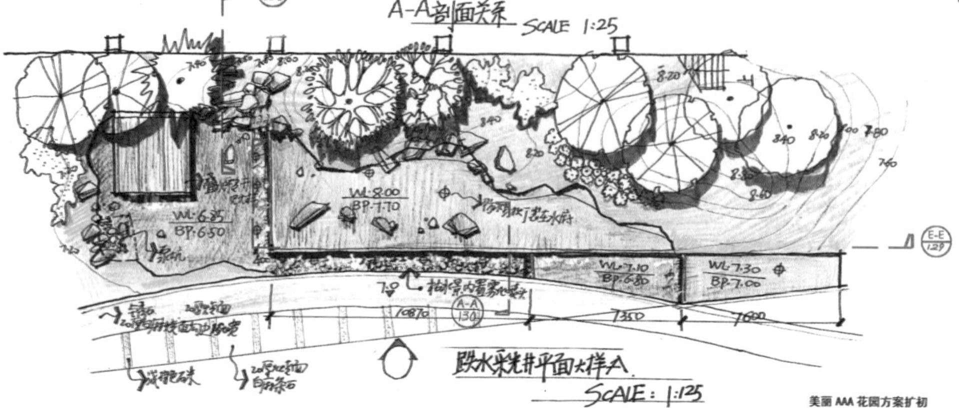

美丽 AAA 花园方案扩初

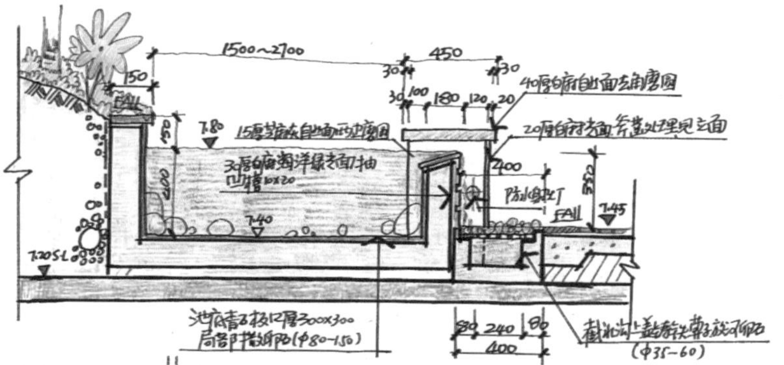

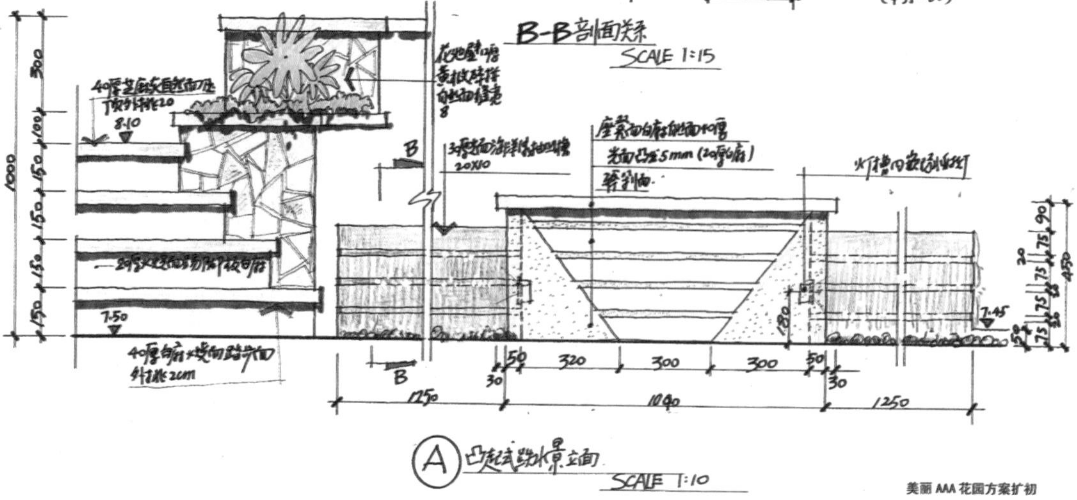

美丽 AAA 花园方案扩初

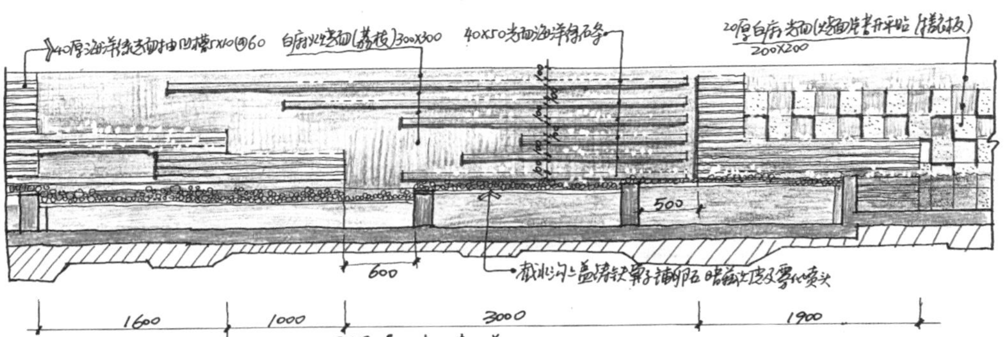

跌水采光景、叠水墙局部大样

SCALE 1:25

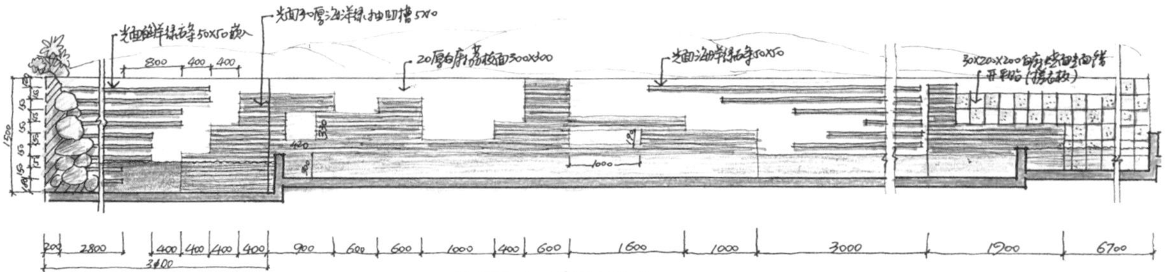

跌水桥井 E-E剖立面展开. 1:40

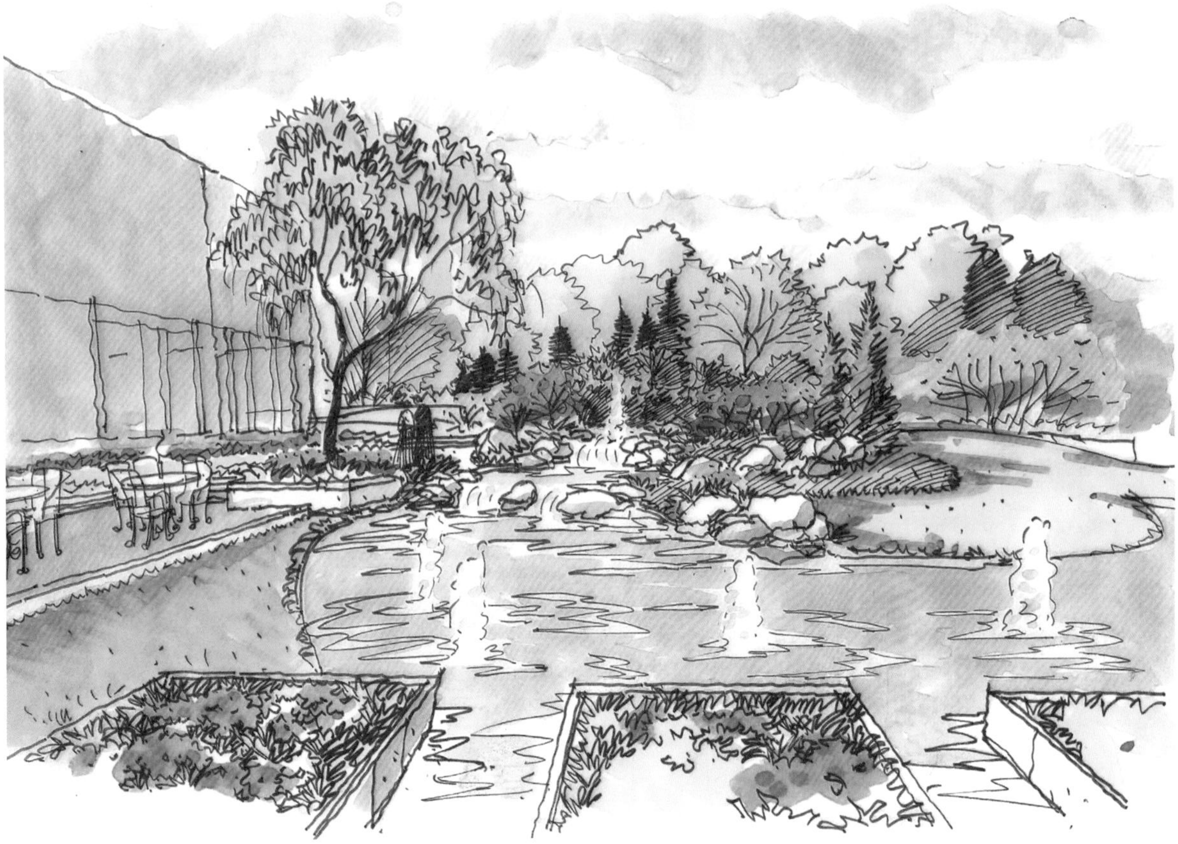

500
600
1600
1000
3000
1900
800
400
400
200
2800
3400
900
600
600
1000
400
600
1600
1000
3000
1900
6700

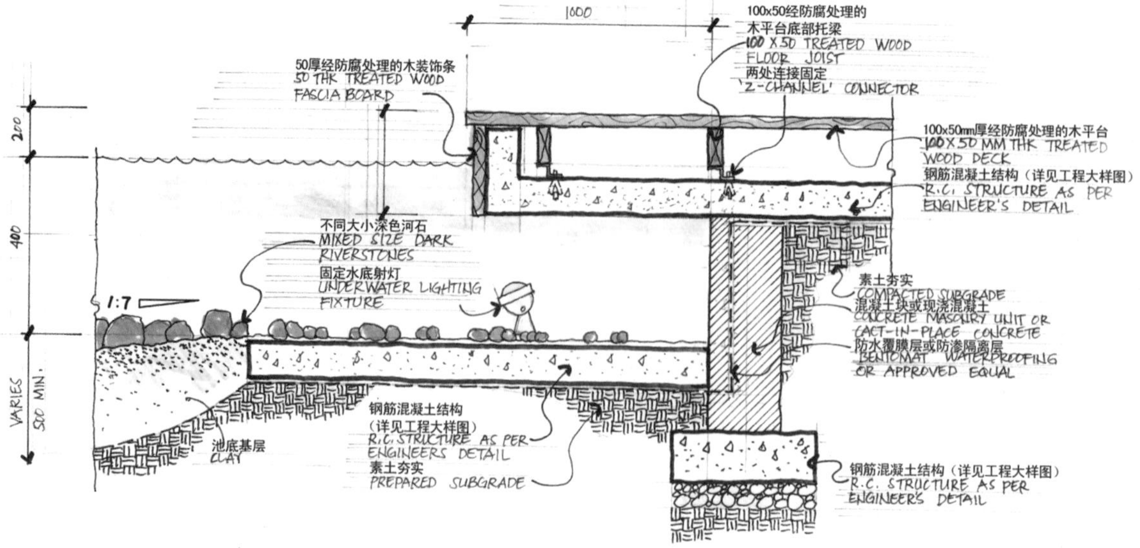
1000
100x50经防腐处理的
木平台底部托梁
100 X 50 TREATED WOOD
FLOOR JOIST
两处连接固定
'Z-CHANNEL' CONNECTOR
50厚经防腐处理的木装饰条
50 THK TREATED WOOD
FASCIA BOARD
100x50mm厚经防腐处理的木平台
100 X 50 MM THK TREATED
WOOD DECK
钢筋混凝土结构（详见工程大样图）
R.C. STRUCTURE AS PER
ENGINEER'S DETAIL
200
400
VARIES
500 MIN.
不同大小深色河石
MIXED SIZE DARK
RIVERSTONES
固定水底射灯
UNDERWATER LIGHTING
FIXTURE
1:7
素土夯实
COMPACTED SUBGRADE
混凝土块或现浇混凝土
CONCRETE MASONRY UNIT OR
CAST-IN-PLACE CONCRETE
防水覆膜层或防渗隔离层
BENTOMAT WATERPROOFING
OR APPROVED EQUAL
池底基层
CLAY
钢筋混凝土结构
（详见工程大样图）
R.C. STRUCTURE AS PER
ENGINEERS DETAIL
素土夯实
PREPARED SUBGRADE
钢筋混凝土结构（详见工程大样图）
R.C. STRUCTURE AS PER
ENGINEER'S DETAIL

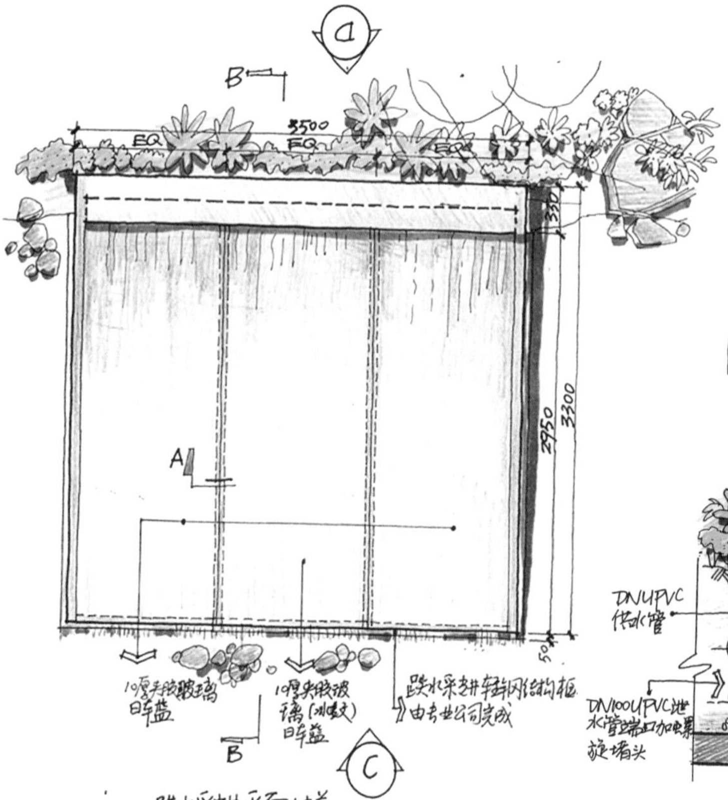
D
B
3500
EQ
EQ
EQ
330
2950
3300
A
50
B
C

跌水采光井平面大样

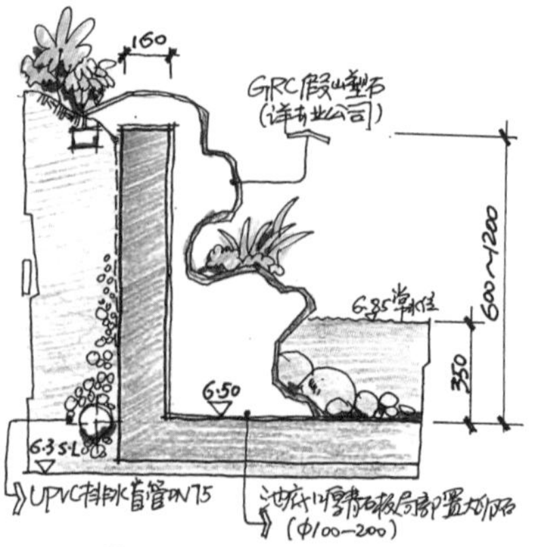
160
GRC假山塑石
(详专业公司)
6.85常水位
600~1200
350
6.50
6.35 S.L
UPVC排水管DN75
(Φ100-200)

① 跌水采光井假山塑石剖面示意

SCALE 1:15

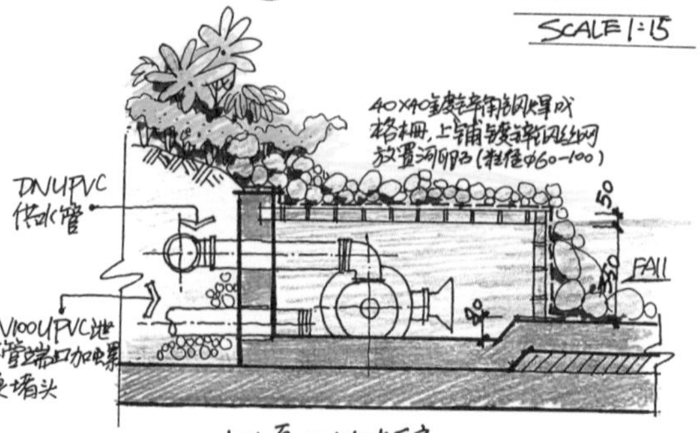
DN UPVC
供水管
50
350
FALL

水池泵坑做法示意

SCALE 1:60

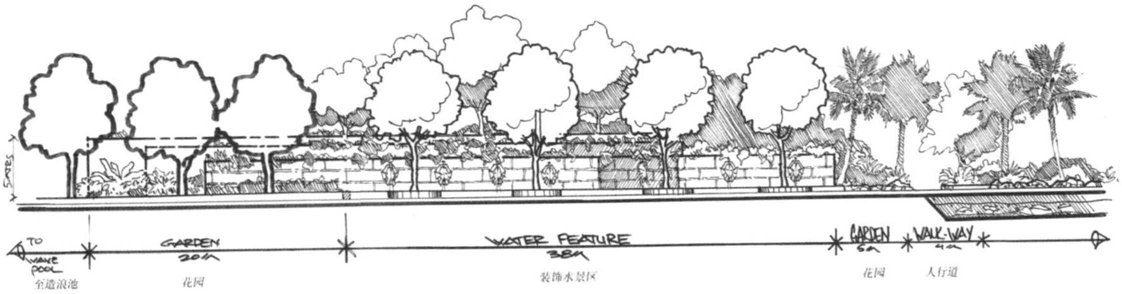
TO WAVE POOL
GARDEN
WATER FEATURE
GARDEN
WALK-WAY
至造浪池
花园
装饰水景区
花园
人行道

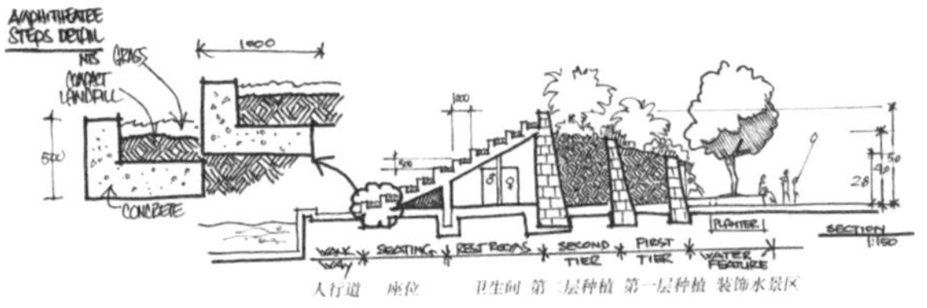
AMPHITHEATRE STEPS DETAIL
NTS
GRASS
COMPACT LANDFILL
CONCRETE
1000
500
PLANTER
SECTION 1:150
WALK WAY
SEATING
REST ROOMS
SECOND TIER
FIRST TIER
WATER FEATURE
人行道
座位
卫生间
第二层种植
第一层种植
装饰水景区

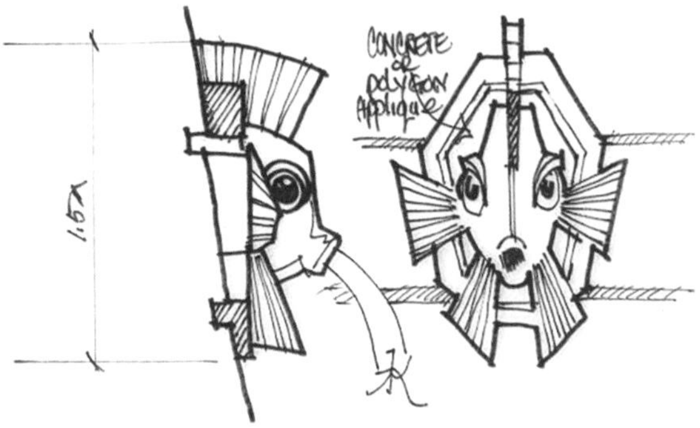
CONCRETE or POLYGON Applique

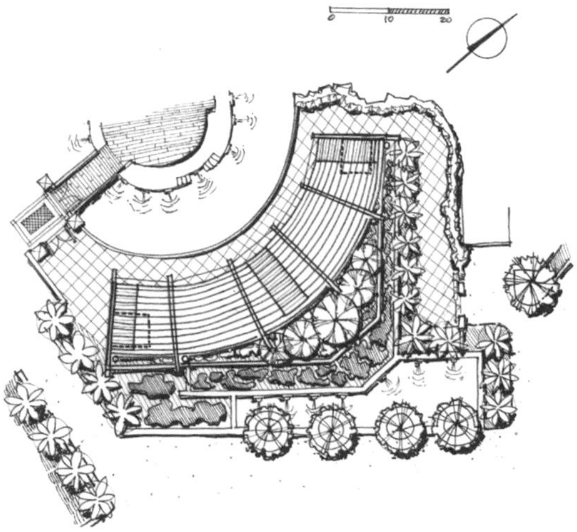

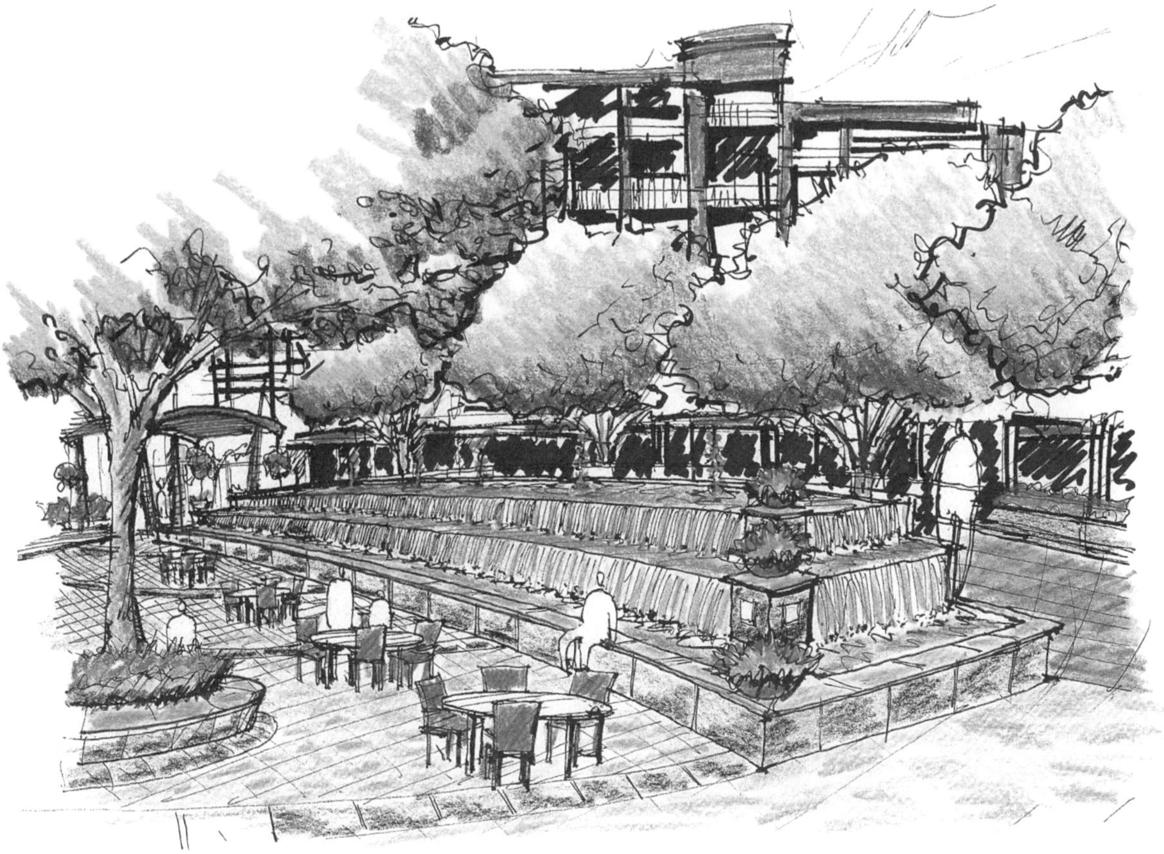

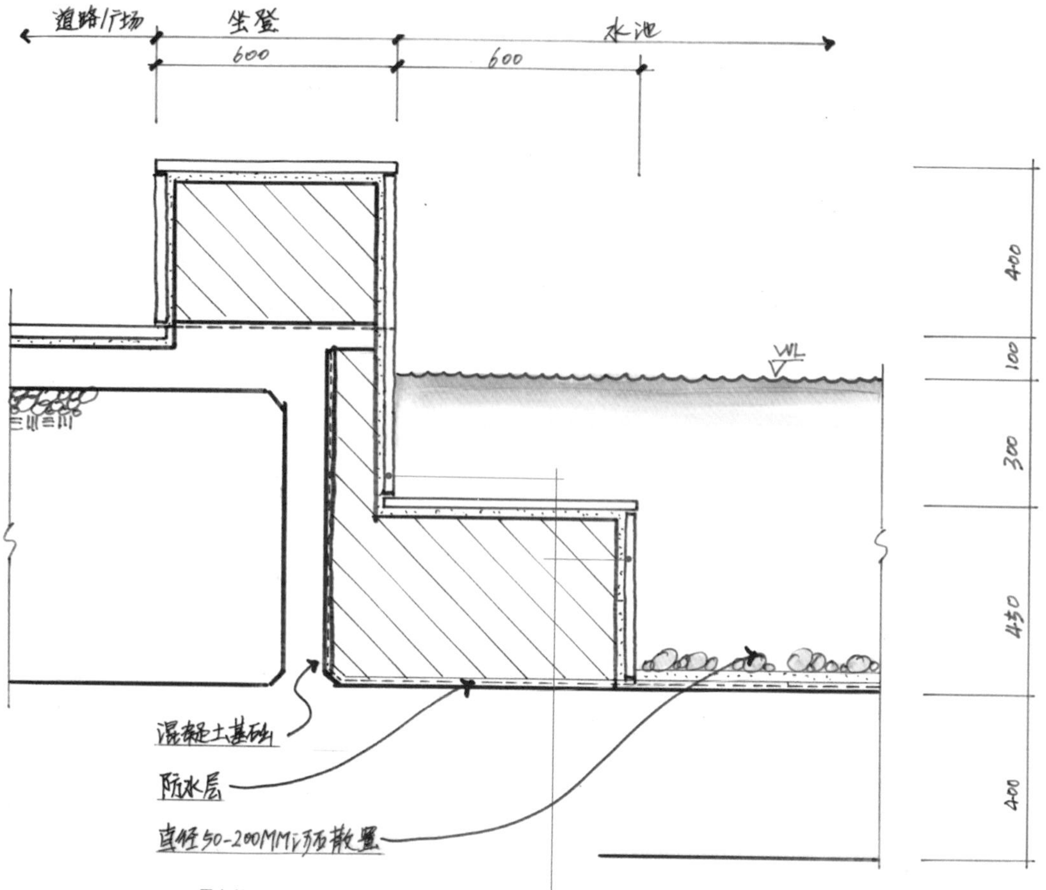
道路/广场
坐凳
水池
600
600
WL
400
100
300
450
400
混凝土基础
防水层
直径50-200MM河石散置
20mm厚自然刀劈面灰色板岩

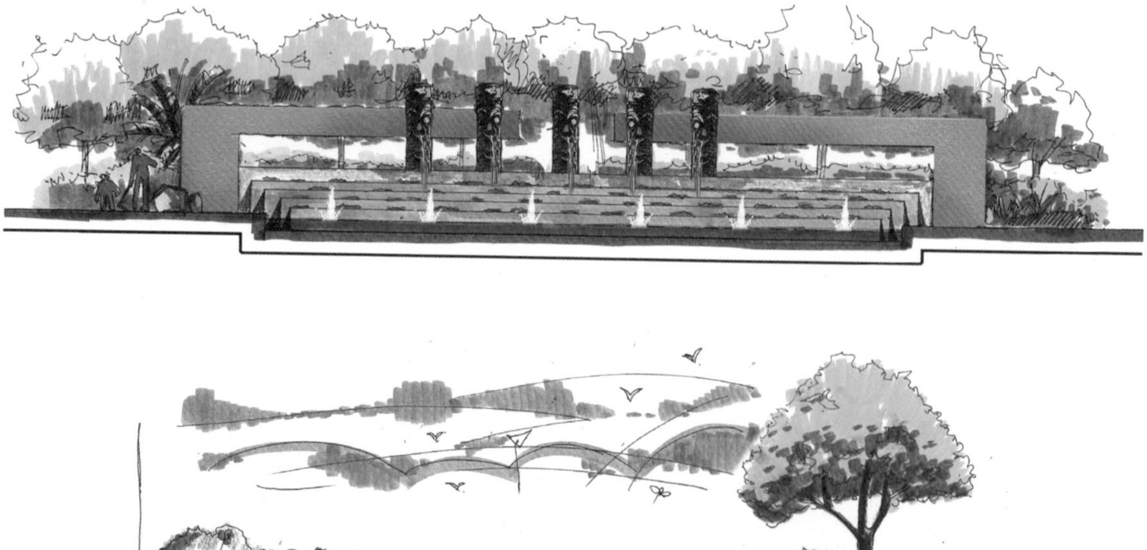

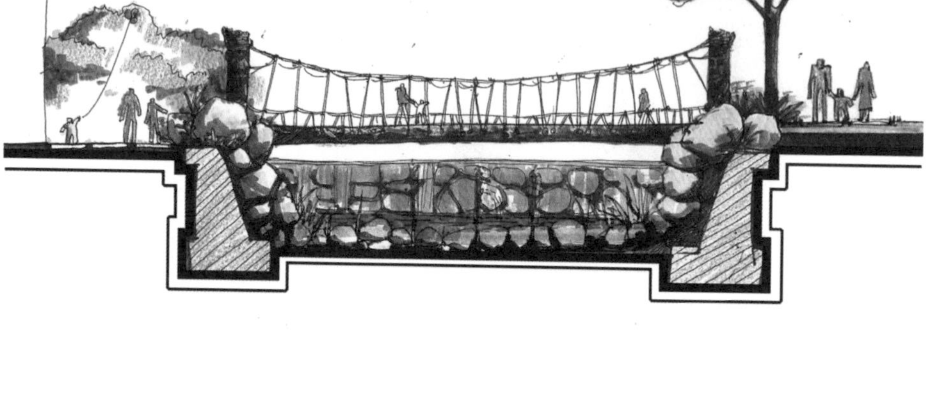

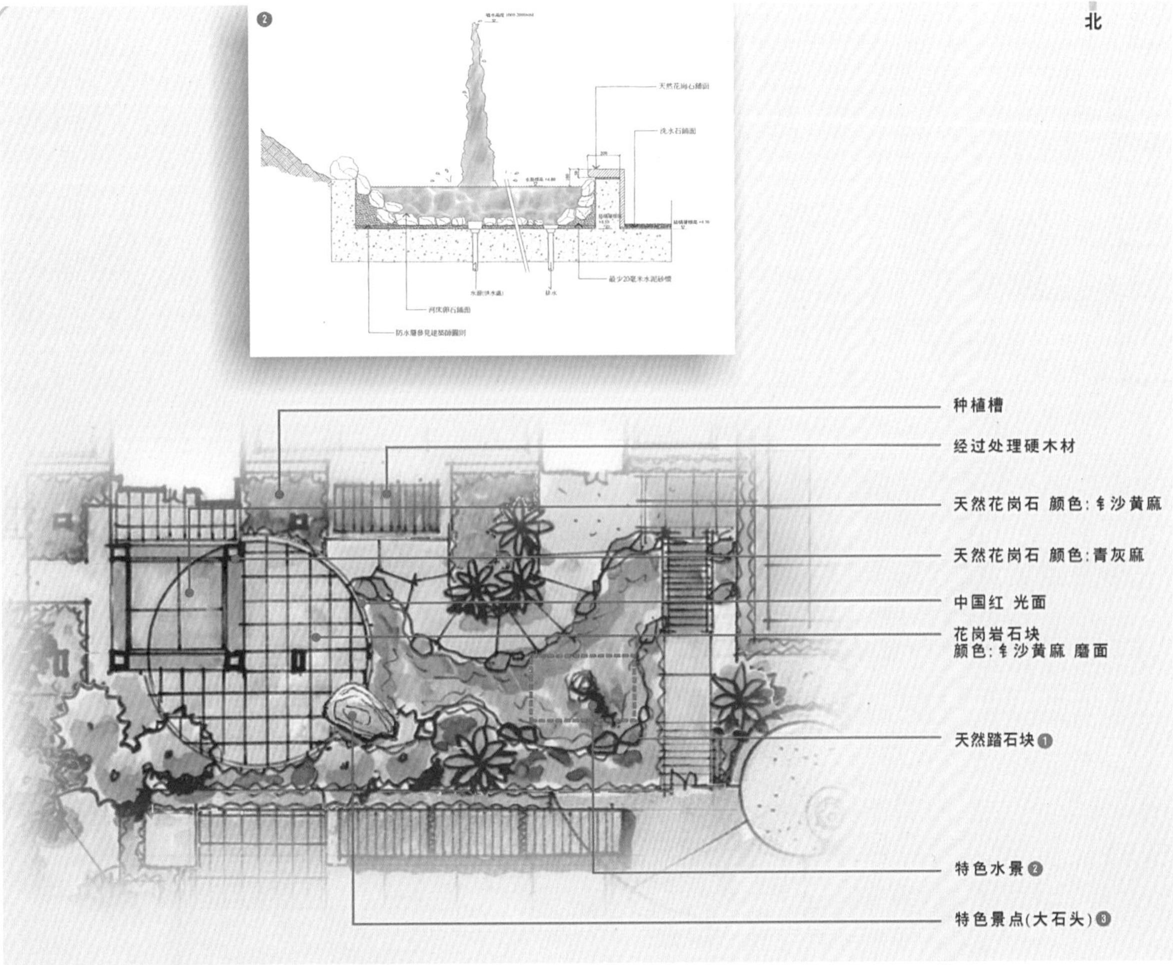
2
天然花崗石鋪面
洗水石鋪面
最少20毫米水泥砂漿
水源(供水處)
河床卵石鋪面
防水層參見建築師圖則
北
种植槽
经过处理硬木材
天然花岗石 颜色:钅沙黄麻
天然花岗石 颜色:青灰麻
中国红 光面
花岗岩石块
颜色:钅沙黄麻 磨面
天然踏石块 1
特色水景 2
特色景点(大石头) 3

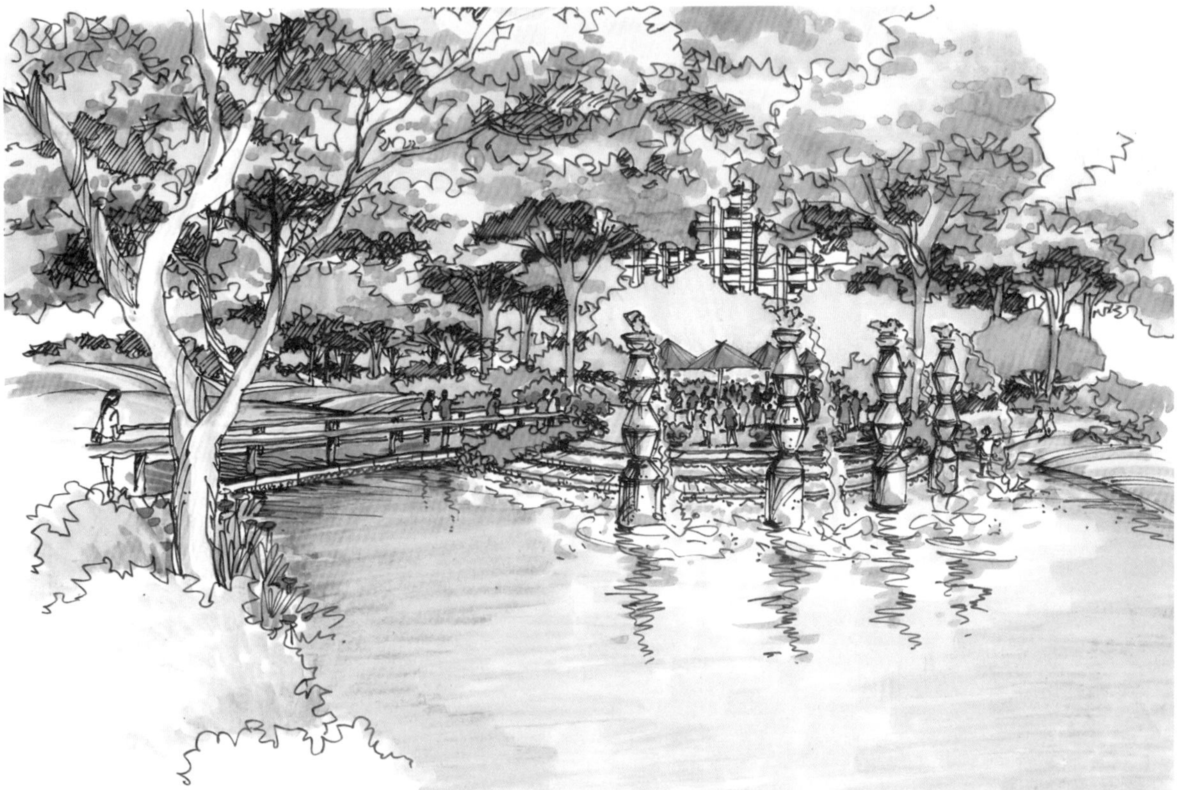

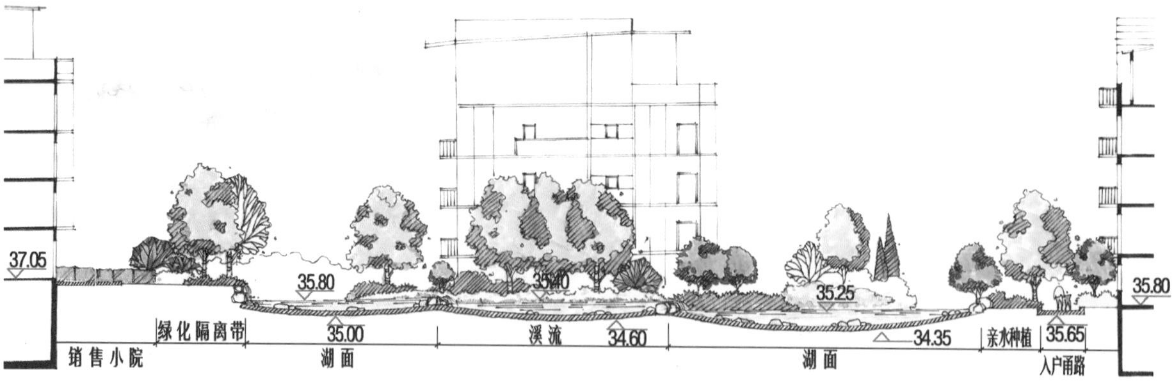
37.05
35.80
35.40
35.25
35.80
绿化隔离带
35.00
溪流
34.60
34.35
亲水种植
35.65
销售小院
湖面
湖面
入户甬路

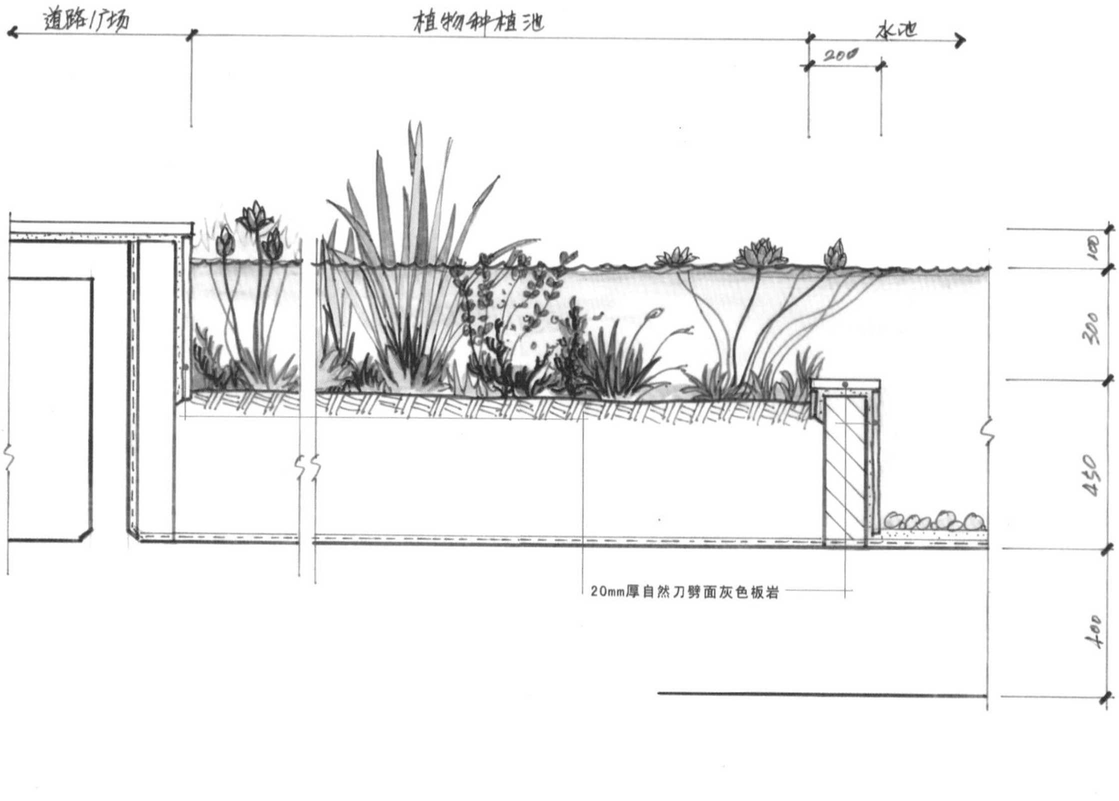
道路/广场
植物种植池
水池
200
100
300
450
400
20mm厚自然刀劈面灰色板岩

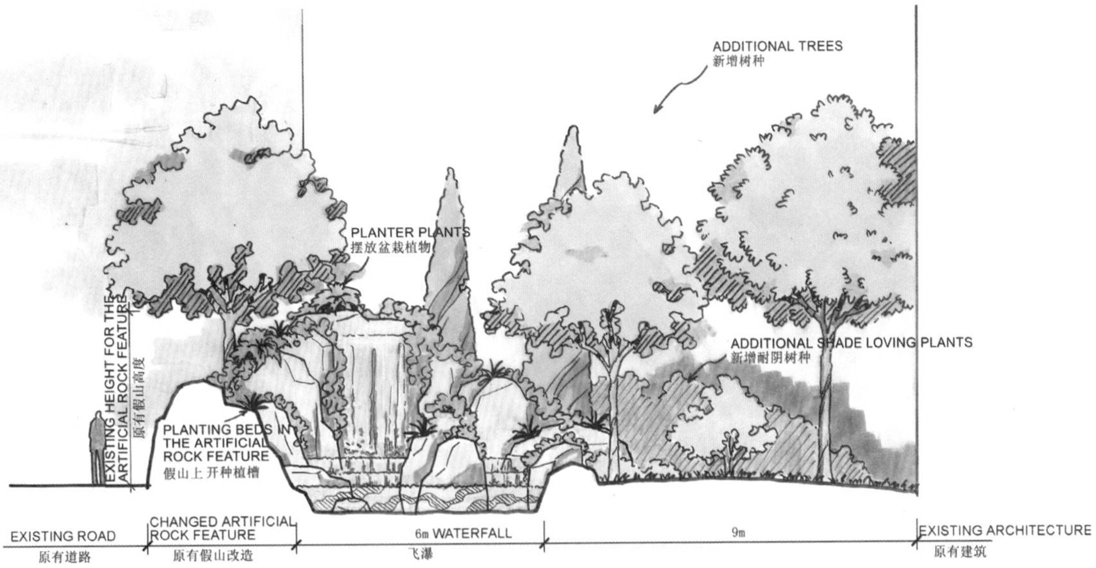
ADDITIONAL TREES
新增树种
PLANTER PLANTS
摆放盆栽植物
EXISTING HEIGHT FOR THE ARTIFICIAL ROCK FEATURE
原有假山高度
ADDITIONAL SHADE LOVING PLANTS
新增耐阴树种
PLANTING BEDS IN THE ARTIFICIAL ROCK FEATURE
假山上开种植槽
EXISTING ROAD
原有道路
CHANGED ARTIFICIAL ROCK FEATURE
原有假山改造
6m WATERFALL
飞瀑
9m
EXISTING ARCHITECTURE
原有建筑

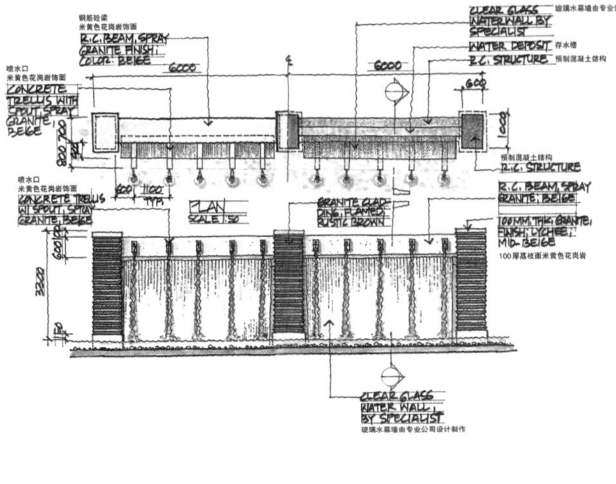
钢筋砼梁
米黄色花岗岩饰面
R.C. BEAM, SPRAY GRANITE FINISH; COLOR: BEIGE
CLEAR GLASS WATERWALL BY SPECIALIST
玻璃水幕墙由专业公司设计制作
WATER DEPOSIT
存水槽
R.C. STRUCTURE
预制混凝土结构
喷水口
米黄色花岗岩饰面
CONCRETE TRELLIS WITH SPOUT, SPRAY GRANITE, BEIGE
6000
6000
600
PLAN
SCALE 1:50
GRANITE CLADDING, FLAMED RUSTIC BROWN
R.C. BEAM, SPRAY GRANITE, BEIGE
100MM THK GRANITE, FINISH: LYCHEE, MID-BEIGE
100厚荔枝面米黄色花岗岩
CONCRETE TRELLIS W/ SPOUT, SPRAY GRANITE, BEIGE
3200
CLEAR GLASS WATER WALL, BY SPECIALIST
玻璃水幕墙由专业公司设计制作

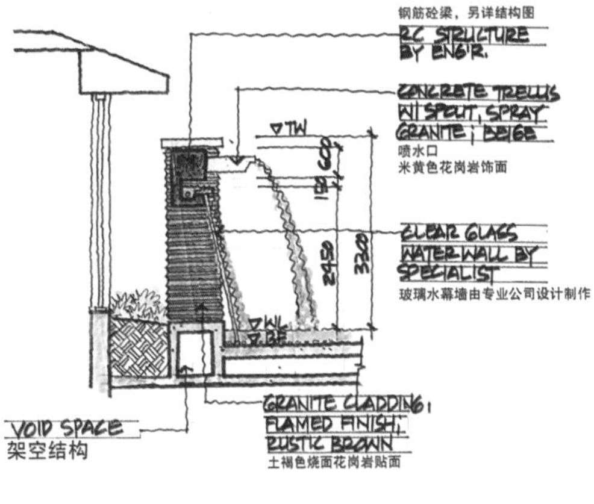
钢筋砼梁，另详结构图
RC STRUCTURE BY ENG'R.
CONCRETE TRELLIS W/ SPOUT, SPRAY GRANITE; BEIGE
喷水口
米黄色花岗岩饰面
TW
WL
BF
2450
3200
CLEAR GLASS WATER WALL BY SPECIALIST
玻璃水幕墙由专业公司设计制作
VOID SPACE
架空结构
GRANITE CLADDING, FLAMED FINISH, RUSTIC BROWN
土褐色烧面花岗岩贴面

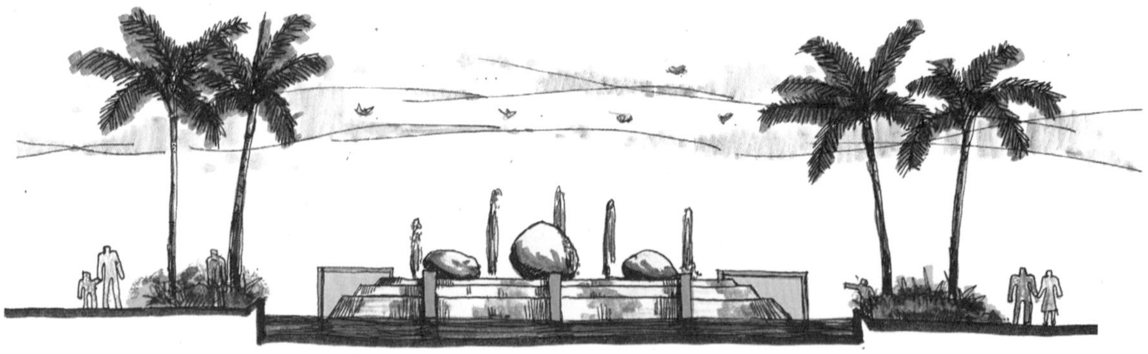

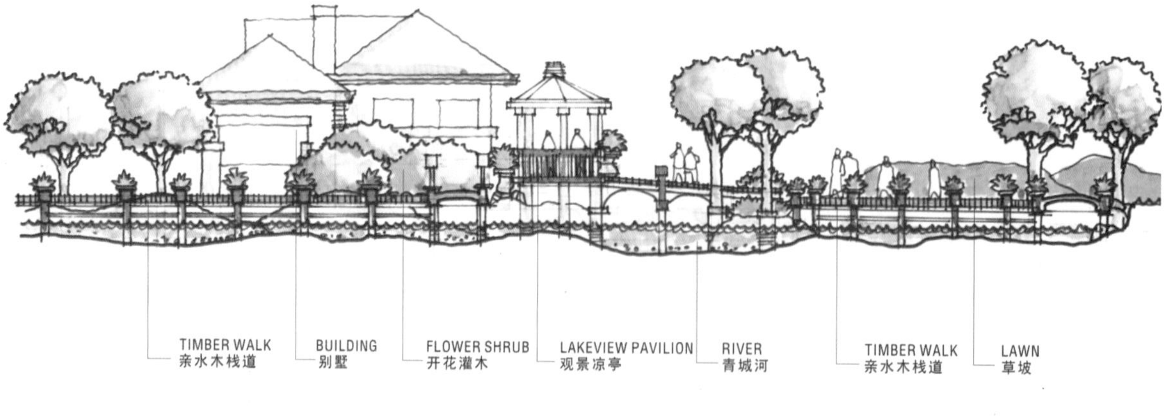
TIMBER WALK
亲水木栈道
BUILDING
别墅
FLOWER SHRUB
开花灌木
LAKEVIEW PAVILION
观景凉亭
RIVER
青城河
TIMBER WALK
亲水木栈道
LAWN
草坡

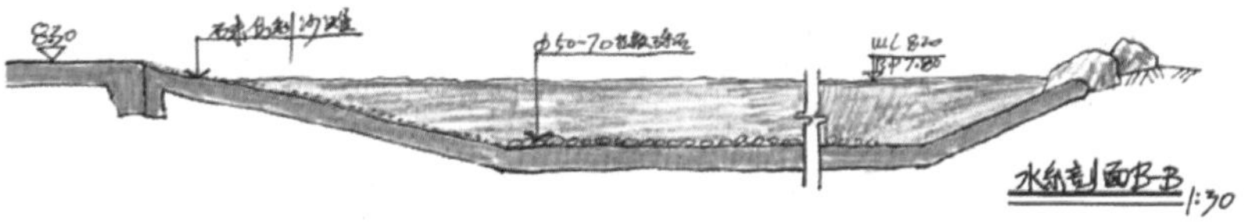
830
WL 8.20
BP 7.80
水系剖面B-B 1:30

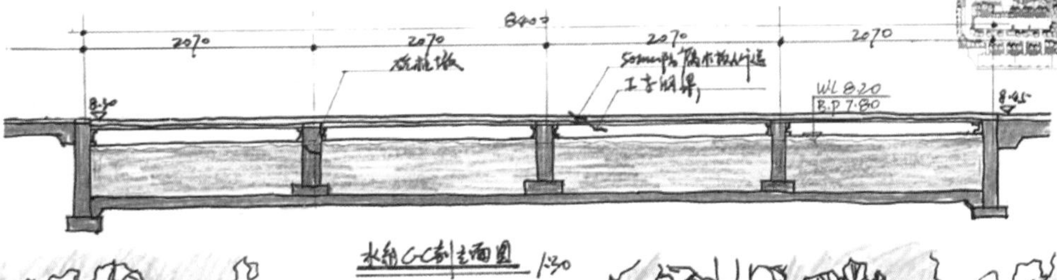
8400
2070
2070
2070
2070
WL 8.20
B.P 7.80
8.45
水系C-C剖立面图 1:30

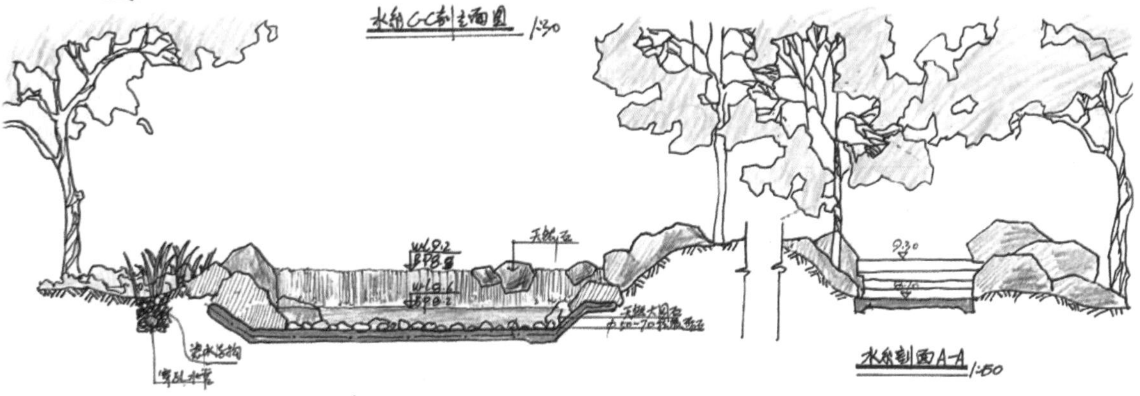
天然石
9.30
水系剖面A-A 1:50

BRANCHING TREES AND SHRUB
喬灌木混合種植
VIEWING BRIDGE
景觀橋
CENTRAL FOUNTAIN
中心涌泉
LEISURE PLAYGROUND
休閒活動場地
STONE PAVING
石材鋪設的地坪
TERRACED BORDER
迭落式花壇
PLANTING HOLE
臨水樹穴
LEISURE SEATING
休閒座椅
TERRACED WATER VIEWING
水景・迭水
STEP TILING WITH GRANITE
石材鋪砌臺階
STONE BOARDWAIK UP WATER
涉水石板

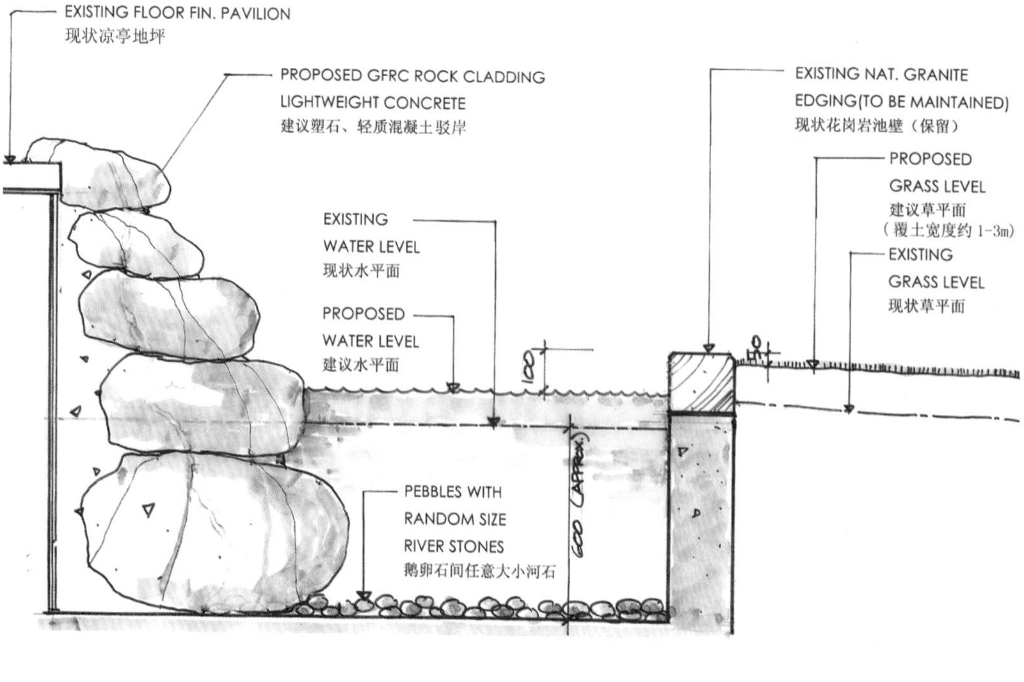
EXISTING FLOOR FIN. PAVILION
现状凉亭地坪
PROPOSED GFRC ROCK CLADDING
LIGHTWEIGHT CONCRETE
建议塑石、轻质混凝土驳岸
EXISTING NAT. GRANITE
EDGING(TO BE MAINTAINED)
现状花岗岩池壁（保留）
PROPOSED
GRASS LEVEL
建议草平面
（覆土宽度约 1-3m）
EXISTING
GRASS LEVEL
现状草平面
EXISTING
WATER LEVEL
现状水平面
PROPOSED
WATER LEVEL
建议水平面
180
600 (APPROX)
PEBBLES WITH
RANDOM SIZE
RIVER STONES
鹅卵石间任意大小河石

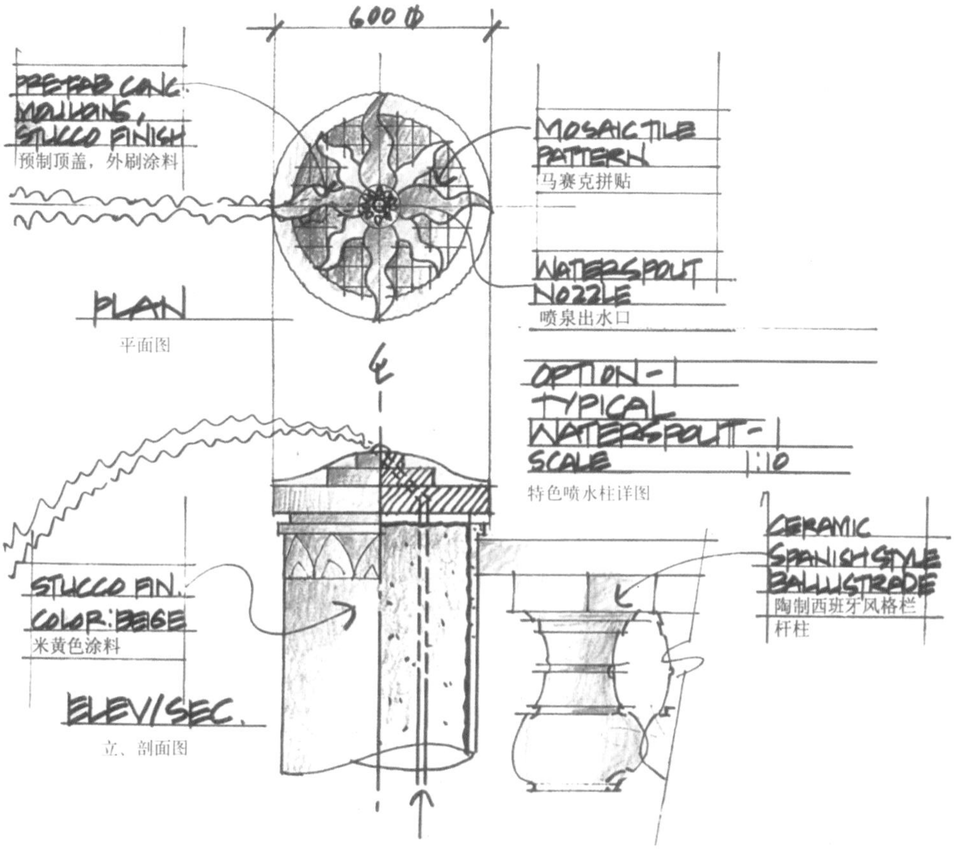
600 Φ
PREFAB CONC. MOULDINGS, STUCCO FINISH
预制顶盖，外刷涂料
MOSAIC TILE PATTERN
马赛克拼贴
WATERSPOUT NOZZLE
喷泉出水口
PLAN
平面图
OPTION-1 TYPICAL WATERSPOUT-SCALE 1:10
特色喷水柱详图
CERAMIC SPANISH STYLE BALLUSTRADE
陶制西班牙风格栏杆柱
STUCCO FIN. COLOR: BEIGE
米黄色涂料
ELEV/SEC.
立、剖面图

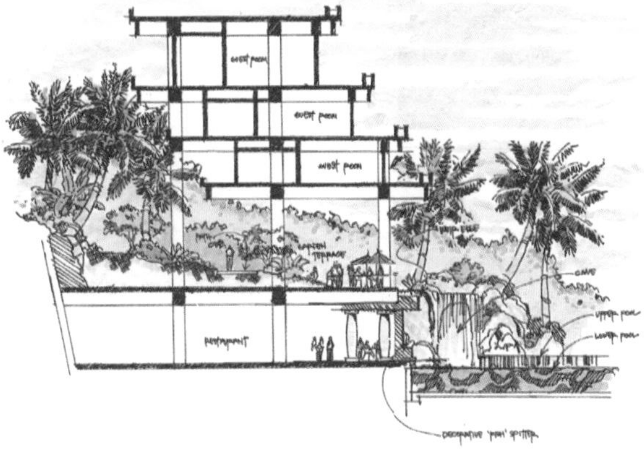

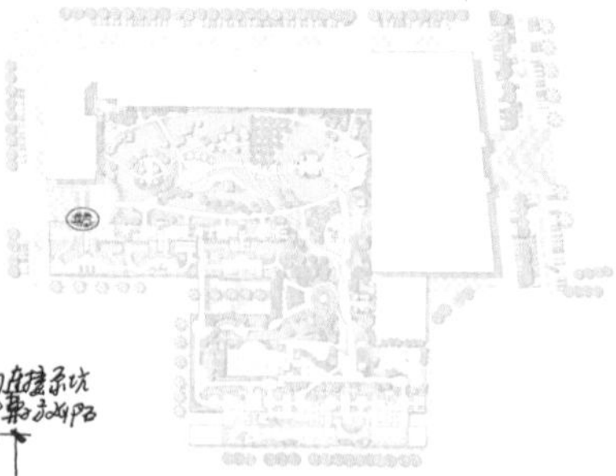

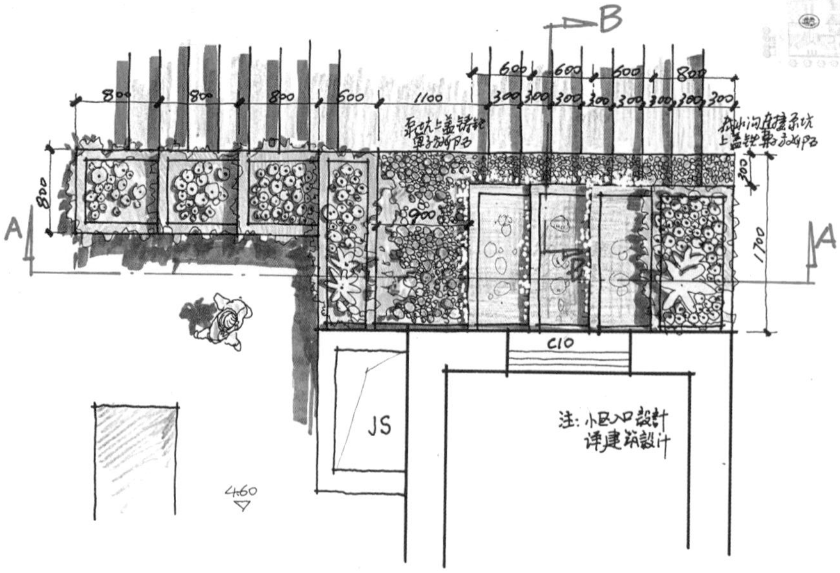
B
A
A
C10
JS
注: 小区入口設計
详建筑設计
4.60

小区主入口叠水花池平面大样

SCALE 1:30

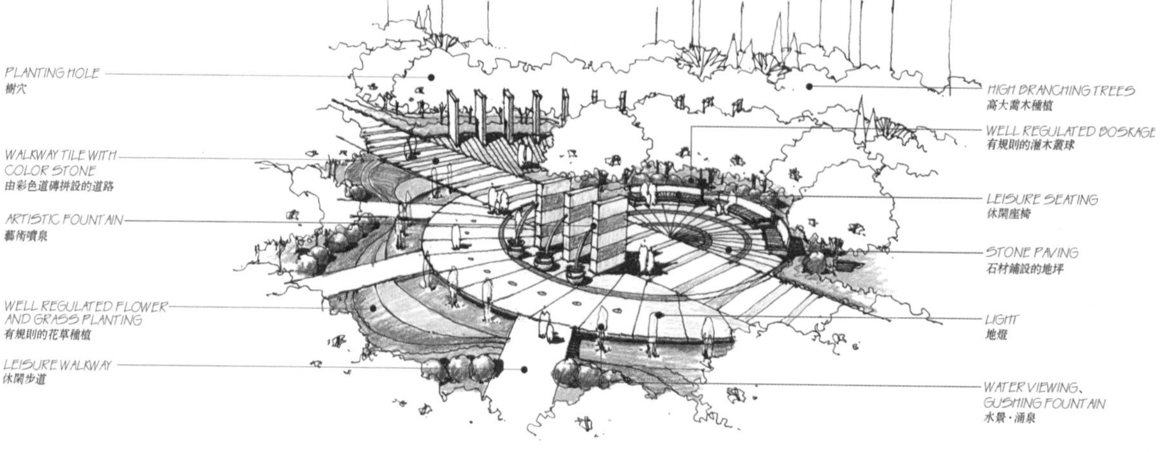
PLANTING HOLE
樹穴
WALKWAY TILE WITH COLOR STONE
由彩色道磚拼設的道路
ARTISTIC FOUNTAIN
藝術噴泉
WELL REGULATED FLOWER AND GRASS PLANTING
有規則的花草種植
LEISURE WALKWAY
休閒步道
HIGH BRANCHING TREES
高大喬木種植
WELL REGULATED BOSKAGE
有規則的灌木叢球
LEISURE SEATING
休閒座椅
STONE PAVING
石材鋪設的地坪
LIGHT
地燈
WATER VIEWING、GUSHING FOUNTAIN
水景・涌泉

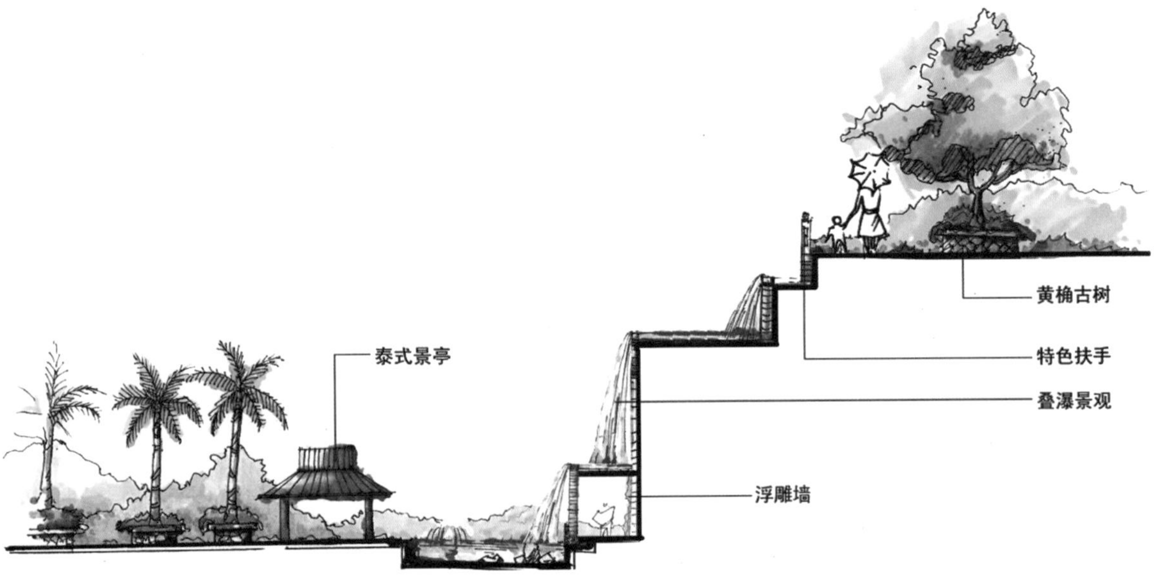
泰式景亭
黄桷古树
特色扶手
叠瀑景观
浮雕墙

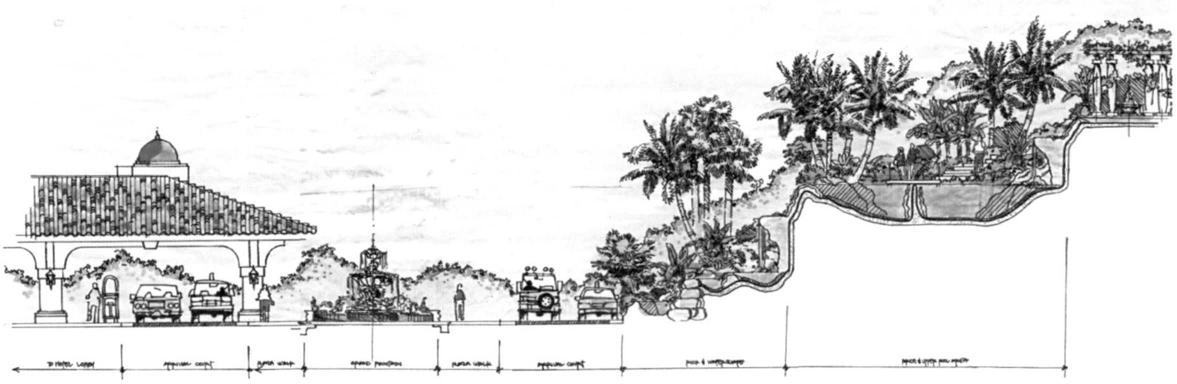

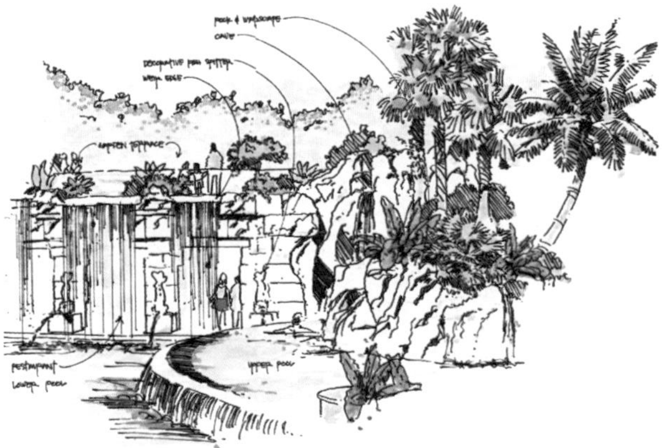

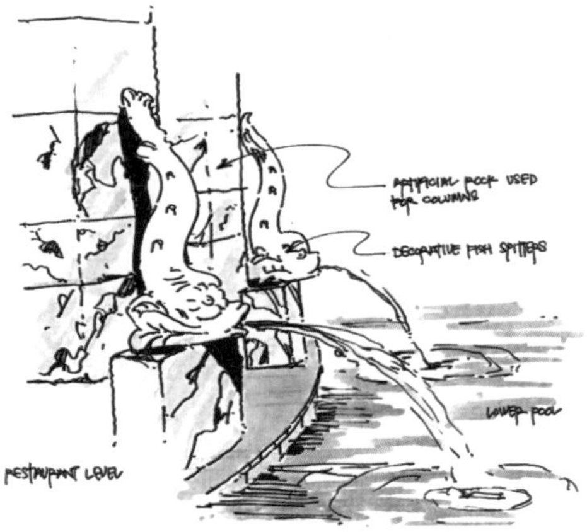
ARTIFICIAL ROCK USED FOR COLUMNS
DECORATIVE FISH SPITTERS
LOWER POOL
RESTAURANT LEVEL

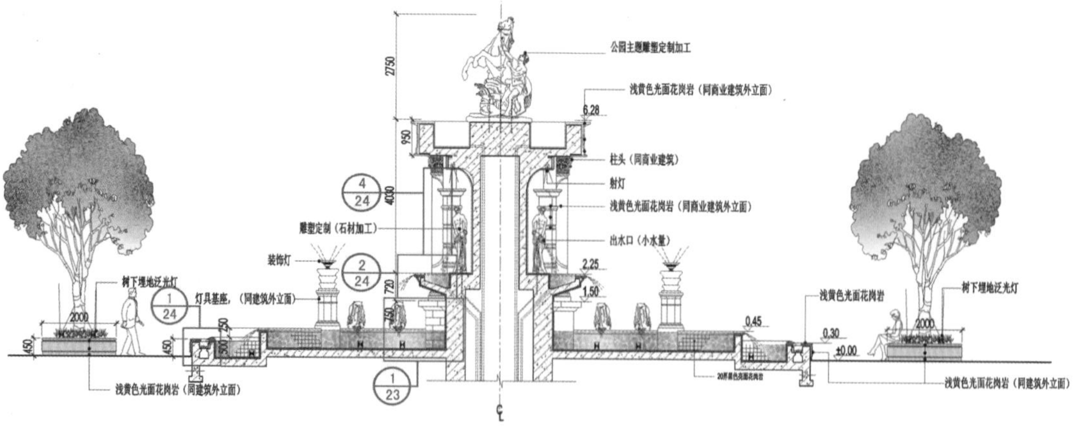
公园主题雕塑定制加工
2750
浅黄色光面花岗岩（同商业建筑外立面）
6.28
950
柱头（同商业建筑）
射灯
4030
浅黄色光面花岗岩（同商业建筑外立面）
雕塑定制（石材加工）
出水口（小水量）
装饰灯
2.25
1.50
720
750
树下埋地泛光灯
灯具基座，（同建筑外立面）
2000
450
250
500
0.45
0.30
±0.00
浅黄色光面花岗岩
树下埋地泛光灯
浅黄色光面花岗岩（同建筑外立面）
20厚黑色光面花岗岩
浅黄色光面花岗岩（同建筑外立面）
4
24
2
24
1
24
1
23

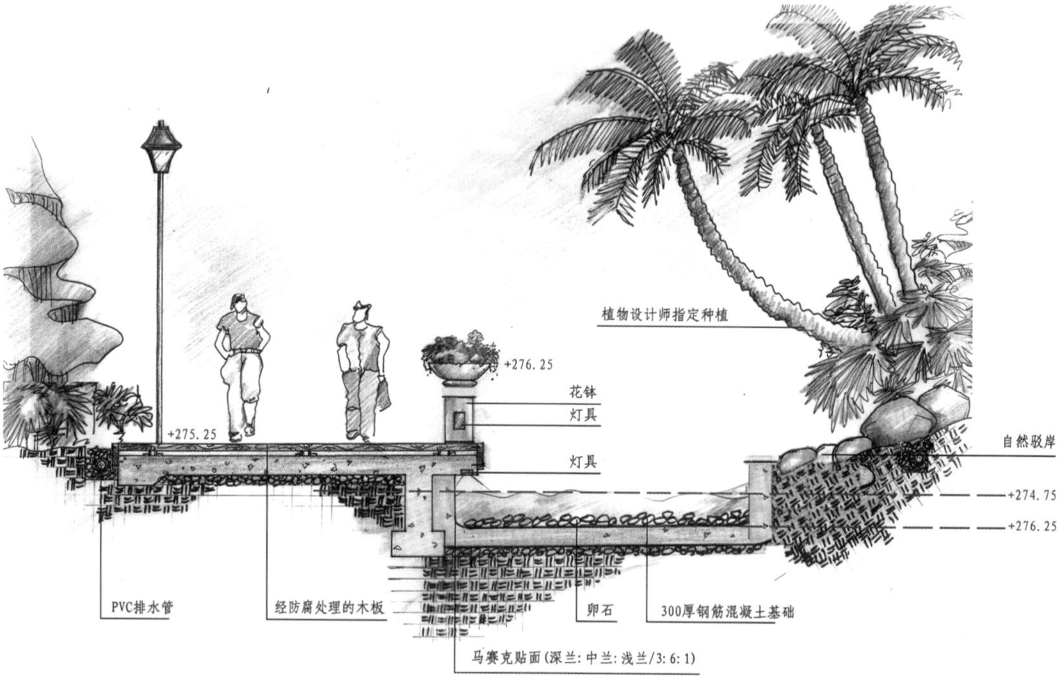
植物设计师指定种植
+276.25
花钵
灯具
+275.25
灯具
自然驳岸
+274.75
+276.25
PVC排水管
经防腐处理的木板
卵石
300厚钢筋混凝土基础
马赛克贴面（深兰：中兰：浅兰/3：6：1）

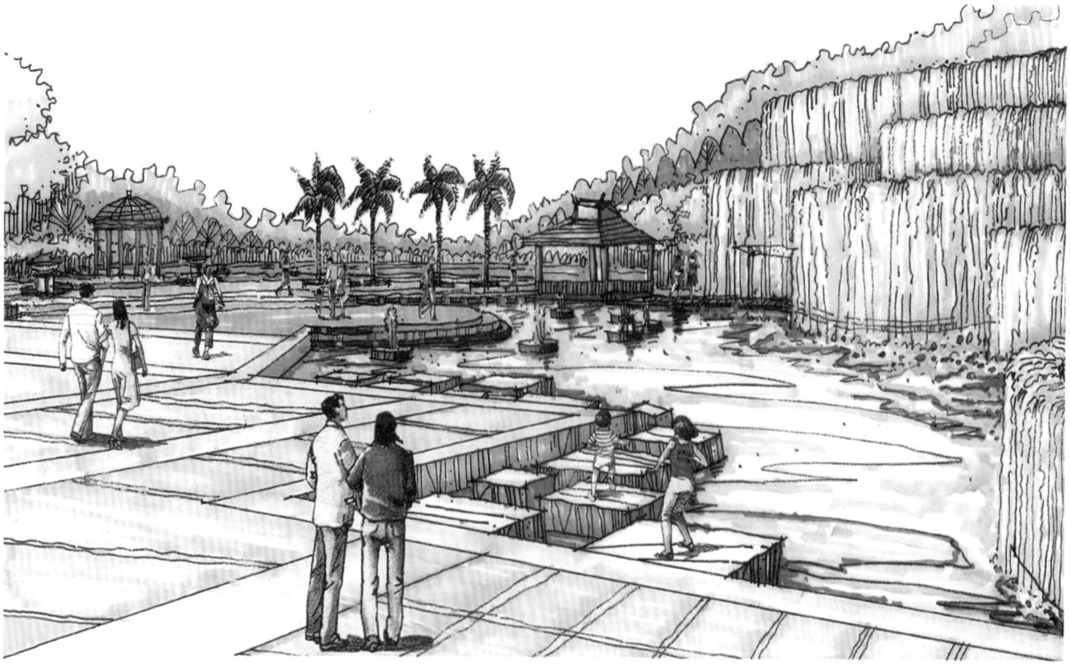

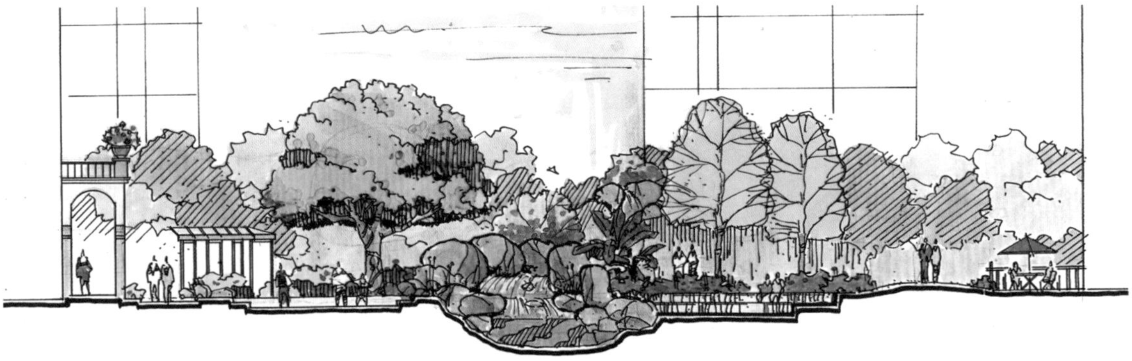

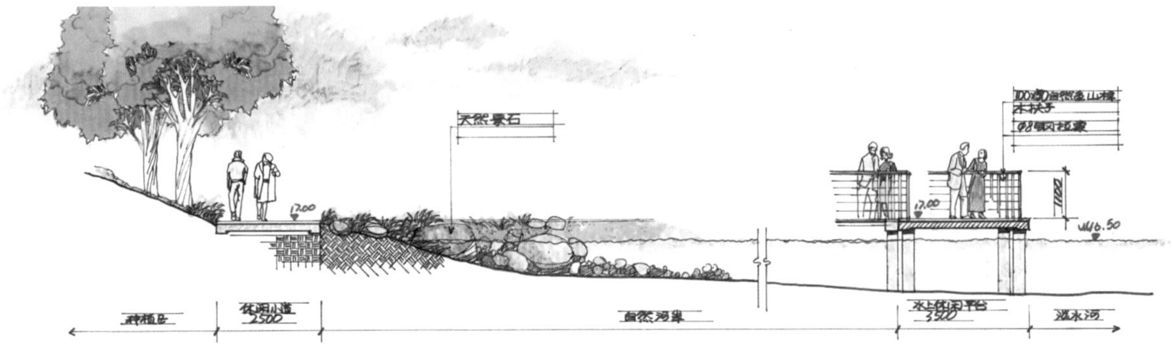
天然景石
1200
1100
种植区
休闲小道
2500
自然河岸
水上休闲平台
3500

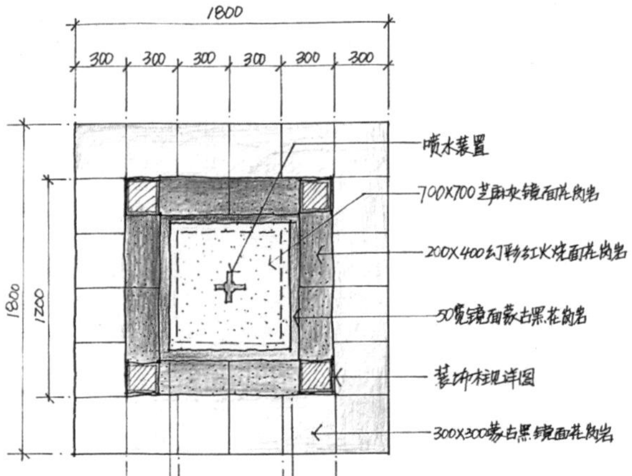
1800
300
300
300
300
300
300
1800
1200
喷水装置
700X700芝麻灰镜面花岗岩
200X400幻彩红火烧面花岗岩
50宽镜面蒙古黑花岗岩
装饰柱见详图
300X300蒙古黑镜面花岗岩
250
250
50
50
600
700

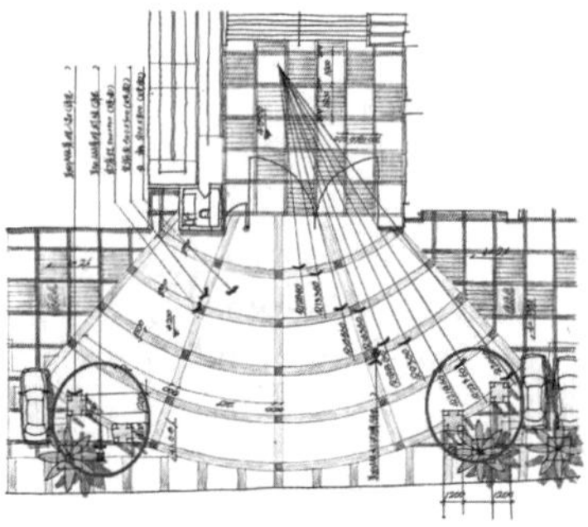

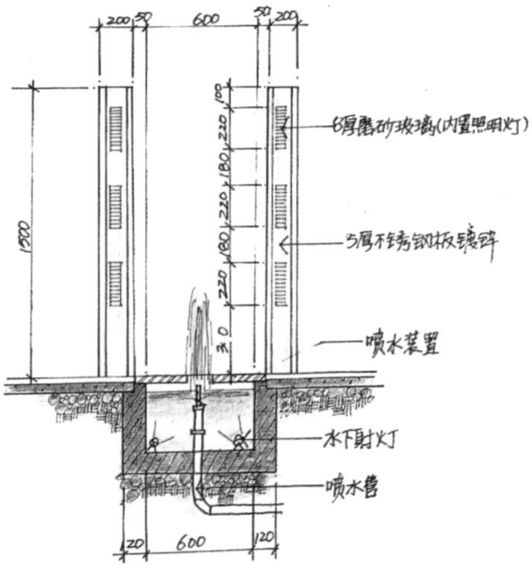
200
50
600
50
200
1500
6厚磨砂玻璃(内置照明灯)
5厚不锈钢板镀锌
喷水装置
水下射灯
喷水管
20
600
120

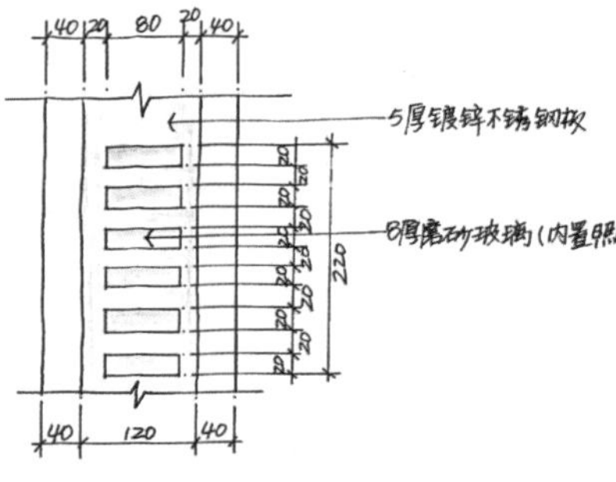
40
20
80
20
40
5厚镀锌不锈钢板
8厚磨砂玻璃(内置照明灯)
220
40
120
40

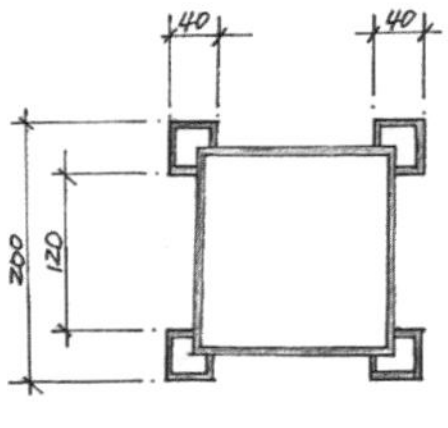
40
40
200
120

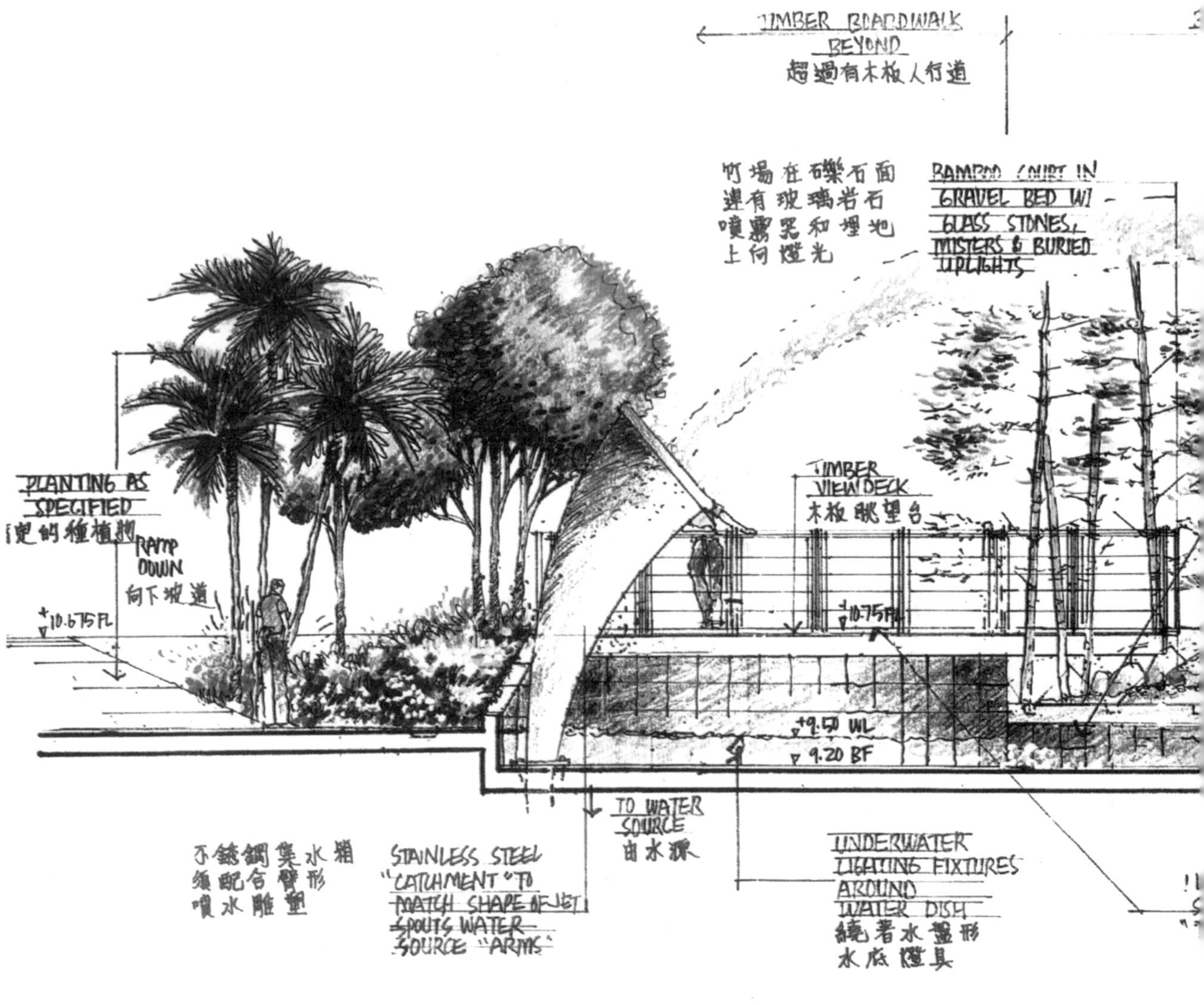
TIMBER BOARDWALK BEYOND
超過有木板人行道
竹場在礫石面
連有玻璃岩石
噴霧器和埋地
上向燈光
BAMBOO COURT IN GRAVEL BED W/ GLASS STONES, MISTERS & BURIED UPLIGHTS
PLANTING AS SPECIFIED
定的種植物
RAMP DOWN
向下坡道
10.675 FL
TIMBER VIEW DECK
木板眺望台
10.75 FL
+9.50 WL
9.20 BF
TO WATER SOURCE
由水源
不銹鋼集水箱
須配合臂形
噴水雕塑
STAINLESS STEEL "CATCHMENT" TO MATCH SHAPE OF JET SPOUTS WATER SOURCE "ARMS"
UNDERWATER LIGHTING FIXTURES AROUND WATER DISH
繞著水盤形
水底燈具

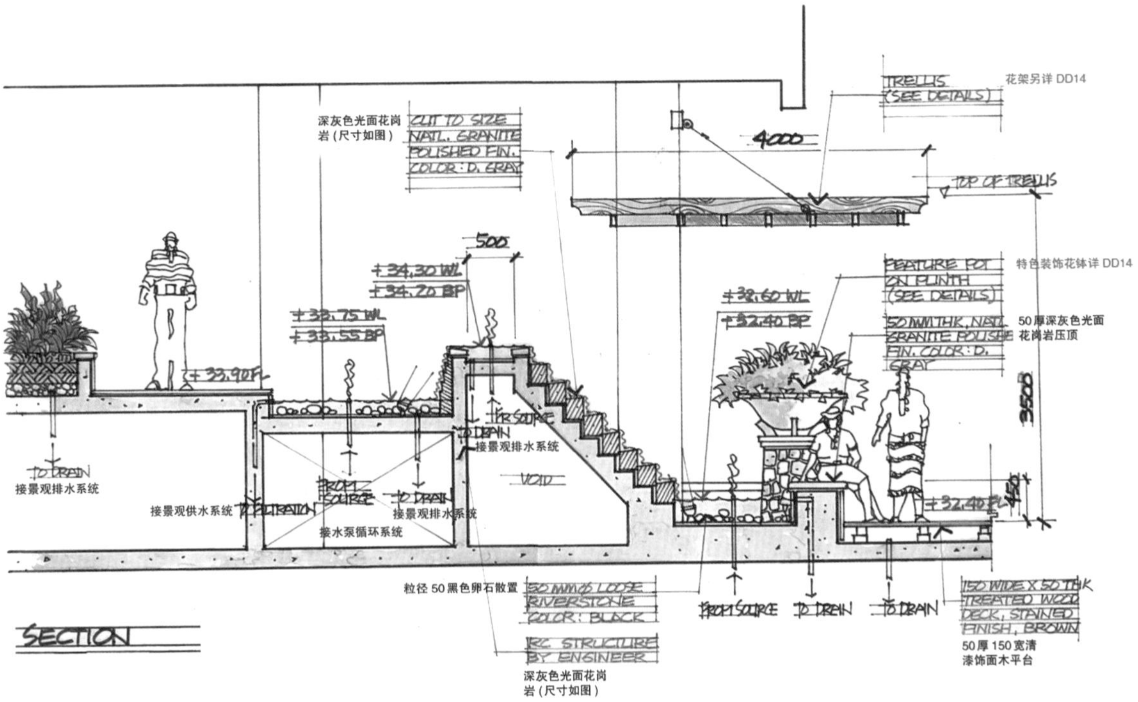

TRELLIS (SEE DETAILS)
花架另详 DD14
深灰色光面花岗岩（尺寸如图）
CUT TO SIZE NATL. GRANITE POLISHED FIN. COLOR: D. GRAY
4000
TOP OF TRELLIS
500
+34.30 WL
+34.20 BP
+33.75 WL
+33.55 BP
+33.90 FL
FEATURE POT ON PLINTH (SEE DETAILS)
特色装饰花钵详 DD14
+32.60 WL
+32.40 BP
50 MM THK. NATL. GRANITE POLISHED FIN. COLOR: D. GRAY
50厚深灰色光面花岗岩压顶
3500
TO DRAIN
接景观排水系统
PUMP SOURCE
TO DRAIN
接景观排水系统
接景观供水系统
TO FILTRATION
接水泵循环系统
VOID
+32.40 FL
粒径 50 黑色卵石散置
50 MM Ø LOOSE RIVERSTONE COLOR: BLACK
RC STRUCTURE BY ENGINEER
深灰色光面花岗岩（尺寸如图）
PUMP SOURCE
TO DRAIN
TO DRAIN
150 WIDE X 50 THK TREATED WOOD DECK, STAINED FINISH, BROWN
50厚 150 宽清漆饰面木平台
SECTION

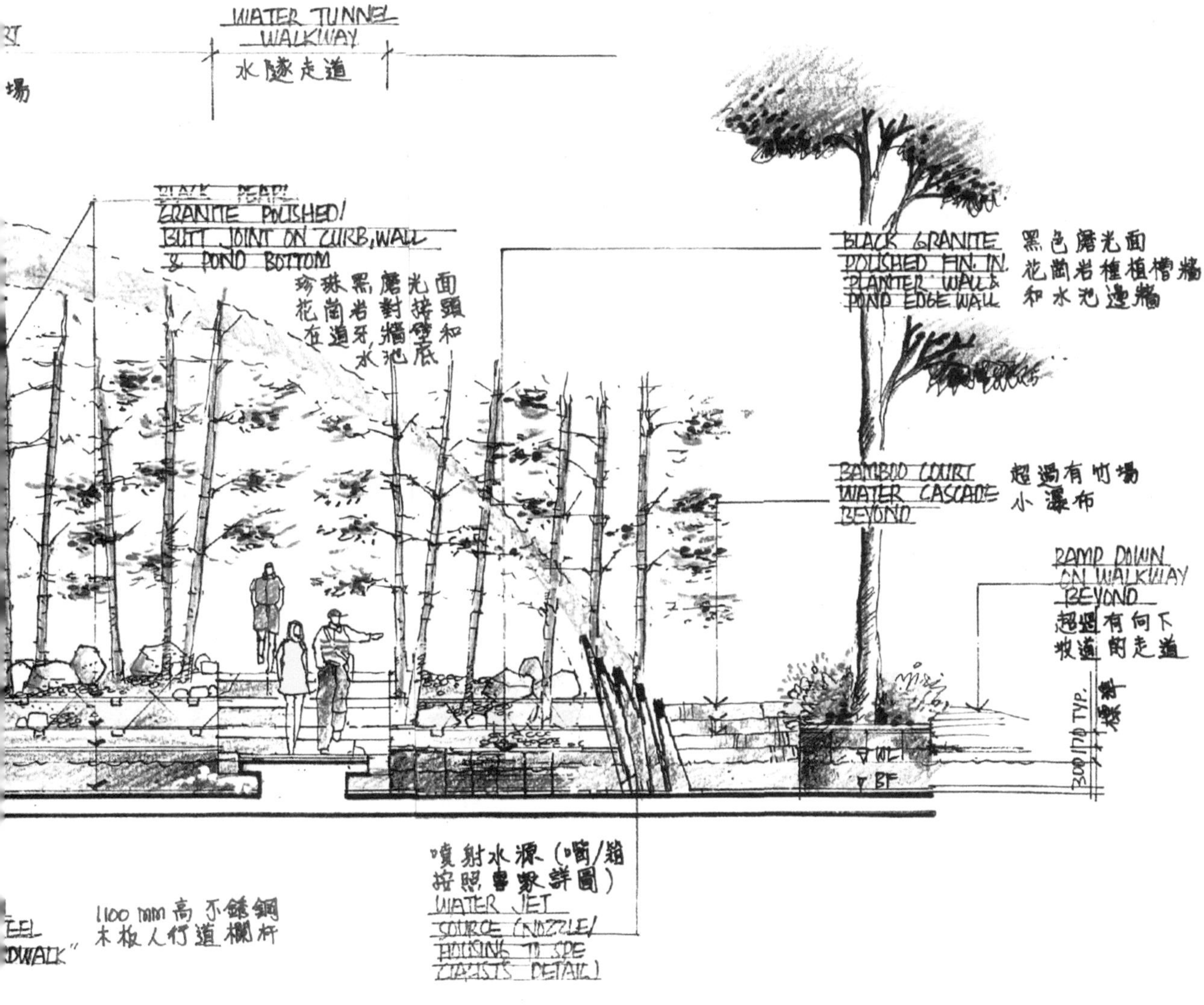

WATER TUNNEL WALKWAY
水隧走道
BLACK PEARL GRANITE POLISHED/ BUTT JOINT ON CURB, WALL & POND BOTTOM
珍珠黑磨光面花崗岩對接頭在道牙,牆壁和水池底
BLACK GRANITE POLISHED FIN. IN. PLANTER WALL & POND EDGE WALL
黑色磨光面花崗岩種植槽牆和水池邊牆
BAMBOO COURT WATER CASCADE BEYOND
超過有竹場小瀑布
RAMP DOWN ON WALKWAY BEYOND
超過有向下坡道的走道
300/170 TYP.
WL
BF
噴射水源（噴/箱按照專家詳圖）
WATER JET SOURCE (NOZZLE/ HOUSING TO SPECIALISTS DETAIL)
1100 mm 高不銹鋼木板人行道欄杆

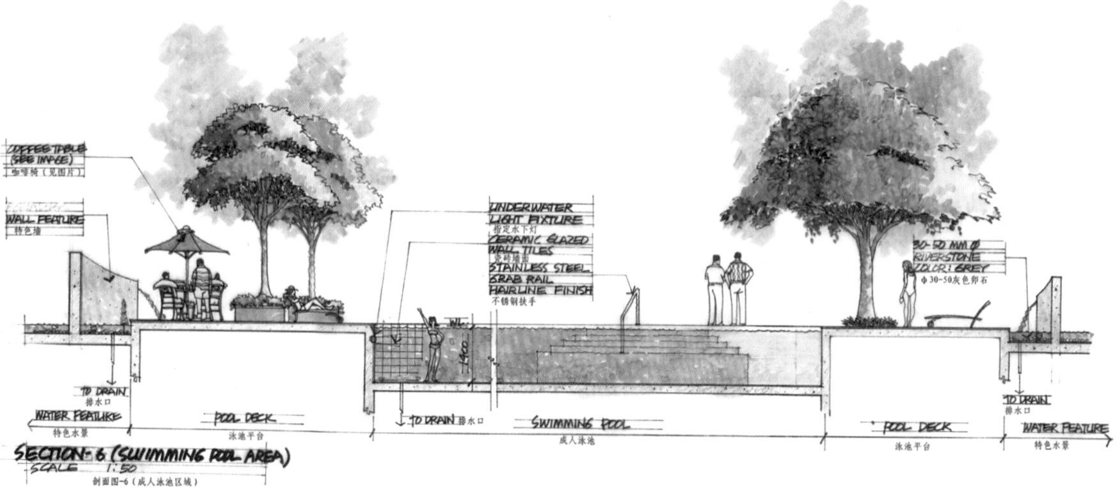
COFFEE TABLE
(SEE IMAGE)
咖啡椅（见图片）
WALL FEATURE
特色墙
UNDERWATER
LIGHT FIXTURE
指定水下灯
CERAMIC GLAZED
WALL TILES
瓷砖墙面
STAINLESS STEEL
GRAB RAIL
HAIRLINE FINISH
不锈钢扶手
WL
1400
30-50 MM Ø
RIVERSTONE
COLOR: GREY
ф30-50灰色卵石
TO DRAIN
排水口
WATER FEATURE
特色水景
POOL DECK
泳池平台
TO DRAIN 排水口
SWIMMING POOL
成人泳池
POOL DECK
泳池平台
TO DRAIN
排水口
WATER FEATURE
特色水景
SECTION-6 (SWIMMING POOL AREA)
SCALE 1:50
剖面图-6（成人泳池区域）

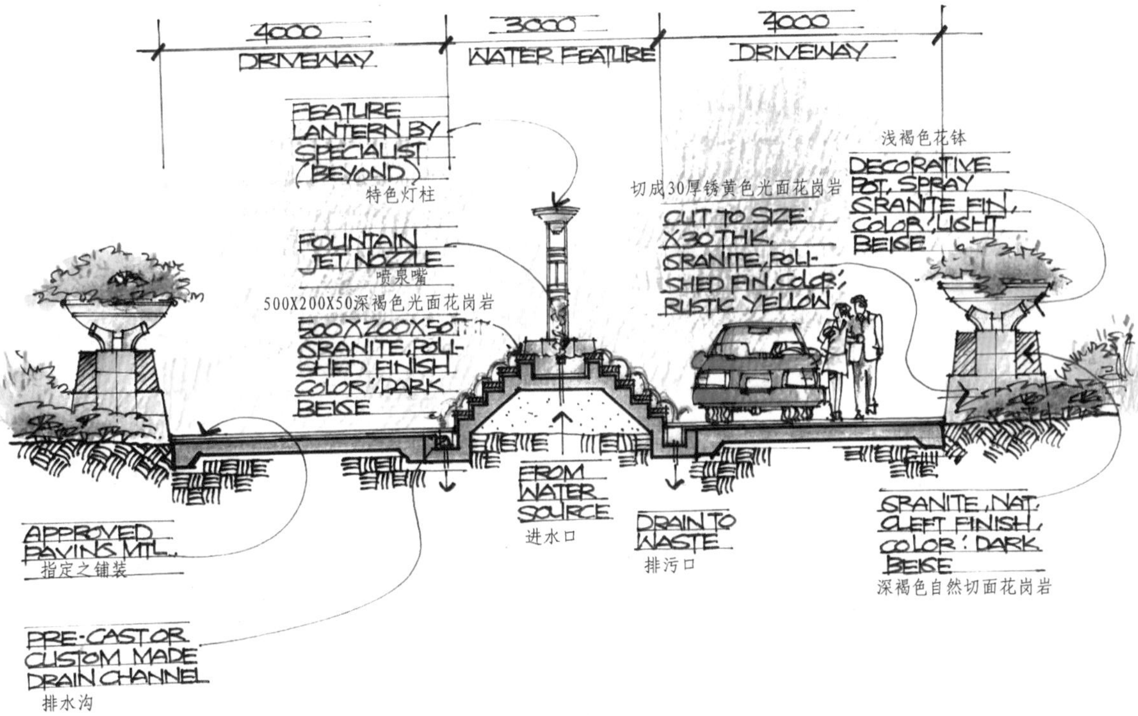
4000
DRIVEWAY
3000
WATER FEATURE
4000
DRIVEWAY
FEATURE LANTERN BY SPECIALIST (BEYOND)
特色灯柱
FOUNTAIN JET NOZZLE
喷泉嘴
500X200X50深褐色光面花岗岩
500X200X50 GRANITE, POLISHED FINISH COLOR: DARK BEIGE
切成30厚锈黄色光面花岗岩
CUT TO SIZE X30 THK. GRANITE, POLISHED FIN. COLOR: RUSTIC YELLOW
浅褐色花钵
DECORATIVE POT, SPRAY GRANITE FIN. COLOR: LIGHT BEIGE
FROM WATER SOURCE
进水口
DRAIN TO WASTE
排污口
APPROVED PAVING MTL.
指定之铺装
PRE-CAST OR CUSTOM MADE DRAIN CHANNEL
排水沟
GRANITE, NAT. CLEFT FINISH, COLOR: DARK BEIGE
深褐色自然切面花岗岩

1E主题水景

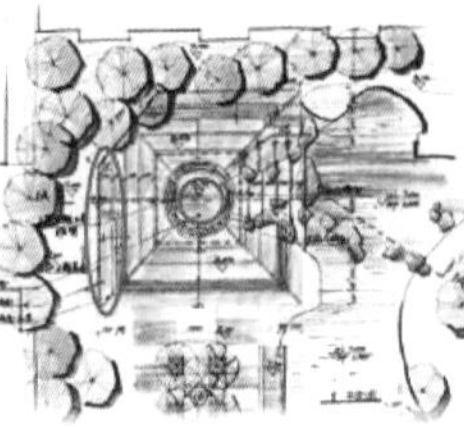

挡土墙
铁艺

1D挡土墙剖立面

水泥本色
铁艺

1D挡土墙

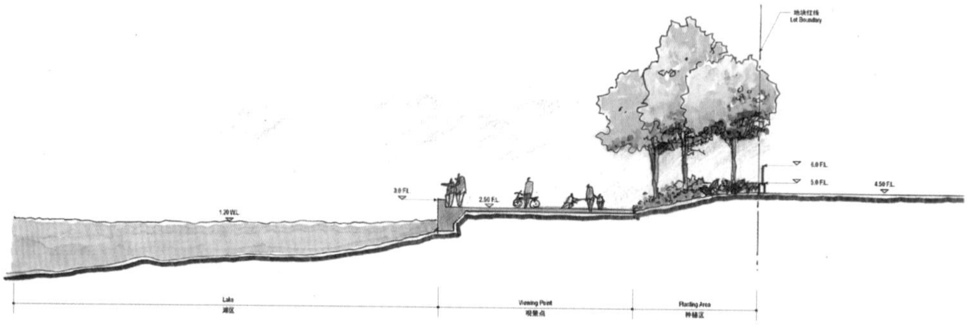
地块红线
Lot Boundary
6.0 F.L.
5.0 F.L.
4.50 F.L.
3.0 F.L.
2.50 F.L.
1.20 W.L.
Lake
湖区
Viewing Point
观景点
Planting Area
种植区

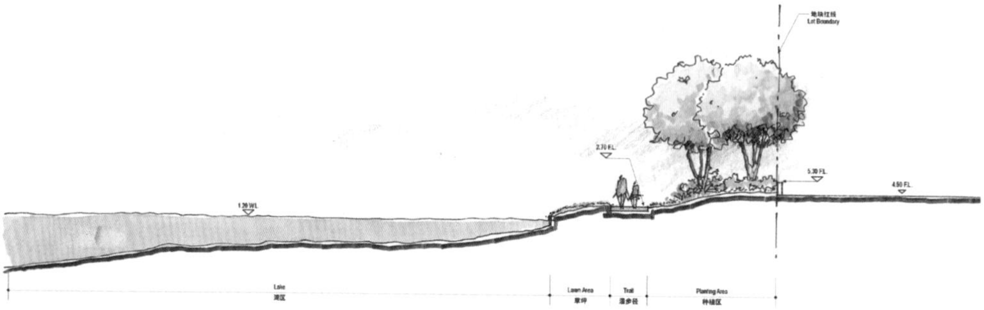
地块红线
Lot Boundary
2.70 F.L.
5.30 F.L.
4.50 F.L.
Lake
湖区
Lawn Area
草坪
Trail
漫步径
Planting Area
种植区

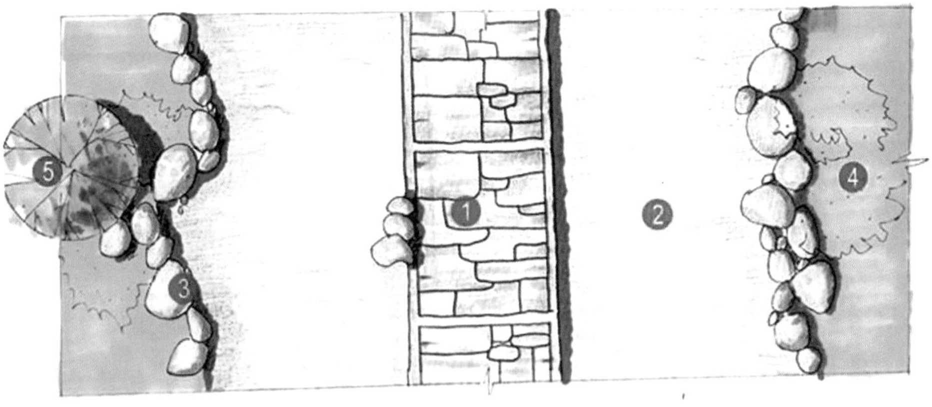

1
2
3
4
5

水中步道
大樹
親水植物
置石

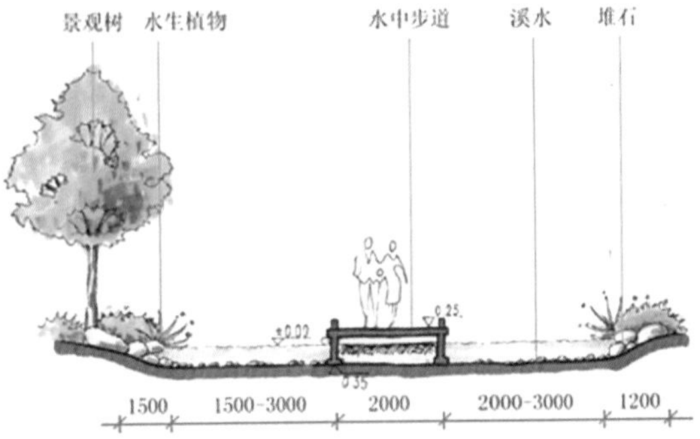

景观树
水生植物
水中步道
溪水
堆石
1500
1500-3000
2000
2000-3000
1200

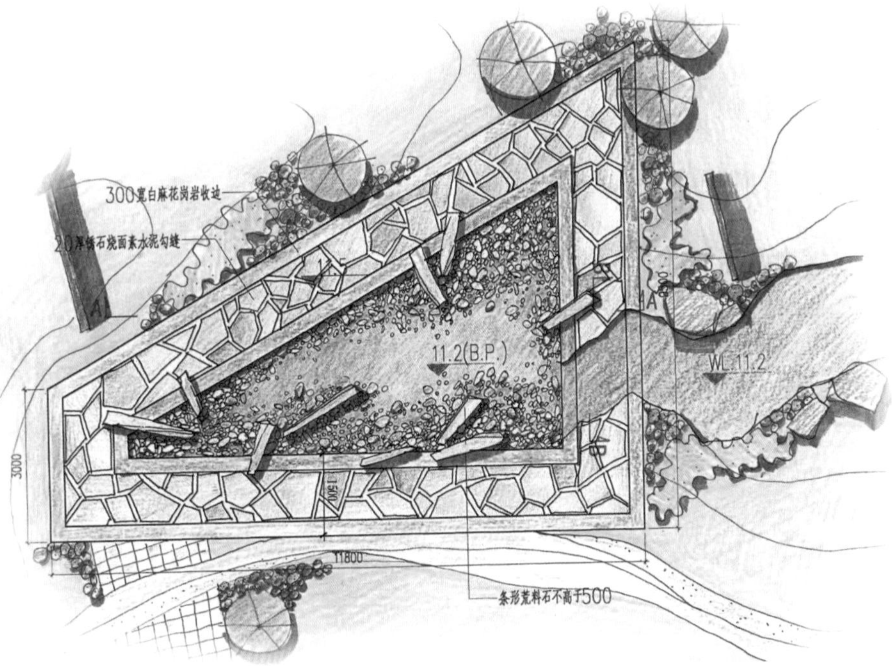

300宽白麻花岗岩收边
11.2(B.P.)
WL.11.2
3000
11800
条形荒料石不高于500

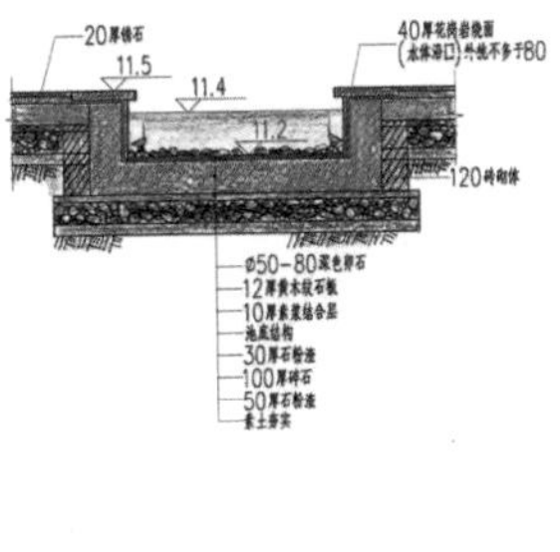

11.5
11.4
11.2

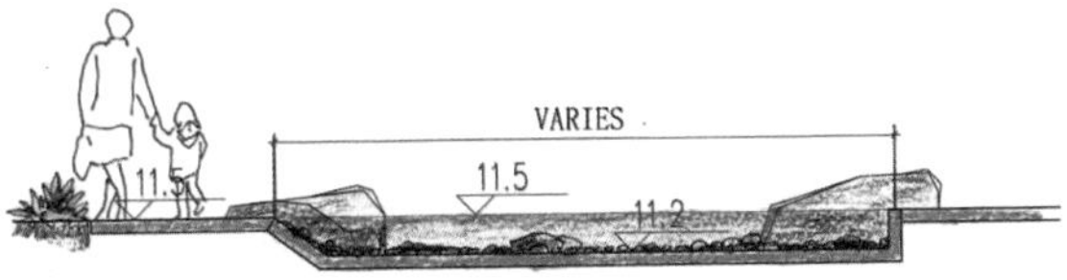

VARIES
11.5
11.5
11.2

入口叠水景观大样
A-A 剖面图
1：50

350 150 350 150 350 150 350 150

600
800
1500
200

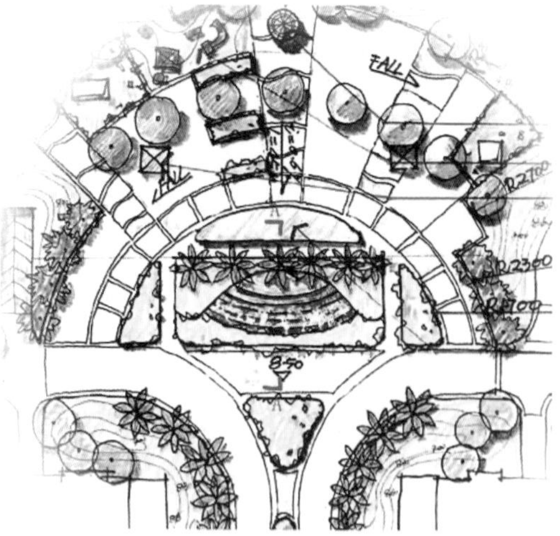

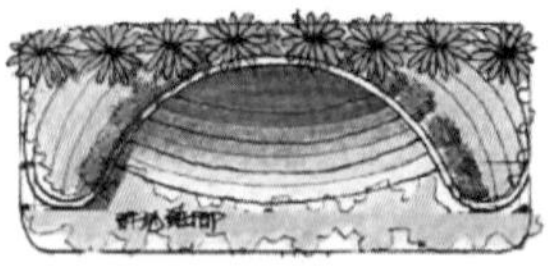

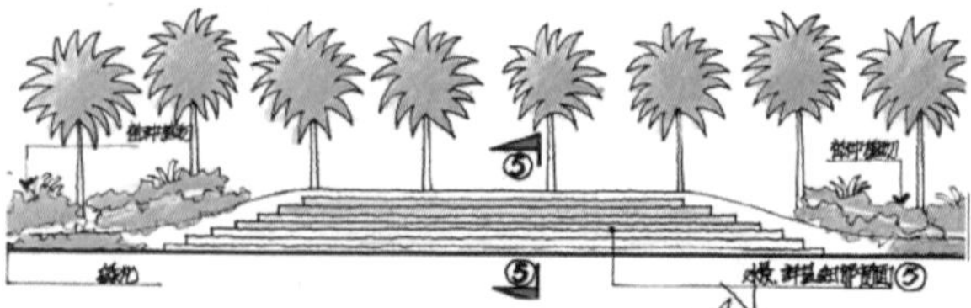

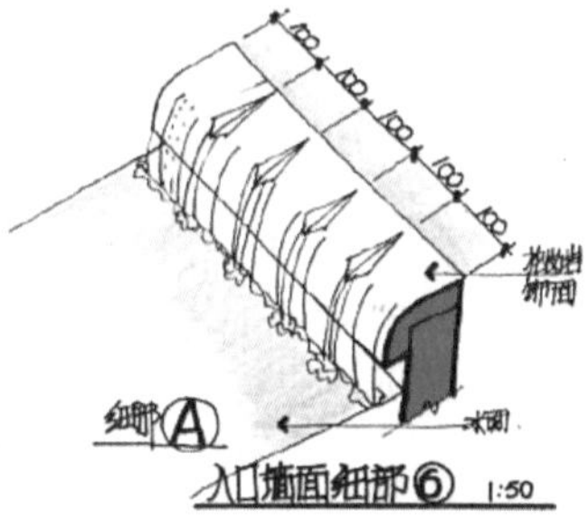

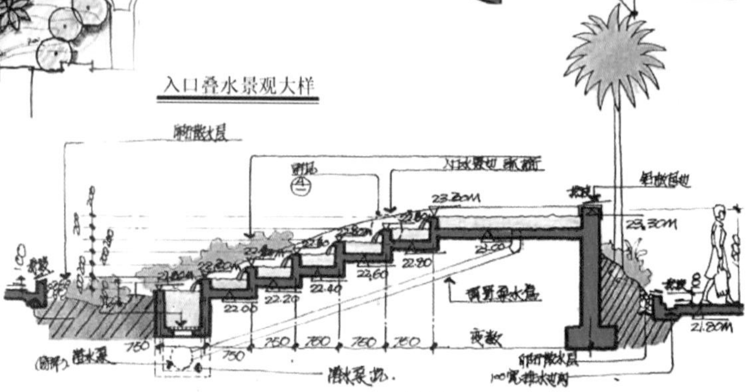

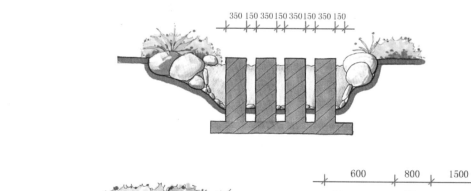

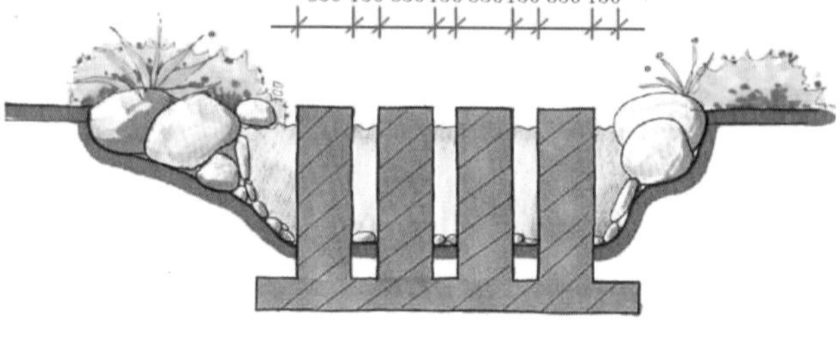

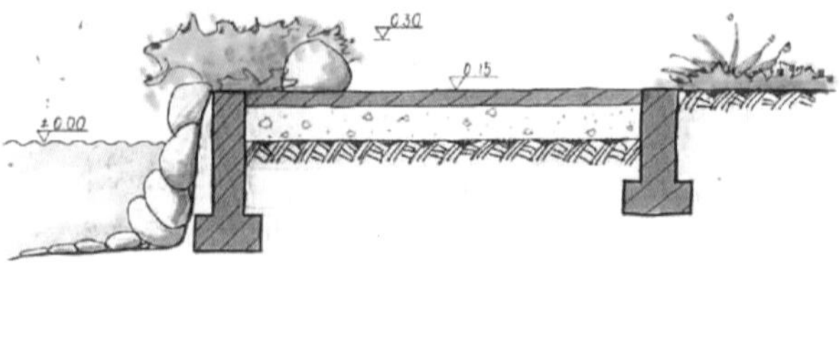

9300
2600
PA
PA
1000
300
1000
WATER JET BUBBLER
起泡器
A
平面圖
PLAN
SCALE 1:100
水平
WATER LEVEL
+12.05
200 X 50 MM THK SOLID YELLOW GRANITE WEIR CUT TO SIZE
实心黄色花岗岩堰按尺寸切割
DAFFODIL YELLOW QUARTZITE
巴西金黄石英板
C
WEIR DET.
SCALE NTS
9300
DAFFODIL YELLOW QUARTZITE
巴西金黄石英板
STRUCTURAL COLUMNS
結構柱
WL
PLANTING AS SPECIFIED
指定的植植物
BEIGE QUARTZITE
灰褐石英板
3000
9.65 TW
9.05 FL
YELLOW QUARTZITE CLADDING
黄石英板牆身飾面
BUBBLER
起泡器
LIGHTING FIXTURES (RECESSED)
燈具(嵌頂)

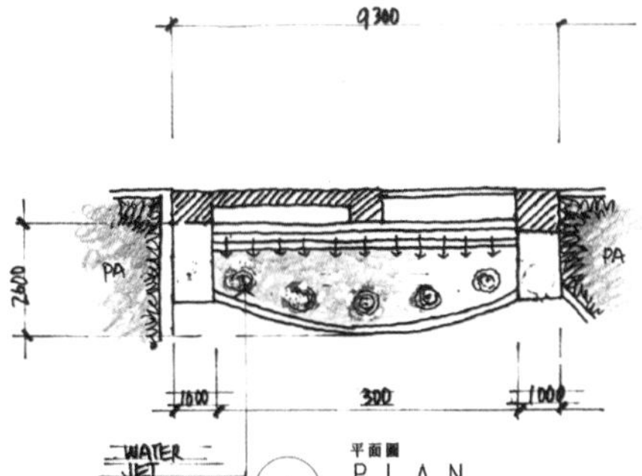

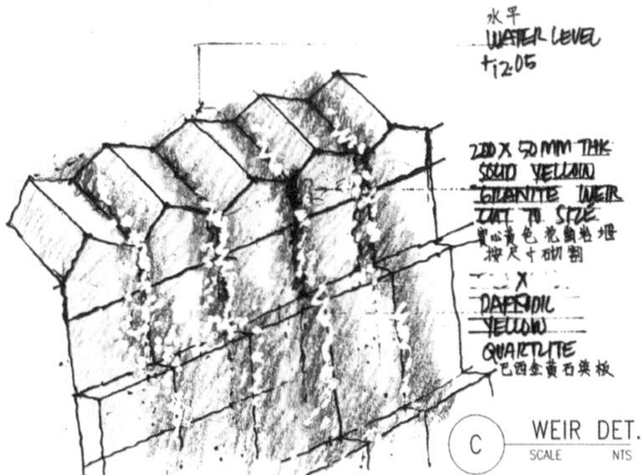

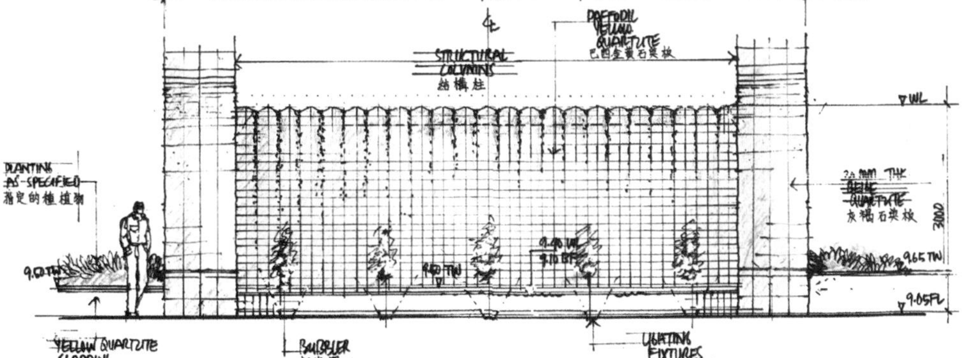

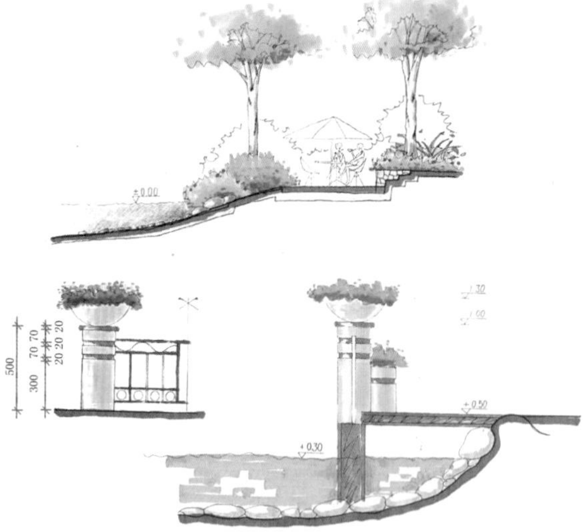

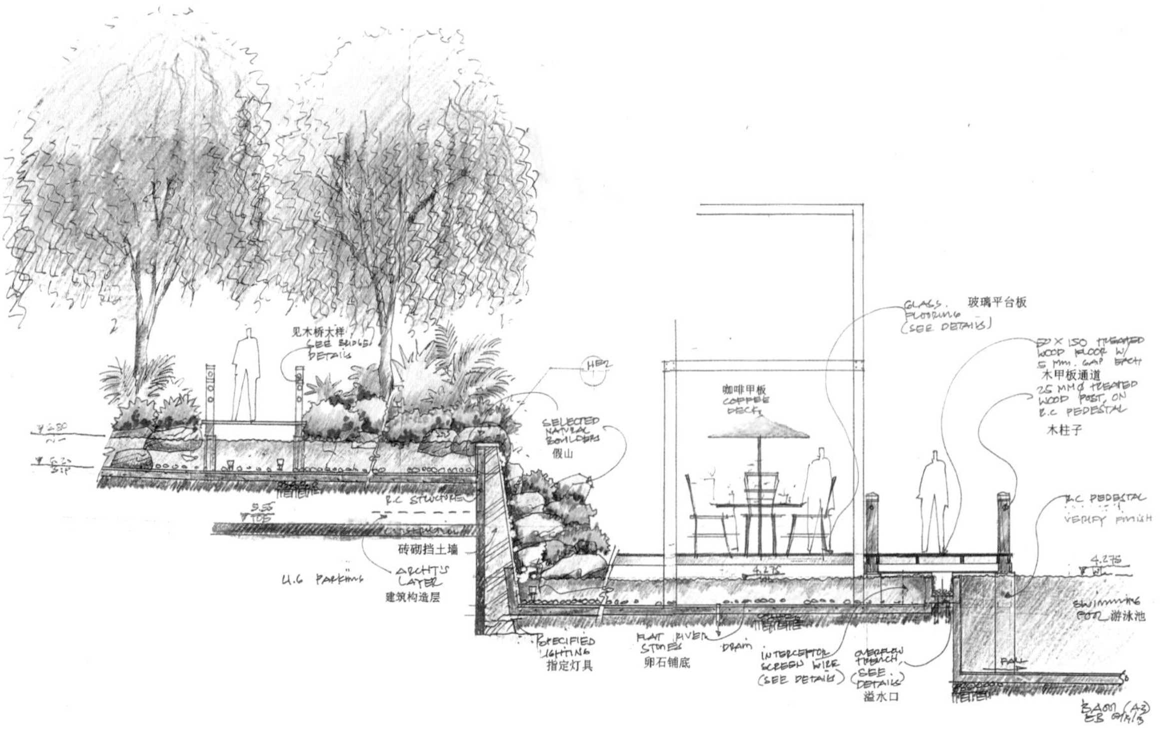
见木桥大样
SEE BRIDGE DETAILS
SELECTED NATURAL BOULDERS
假山
R.C STRUCTURE
砖砌挡土墙
U.G PARKING
ARCHT'S LAYER
建筑构造层
SPECIFIED LIGHTING
指定灯具
FLAT RIVER STONES
卵石铺底
DRAIN
咖啡甲板
COFFEE DECK
GLASS. FLOORING (SEE DETAILS)
玻璃平台板
木甲板通道
木柱子
INTERCEPTOR SCREEN WIRE (SEE DETAILS)
OVERFLOW TRENCH, (SEE DETAILS)
溢水口
R.C PEDESTAL
VERIFY FINISH
4.275
SWIMMING POOL 游泳池
FALL

VENTILATION WOOD TRELLIS
SUBWAY ENTRY.
WATER WALL
PLANTERS
MAIN AXIS
DECOMPOSED GRANITE MULCH.

水系平面 1:250

HIGH BRANCHING TREE (GINKGO TREE)
高大喬木(銀杏)
SHRUB AND ARBOR
喬灌木混合種植
WATER SURFACE ON A ARTIFICIAL LAKE
人工湖水面
CURVATURE WATERBANK
曲綫水岸綫
PLANTINGHOLE WITH HIGH BRANCHING TREE SHRUB AROUND
由大型灌木圍合種植在樹穴下部
TREE CAVE IN WATER WITH SYMETRICAL ARRANGEMENT
對稱排列的水中樹穴

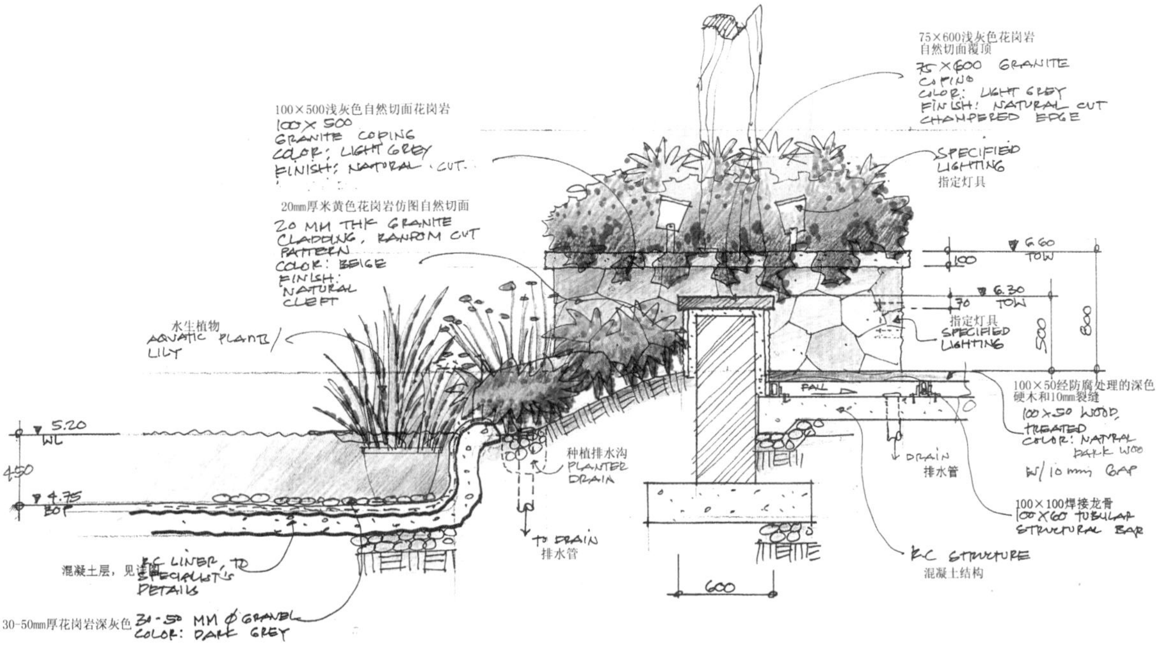
75×600浅灰色花岗岩
自然切面覆顶
75×600 GRANITE
COPING
COLOR: LIGHT GREY
FINISH: NATURAL CUT
CHAMFERED EDGE
100×500浅灰色自然切面花岗岩
100×500
GRANITE COPING
COLOR: LIGHT GREY
FINISH: NATURAL CUT.
SPECIFIED
LIGHTING
指定灯具
20mm厚米黄色花岗岩仿图自然切面
20 MM THK GRANITE
CLADDING, RANDOM CUT
PATTERN
COLOR: BEIGE
FINISH:
NATURAL
CLEFT
6.60
TOW
100
6.30
TOW
70
指定灯具
SPECIFIED
LIGHTING
500
800
水生植物
AQUATIC PLANTS/
LILY
FALL
100×50经防腐处理的深色
硬木和10mm裂缝
100×50 WOOD,
TREATED
COLOR: NATURAL
DARK WOO
W/10mm GAP
5.20
WL
450
4.75
BOP
种植排水沟
PLANTER
DRAIN
DRAIN
排水管
100×100焊接龙骨
100×60 TUBULAR
STRUCTURAL BAR
TO DRAIN
排水管
RC STRUCTURE
混凝土结构
混凝土层，见详图
RC LINER, TO
SPECIALIST'S
DETAILS
600
30-50mm厚花岗岩深灰色
30-50 MM Ø GRAVEL
COLOR: DARK GREY

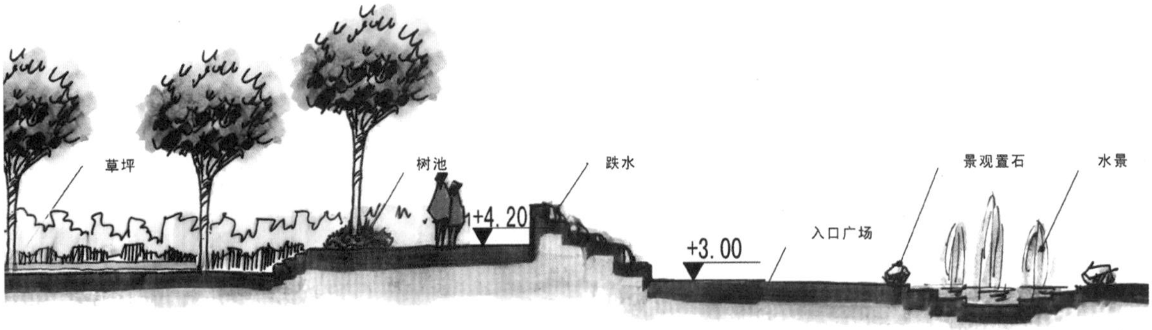
草坪
树池
+4.20
跌水
+3.00
入口广场
景观置石
水景

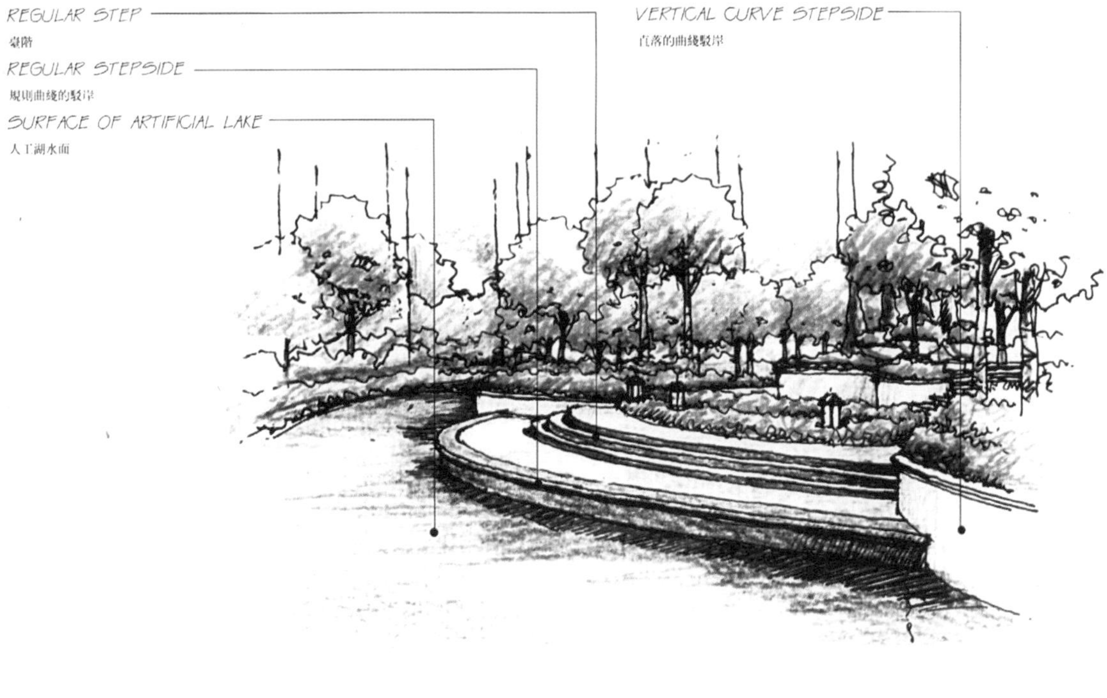
REGULAR STEP
台阶
REGULAR STEPSIDE
规则曲线的驳岸
SURFACE OF ARTIFICIAL LAKE
人工湖水面
VERTICAL CURVE STEPSIDE
直落的曲线驳岸

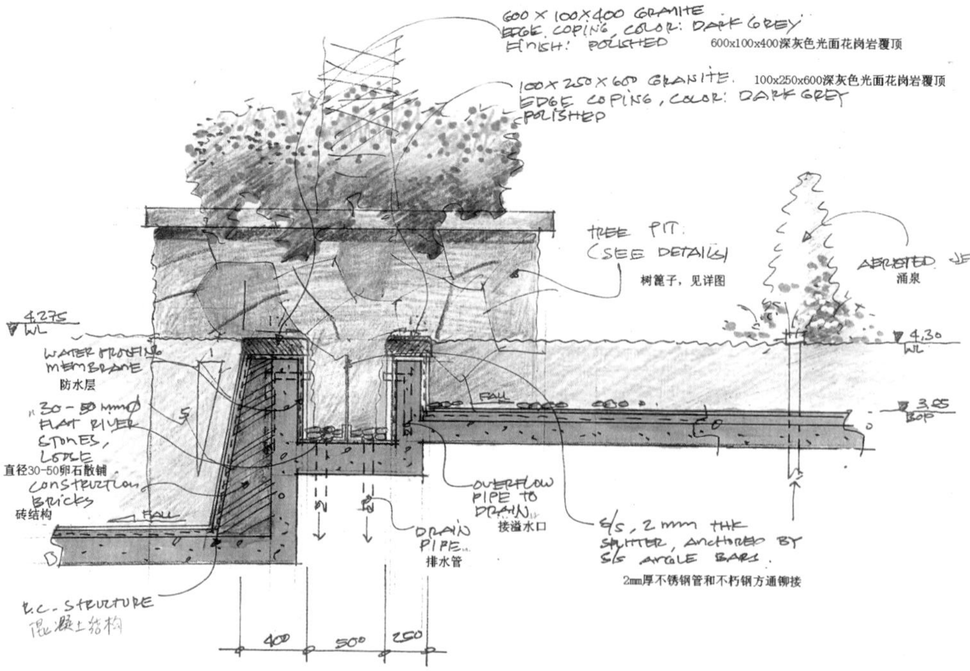
600 X 100 X 400 GRANITE EDGE COPING, COLOR: DARK GREY FINISH: POLISHED
600x100x400深灰色光面花岗岩覆顶
100 X 250 X 600 GRANITE EDGE COPING, COLOR: DARK GREY POLISHED
100x250x600深灰色光面花岗岩覆顶
TREE PIT (SEE DETAILS)
树篦子，见详图
AERATED JETS
涌泉
4.275 WL
4.30 WL
WATERPROOFING MEMBRANE
防水层
30-50 MMØ FLAT RIVER STONES, LOOSE
直径30-50卵石散铺
CONSTRUCTION BRICKS
砖结构
FALL
3.95 BOP
OVERFLOW PIPE TO DRAIN
接溢水口
DRAIN PIPE
排水管
S/S, 2 mm THK SPLITTER, ANCHORED BY S/S ANGLE BARS
2mm厚不锈钢管和不朽钢方通铆接
R.C. STRUCTURE
混凝土结构
400
500
250

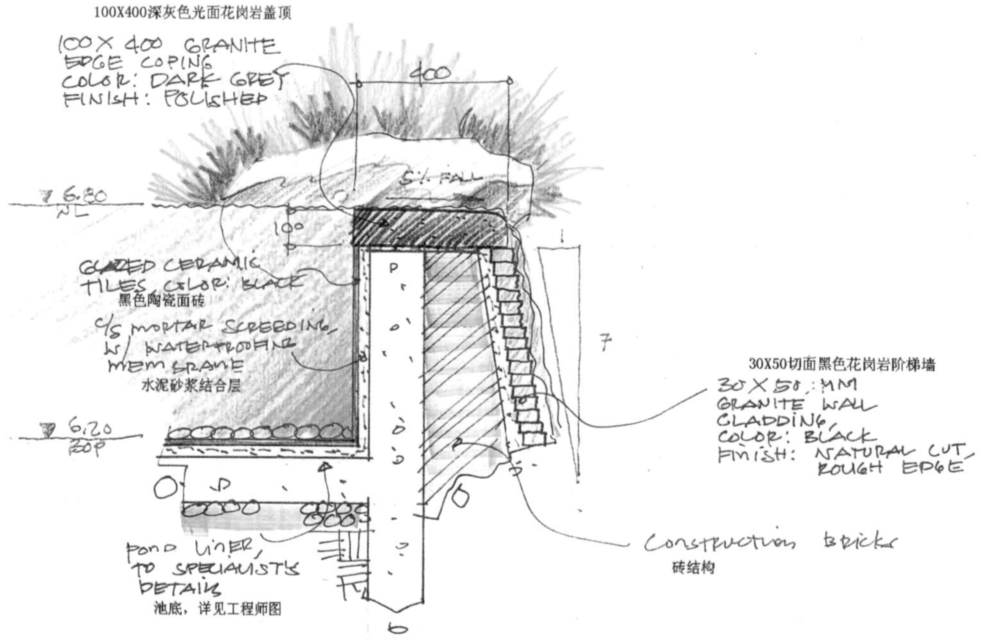
100X400深灰色光面花岗岩盖顶
100 X 400 GRANITE EDGE COPING COLOR: DARK GREY FINISH: POLISHED
400
5% FALL
6.80 WL
100
GLAZED CERAMIC TILES COLOR: BLACK
黑色陶瓷面砖
C/S MORTAR SCREEDING W/ WATERPROOFING MEMBRANE
水泥砂浆结合层
6.20 BOP
30X50切面黑色花岗岩阶梯墙
30 X 50 MM GRANITE WALL CLADDING, COLOR: BLACK FINISH: NATURAL CUT, ROUGH EDGE
CONSTRUCTION BRICKS
砖结构
POND LINER, TO SPECIALIST'S DETAIL
池底，详见工程师图

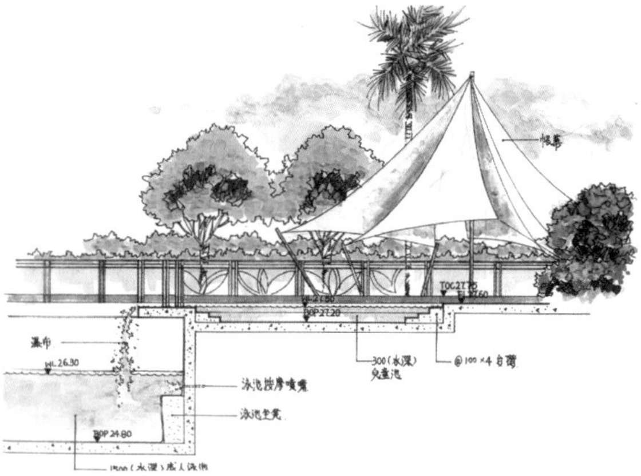
瀑布
WL.26.30
泳池按摩喷嘴
泳池坐凳
BOP 24.80
BOP 27.20
300(水深)
儿童池
@100×4台阶

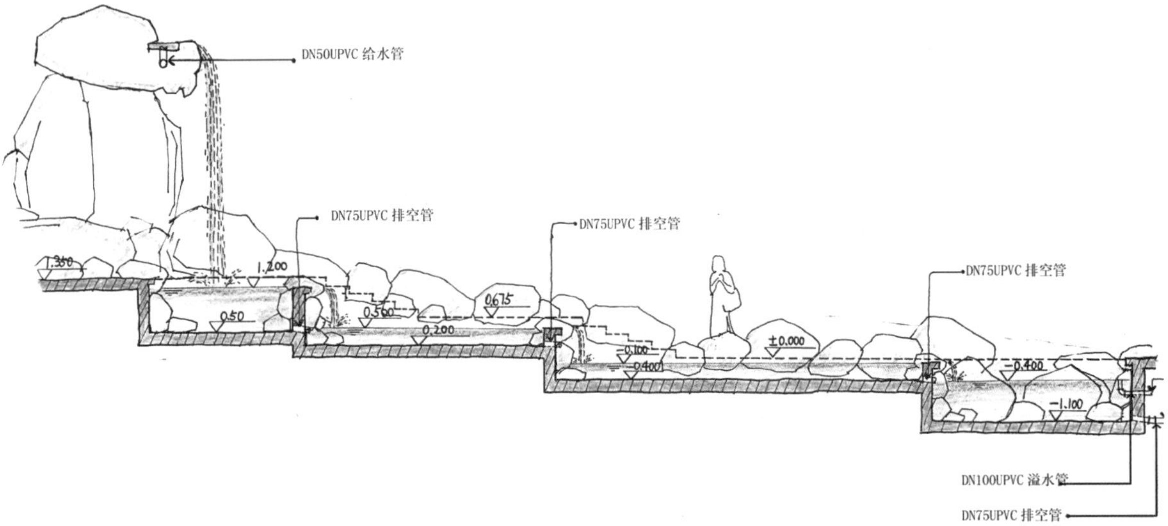
DN50UPVC 给水管
DN75UPVC 排空管
DN75UPVC 排空管
DN75UPVC 排空管
1.380
1.200
0.50
0.560
0.675
0.200
±0.000
0.100
0.400
-0.400
-1.100
DN100UPVC 溢水管
DN75UPVC 排空管

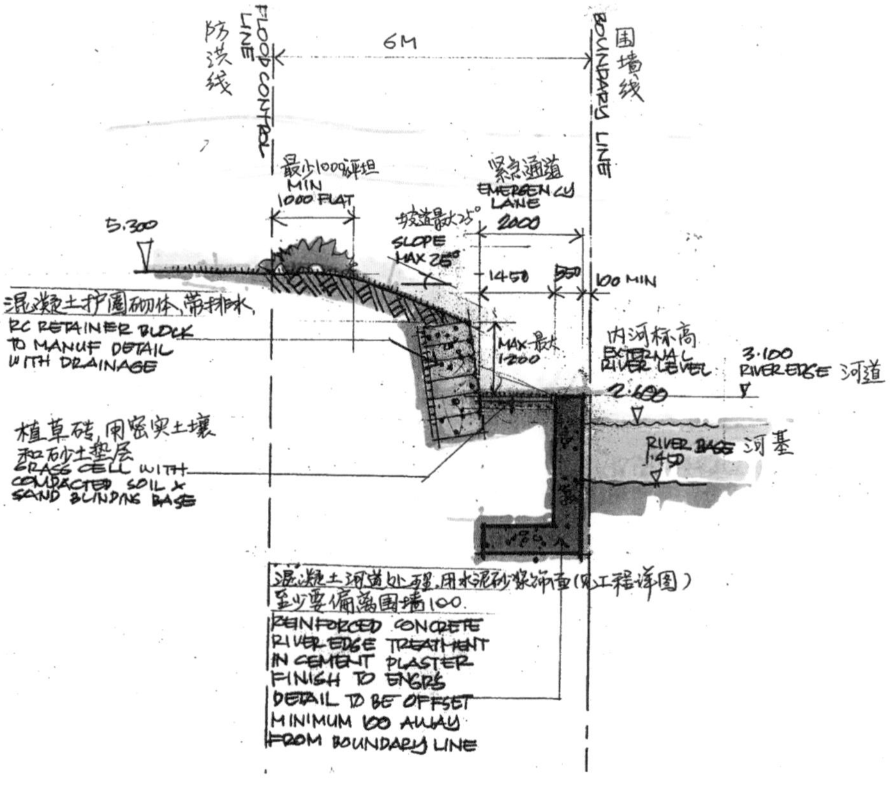
防洪线
FLOOD CONTROL LINE
6M
围墙线
BOUNDARY LINE
最少1000平坦
MIN 1000 FLAT
紧急通道
EMERGENCY LANE
5.300
坡道最大25°
SLOPE MAX 25°
2000
1450
550
100 MIN
混凝土护围砌体,带排水
RC RETAINER BLOCK TO MANUF DETAIL WITH DRAINAGE
MAX-最大 1200
内河标高
EXTERNAL RIVER LEVEL
3.100
RIVER EDGE 河道
2.600
植草砖,用密实土壤和砂土垫层
GRASS CELL WITH COMPACTED SOIL & SAND BLINDING BASE
RIVER BASE 河基
1.450
混凝土河道处理,用水泥砂浆饰面(见工程详图)至少要偏离围墙100.
REINFORCED CONCRETE RIVER EDGE TREATMENT IN CEMENT PLASTER FINISH TO ENGR'S DETAIL TO BE OFFSET MINIMUM 100 AWAY FROM BOUNDARY LINE

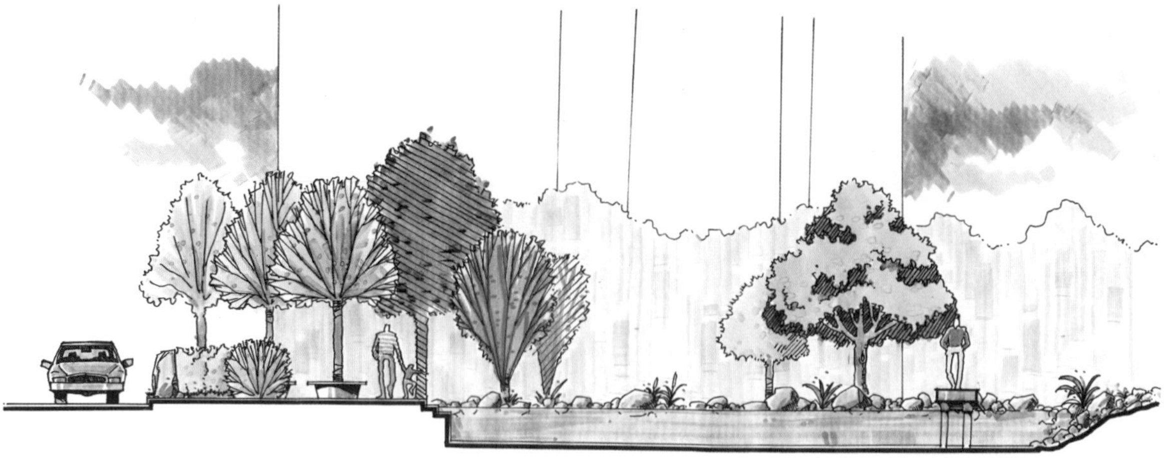

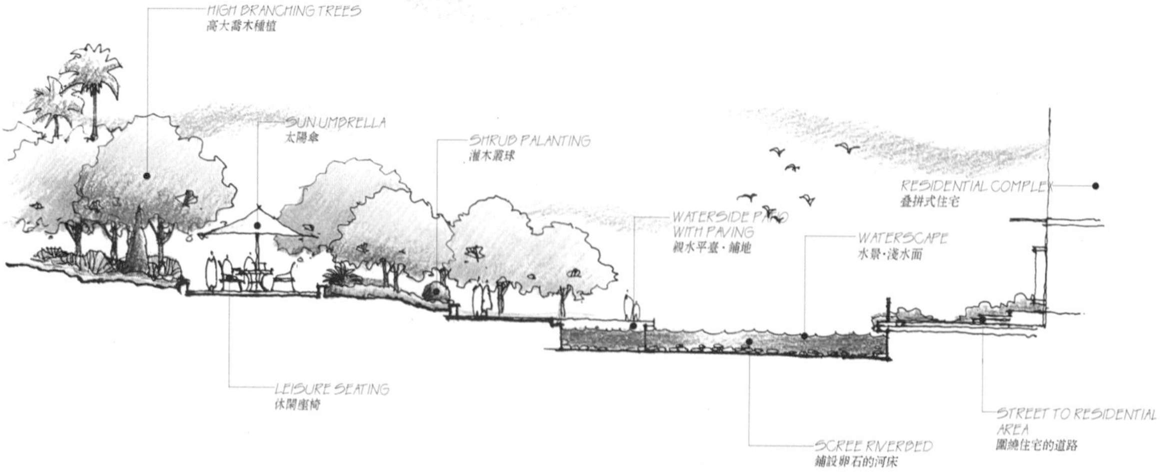
HIGH BRANCHING TREES
高大喬木種植
SUN UMBRELLA
太陽傘
SHRUB PALANTING
灌木叢球
WATERSIDE PA
WITH PAVING
親水平臺・鋪地
WATERSCAPE
水景・淺水面
RESIDENTIAL COMPLEX
叠拼式住宅
LEISURE SEATING
休閑座椅
SCREE RIVERBED
鋪設卵石的河床
STREET TO RESIDENTIAL AREA
圍繞住宅的道路

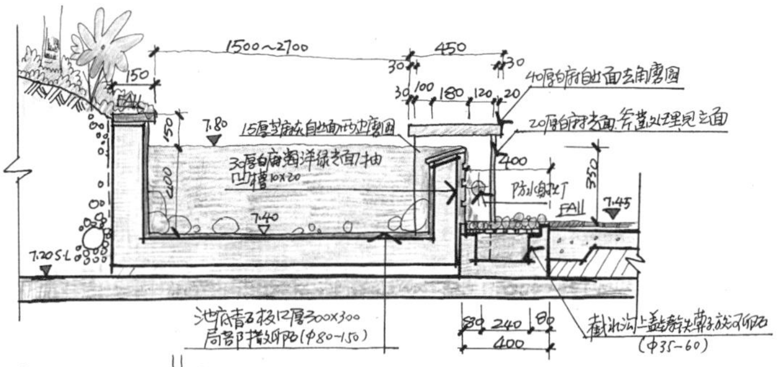

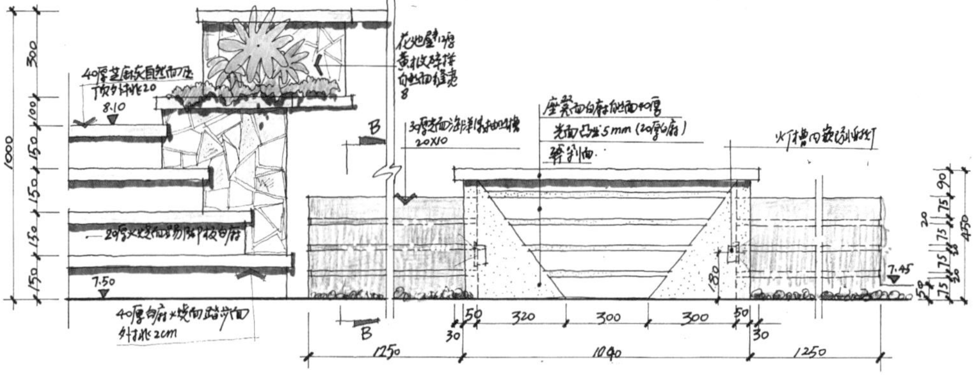

山石叠水

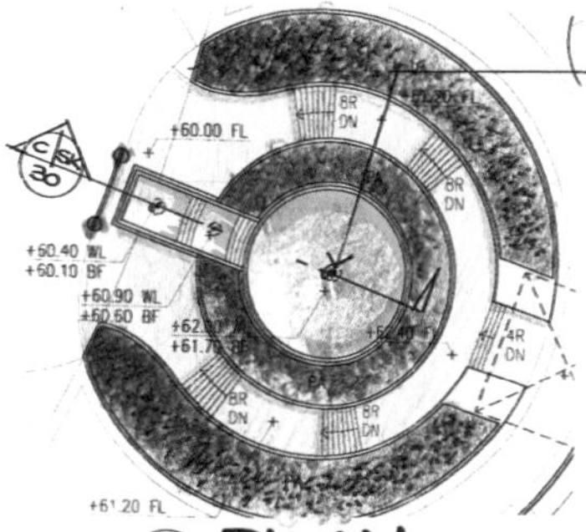
BUBBLER JET NOZZLE BY SPECIALIST
起泡器噴嘴 由專家擔任
+60.00 FL
+60.40 WL
+60.10 BF
+60.90 WL
+60.60 BF
+61.20 FL
8R DN
4R DN

A PLAN 平面圖 SCALE 1:600

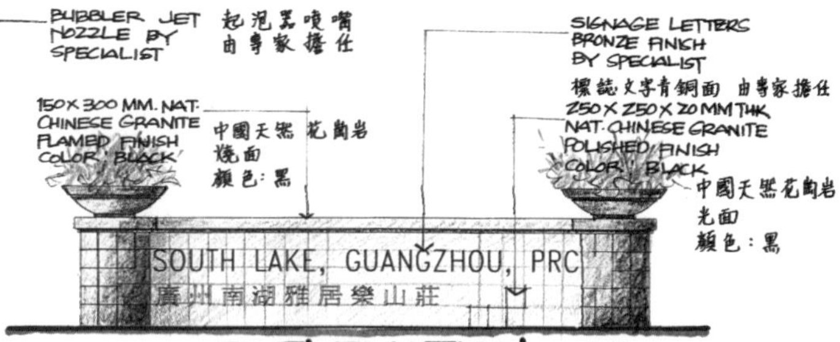
150 X 300 MM. NAT. CHINESE GRANITE FLAMED FINISH COLOR: BLACK
中國天然花崗岩 燒面 顏色：黑
SIGNAGE LETTERS BRONZE FINISH BY SPECIALIST
標誌文字青銅面 由專家擔任
250 X 250 X 20 MM THK NAT. CHINESE GRANITE POLISHED FINISH COLOR: BLACK
中國天然花崗岩 光面 顏色：黑
SOUTH LAKE, GUANGZHOU, PRC
廣州南湖雅居樂山莊

B ELEVATION 立面圖 SCALE 1:50

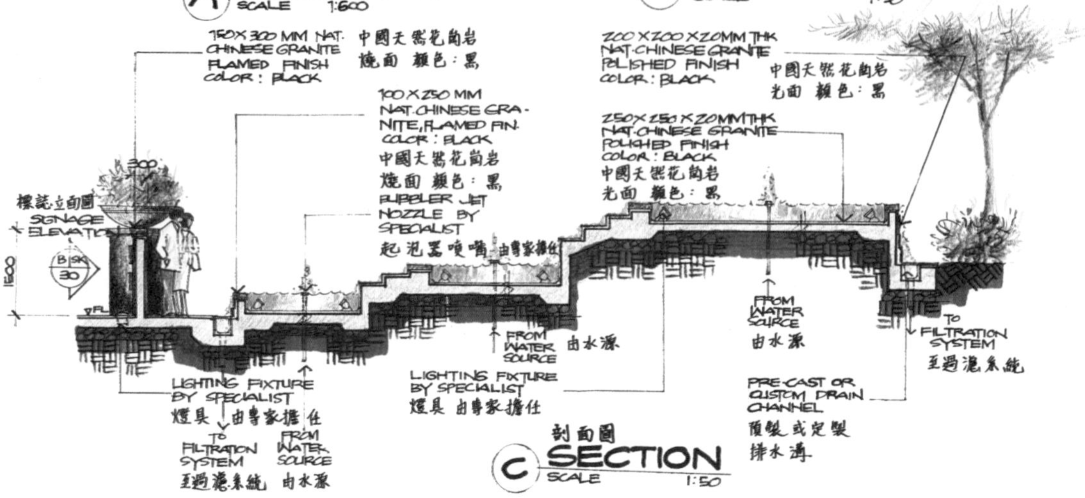
150 X 300 MM NAT. CHINESE GRANITE FLAMED FINISH COLOR: BLACK
中國天然花崗岩 燒面 顏色：黑
100 X 250 MM NAT. CHINESE GRANITE, FLAMED FIN. COLOR: BLACK
中國天然花崗岩 燒面 顏色：黑
BUBBLER JET NOZZLE BY SPECIALIST
起泡器噴嘴 由專家擔任
200 X 200 X 20MM THK NAT. CHINESE GRANITE POLISHED FINISH COLOR: BLACK
中國天然花崗岩 光面 顏色：黑
250 X 250 X 20MM THK NAT. CHINESE GRANITE POLISHED FINISH COLOR: BLACK
中國天然花崗岩 光面 顏色：黑
標誌立面圖 SIGNAGE ELEVATION
1500
300
FROM WATER SOURCE 由水源
TO FILTRATION SYSTEM 至過濾系統
LIGHTING FIXTURE BY SPECIALIST
燈具 由專家擔任
PRE-CAST OR CUSTOM DRAIN CHANNEL
預製或定製排水溝

C SECTION 剖面圖 SCALE 1:50

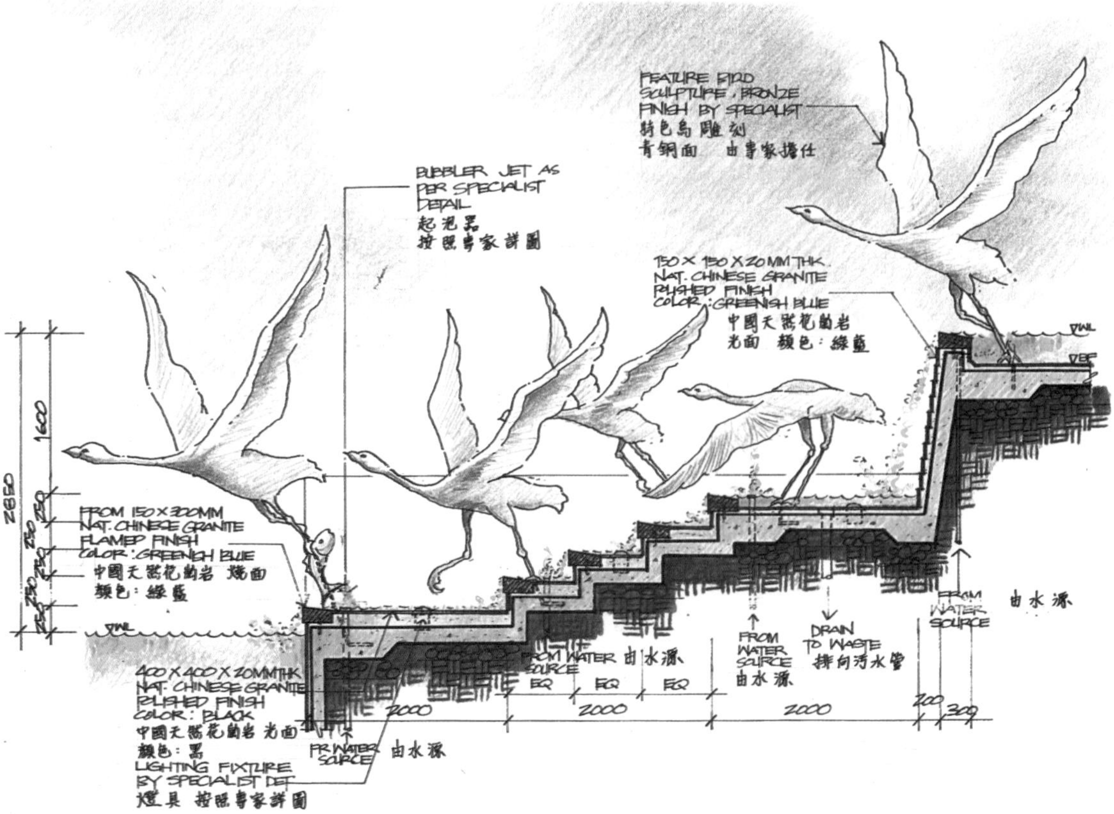
FEATURE BIRD SCULPTURE, BRONZE FINISH BY SPECIALIST
特色鳥雕刻 青銅面 由專家擔任
BUBBLER JET AS PER SPECIALIST DETAIL
起泡器 按照專家詳圖
150 X 150 X 20 MM THK NAT. CHINESE GRANITE POLISHED FINISH COLOR: GREENISH BLUE
中國天然花崗岩 光面 顏色：綠藍
FROM 150 X 300MM NAT. CHINESE GRANITE FLAMED FINISH COLOR: GREENISH BLUE
中國天然花崗岩 燒面 顏色：綠藍
400 X 400 X 20MM THK NAT. CHINESE GRANITE POLISHED FINISH COLOR: BLACK
中國天然花崗岩 光面 顏色：黑
LIGHTING FIXTURE BY SPECIALIST DET
燈具 按照專家詳圖
FR WATER SOURCE 由水源
FROM WATER SOURCE 由水源
DRAIN TO WASTE 排向污水管
EQ
2000
2850
1600
200
300

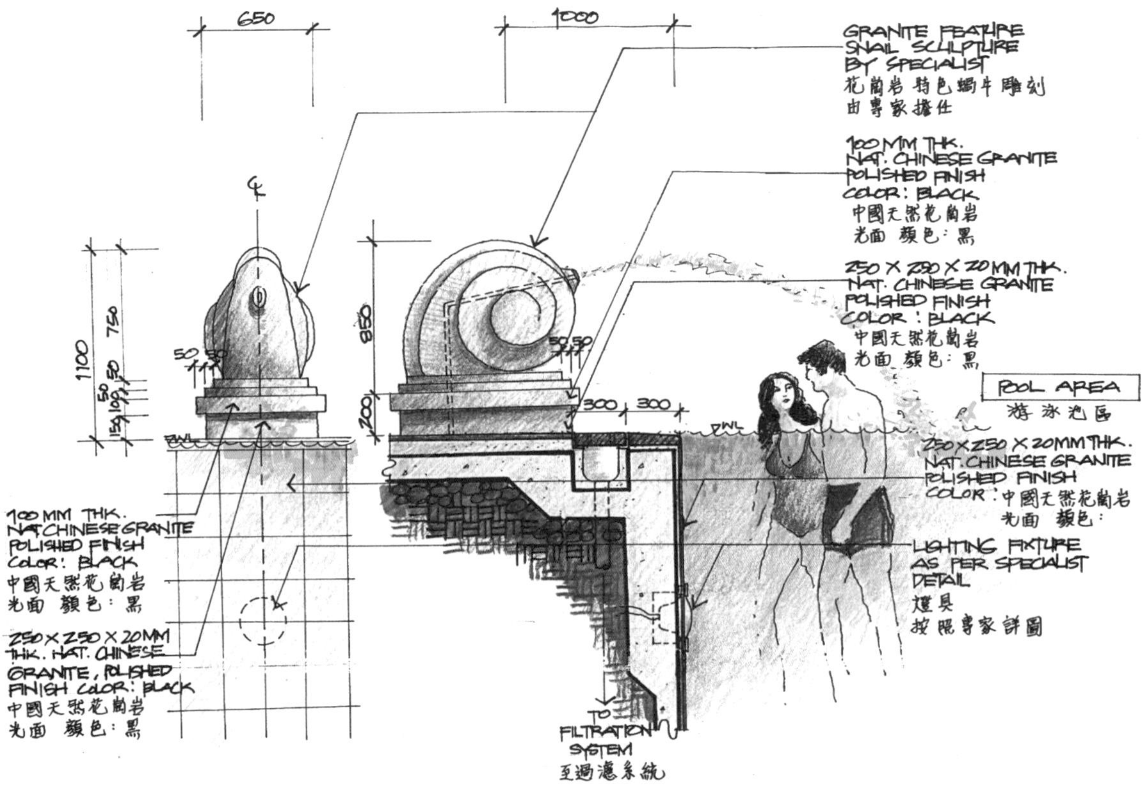
650
1000
GRANITE FEATURE SNAIL SCULPTURE BY SPECIALIST
花崗岩特色蝸牛雕刻
由專家擔任
100MM THK. NAT. CHINESE GRANITE POLISHED FINISH COLOR: BLACK
中國天然花崗岩
光面 顏色：黑
250 X 250 X 20MM THK. NAT. CHINESE GRANITE POLISHED FINISH COLOR: BLACK
中國天然花崗岩
光面 顏色：黑
POOL AREA
游泳池區
250 X 250 X 20MM THK. NAT. CHINESE GRANITE POLISHED FINISH COLOR: 中國天然花崗岩
光面 顏色：
LIGHTING FIXTURE AS PER SPECIALIST DETAIL
燈具
按照專家詳圖
100MM THK. NAT. CHINESE GRANITE POLISHED FINISH COLOR: BLACK
中國天然花崗岩
光面 顏色：黑
250 X 250 X 20MM THK. NAT. CHINESE GRANITE, POLISHED FINISH COLOR: BLACK
中國天然花崗岩
光面 顏色：黑
TO FILTRATION SYSTEM
至過濾系統
1100
750
880
200
300
300
DWL

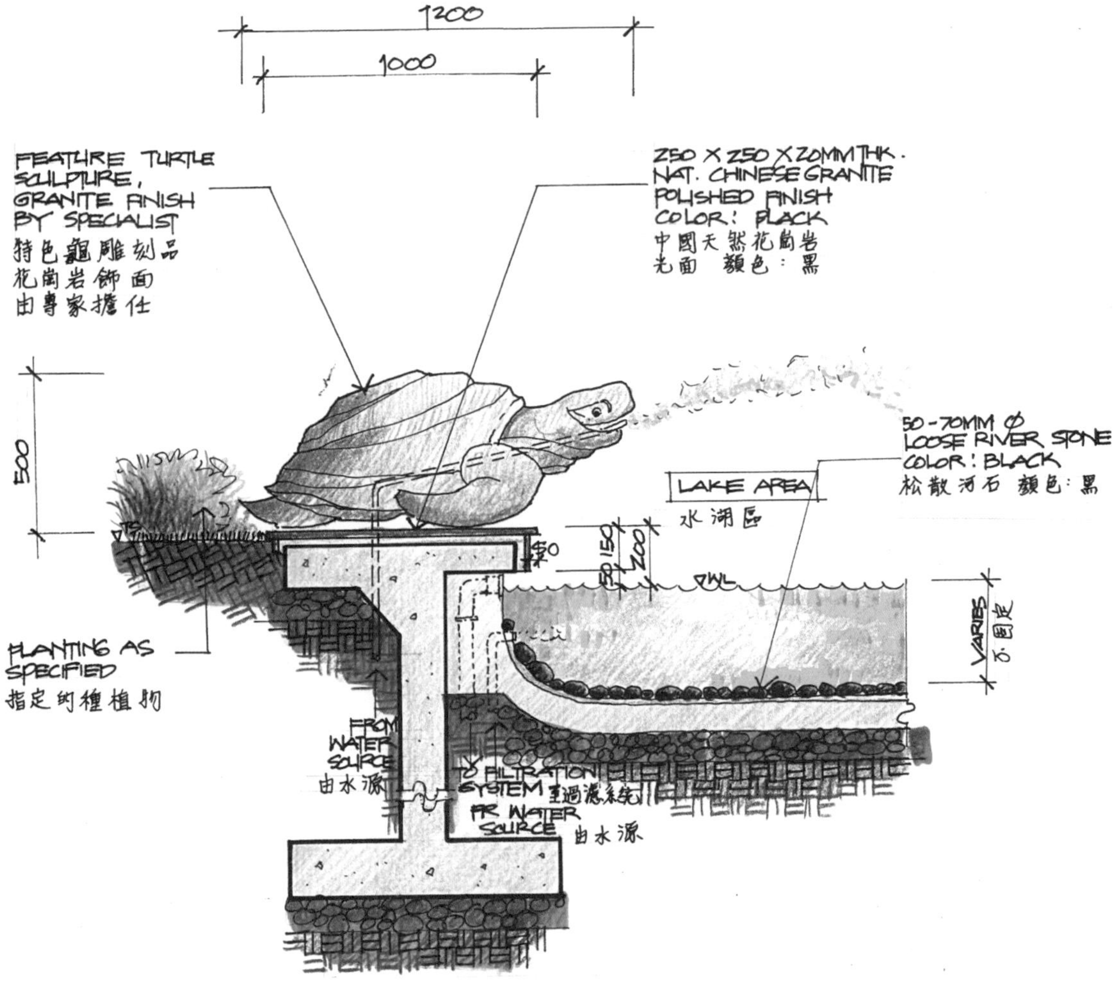
1200
1000
FEATURE TURTLE SCULPTURE, GRANITE FINISH BY SPECIALIST
特色龜雕刻品
花崗岩飾面
由專家擔任
250 X 250 X 20MM THK. NAT. CHINESE GRANITE POLISHED FINISH COLOR: BLACK
中國天然花崗岩
光面 顏色：黑
500
50-70MM Ø LOOSE RIVER STONE COLOR: BLACK
松散河石 顏色：黑
LAKE AREA
水湖區
150
200
DWL
VARIES
不固定
PLANTING AS SPECIFIED
指定的種植物
FROM WATER SOURCE
由水源
TO FILTRATION SYSTEM 至過濾系統
FR WATER SOURCE 由水源

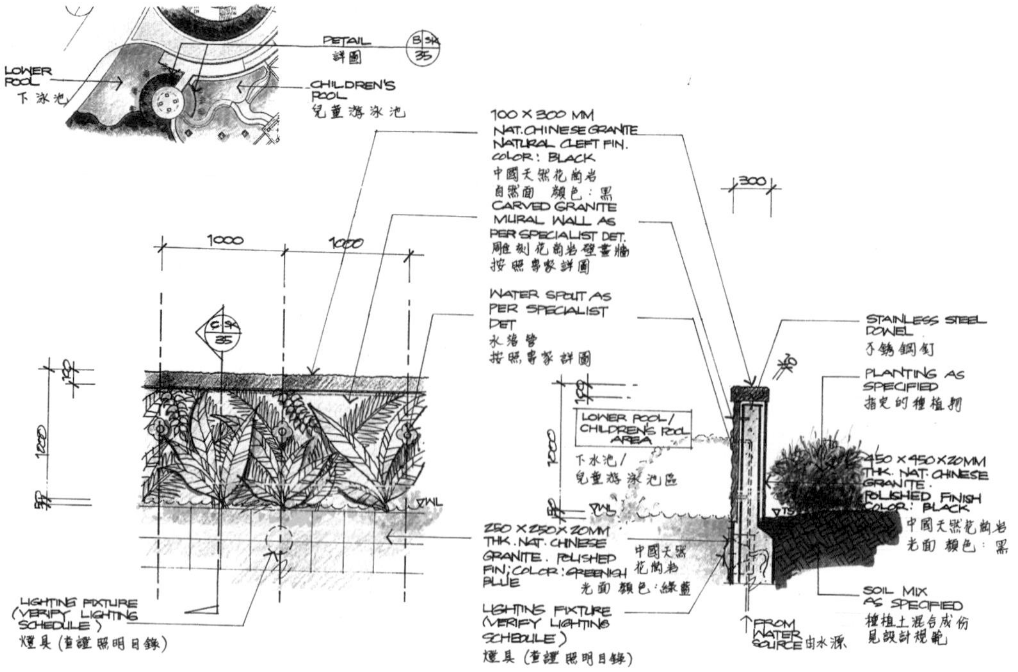
DETAIL
詳圖
B SK
35
LOWER POOL
下泳池
CHILDREN'S POOL
兒童游泳池
100 X 300 MM
NAT. CHINESE GRANITE
NATURAL CLEFT FIN.
COLOR: BLACK
中國天然花崗岩
自然面 顏色：黑
CARVED GRANITE
MURAL WALL AS
PER SPECIALIST DET.
雕刻花崗岩壁畫牆
按照專家詳圖
1000
1000
300
C SK
35
WATER SPOUT AS
PER SPECIALIST
DET
水溶管
按照專家詳圖
1200
LOWER POOL/
CHILDREN'S POOL
AREA
下水池/
兒童游泳池區
1000
WL
STAINLESS STEEL
DOWEL
不銹鋼釘
PLANTING AS
SPECIFIED
指定的種植物
450 X 450 X 20MM
THK. NAT. CHINESE
GRANITE.
POLISHED FINISH
COLOR: BLACK
中國天然花崗岩
光面 顏色：黑
250 X 250 X 20MM
THK. NAT. CHINESE
GRANITE. POLISHED
FIN; COLOR: GREENISH
BLUE
中國天然
花崗岩
光面 顏色：綠藍
LIGHTING FIXTURE
(VERIFY LIGHTING
SCHEDULE)
燈具(查證照明目錄)
LIGHTING FIXTURE
(VERIFY LIGHTING
SCHEDULE)
燈具(查證照明目錄)
SOIL MIX
AS SPECIFIED
種植土混合成份
見設計規範
FROM
WATER
SOURCE 由水源

ROMAN

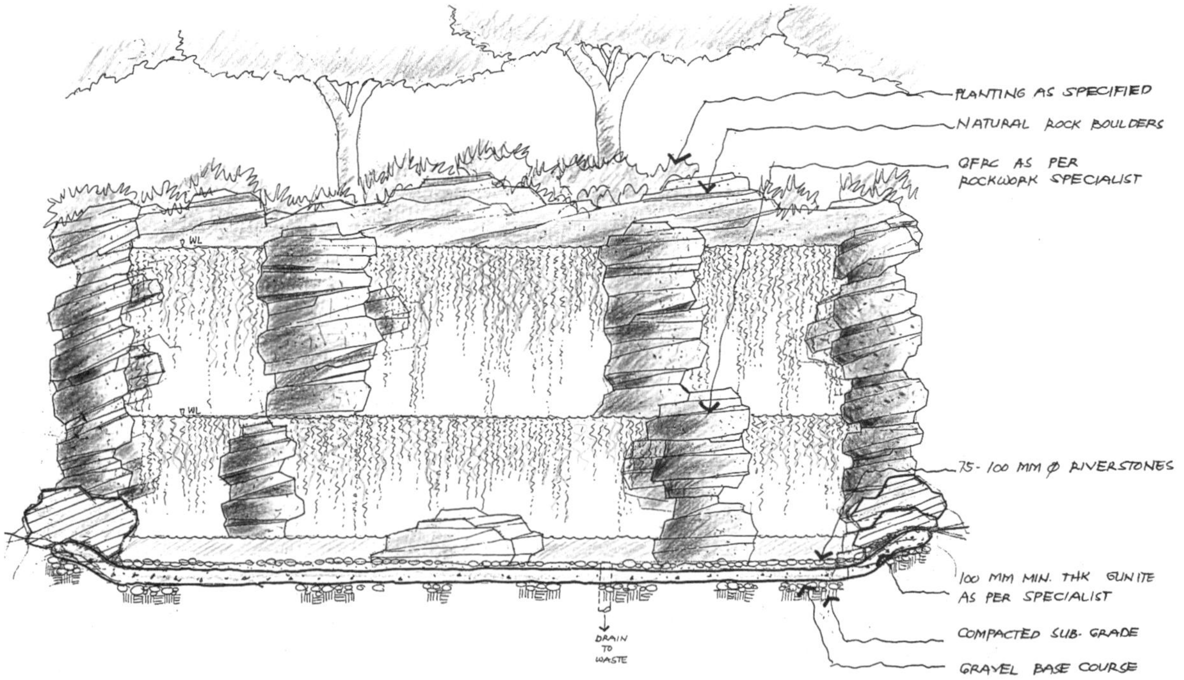
WL
WL
PLANTING AS SPECIFIED
NATURAL ROCK BOULDERS
GFRC AS PER ROCKWORK SPECIALIST
75-100 MM Ø RIVERSTONES
100 MM MIN. THK GUNITE AS PER SPECIALIST
COMPACTED SUB. GRADE
GRAVEL BASE COURSE
DRAIN TO WASTE

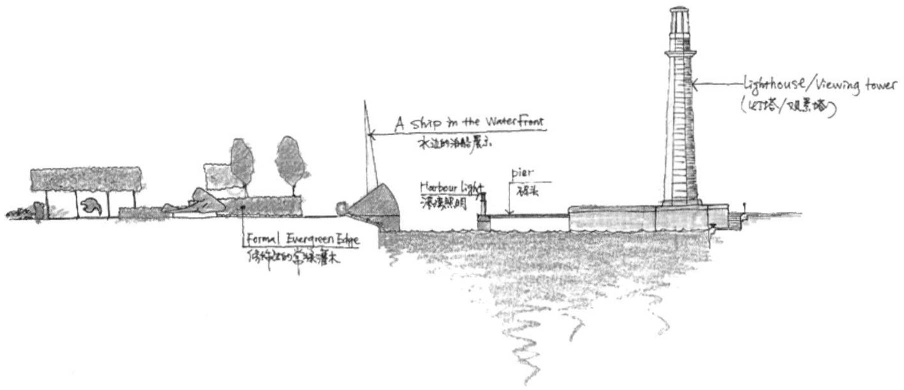
Lighthouse/Viewing tower
(灯塔/观景塔)
A ship in the Waterfront
Harbour light
港湾照明
pier
码头
Formal Evergreen Edge

WATER BUBBLER JETS, 500 MM HIGH
BUILDING COLUMN
建筑柱位
POOL
PLANTER WALL SEE
PLAN 平面
SCALE 1:50
R.C. STRUCTURE
30MM THK. GRANITE CLADDING
CEMENT MORTAR ADHESIVE
50X50MM GRANITE CORNER TRIM, FINISH & COLOR TO MATCH CLADDING
TYPICAL CORNER TRIM DETAIL
SCALE 1:5
转角大样
300X100MM THK. SANDSTONE COPING, HONED FINISH, BEIGE COLOR
300X100MM THK. POLISHED BLACK GRANITE COPING
30MM THK. GRANITE CLADDING, RANDOM CUT, BUTT JOINT, BEIGE COLOR
WATER BUBBLER JETS,500MM HIGH
TYPICAL CORNER TRIM DETAIL
250X100MM THK. POLISHED GRANITE STEPS, BLACK COLOR
20MM THK. POLISHED BLACK GRANITE PAVING
TIMBER FOOTBRIDGE SEE
FEATURE POT & PLINTH DETAIL
NATURAL ROCK BOULDERS BY ROCKWORKS SPECIALIST
300X300MM RED GRANITE ACCENT, NATURAL CLEFT FINISH
WATERSIDE CAFE STRUCTURE, REFER TO ARCHITECTURAL DETAILS
LAKE
湖
300X300MM FLAMED GRANITE BAND, GREY COLOR
TYPICAL ENTRANCE STEPS DETAIL, SEE
600X300MM FLAMED GRANITE, BEIGE COLOR
600X600MM FLAMED GRANITE PAVING WITH 100X100MM POLISHED BLACK GRANITE ACCENT
300X300MM FLAMED GRANITE PAVING, BEIGE COLOR
50MM WIDTH POLISHED BLACK GRANITE STRIPE
100X100X100MM THK. GRANITE SETTS, NATURAL CLEFT FINISH, YELLOW COLOR
WATER FEATURE SEE
水景,参见
PAVING PATTERN
SCALE 1:100
600X600MM FLAMED GRANITE PAVING, BEIGE COLOR
BUILDING ENTRANCE-1 PAVING DETAIL
建筑入口-1
铺装大样

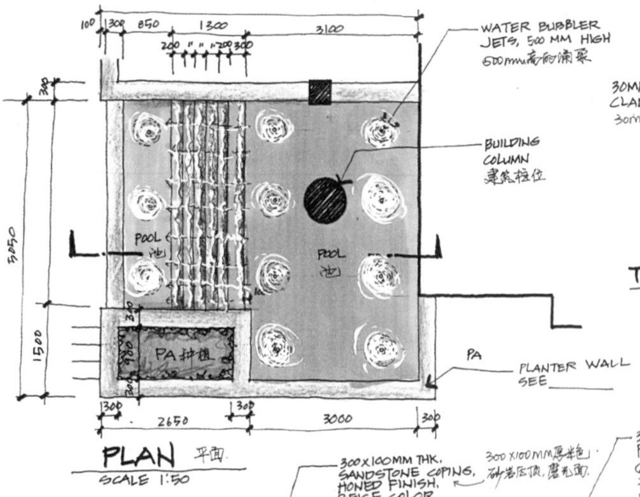

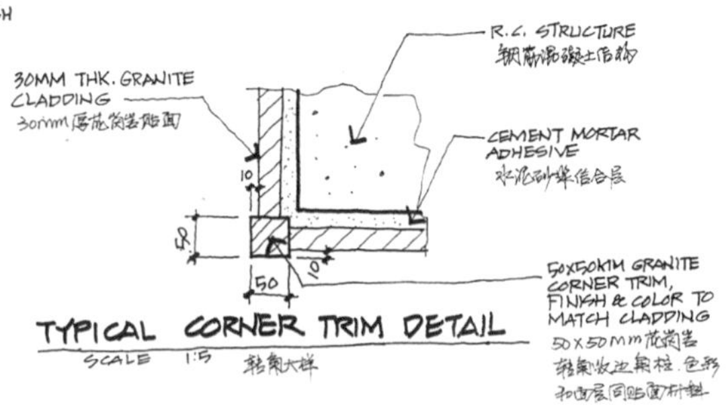

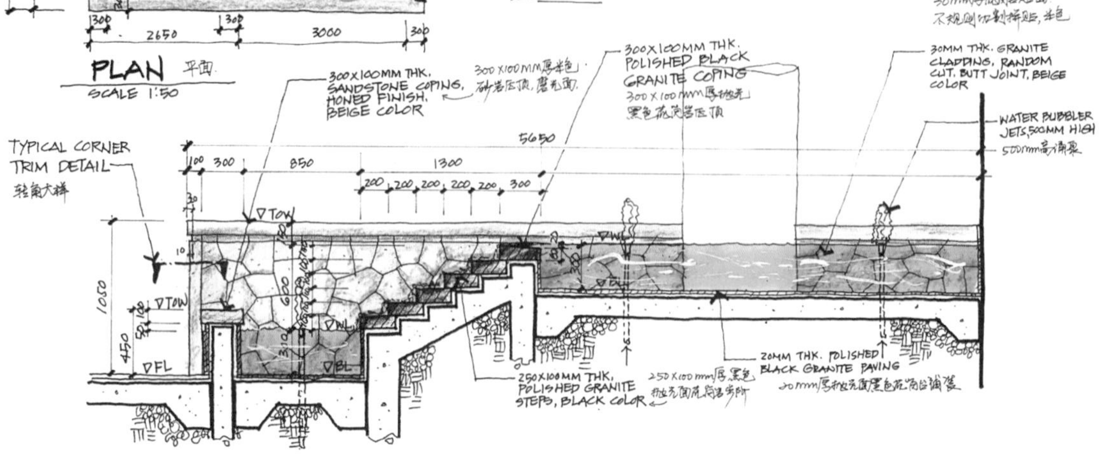

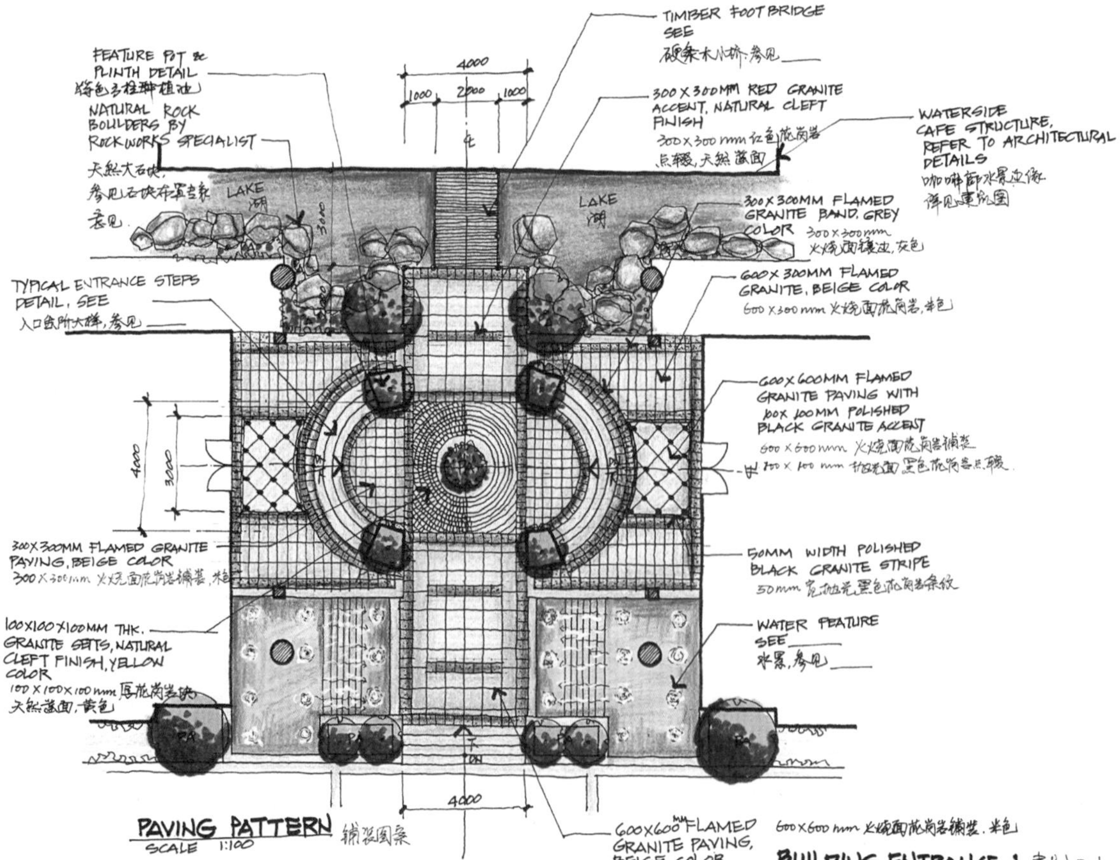

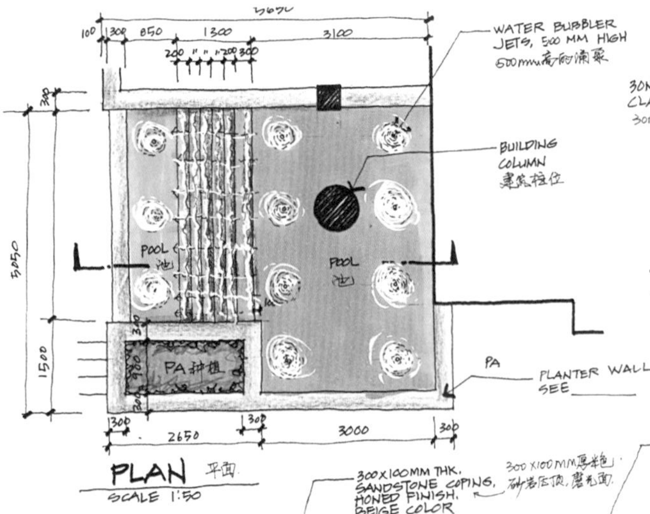
WATER BUBBLER JETS, 500 MM HIGH
500mm高的涌泉
BUILDING COLUMN
建筑柱位
POOL 池
PA 种植
PLANTER WALL SEE
PLAN 平面
SCALE 1:50

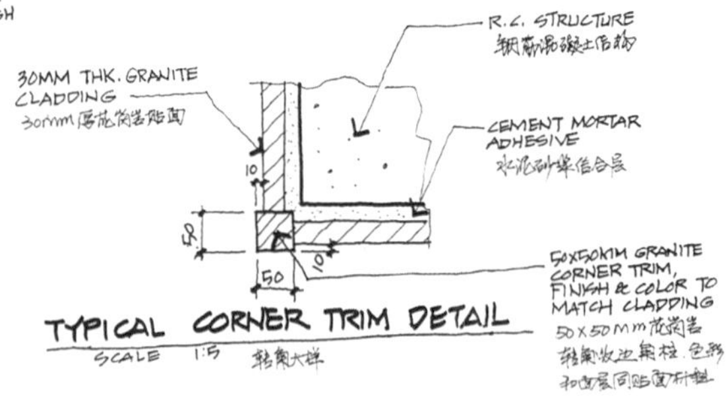
30MM THK. GRANITE CLADDING
R.C. STRUCTURE
CEMENT MORTAR ADHESIVE
50X50MM GRANITE CORNER TRIM, FINISH & COLOR TO MATCH CLADDING
TYPICAL CORNER TRIM DETAIL
SCALE 1:5

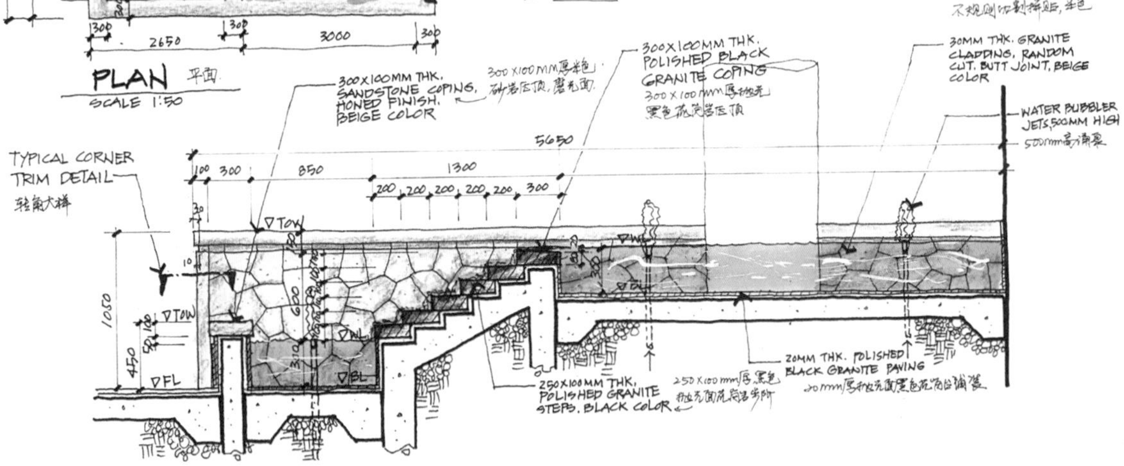
TYPICAL CORNER TRIM DETAIL
300X100MM THK. SANDSTONE COPING, HONED FINISH, BEIGE COLOR
300X100MM THK. POLISHED BLACK GRANITE COPING
30MM THK. GRANITE CLADDING, RANDOM CUT, BUTT JOINT, BEIGE COLOR
WATER BUBBLER JETS,500MM HIGH
250X100MM THK. POLISHED GRANITE STEPS, BLACK COLOR
20MM THK. POLISHED BLACK GRANITE PAVING

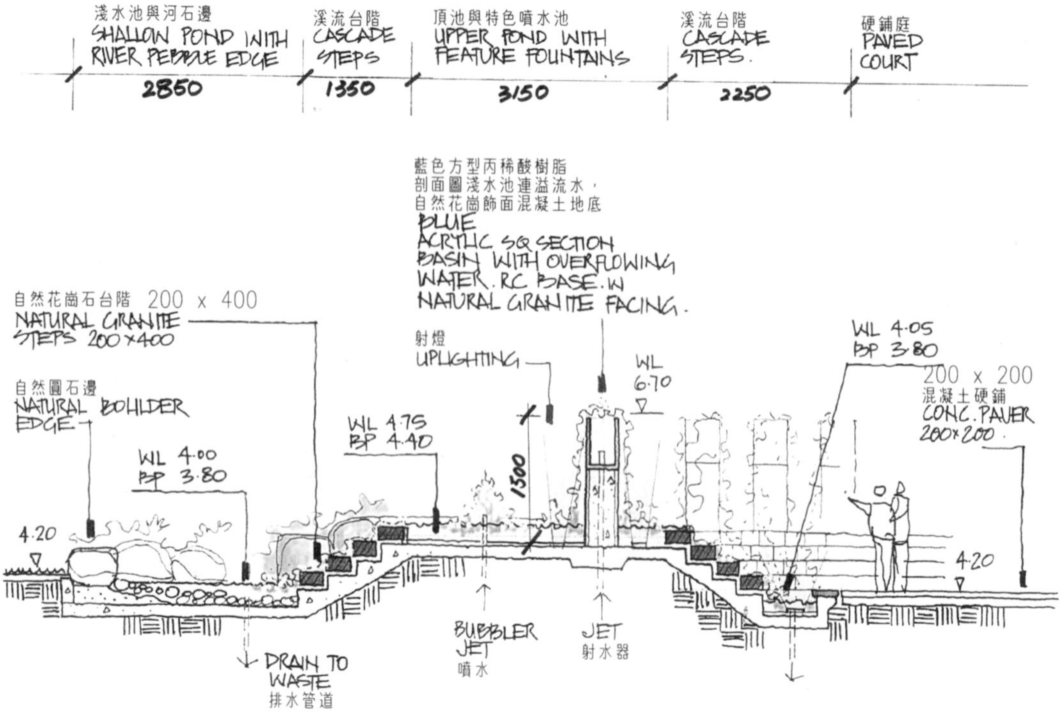
浅水池與河石邊
SHALLOW POND WITH RIVER PEBBLE EDGE
2850
溪流台階
CASCADE STEPS
1350
頂池與特色噴水池
UPPER POND WITH FEATURE FOUNTAINS
3150
溪流台階
CASCADE STEPS.
2250
硬鋪庭
PAVED COURT
藍色方型丙稀酸樹脂
剖面圖淺水池連溢流水，
自然花崗飾面混凝土地底
BLUE ACRYLIC SQ SECTION BASIN WITH OVERFLOWING WATER. RC BASE. W NATURAL GRANITE FACING.
自然花崗石台階 200 x 400
NATURAL GRANITE STEPS 200×400
射燈
UPLIGHTING
WL 6.70
WL 4.05
BP 3.80
200 x 200
混凝土硬鋪
CONC. PAVER 200×200.
自然圓石邊
NATURAL BOULDER EDGE
WL 4.75
BP 4.40
WL 4.00
BP 3.80
1500
4.20
4.20
BUBBLER JET
噴水
JET
射水器
DRAIN TO WASTE
排水管道

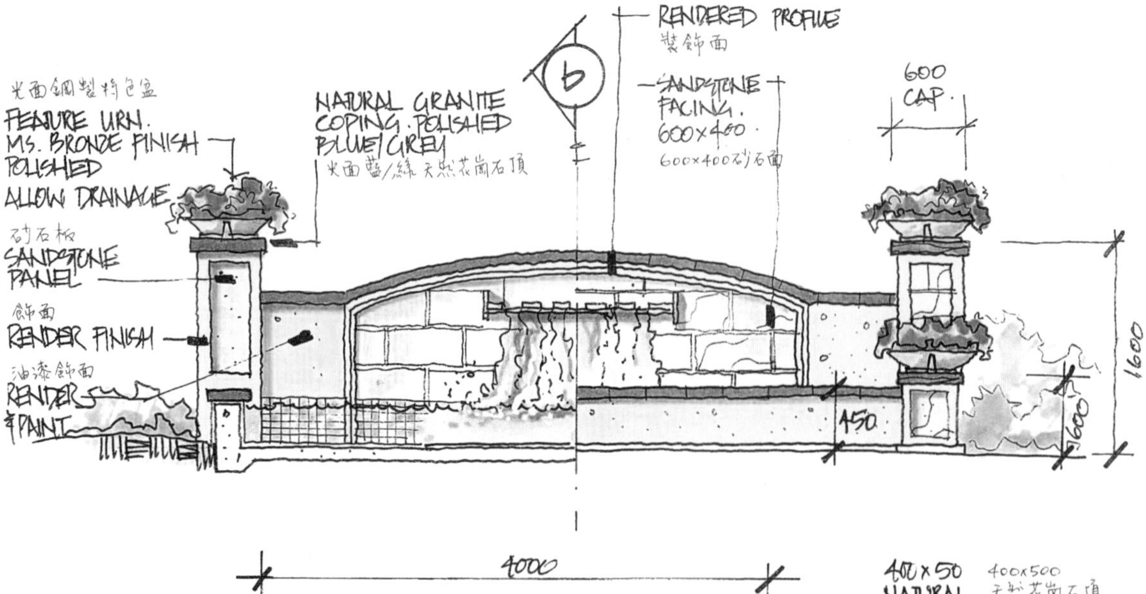
RENDERED PROFILE
裝飾面
光面銅製特色盆
FEATURE URN. M.S. BRONZE FINISH POLISHED ALLOW DRAINAGE
NATURAL GRANITE COPING. POLISHED BLUE/GREY
光面藍/綠天然花崗石頂
SANDSTONE FACING. 600×400.
600×400砂石面
600 CAP.
砂石板
SANDSTONE PANEL
飾面
RENDER FINISH
油漆飾面
RENDER & PAINT
450
1600
600
4000

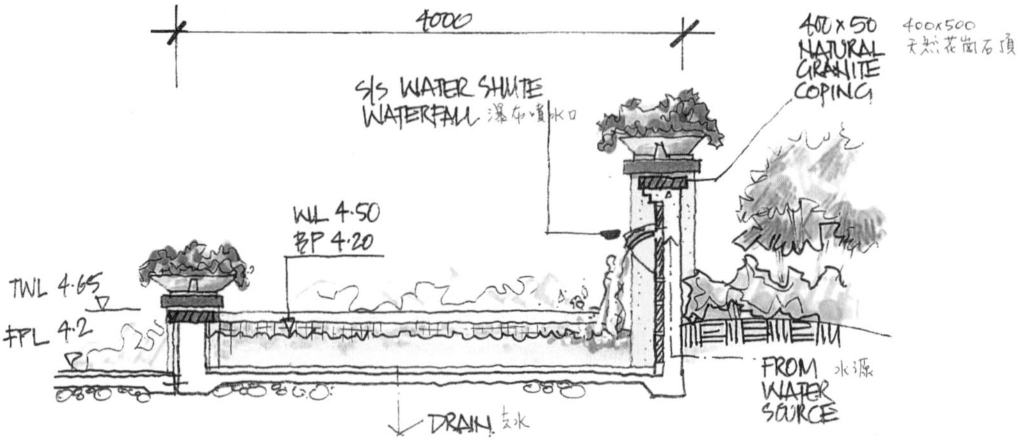
400×50 NATURAL GRANITE COPING
400×500 天然花崗石頂
S/S WATER SHUTE WATERFALL 瀑布噴水口
WL 4.50
BP 4.20
TWL 4.65
FPL 4.2
FROM WATER SOURCE 水源
DRAIN 去水

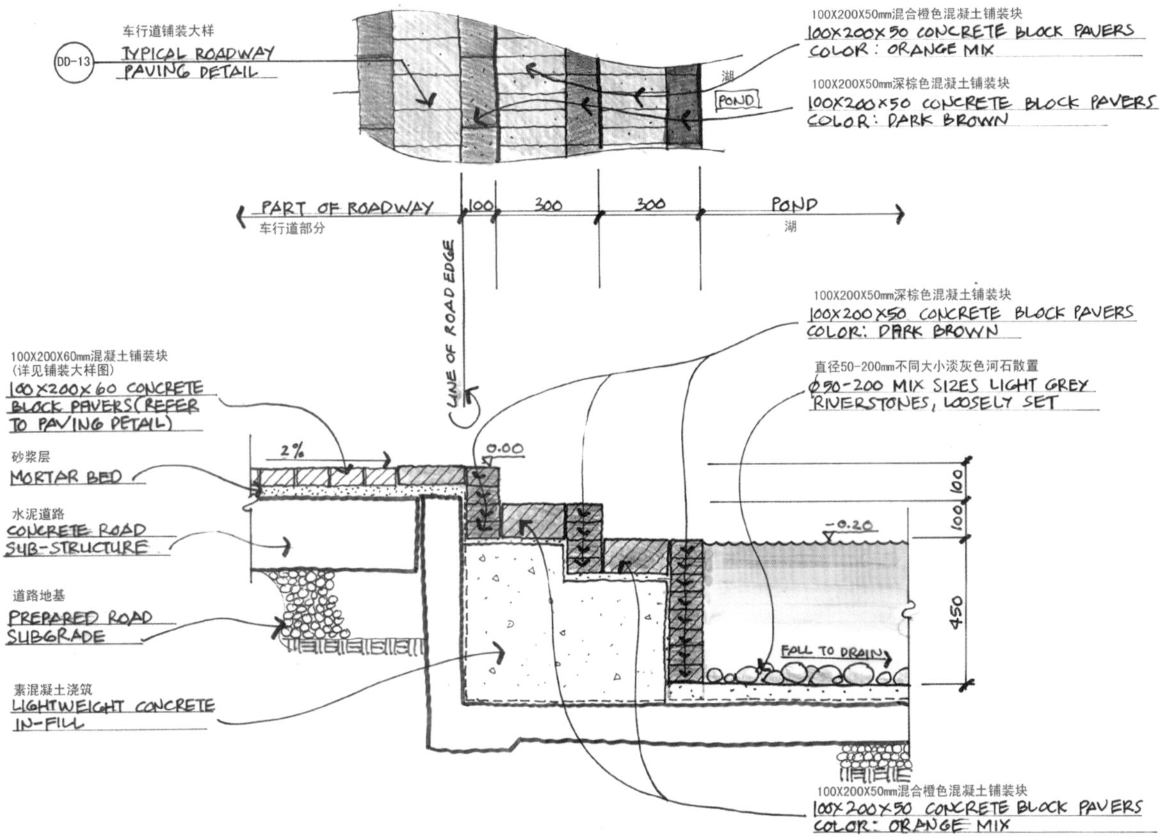

车行道铺装大样
DD-13
TYPICAL ROADWAY PAVING DETAIL
100X200X50mm混合橙色混凝土铺装块
100X200X50 CONCRETE BLOCK PAVERS COLOR: ORANGE MIX
湖
POND
100X200X50mm深棕色混凝土铺装块
100X200X50 CONCRETE BLOCK PAVERS COLOR: DARK BROWN
PART OF ROADWAY
车行道部分
100
300
300
POND
湖
LINE OF ROAD EDGE
100X200X50mm深棕色混凝土铺装块
100X200X50 CONCRETE BLOCK PAVERS COLOR: DARK BROWN
100X200X60mm混凝土铺装块
(详见铺装大样图)
100X200X60 CONCRETE BLOCK PAVERS (REFER TO PAVING DETAIL)
直径50-200mm不同大小淡灰色河石散置
Ø50-200 MIX SIZES LIGHT GREY RIVERSTONES, LOOSELY SET
2%
0.00
砂浆层
MORTAR BED
100
100
-0.20
水泥道路
CONCRETE ROAD SUB-STRUCTURE
道路地基
PREPARED ROAD SUBGRADE
450
FALL TO DRAIN
素混凝土浇筑
LIGHTWEIGHT CONCRETE IN-FILL
100X200X50mm混合橙色混凝土铺装块
100X200X50 CONCRETE BLOCK PAVERS COLOR: ORANGE MIX

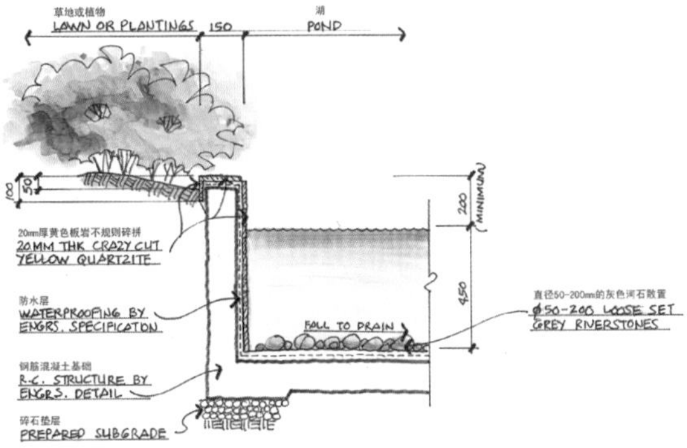

草地或植物
LAWN OR PLANTINGS
150
湖
POND
100
50
200 (MINIMUM)
450
20mm厚黄色板岩不规则碎拼
20MM THK CRAZY CUT YELLOW QUARTZITE
防水层
WATERPROOFING BY ENGRS. SPECIFICATION
FALL TO DRAIN
直径50-200mm的灰色河石散置
Ø50-200 LOOSE SET GREY RIVERSTONES
钢筋混凝土基础
R.C. STRUCTURE BY ENGRS. DETAIL
碎石垫层
PREPARED SUBGRADE

特色植物
PLANTINGS AS SPECIFIED
90度直角
90° RIGHT ANGLE CORNER
改良种植土
AMENDED SOIL
20mm厚黄色板岩不规则碎拼
20MM THK CRAZY CUT YELLOW QUARTZITE
缝与缝之间连成线
JOINTS ARE ALIGNED

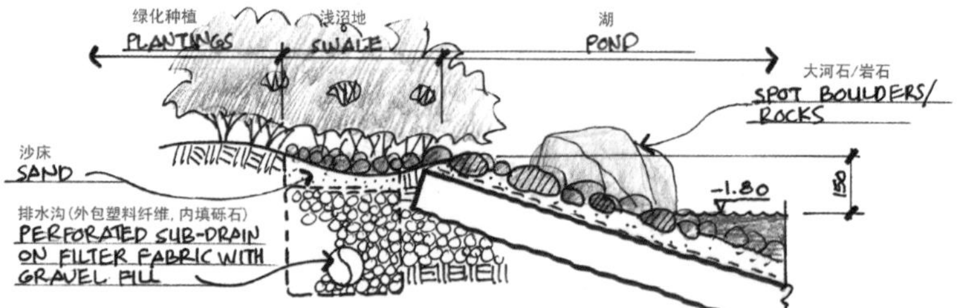
绿化种植
PLANTINGS
浅沼地
SWALE
湖
POND
大河石/岩石
SPOT BOULDERS/ ROCKS
沙床
SAND
排水沟(外包塑料纤维，内填砾石)
PERFORATED SUB-DRAIN ON FILTER FABRIC WITH GRAVEL FILL
-1.80
150

PLANTING EDGE 软质（植物）驳岸做法

SCALE 1:10

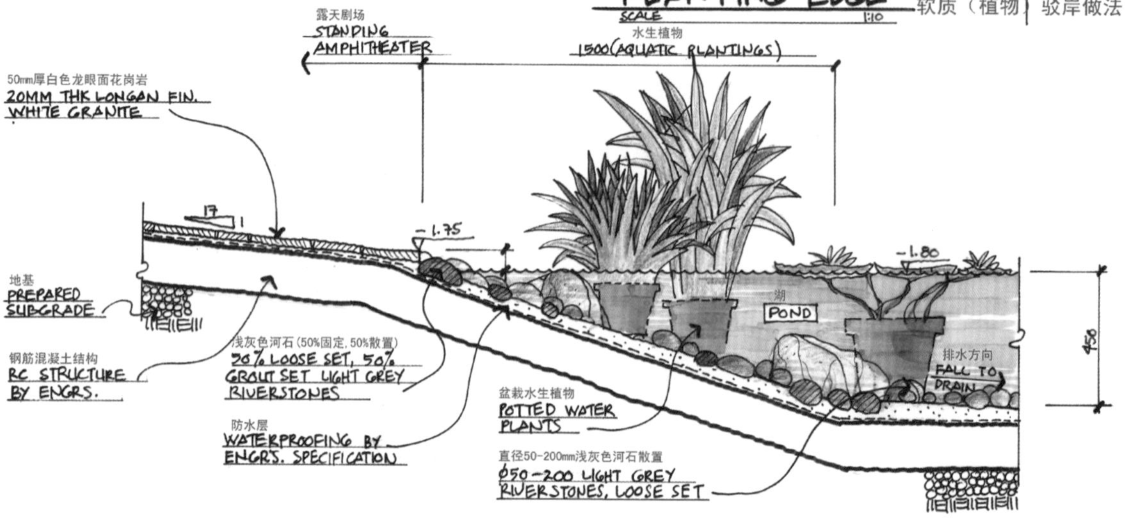
露天剧场
STANDING AMPHITHEATER
水生植物
1500(AQUATIC PLANTINGS)
50mm厚白色龙眼面花岗岩
20MM THK LONGAN FIN. WHITE GRANITE
17
1
-1.75
-1.80
地基
PREPARED SUBGRADE
湖
POND
450
钢筋混凝土结构
RC STRUCTURE BY ENGRS.
浅灰色河石(50%固定，50%散置)
50% LOOSE SET, 50% GROUT SET LIGHT GREY RIVERSTONES
排水方向
FALL TO DRAIN
盆栽水生植物
POTTED WATER PLANTS
防水层
WATERPROOFING BY ENGR'S. SPECIFICATION
直径50-200mm浅灰色河石散置
Ø50-200 LIGHT GREY RIVERSTONES, LOOSE SET

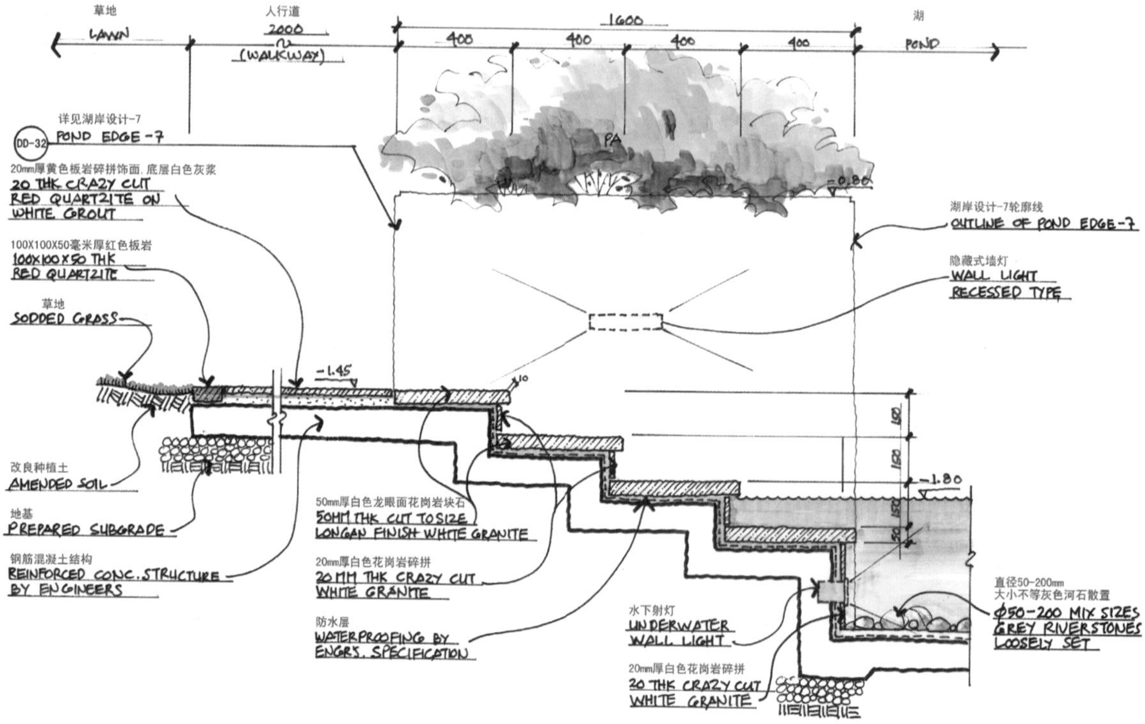
草地
LAWN
人行道
2000
(WALKWAY)
1600
400
400
400
400
湖
POND
DD-32
详见湖岸设计-7
POND EDGE -7
PA
-0.80
20mm厚黄色板岩碎拼饰面，底层白色灰浆
20 THK CRAZY CUT RED QUARTZITE ON WHITE GROUT
湖岸设计-7轮廓线
OUTLINE OF POND EDGE-7
100X100X50毫米厚红色板岩
100X100X50 THK RED QUARTZITE
隐藏式墙灯
WALL LIGHT RECESSED TYPE
草地
SODDED GRASS
-1.45
10
150
150
-1.80
150
50
改良种植土
AMENDED SOIL
地基
PREPARED SUBGRADE
50mm厚白色龙眼面花岗岩块石
50MM THK CUT TO SIZE LONGAN FINISH WHITE GRANITE
钢筋混凝土结构
REINFORCED CONC. STRUCTURE BY ENGINEERS
20mm厚白色花岗岩碎拼
20 MM THK CRAZY CUT WHITE GRANITE
直径50-200mm 大小不等灰色河石散置
Ø50-200 MIX SIZES GREY RIVERSTONES LOOSELY SET
水下射灯
UNDERWATER WALL LIGHT
防水层
WATERPROOFING BY ENGRS. SPECIFICATION
20mm厚白色花岗岩碎拼
20 THK CRAZY CUT WHITE GRANITE

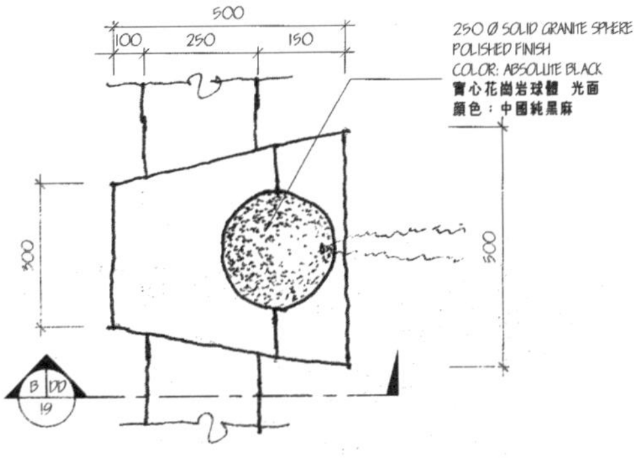

250 Ø SOLID GRANITE SPHERE
POLISHED FINISH
COLOR: ABSOLUTE BLACK
實心花崗岩球體 光面
顏色：中國純黑麻
A
平面圖
PLAN
SCALE 1:10
250 Ø SOLID GRANITE SPHERE
FLAMED FINISH
COLOR: ABSOLUTE BLACK
實心花崗岩球體 燒面
顏色：中國純黑麻
PEBBLE WASHED, COLOR: BEIGE
洗水石 顏色：灰黃
C
遠景圖
PERSPECTIVE
SCALE NTS
PEBBLE WASHED, COLOR: BEIGE
洗水石 顏色：灰黃
R.C. STRUCTURE BY ENGINEER
鋼筋混凝土結構
由工程師擔任
LIGHTING FIXTURE AS SPECIFIED
(REFER TO LIGHTING PLAN)
指定燈具（參照照明配置圖）
B
剖面圖
SECTION
SCALE 1:10
TYPICAL TREE COLLAR
SEE DETAIL
AERATED WATER JETS
50 MM THK NATURAL GRANITE
FLAMED FINISH; COLOUR: RUSTIC YELLOW
STAINLESS STEEL TRAY WITH
METAL SUPPORT
APPROVED PAVING MATERIAL
WATERPROOFING AS PER ARCHITECTS
SPECIFICATIONS
R.C. STRUCTURE AS PER ENGINEER'S
DETAIL
TO
FILTRATION
DRAIN TO
WASTE
AERATED WATER JETS
TYPICAL TREE COLLAR
SEE DETAIL
SEE DETAIL
FALL
FALL
FROM
WATER
SOURCE
FROM
WATER SOURCE
FROM
WATER
SOURCE

水景

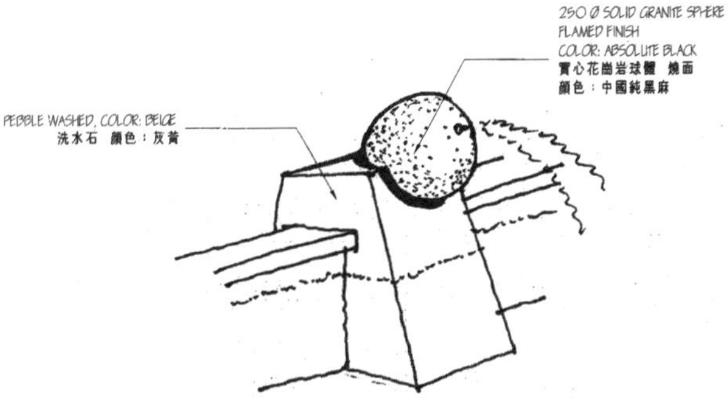

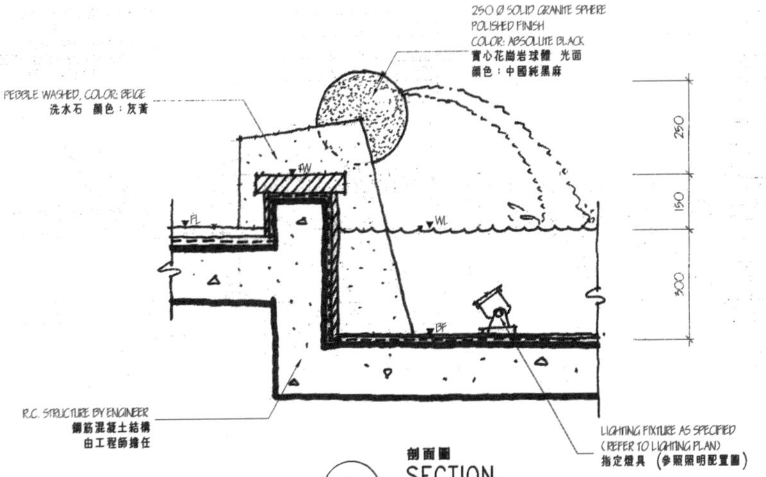

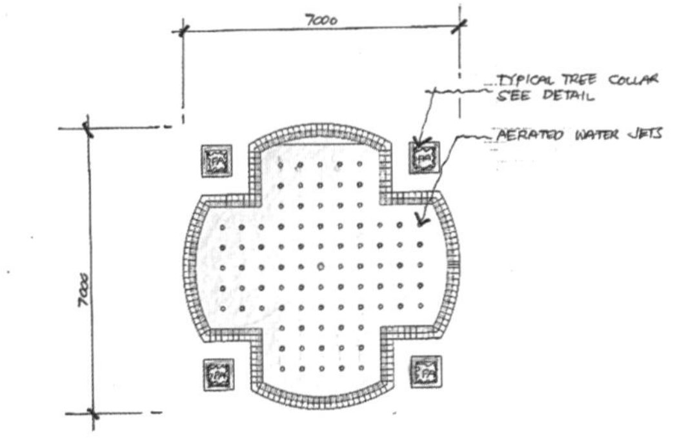

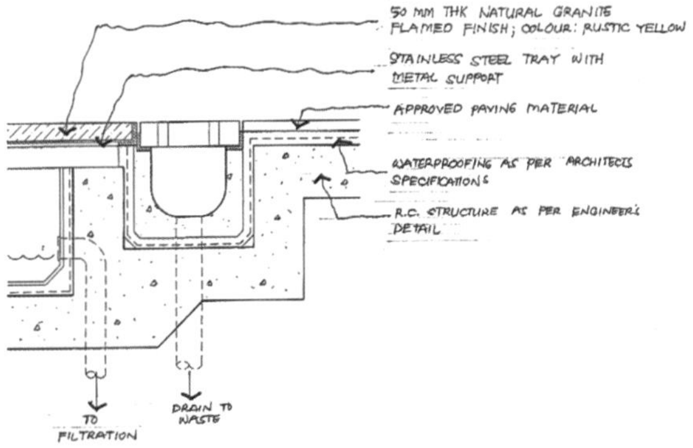

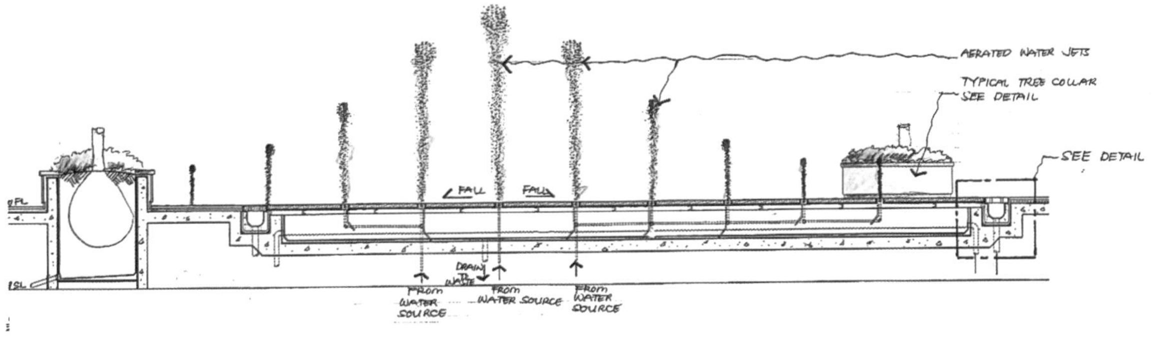

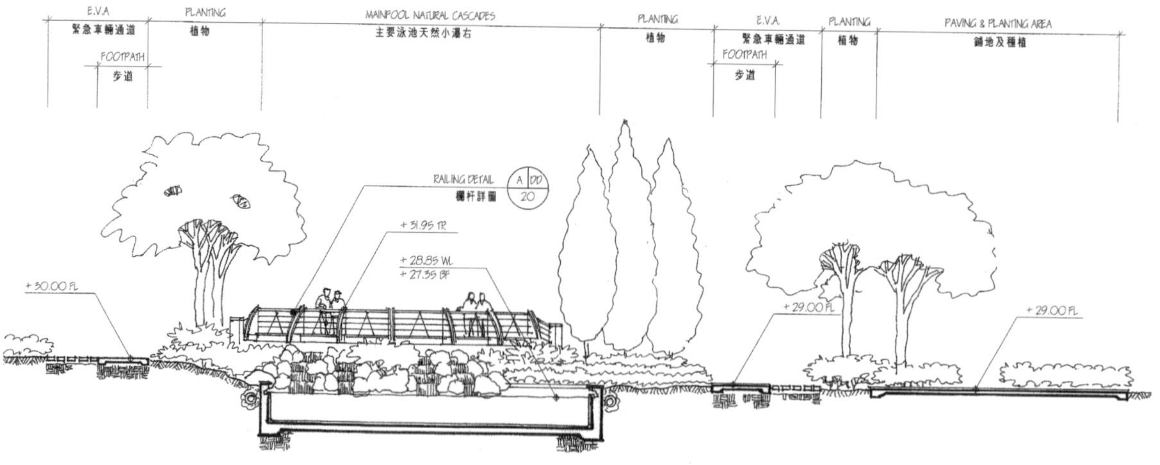
E.V.A
緊急車輛通道
FOOTPATH
步道
PLANTING
植物
MAINPOOL NATURAL CASCADES
主要泳池天然小瀑右
PLANTING
植物
E.V.A
緊急車輛通道
FOOTPATH
步道
PLANTING
植物
PAVING & PLANTING AREA
鋪地及種植
RAILING DETAIL
欄杆詳圖
A
DD
20
+ 31.95 TR
+ 28.85 WL
+ 27.35 BF
+ 30.00 FL
+ 29.00 FL
+ 29.00 FL

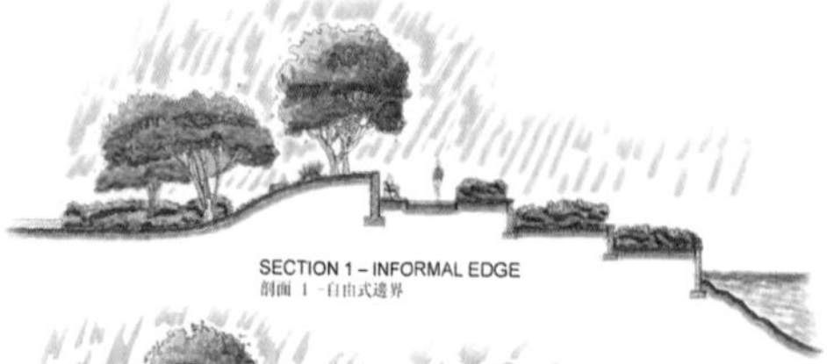
SECTION 1 – INFORMAL EDGE
剖面 1 -自由式邊界

SECTION 4 – INFORMAL EDGE
剖面 4 自由式邊界

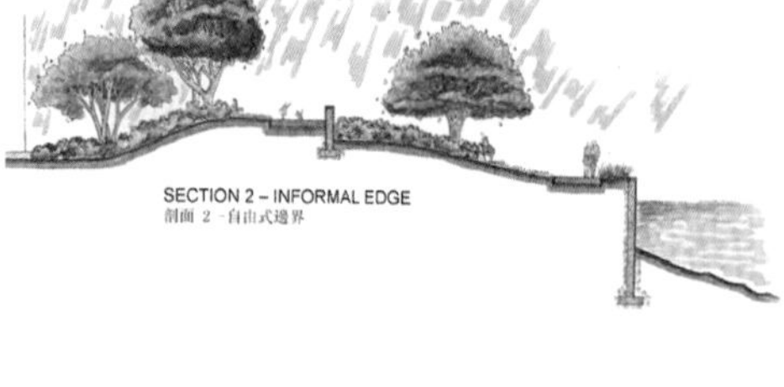
SECTION 2 – INFORMAL EDGE
剖面 2 -自由式邊界

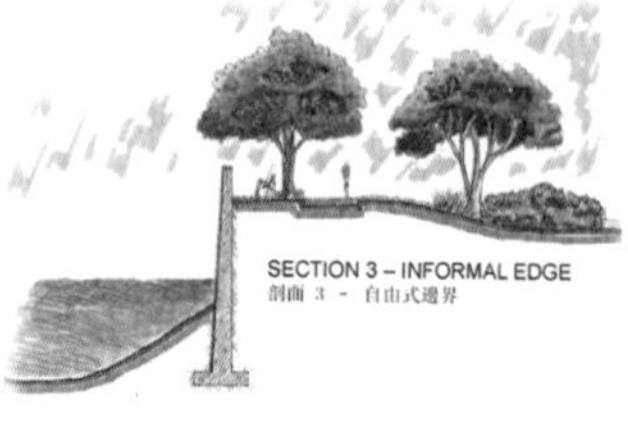
SECTION 3 – INFORMAL EDGE
剖面 3 - 自由式邊界

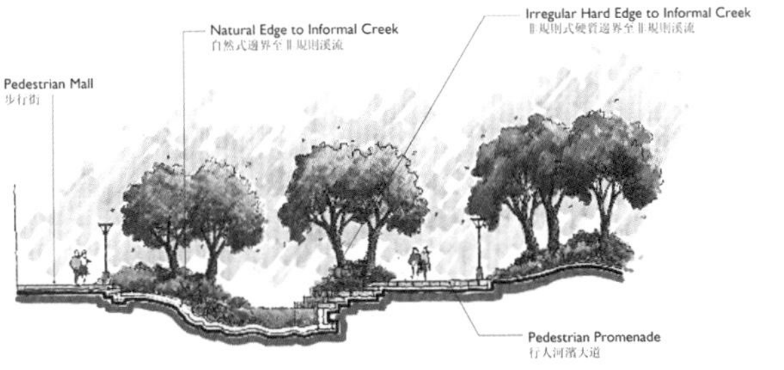
Natural Edge to Informal Creek
自然式邊界至非規則溪流
Irregular Hard Edge to Informal Creek
非規則式硬質邊界至非規則溪流
Pedestrian Mall
步行街
Pedestrian Promenade
行人河濱大道

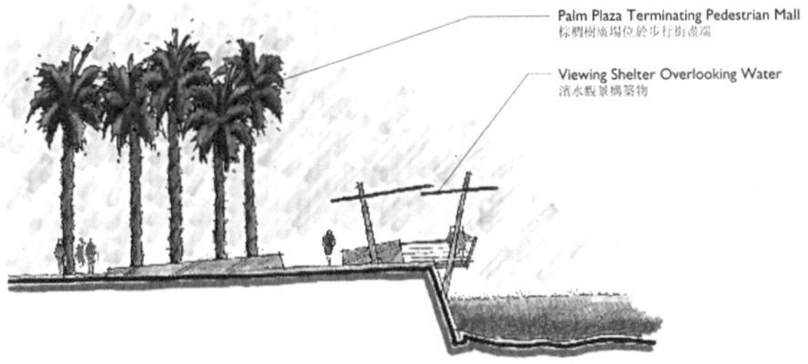
Palm Plaza Terminating Pedestrian Mall
棕櫚樹廣場位於步行街盡端
Viewing Shelter Overlooking Water
濱水觀景構築物

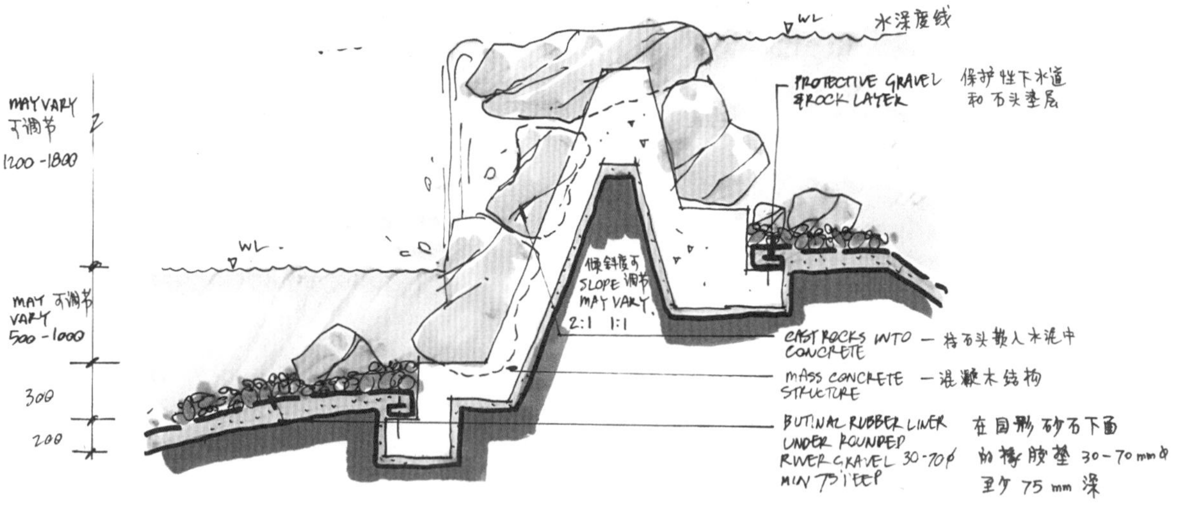
WL
水深度线
MAY VARY
可调节
1200-1800
WL
MAY VARY
可调节
500-1000
300
200
PROTECTIVE GRAVEL & ROCK LAYER
保护性下水道和石头垫层
倾斜度可调节
SLOPE MAY VARY
2:1 1:1
CAST ROCKS INTO CONCRETE
将石头散入水泥中
MASS CONCRETE STRUCTURE
混凝土结构
BUTINAL RUBBER LINER UNDER ROUNDED RIVER GRAVEL 30-70ϕ MIN 75 DEEP
在圆形砂石下面的橡胶垫 30-70mmϕ 至少75mm深

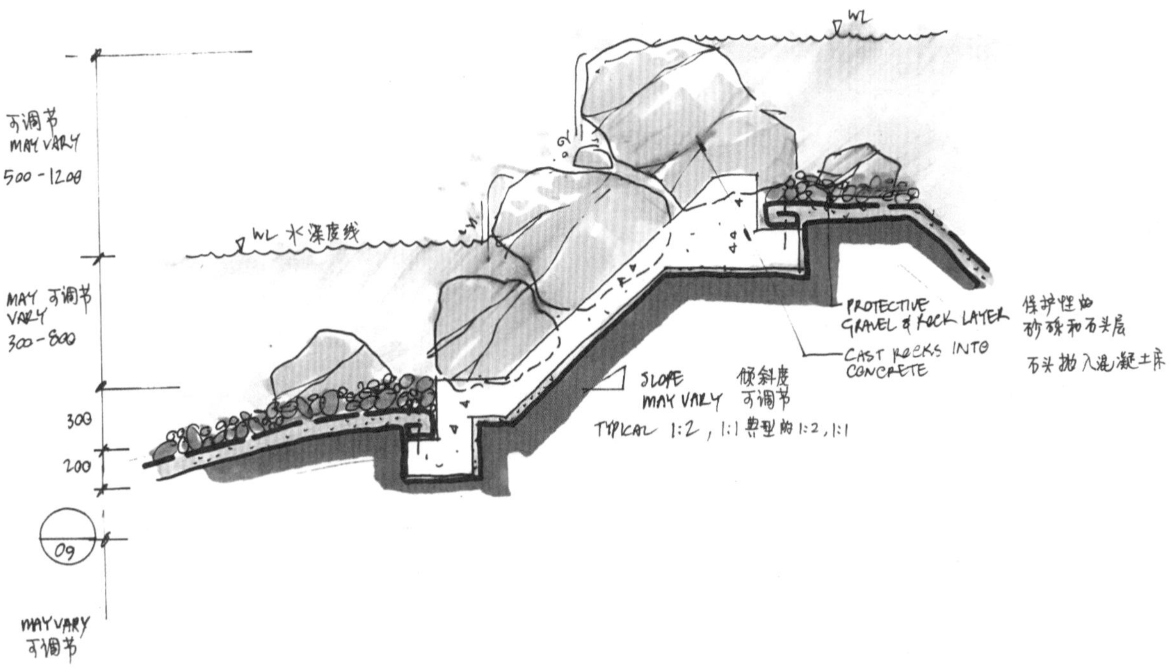
WL
可调节
MAY VARY
500-1200
WL 水深度线
MAY VARY
可调节
300-800
300
200
09
MAY VARY
可调节
PROTECTIVE GRAVEL & ROCK LAYER
保护性的砂砾和石头层
CAST ROCKS INTO CONCRETE
石头抛入混凝土床
SLOPE MAY VARY
倾斜度可调节
TYPICAL 1:2, 1:1 典型的1:2, 1:1

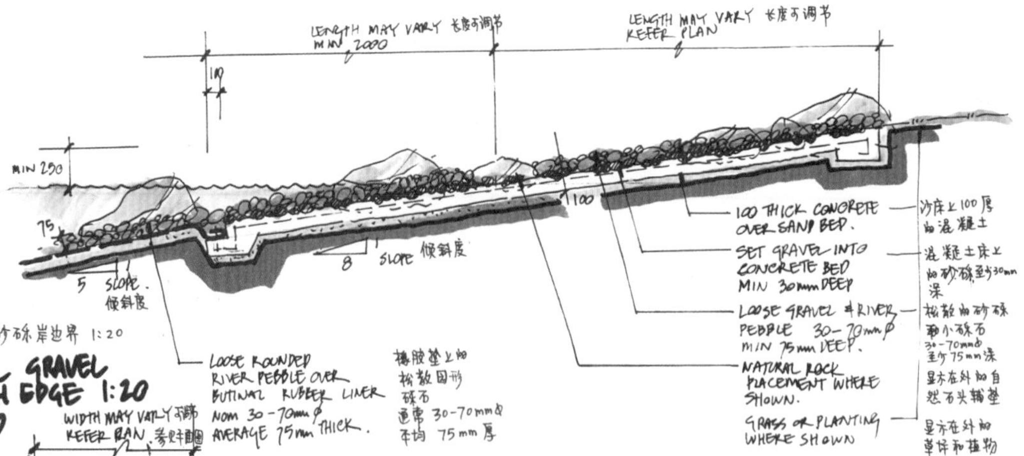

LENGTH MAY VARY 长度可调节
MIN 2000
LENGTH MAY VARY 长度可调节
REFER PLAN
MIN 250
75
100
110
5 SLOPE 倾斜度
8 SLOPE 倾斜度
100 THICK CONCRETE OVER SAND BED. 沙床上100厚的混凝土
SET GRAVEL INTO CONCRETE BED MIN 30mm DEEP 混凝土床上的砂砾至少30mm深
LOOSE GRAVEL & RIVER PEBBLE 30–70mm ϕ MIN 75mm DEEP. 松散的砂砾和小砾石 30–70mmϕ 至少75mm深
NATURAL ROCK PLACEMENT WHERE SHOWN. 显示在外的自然石头铺垫
GRASS OR PLANTING WHERE SHOWN 显示在外的草坪和植物
LOOSE ROUNDED RIVER PEBBLE OVER BUTINAL RUBBER LINER Nom 30–70mm ϕ AVERAGE 75mm THICK. 橡胶垫上的松散圆形砾石 通常30–70mmϕ 平均75mm厚
典型的砂砾岸边界 1:20
07 TYPICAL GRAVEL BEACH EDGE 1:20 TYPE D
D型
WIDTH MAY VARY REFER PLAN
200
250–350
5
8 1 SLOPE 倾斜度
100 THICK CONCRETE SLAB OVER SAND BED. 在沙床上的100厚度的混凝土板
SET GRAVEL & RIVER PEBBLE INTO CONCRETE MIN 30mm DEEP 放置在混凝土上的砂砾和小砾石，至少30mm深
LOOSE RIVER PEBBLE MULCH 75mm DEEP OVER AQUATIC PLANTING SOIL 在水生植物土壤上的松散护根砂砾，75mm深
08 GRAVEL BEACH WITH AQUATIC PLANTING. 1:20 TYPE D1.
有水生植物的砂砾岸边 1:20
D1型

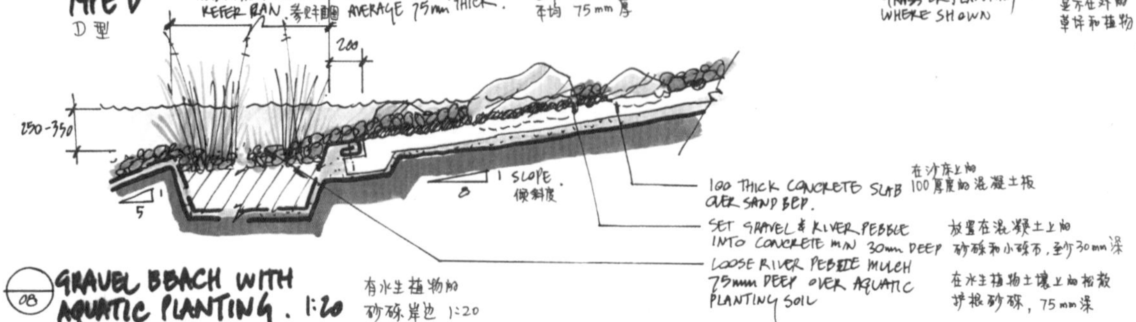

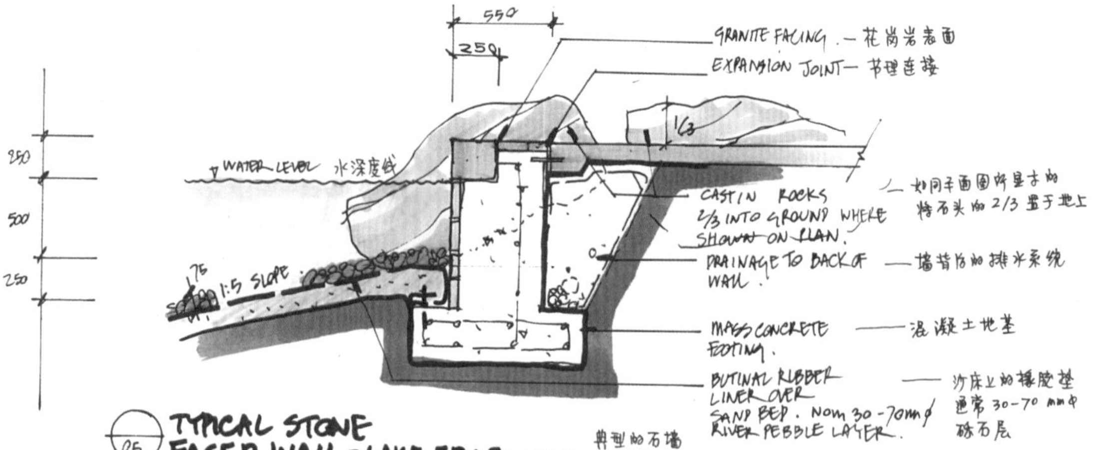

550
250
1/3
WATER LEVEL 水深度线
250
500
250
75
1:5 SLOPE
GRANITE FACING. —花岗岩表面
EXPANSION JOINT—节理连接
CAST IN ROCKS 2/3 INTO GROUND WHERE SHOWN ON PLAN. 如同平面图所显示的将石头的2/3置于地上
DRAINAGE TO BACK OF WALL —墙背后的排水系统
MASS CONCRETE FOOTING. —混凝土地基
BUTINAL RUBBER LINER OVER SAND BED. Nom 30–70mm ϕ RIVER PEBBLE LAYER. —沙床上的橡胶垫 通常30–70mmϕ 砾石层
05 TYPICAL STONE FACED WALL – LAKE EDGE 1:20 TYPE C.
典型的石墙湖边界 1:20
C型
550
250
200
250
500
75
400
50
EXTEND GRANITE FACING BELOW GRADE. —延伸花岗岩向下程度
CAST BOULDERS INTO WALL—将圆石嵌入墙中
MASS CONCRETE WALL WITH SPLIT FACE GRANITE —有裂面花岗石的混凝土墙
DRAINAGE TO BACK OF WALL —墙后背的排水系统
AQUATIC PLANTING SOIL WITH GRAVEL MULCH COVER Nom 30–70mm ϕ. —有砂砾护根的水生植物土壤 通常30–70mmϕ
06 TYPICAL STONE FACED LAKE EDGE WITH AQUATIC PLANTING. 1:20 TYPE C1
典型的水生植物石面湖边界 1:20
C1型

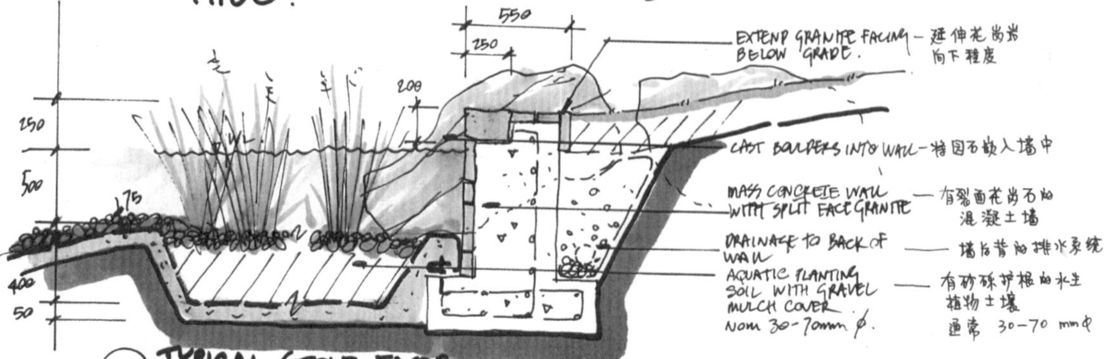

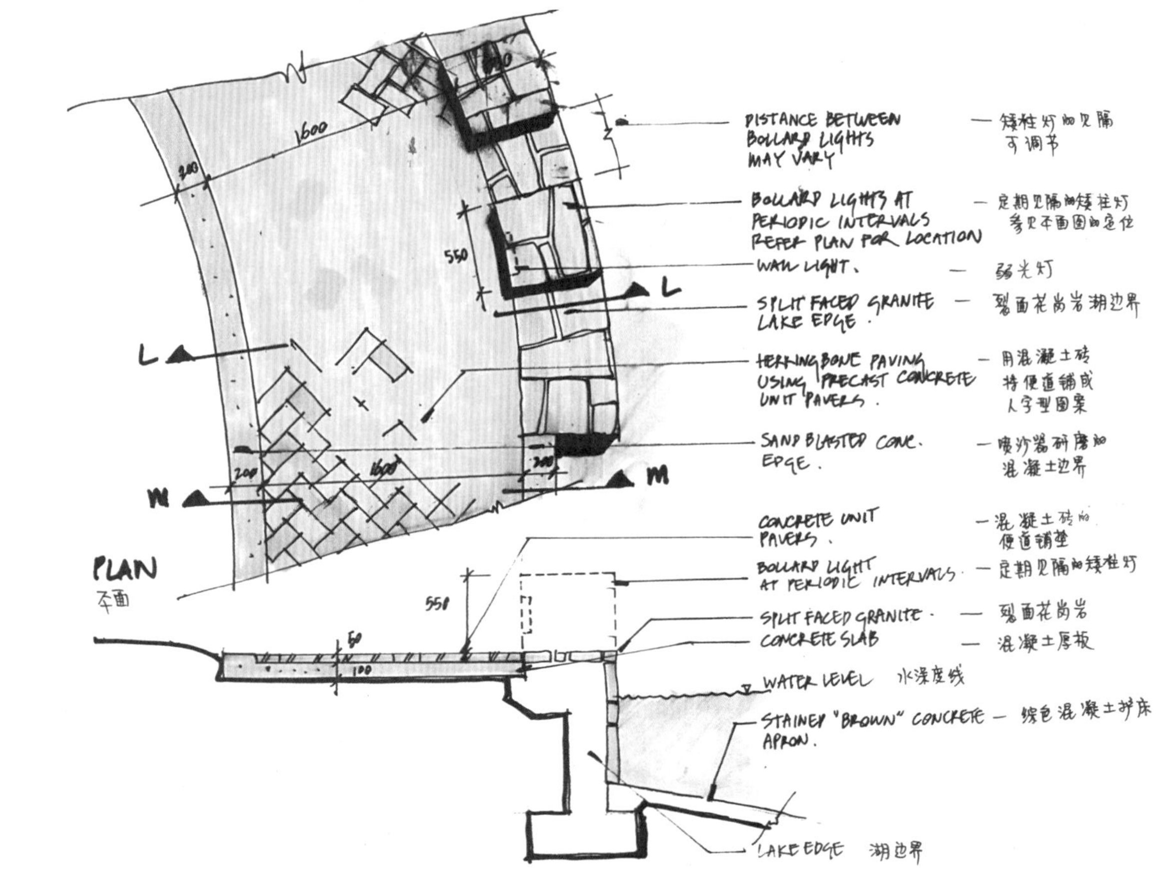

05A TYPICAL STONE LAKE EDGE SECTION LL 1:20 典型的石湖边界 LL断面图 1:20

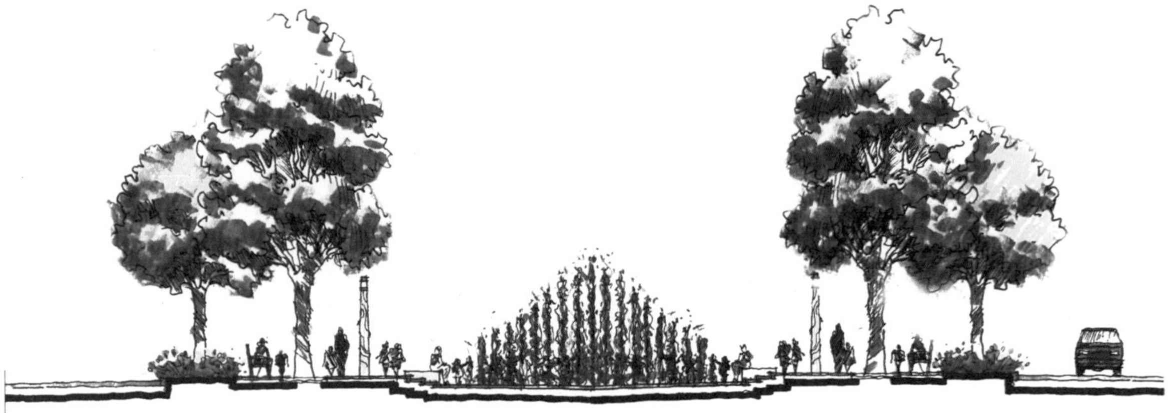

林荫大道　雕塑灯柱　管风琴音乐喷泉　休憩广场　入口大道

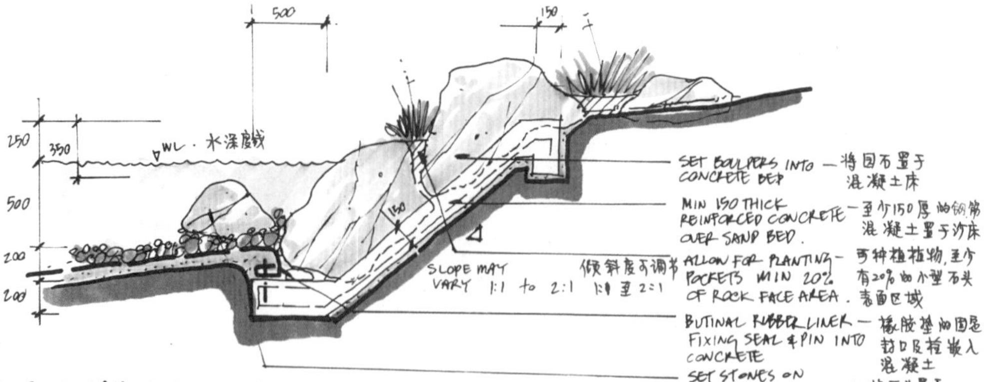

(03) TYPICAL ROCK LAKE EDGE 1:20 TYPE B
典型石头湖边界 1:20
B型

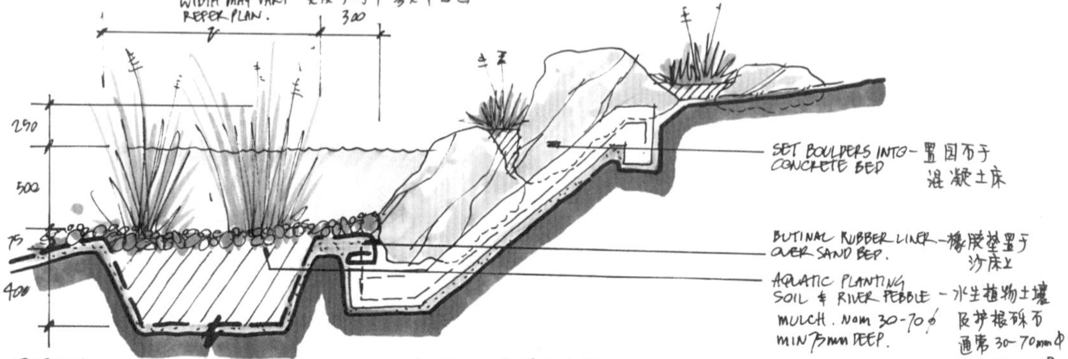

(04) TYPICAL ROCK LAKE EDGE WITH AQUATIC PLANTING 1:20 TYPE B1.
典型的水生植物石头湖边界 1:20
B1型

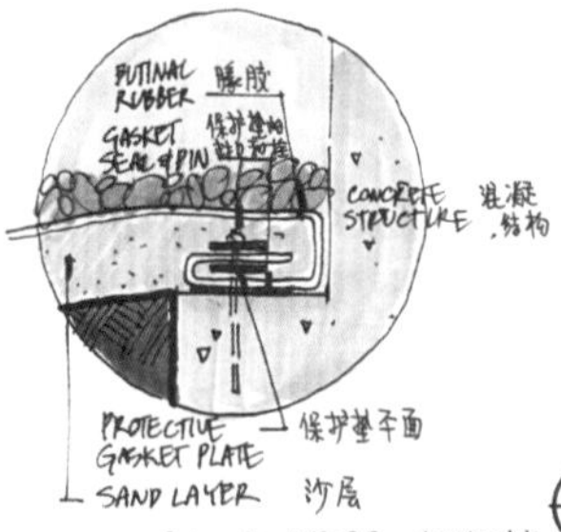

TYPICAL RUBBER LINER FIXING DETAIL NTS.
典型橡胶垫的固定详细图

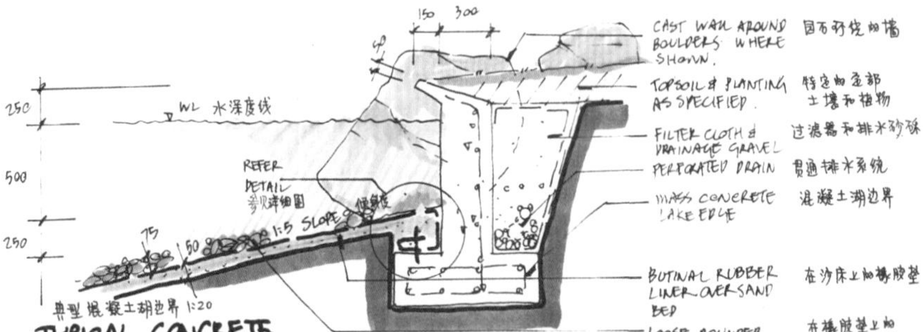

典型混凝土湖边界 1:20
(01) TYPICAL CONCRETE LAKE EDGE 1:20 TYPE A
A型

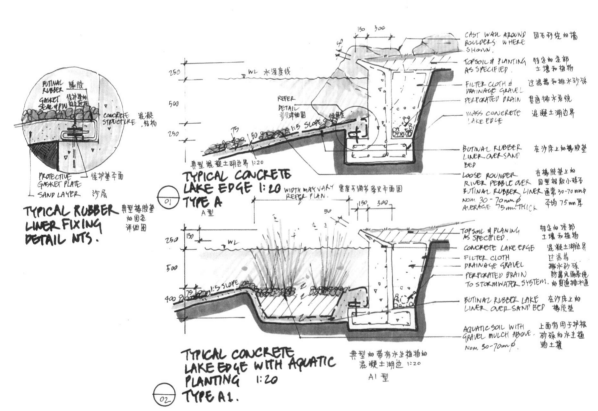

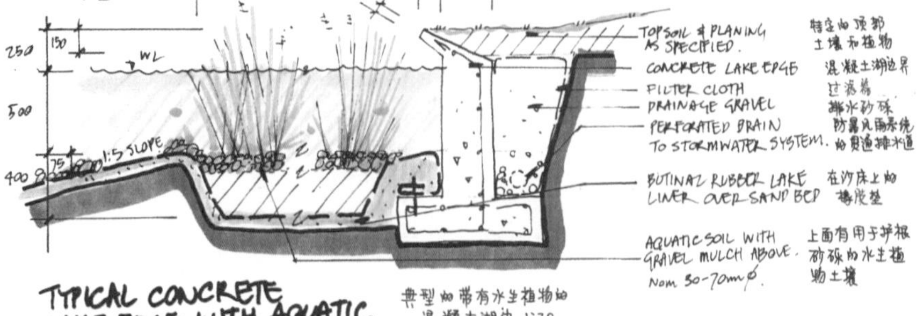

TYPICAL CONCRETE LAKE EDGE WITH AQUATIC PLANTING 1:20
(02) TYPE A1.
典型的带有水生植物的混凝土湖边 1:20
A1型

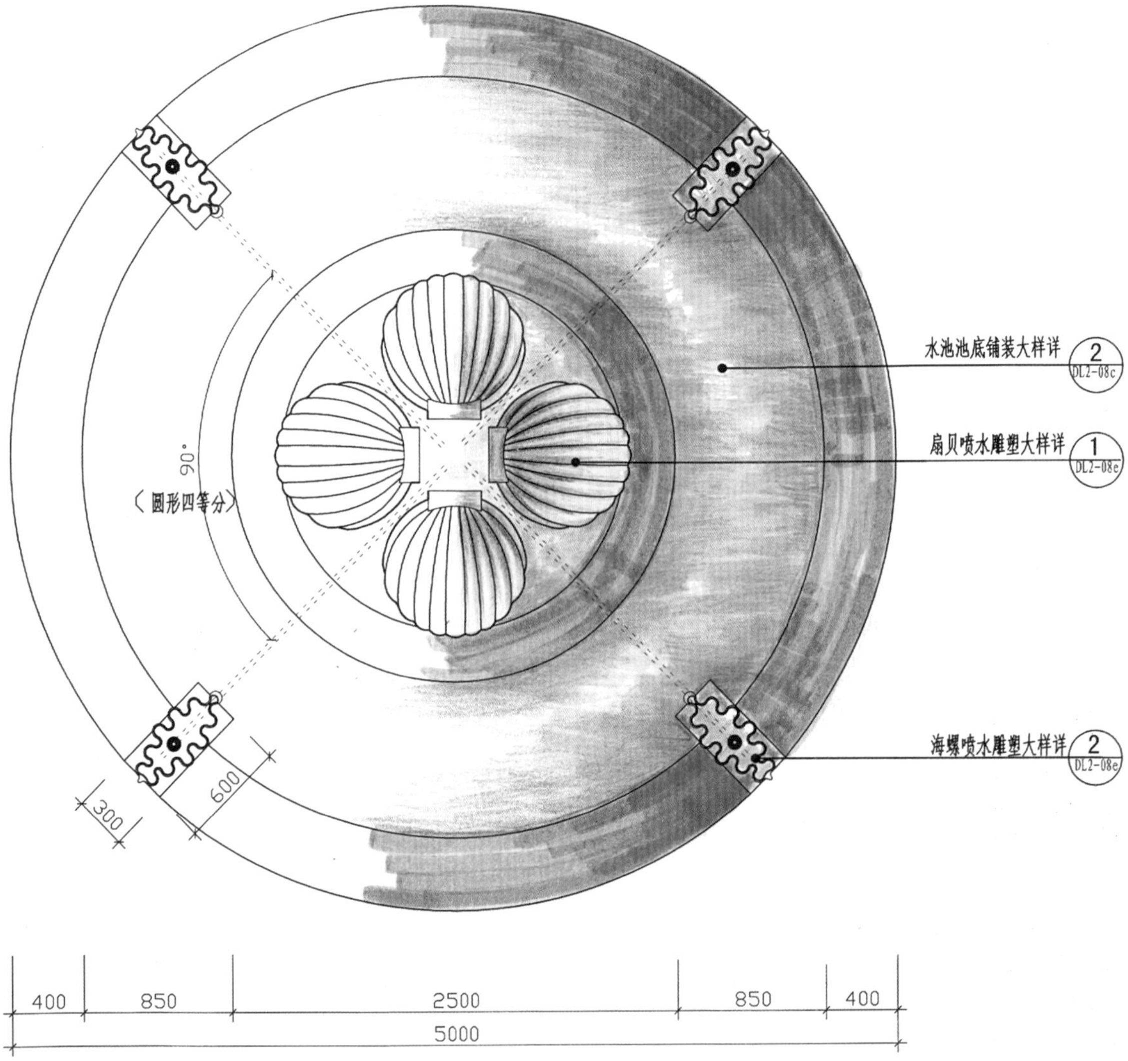
水池池底铺装大样详
2
DL2-08c
扇贝喷水雕塑大样详
1
DL2-08e
海螺喷水雕塑大样详
2
DL2-08e
90°
(圆形四等分)
600
300
400
850
2500
850
400
5000

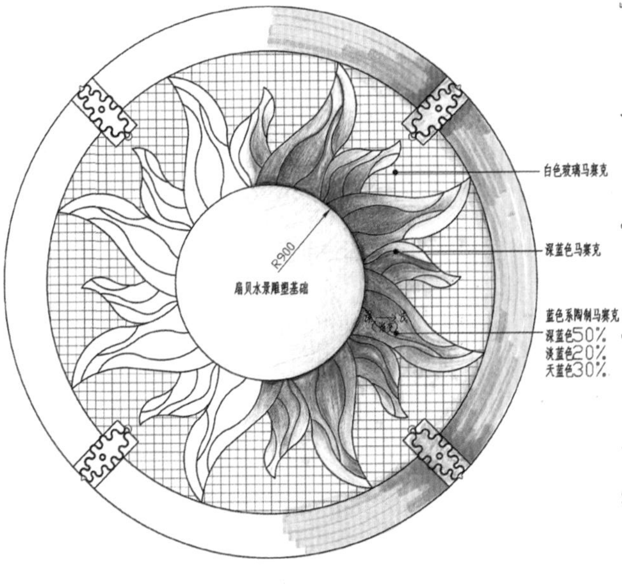
白色玻璃马赛克
深蓝色马赛克
蓝色系陶制马赛克
深蓝色50%
淡蓝色20%
天蓝色30%
扇贝水景雕塑基础
R900

0
0.5
1
2
3
4
5

网格间距为0.1米。

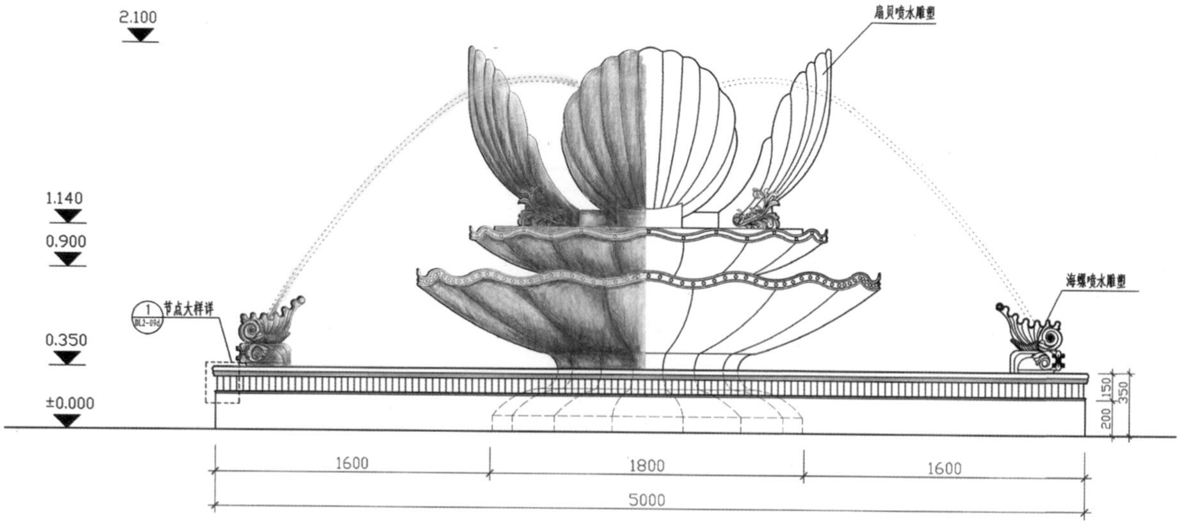
2.100
扇贝喷水雕塑
1.140
0.900
1
节点大样详
0.350
±0.000
海螺喷水雕塑
150
350
200
1600
1800
1600
5000

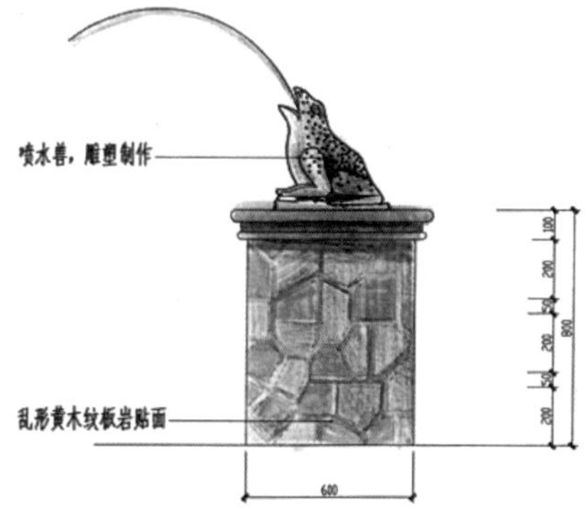
喷水兽，雕塑制作
乱形黄木纹板岩贴面
100
200
50
200
50
200
800
600

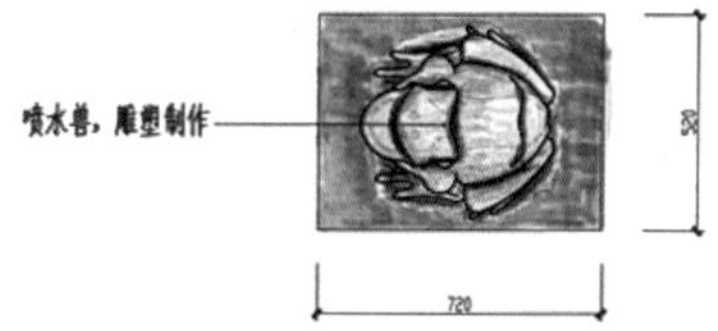
喷水兽，雕塑制作
520
720

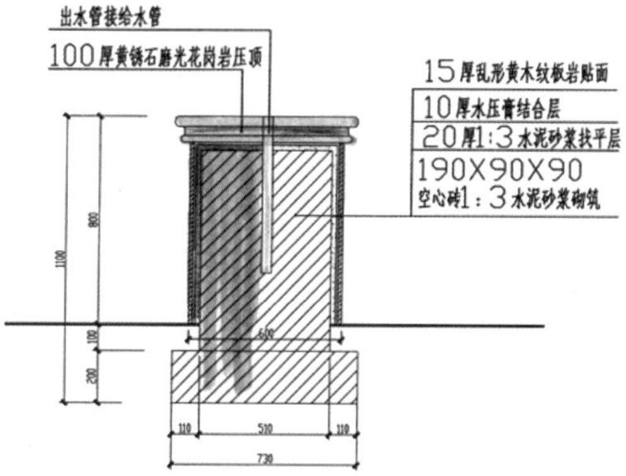
出水管接给水管
100厚黄锈石磨光花岗岩压顶
15厚乱形黄木纹板岩贴面
10厚水压膏结合层
20厚1:3水泥砂浆找平层
190X90X90
空心砖1:3水泥砂浆砌筑
1100
800
100
200
600
110
510
110
730

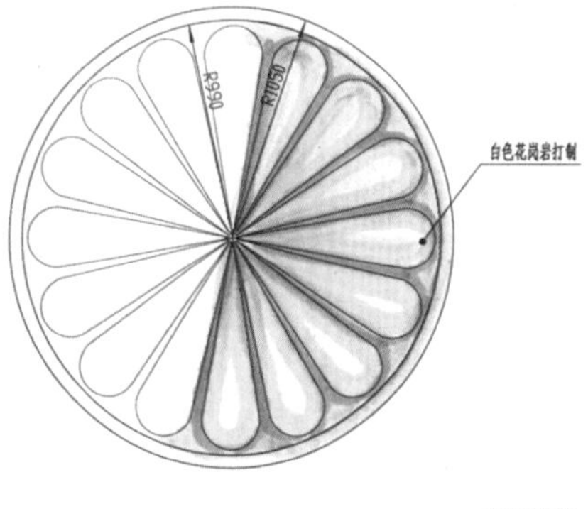
白色花岗岩打制
R1050

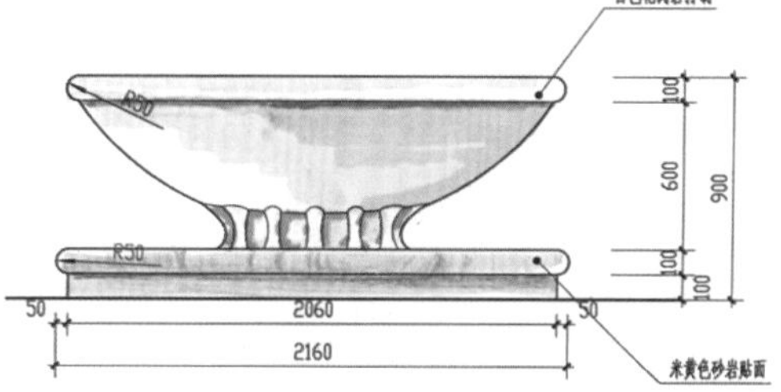
白色花岗岩打制
R50
R50
100
600
100
900
100
50
2060
50
2160
米黄色砂岩贴面

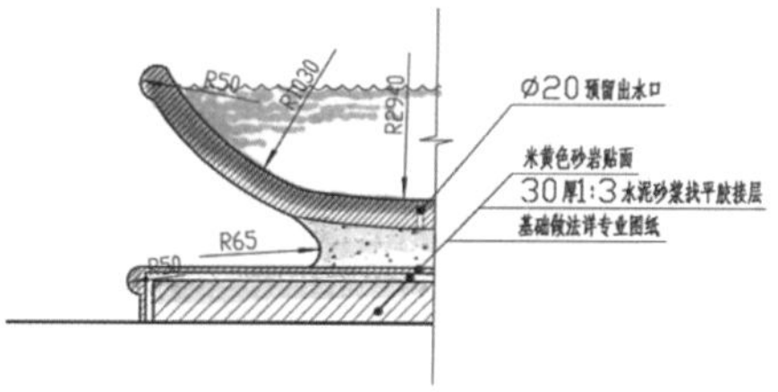
R50
R1030
R2940
Ø20 预留出水口
米黄色砂岩贴面
30厚1:3水泥砂浆找平找坡层
基础做法详专业图纸
R65
R50

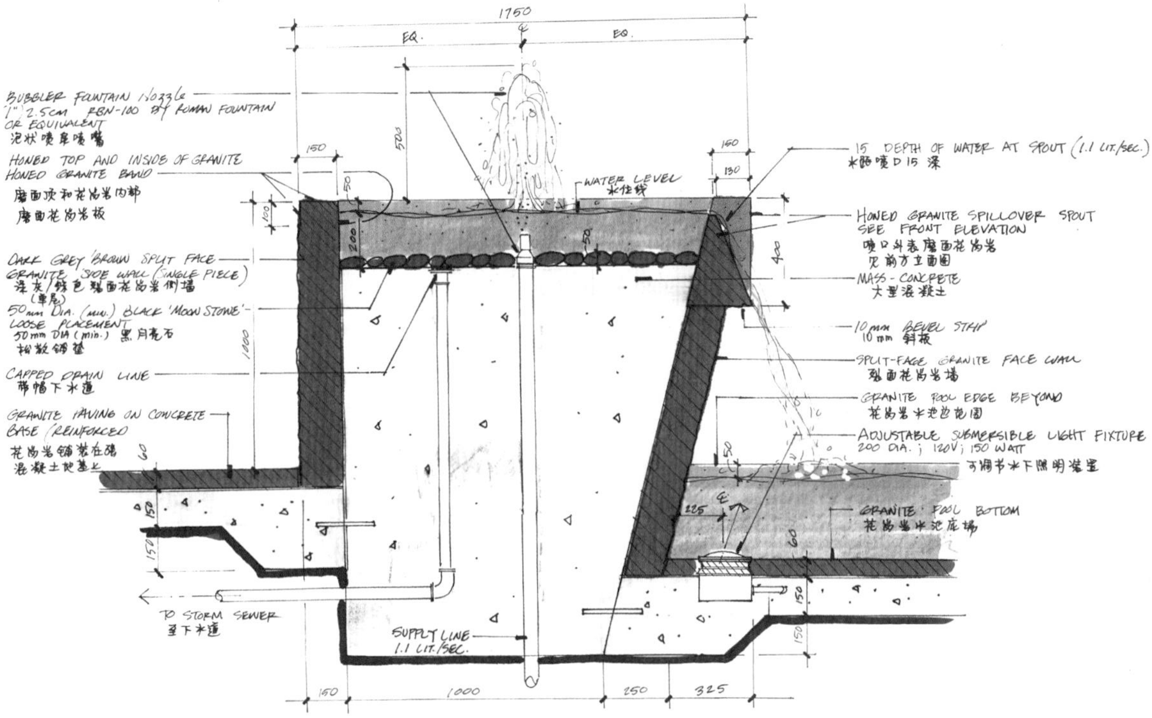
1750
EQ.
EQ.
BUBBLER FOUNTAIN NOZZLE 1" 2.5cm FBN-100 BY ROMAN FOUNTAIN OR EQUIVALENT
泡状喷泉喷嘴
HONED TOP AND INSIDE OF GRANITE HONED GRANITE BAND
磨面顶和花岗岩内部
磨面花岗岩板
DARK GREY/BROWN SPLIT FACE GRANITE SIDE WALL (SINGLE PIECE)
深灰/棕色裂面花岗岩侧墙（单片）
50mm DIA. (min.) BLACK 'MOON STONE' LOOSE PLACEMENT
50mm DIA (min.) 黑月亮石松散铺设
CAPPED DRAIN LINE
带帽下水道
GRANITE PAVING ON CONCRETE BASE/REINFORCED
花岗岩铺装在强混凝土地基上
TO STORM SEWER
至下水道
SUPPLY LINE 1.1 LIT./SEC.
WATER LEVEL
水位线
15 DEPTH OF WATER AT SPOUT (1.1 LIT./SEC.)
水距喷口15深
HONED GRANITE SPILLOVER SPOUT SEE FRONT ELEVATION
喷口外表磨面花岗岩
见前方立面图
MASS-CONCRETE
大型混凝土
10mm BEVEL STRIP
10mm 斜板
SPLIT-FACE GRANITE FACE WALL
裂面花岗岩墙
GRANITE POOL EDGE BEYOND
花岗岩水池边范围
ADJUSTABLE SUBMERSIBLE LIGHT FIXTURE 200 DIA.; 120V; 150 WATT
可调节水下照明装置
GRANITE POOL BOTTOM
花岗岩水池底部
150
1000
250
325

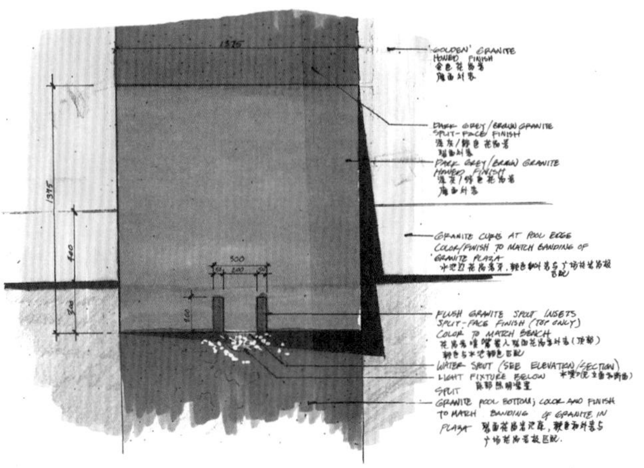
1375
'GOLDEN' GRANITE HONED FINISH
DARK GREY/BROWN GRANITE SPLIT-FACE FINISH
DARK GREY/BROWN GRANITE HONED FINISH
GRANITE CURB AT POOL EDGE COLOR/FINISH TO MATCH BANDING OF GRANITE PLAZA
FLUSH GRANITE SPOUT INSETS SPLIT-FACE FINISH (TOP ONLY) COLOR TO MATCH BENCH
WATER SPOUT (SEE ELEVATION/SECTION)
LIGHT FIXTURE BELOW
SPOUT
GRANITE POOL BOTTOM; COLOR AND FINISH TO MATCH BANDING OF GRANITE IN PLAZA

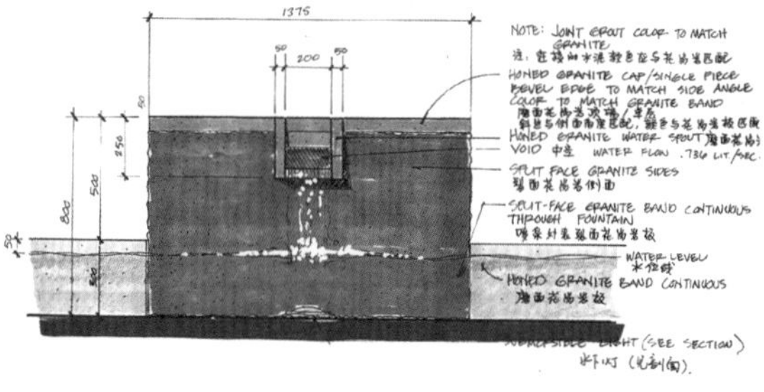
1375
NOTE: JOINT GROUT COLOR TO MATCH GRANITE.
HONED GRANITE CAP/SINGLE PIECE BEVEL EDGE TO MATCH SIDE ANGLE COLOR TO MATCH GRANITE BAND
HONED GRANITE WATER SPOUT
VOID 中空 WATER FLOW
SPLIT FACE GRANITE SIDES
SPLIT-FACE GRANITE BAND CONTINUOUS THROUGH FOUNTAIN
WATER LEVEL
水位线
HONED GRANITE BAND CONTINUOUS

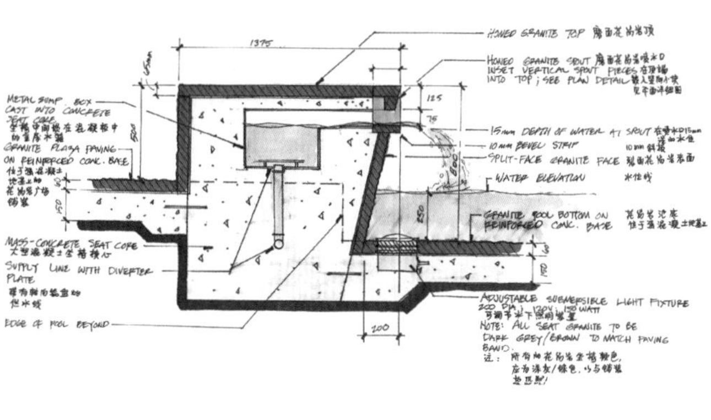
1375
HONED GRANITE TOP
METAL SUMP BOX CAST INTO CONCRETE SEAT CORE
GRANITE PLAZA PAVING ON REINFORCED CONC. BASE
MASS-CONCRETE SEAT CORE
SUPPLY LINE WITH DIVERTER PLATE
EDGE OF POOL BEYOND
15mm DEPTH OF WATER AT SPOUT
10mm BEVEL STRIP
SPLIT-FACE GRANITE FACE
WATER ELEVATION
GRANITE POOL BOTTOM ON REINFORCED CONC. BASE
ADJUSTABLE SUBMERSIBLE LIGHT FIXTURE
NOTE: ALL SEAT GRANITE TO BE DARK GREY/BROWN TO MATCH PAVING BAND.

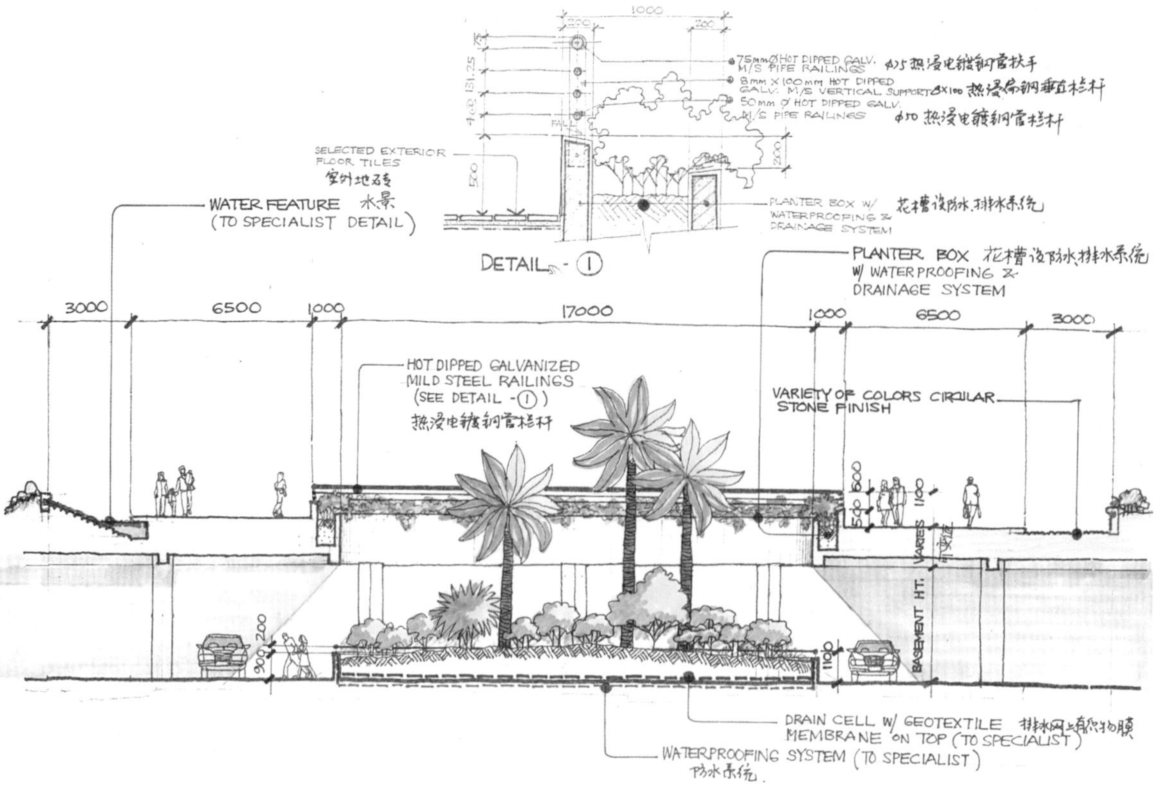
1000
200
200
75mm Ø HOT DIPPED GALV. M/S PIPE RAILINGS
8mm X 100mm HOT DIPPED GALV. M/S VERTICAL SUPPORT
50mm Ø HOT DIPPED GALV. M/S PIPE RAILINGS
热浸电镀钢管栏杆
SELECTED EXTERIOR FLOOR TILES
室外地石砖
PLANTER BOX W/ WATERPROOFING & DRAINAGE SYSTEM
DETAIL - 1
WATER FEATURE 水景
(TO SPECIALIST DETAIL)
PLANTER BOX 花槽设防水、排水系统
W/ WATERPROOFING & DRAINAGE SYSTEM
3000
6500
1000
17000
1000
6500
3000
HOT DIPPED GALVANIZED MILD STEEL RAILINGS
(SEE DETAIL - 1)
热浸电镀钢管栏杆
VARIETY OF COLORS CIRCULAR STONE FINISH
BASEMENT HT. VARIES
DRAIN CELL W/ GEOTEXTILE MEMBRANE ON TOP (TO SPECIALIST)
WATERPROOFING SYSTEM (TO SPECIALIST)
防水系统

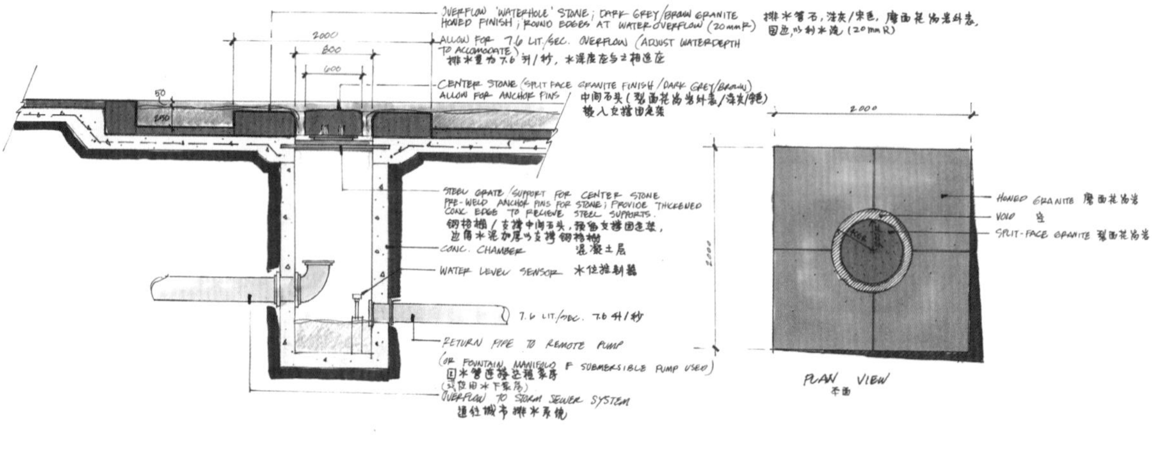
OVERFLOW 'WATERHOLE' STONE; DARK GREY/BROWN GRANITE HONED FINISH; ROUND EDGES AT WATER OVERFLOW (20mm R)
ALLOW FOR 7.6 LIT./SEC. OVERFLOW (ADJUST WATERDEPTH TO ACCOMODATE)
2000
800
600
CENTER STONE (SPLIT FACE GRANITE FINISH / DARK GREY/BROWN)
ALLOW FOR ANCHOR PINS
STEEL GRATE / SUPPORT FOR CENTER STONE
PRE-WELD ANCHOR PINS FOR STONE; PROVIDE THICKENED CONC. EDGE TO RECIEVE STEEL SUPPORTS.
CONC. CHAMBER 混凝土层
WATER LEVEL SENSOR
7.6 LIT./SEC.
RETURN PIPE TO REMOTE PUMP
(OR FOUNTAIN MANIFOLD IF SUBMERSIBLE PUMP USED)
OVERFLOW TO STORM SEWER SYSTEM
2000
HONED GRANITE
VOID 空
SPLIT-FACE GRANITE
PLAN VIEW
平面

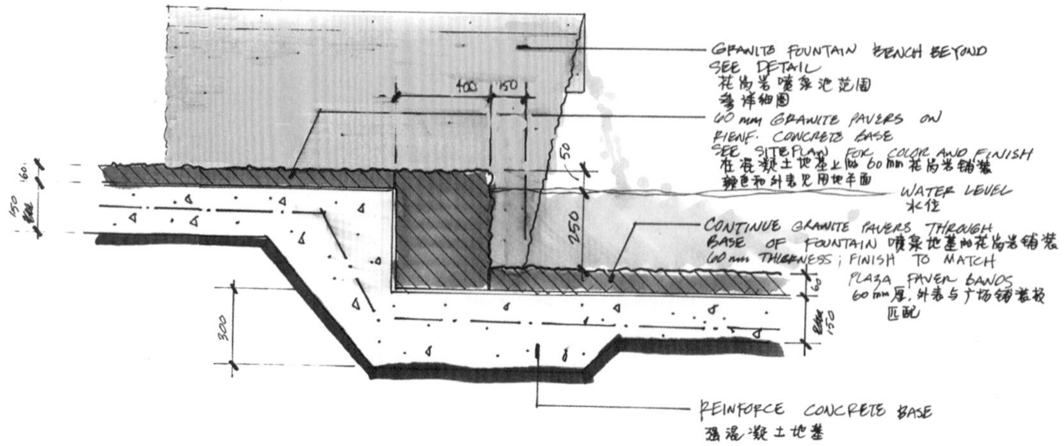
GRANITE FOUNTAIN BENCH BEYOND
SEE DETAIL
400
150
60 mm GRANITE PAVERS ON REINF. CONCRETE BASE
SEE SITE PLAN FOR COLOR AND FINISH
WATER LEVEL
水位
50
250
CONTINUE GRANITE PAVERS THROUGH BASE OF FOUNTAIN
60 mm THICKNESS; FINISH TO MATCH PLAZA PAVER BANDS
60
150
300
REINFORCE CONCRETE BASE
强混凝土地基

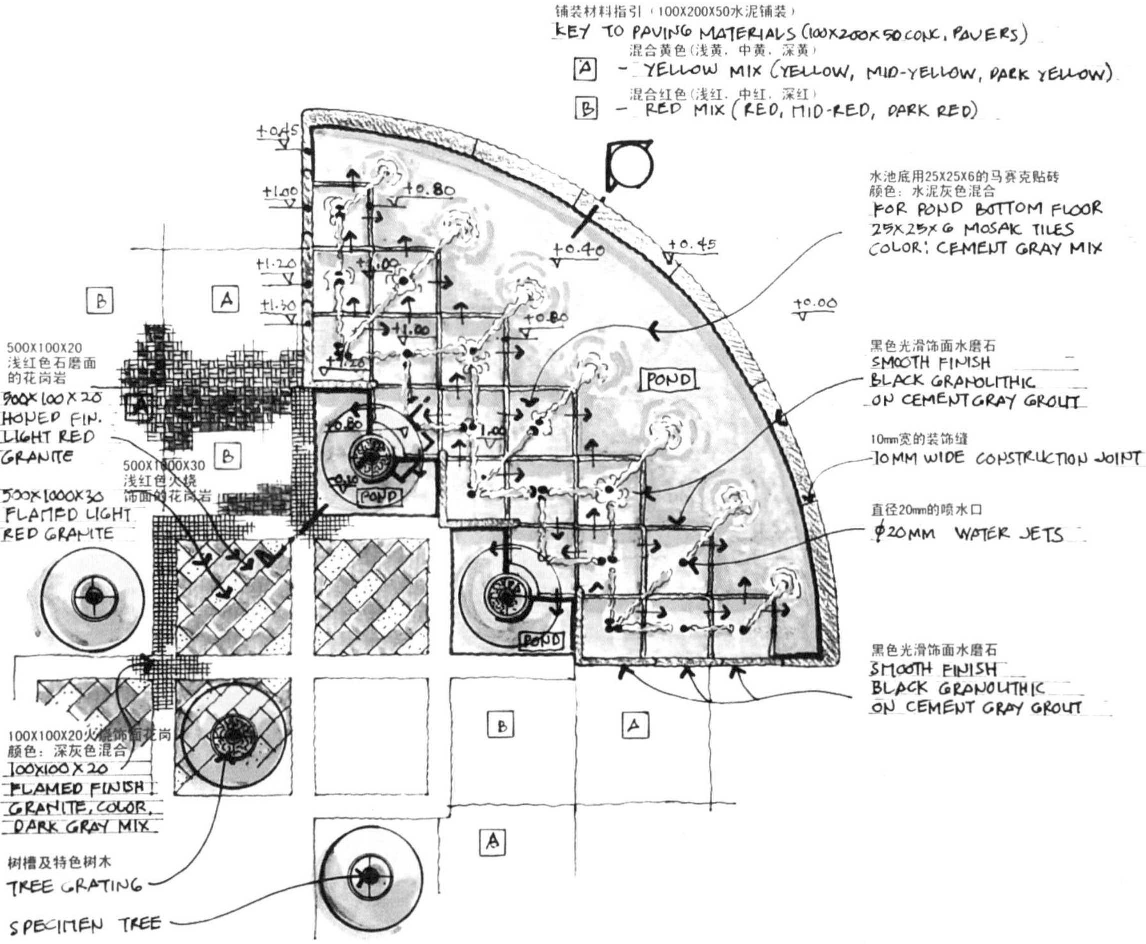

铺装材料指引（100X200X50水泥铺装）
KEY TO PAVING MATERIALS (100X200X50 CONC. PAVERS)
混合黄色(浅黄、中黄、深黄)
A - YELLOW MIX (YELLOW, MID-YELLOW, DARK YELLOW)
混合红色(浅红、中红、深红)
B - RED MIX (RED, MID-RED, DARK RED)
水池底用25X25X6的马赛克贴砖
颜色：水泥灰色混合
FOR POND BOTTOM FLOOR
25X25X6 MOSAIC TILES
COLOR: CEMENT GRAY MIX
黑色光滑饰面水磨石
SMOOTH FINISH
BLACK GRANOLITHIC
ON CEMENT GRAY GROUT
10mm宽的装饰缝
10MM WIDE CONSTRUCTION JOINT
直径20mm的喷水口
ϕ20MM WATER JETS
黑色光滑饰面水磨石
SMOOTH FINISH
BLACK GRANOLITHIC
ON CEMENT GRAY GROUT
500X100X20
浅红色石磨面
的花岗岩
500X100X20
HONED FIN.
LIGHT RED
GRANITE
500X1000X30
浅红色火烧
饰面的花岗岩
500X1000X30
FLAMED LIGHT
RED GRANITE
100X100X20火烧饰面花岗
颜色：深灰色混合
100X100X20
FLAMED FINISH
GRANITE, COLOR,
DARK GRAY MIX
树槽及特色树木
TREE GRATING
SPECIMEN TREE
POND
+0.45
+1.00
+1.20
+1.30
+0.80
+0.40
+0.00
+1.00

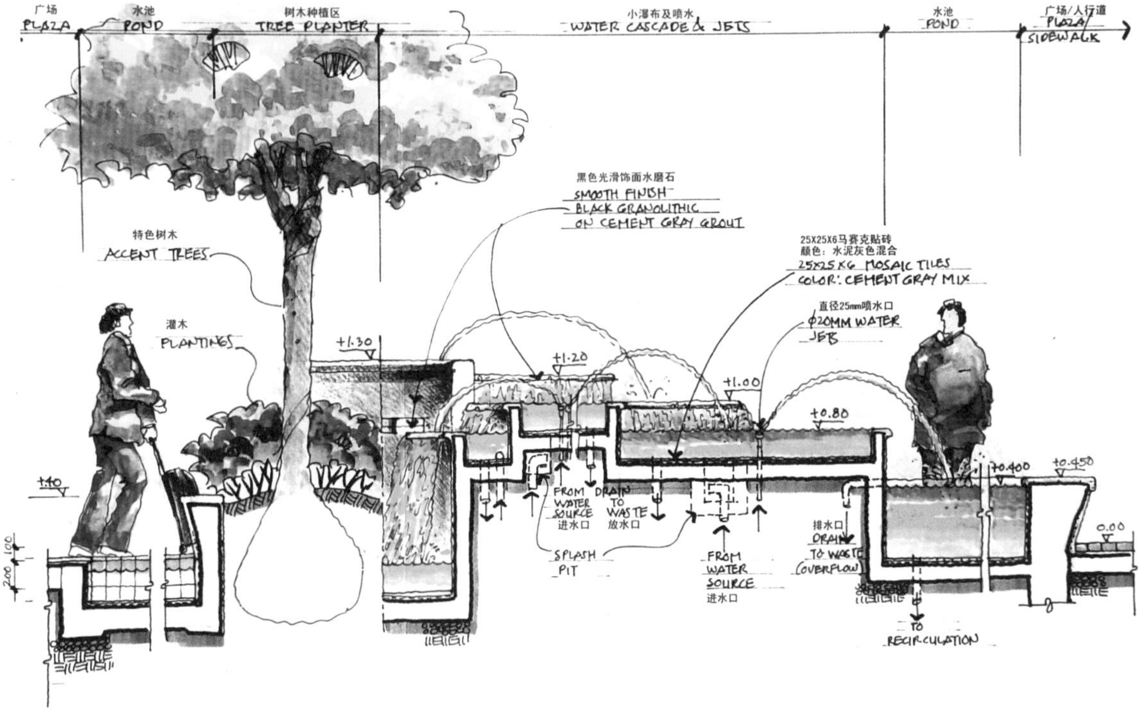

广场
PLAZA
水池
POND
树木种植区
TREE PLANTER
小瀑布及喷水
WATER CASCADE & JETS
水池
POND
广场/人行道
PLAZA/SIDEWALK
黑色光滑饰面水磨石
SMOOTH FINISH
BLACK GRANOLITHIC
ON CEMENT GRAY GROUT
特色树木
ACCENT TREES
25X25X6马赛克贴砖
颜色：水泥灰色混合
25X25X6 MOSAIC TILES
COLOR: CEMENT GRAY MIX
直径25mm喷水口
ϕ20MM WATER JETS
灌木
PLANTINGS
+1.30
+1.20
+1.00
+0.80
+0.400
+0.450
+.40
0.00
FROM WATER SOURCE
进水口
DRAIN TO WASTE
放水口
SPLASH PIT
FROM WATER SOURCE
进水口
排水口
DRAIN TO WASTE OVERFLOW
TO RECIRCULATION

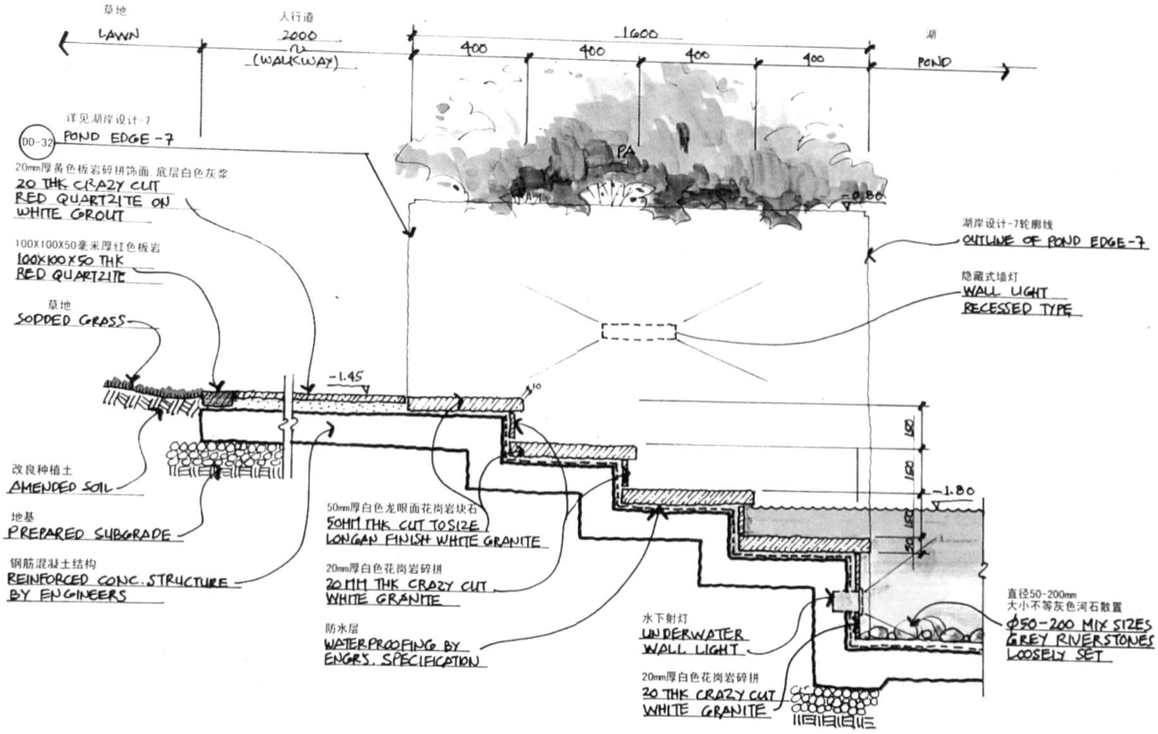
草地
LAWN
人行道
2000
(WALKWAY)
1600
400
400
400
400
湖
POND
详见湖岸设计-7
DD-32
POND EDGE-7
20mm厚黄色板岩碎拼饰面，底层白色灰浆
20 THK CRAZY CUT
RED QUARTZITE ON
WHITE GROUT
100X100X50毫米厚红色板岩
100X100X50 THK
RED QUARTZITE
草地
SODDED GRASS
-1.45
-0.80
湖岸设计-7轮廓线
OUTLINE OF POND EDGE-7
隐藏式墙灯
WALL LIGHT
RECESSED TYPE
改良种植土
AMENDED SOIL
地基
PREPARED SUBGRADE
钢筋混凝土结构
REINFORCED CONC. STRUCTURE
BY ENGINEERS
50mm厚白色龙眼面花岗岩块石
50MM THK CUT TO SIZE
LONGAN FINISH WHITE GRANITE
20mm厚白色花岗岩碎拼
20 MM THK CRAZY CUT
WHITE GRANITE
防水层
WATERPROOFING BY
ENGRS. SPECIFICATION
-1.80
直径50-200mm
大小不等灰色河石散置
Φ50-200 MIX SIZES
GREY RIVERSTONES
LOOSELY SET
水下射灯
UNDERWATER
WALL LIGHT
20mm厚白色花岗岩碎拼
20 THK CRAZY CUT
WHITE GRANITE

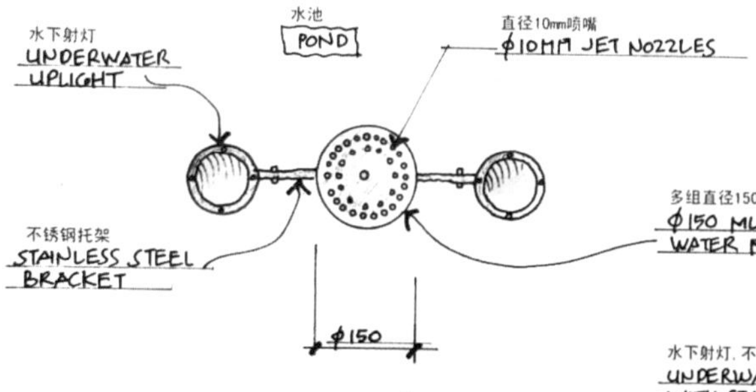
水下射灯
UNDERWATER
UPLIGHT
水池
POND
直径10mm喷嘴
Φ10MM JET NOZZLES
不锈钢托架
STAINLESS STEEL
BRACKET
Φ150

PLAN 平面图

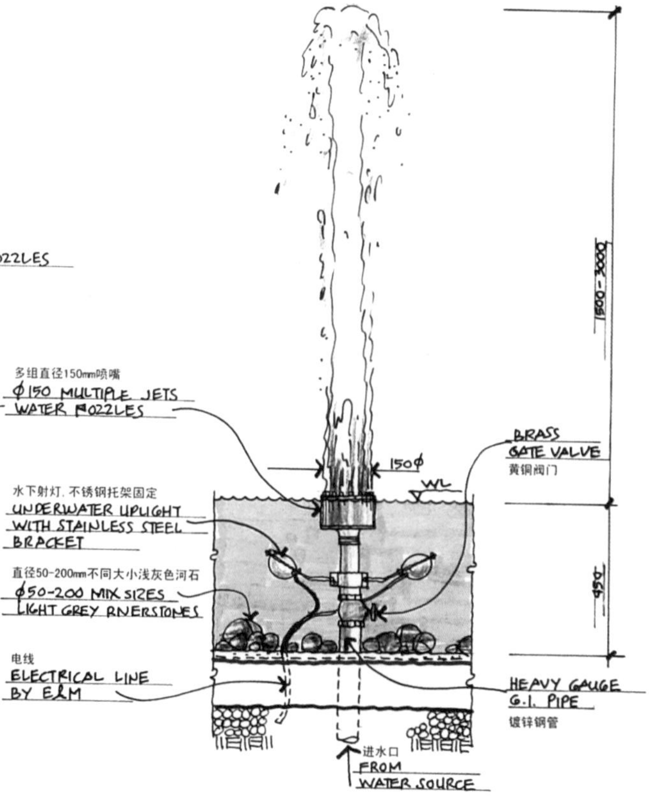
1500-3000
多组直径150mm喷嘴
Φ150 MULTIPLE JETS
WATER NOZZLES
BRASS
GATE VALVE
黄铜阀门
150Φ
WL
水下射灯，不锈钢托架固定
UNDERWATER UPLIGHT
WITH STAINLESS STEEL
BRACKET
450
直径50-200mm不同大小浅灰色河石
Φ50-200 MIX SIZES
LIGHT GREY RIVERSTONES
电线
ELECTRICAL LINE
BY E&M
HEAVY GAUGE
G.I. PIPE
镀锌钢管
进水口
FROM
WATER SOURCE

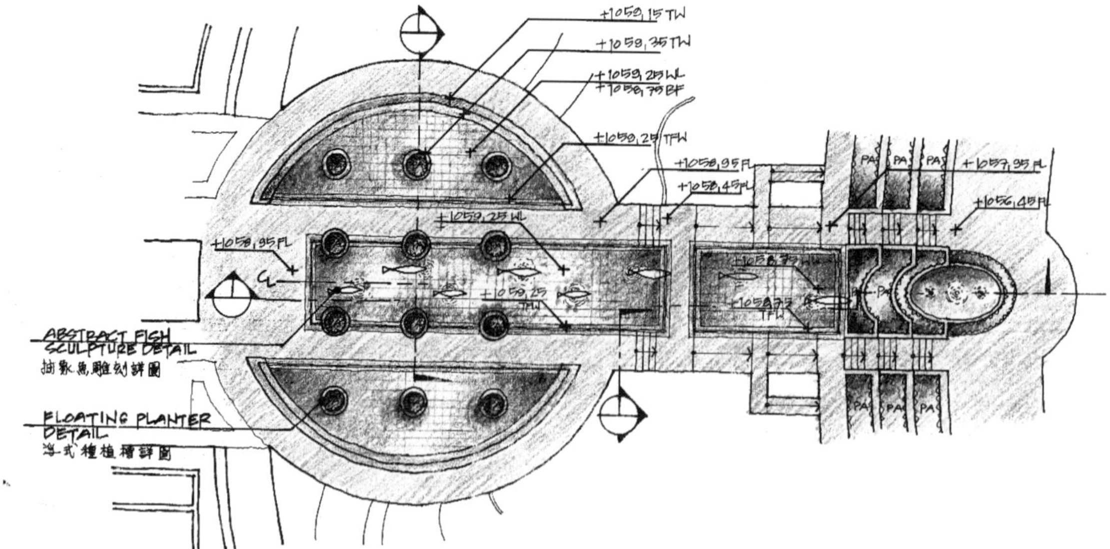
+1059.15 TW
+1059.35 TW
+1059.25 WL
+1058.75 BF
+1059.25 TFW
+1058.95 FL
+1058.45 FL
+1057.95 FL
+1056.45 FL
ABSTRACT FISH SCULPTURE DETAIL
抽象鱼雕刻详图
FLOATING PLANTER DETAIL
浮式种植槽详图

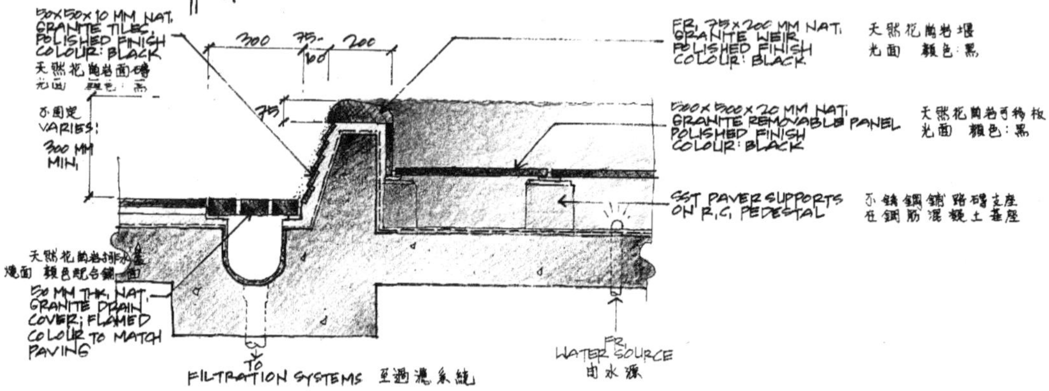
50×50×10 MM NAT. GRANITE TILES, POLISHED FINISH COLOUR: BLACK
天然花岗岩面砖
光面 颜色：黑
300
75
100
200
不固定
VARIES:
300 MM MIN.
FR. 75×200 MM NAT. GRANITE WEIR, POLISHED FINISH COLOUR: BLACK
天然花岗岩堰
光面 颜色：黑
500×500×20 MM NAT. GRANITE REMOVABLE PANEL POLISHED FINISH COLOUR: BLACK
天然花岗岩可移板
光面 颜色：黑
SST PAVER SUPPORTS ON R.C. PEDESTAL
不锈钢铺路砖支座
在钢筋混凝土基座
天然花岗岩排水盖
烧面 颜色配合铺面
50 MM THK. NAT. GRANITE DRAIN COVER; FLAMED COLOUR TO MATCH PAVING
TO FILTRATION SYSTEMS 至过滤系统
FR. WATER SOURCE
由水源

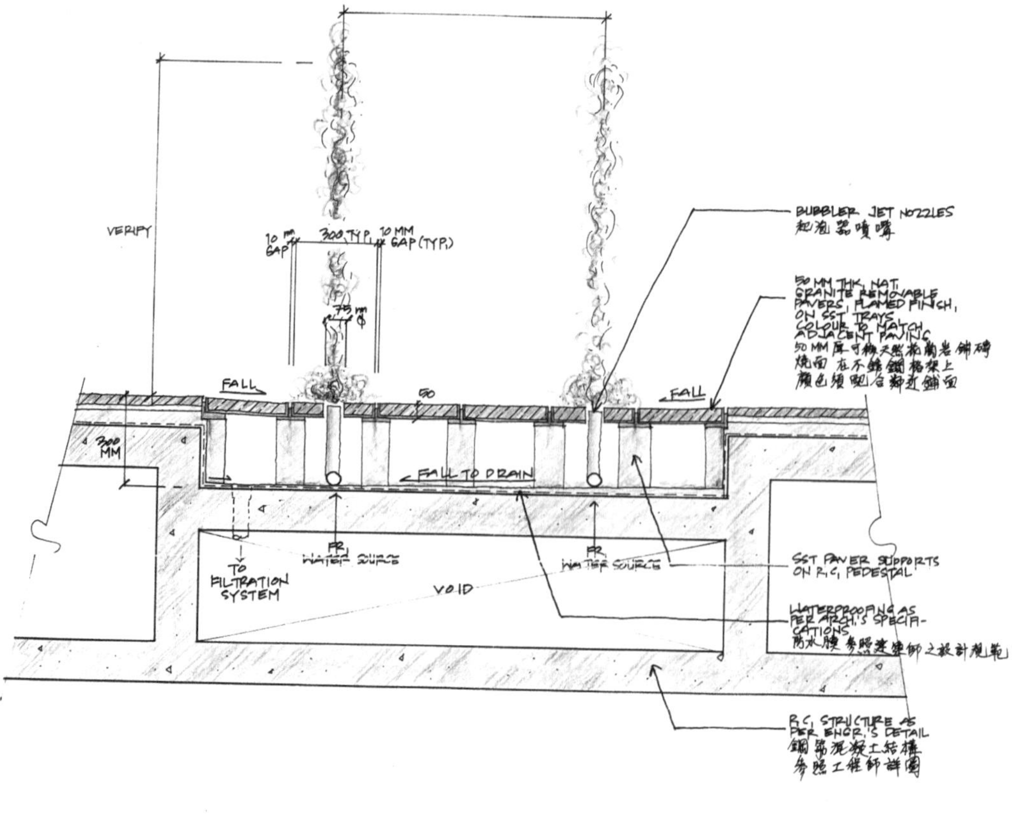
VERIFY
10 MM GAP
300 TYP.
10 MM GAP (TYP.)
75 MM Ø
BUBBLER JET NOZZLES
起泡器喷嘴
50 MM THK. NAT. GRANITE REMOVABLE PAVERS, FLAMED FINISH, ON SST TRAYS, COLOUR TO MATCH ADJACENT PAVING
50 MM 厚可移天然花岗岩铺砖
烧面 在不锈钢格架上
颜色须配合邻近铺面
FALL
50
FALL
300 MM
FALL TO DRAIN
FR. WATER SOURCE
FR. WATER SOURCE
TO FILTRATION SYSTEM
VOID
SST PAVER SUPPORTS ON R.C. PEDESTAL
WATERPROOFING AS PER ARCH.'S SPECIFICATIONS
防水膜参照建筑师之设计规范
R.C. STRUCTURE AS PER ENGR.'S DETAIL
钢筋混凝土结构
参照工程师详图

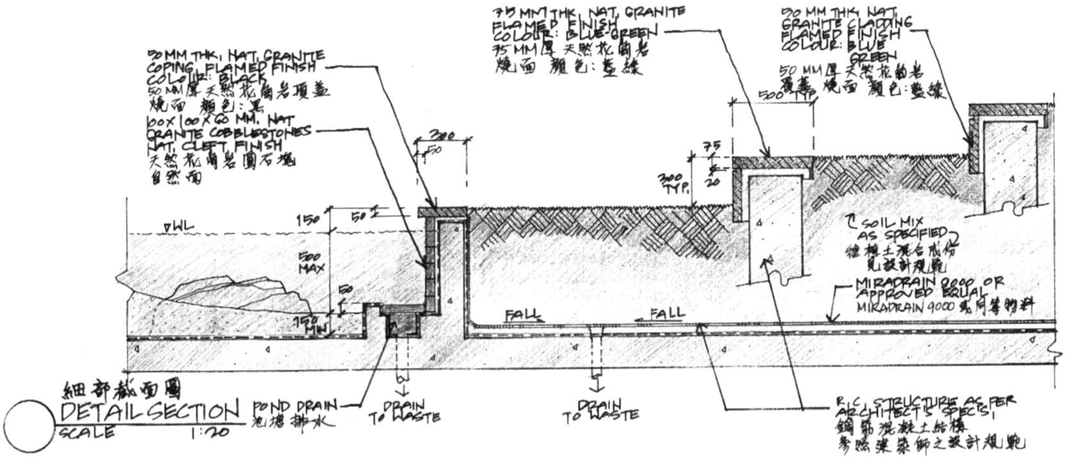
50 MM THK. NAT. GRANITE COPING, FLAMED FINISH COLOUR: BLACK
50 MM厚 天然花崗岩頂蓋 燒面 顏色：黑
100x100x60 MM. NAT. GRANITE COBBLESTONES NAT. CLEFT FINISH
天然花崗岩圓石塊 自然面
75 MM THK. NAT. GRANITE FLAMED FINISH COLOUR: BLUE-GREEN
75 MM厚 天然花崗岩 燒面 顏色：藍綠
50 MM THK. NAT. GRANITE CLADDING FLAMED FINISH COLOUR: BLUE GREEN
50 MM厚 天然花崗岩 燒面 顏色：藍綠
SOIL MIX AS SPECIFIED
種植土混合成份 見設計規範
MIRADRAIN 9000 OR APPROVED EQUAL
MIRADRAIN 9000 或同等物料
WL
150
500 MAX
50
150 MIN
300
50
500
75
300 TYP.
20
FALL
FALL
POND DRAIN
池塘排水
DRAIN TO WASTE
DRAIN TO WASTE
R.C. STRUCTURE AS PER ARCHITECT'S SPEC'S.
鋼筋混凝土結構 參照建築師之設計規範

細部截面圖
DETAIL SECTION
SCALE 1:20

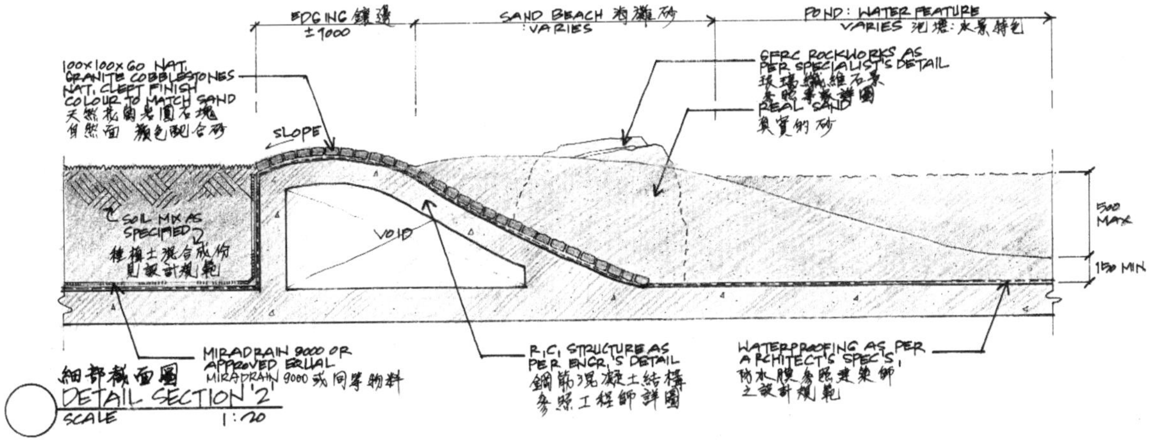
EDGING 鑲邊 ±1000
SAND BEACH 沙灘砂 VARIES
POND: WATER FEATURE VARIES 池塘：水景特色
100x100x60 NAT. GRANITE COBBLESTONES NAT. CLEFT FINISH COLOUR TO MATCH SAND
天然花崗岩圓石塊 自然面 顏色配合砂
GFRC ROCKWORKS AS PER SPECIALIST'S DETAIL
REAL SAND
真實的砂
SLOPE
SOIL MIX AS SPECIFIED
種植土混合成份 見設計規範
VOID
500 MAX
150 MIN
MIRADRAIN 9000 OR APPROVED EQUAL
MIRADRAIN 9000 或同等物料
R.C. STRUCTURE AS PER ENGR.'S DETAIL
鋼筋混凝土結構 參照工程師詳圖
WATERPROOFING AS PER ARCHITECT'S SPEC'S.
防水膜參照建築師之設計規範

細部截面圖
DETAIL SECTION '2'
SCALE 1:20

3450
1000
R.C. STRUCTURE AS PER ENGINEER'S DETAIL
鋼筋混凝土結構參照工程師詳圖

立面圖
ELEVATION
SCALE 1:100

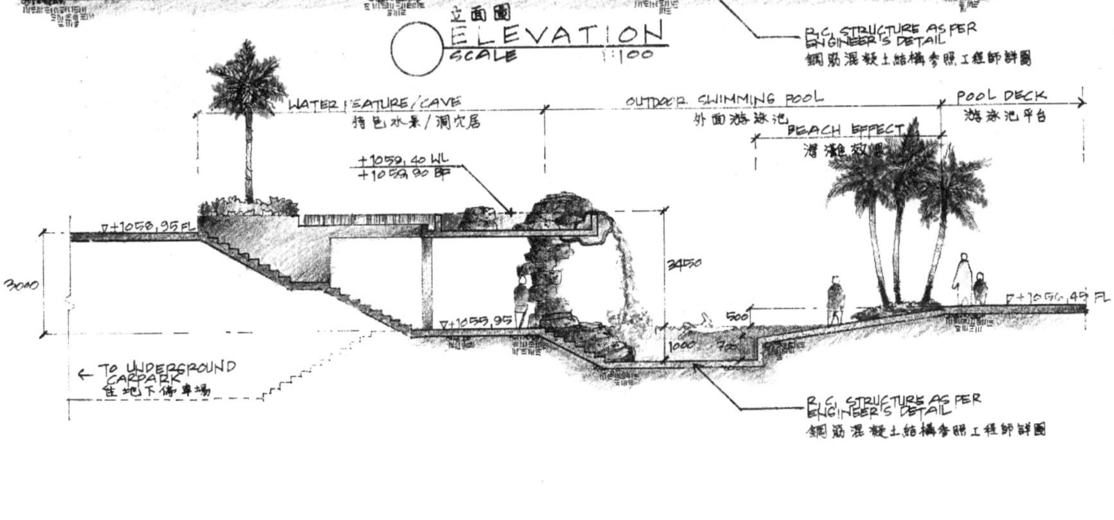
WATER FEATURE/CAVE
特色水景/洞穴居
OUTDOOR SWIMMING POOL
外面游泳池
POOL DECK
游泳池平台
BEACH EFFECT
+1059.40 WL
+1059.90 BF
+1058.95 FL
+1055.95
+1056.45 FL
3000
3450
500
1000
← TO UNDERGROUND CARPARK
往地下停車場
R.C. STRUCTURE AS PER ENGINEER'S DETAIL
鋼筋混凝土結構參照工程師詳圖

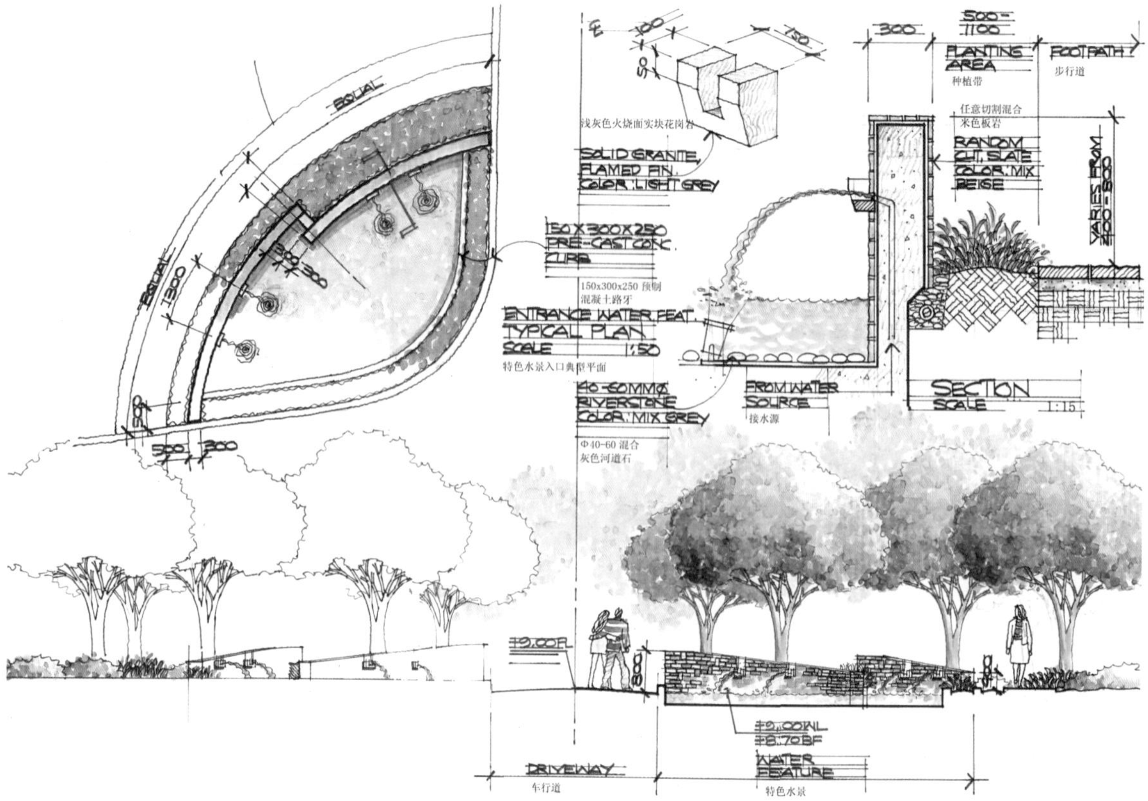
EQUAL
EQUAL
1300
300
300x300
500 300
浅灰色火烧面实块花岗岩
SOLID GRANITE,
FLAMED FIN.
COLOR: LIGHT GREY
150 X 300 X 250
PRE-CAST CONC.
CURB
150x300x250 预制
混凝土路牙
ENTRANCE WATER FEAT.
TYPICAL PLAN
SCALE 1:50
特色水景入口典型平面
40-60MMØ
RIVERSTONE
COLOR: MIX GREY
Φ40-60 混合
灰色河道石
FROM WATER
SOURCE
接水源
300
500-1100
PLANTING
AREA
种植带
FOOTPATH
步行道
任意切割混合
米色板岩
RANDOM
CUT, SLATE
COLOR: MIX
BEIGE
VARIES FROM
400-800
SECTION
SCALE 1:15
+9.00FF
800
+9.00WL
+8.70BF
DRIVEWAY
车行道
WATER
FEATURE
特色水景

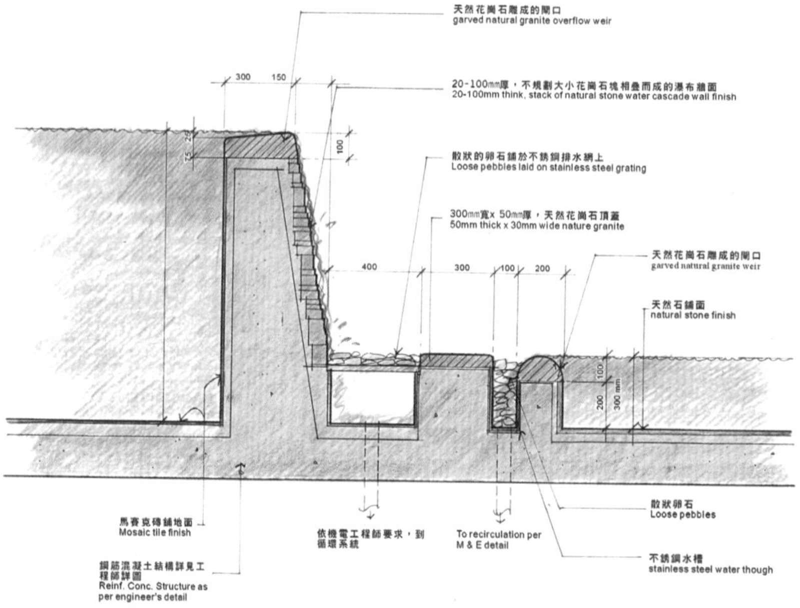
天然花崗石雕成的閘口
garved natural granite overflow weir
300
150
20-100mm厚，不規劃大小花崗石塊相疊而成的瀑布牆面
20-100mm think, stack of natural stone water cascade wall finish
100
散狀的卵石鋪於不銹鋼排水網上
Loose pebbles laid on stainless steel grating
300mm寬x 50mm厚，天然花崗石頂蓋
50mm thick x 30mm wide nature granite
400
300
100
200
天然花崗石雕成的閘口
garved natural granite weir
天然石鋪面
natural stone finish
100
200
300 mm
散狀卵石
Loose pebbles
馬賽克磚鋪地面
Mosaic tile finish
依機電工程師要求，到循環系統
To recirculation per
M & E detail
不銹鋼水槽
stainless steel water though
鋼筋混凝土結構詳見工程師詳圖
Reinf. Conc. Structure as
per engineer's detail

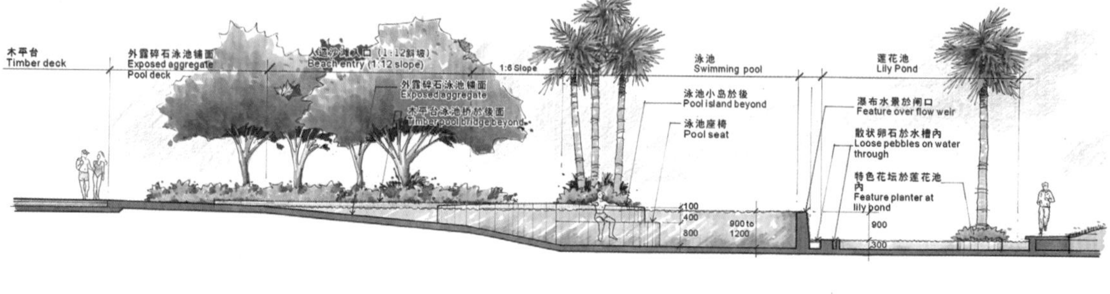
木平台
Timber deck
外露碎石泳池鋪面
Exposed aggregate
Pool deck
人造沙灘入口（1:12斜坡）
Beach entry (1:12 slope)
1:6 Slope
外露碎石泳池鋪面
Exposed aggregate
木平台泳池橋於後面
Timber pool bridge beyond
泳池
Swimming pool
泳池小島於後
Pool island beyond
泳池座椅
Pool seat
100
400
800
900 to
1200
蓮花池
Lily Pond
瀑布水景於閘口
Feature over flow weir
散狀卵石於水槽內
Loose pebbles on water
through
特色花坛於蓮花池內
Feature planter at
lily pond
900
300

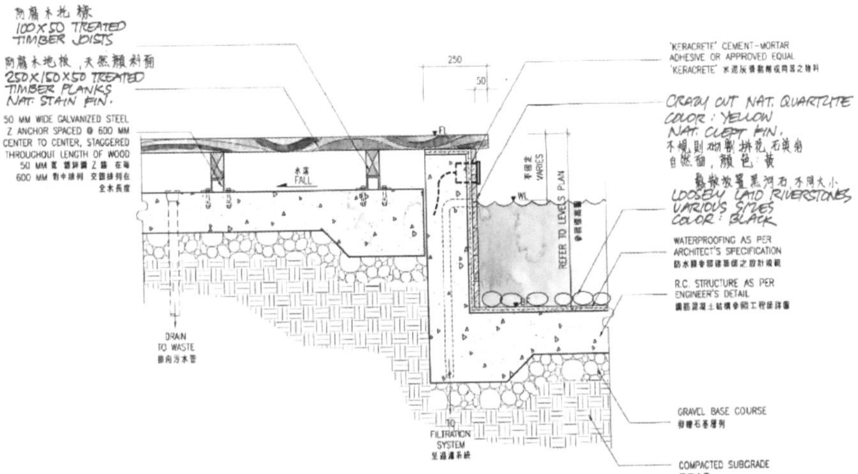
100X50 TREATED TIMBER JOISTS
250X150X50 TREATED TIMBER PLANKS NAT. STAIN FIN.
50 MM WIDE GALVANIZED STEEL Z ANCHOR SPACED @ 600 MM CENTER TO CENTER, STAGGERED THROUGHOUT LENGTH OF WOOD
FALL
DRAIN TO WASTE
250
50
FL
'KERACRETE' CEMENT-MORTAR ADHESIVE OR APPROVED EQUAL
CRAZY CUT NAT. QUARTZITE COLOR : YELLOW NAT. CLEFT FIN.
VARIES
REFER TO LEVELS PLAN
WL
LOOSELY LAID RIVERSTONES VARIOUS SIZES COLOR : BLACK
WATERPROOFING AS PER ARCHITECT'S SPECIFICATION
R.C. STRUCTURE AS PER ENGINEER'S DETAIL
TO FILTRATION SYSTEM
GRAVEL BASE COURSE
COMPACTED SUBGRADE

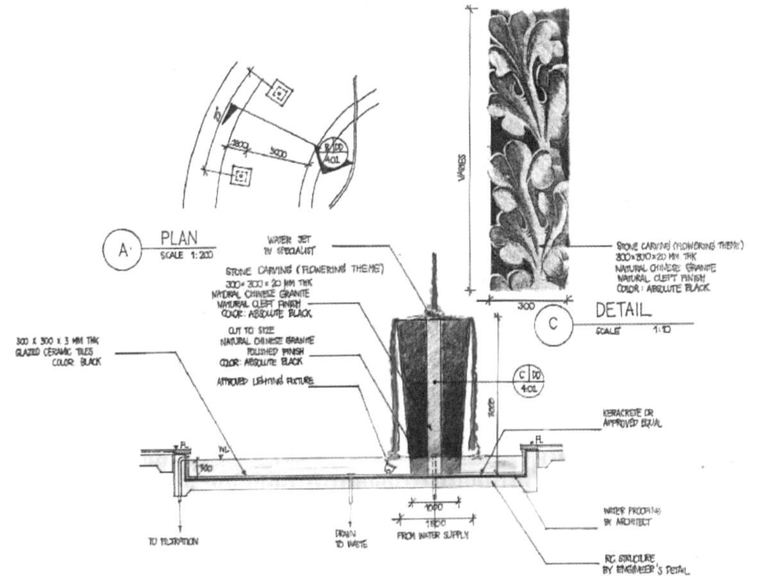
A PLAN SCALE 1:200
VARIES
300
C DETAIL SCALE 1:10
STONE CARVING (FLOWERING THEME) 300×300×20 MM THK NATURAL CHINESE GRANITE NATURAL CLEFT FINISH COLOR : ABSOLUTE BLACK
WATER JET BY SPECIALIST
STONE CARVING (FLOWERING THEME) 300×300×20 MM THK NATURAL CHINESE GRANITE NATURAL CLEFT FINISH COLOR: ABSOLUTE BLACK
CUT TO SIZE NATURAL CHINESE GRANITE POLISHED FINISH COLOR: ABSOLUTE BLACK
APPROVED LIGHTING FIXTURE
300 X 300 X 3 MM THK GLAZED CERAMIC TILES COLOR BLACK
KERACRETE OR APPROVED EQUAL
WATER PROOFING BY ARCHITECT
RC STRUCTURE BY ENGINEER'S DETAIL
TO FILTRATION
DRAIN TO WASTE
1000
1800
FROM WATER SUPPLY

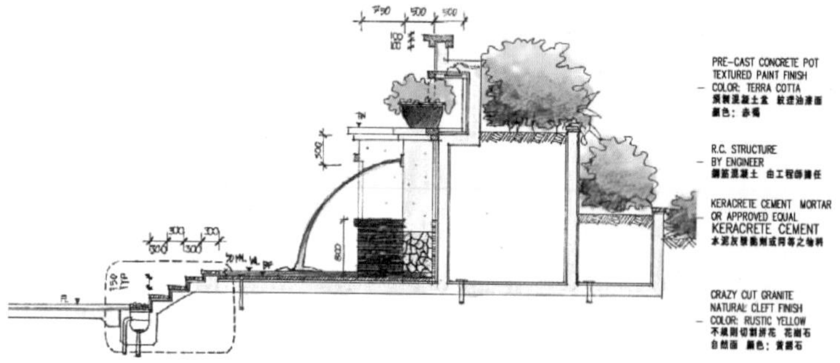
PRE-CAST CONCRETE POT
TEXTURED PAINT FINISH
COLOR: TERRA COTTA
預製混凝土盆 紋理油漆面
顏色：赤褐
R.C. STRUCTURE
BY ENGINEER
鋼筋混凝土 由工程師擔任
KERACRETE CEMENT MORTAR
OR APPROVED EQUAL
KERACRETE CEMENT
水泥灰漿黏劑或同等之物料
CRAZY CUT GRANITE
NATURAL CLEFT FINISH
COLOR: RUSTIC YELLOW
不規則切割拼花 花崗石
自然面 顏色：黃銹石

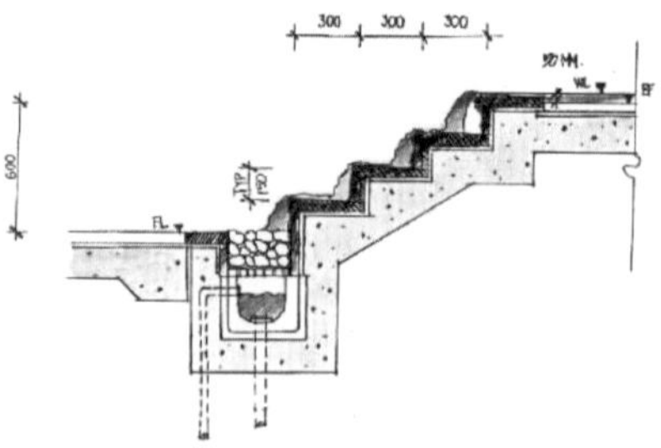
300
300
300
600

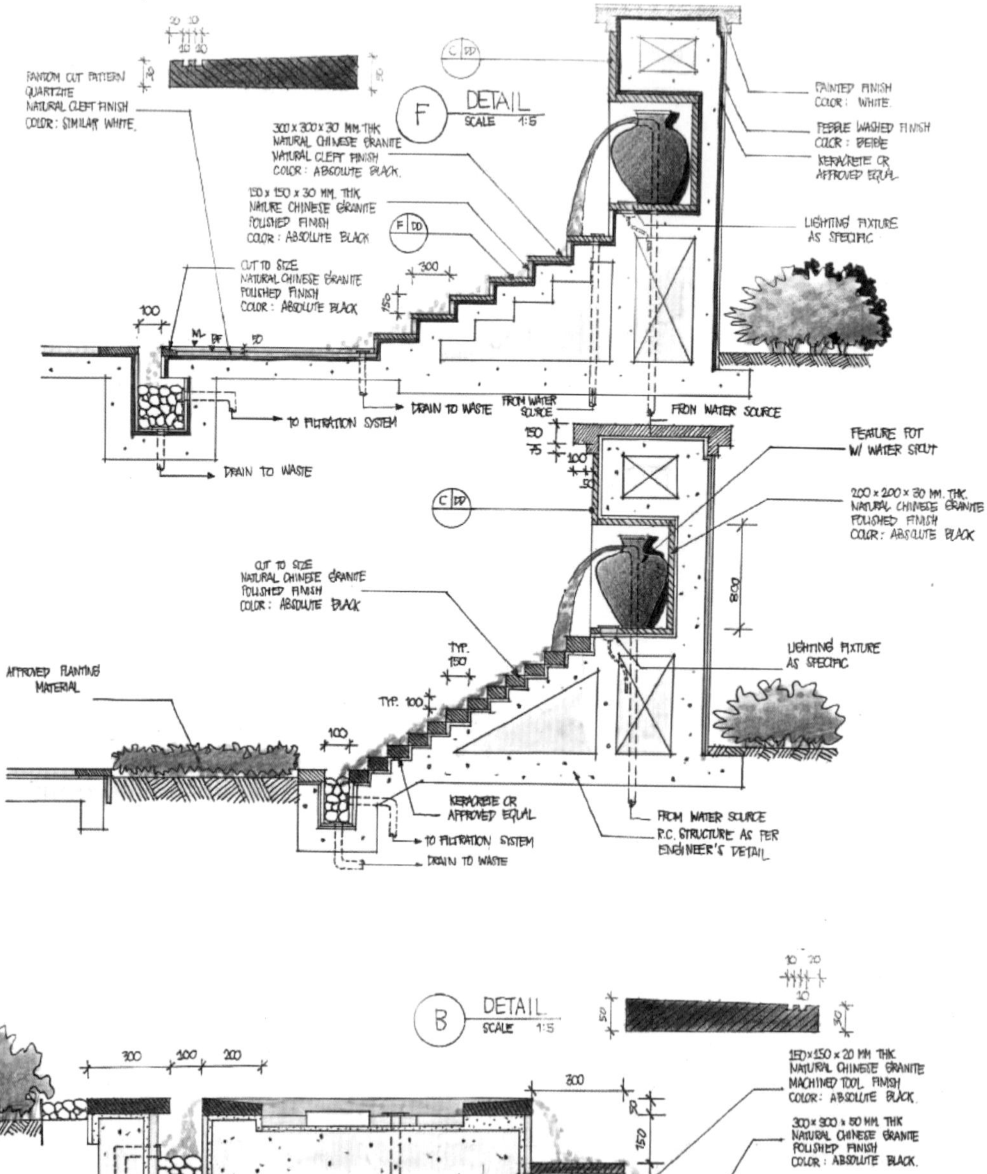
F
DETAIL
SCALE 1:5
RANDOM CUT PATTERN
QUARTZITE
NATURAL CLEFT FINISH
COLOR: SIMILAR WHITE
300 x 300 x 30 MM THK
NATURAL CHINESE GRANITE
NATURAL CLEFT FINISH
COLOR: ABSOLUTE BLACK
150 x 150 x 30 MM THK
NATURE CHINESE GRANITE
POLISHED FINISH
COLOR: ABSOLUTE BLACK
CUT TO SIZE
NATURAL CHINESE GRANITE
POLISHED FINISH
COLOR: ABSOLUTE BLACK
PAINTED FINISH
COLOR: WHITE
PEBBLE WASHED FINISH
COLOR: BEIGE
KERACRETE OR
APPROVED EQUAL
LIGHTING FIXTURE
AS SPECIFIC
TO FILTRATION SYSTEM
DRAIN TO WASTE
FROM WATER SOURCE
FEATURE POT
W/ WATER SPOUT
200 x 200 x 30 MM THK
NATURAL CHINESE GRANITE
POLISHED FINISH
COLOR: ABSOLUTE BLACK
CUT TO SIZE
NATURAL CHINESE GRANITE
POLISHED FINISH
COLOR: ABSOLUTE BLACK
APPROVED PLANTING
MATERIAL
LIGHTING FIXTURE
AS SPECIFIC
KERACRETE OR
APPROVED EQUAL
TO FILTRATION SYSTEM
DRAIN TO WASTE
FROM WATER SOURCE
R.C. STRUCTURE AS PER
ENGINEER'S DETAIL

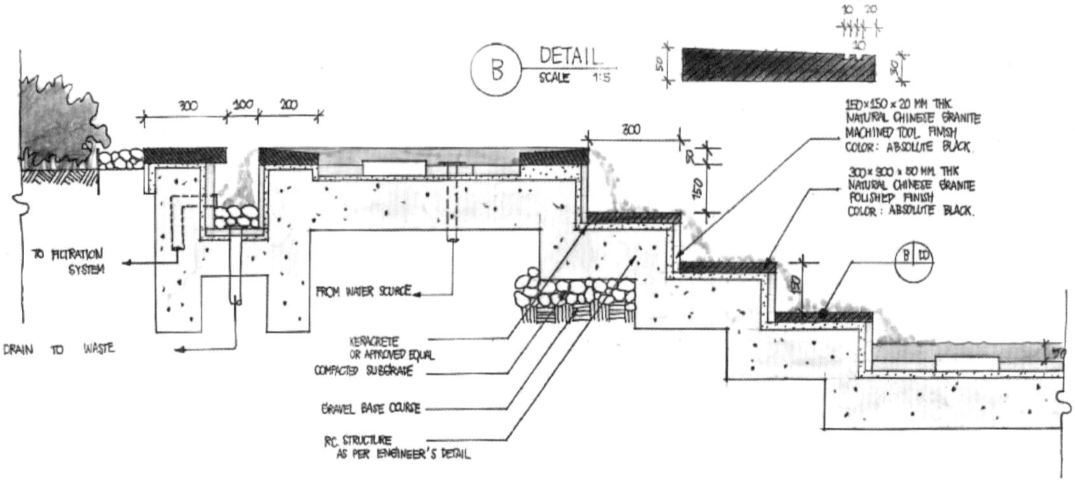
B
DETAIL
SCALE 1:5
150 x 150 x 20 MM THK
NATURAL CHINESE GRANITE
MACHINED TOOL FINISH
COLOR: ABSOLUTE BLACK
300 x 300 x 50 MM THK
NATURAL CHINESE GRANITE
POLISHED FINISH
COLOR: ABSOLUTE BLACK
TO FILTRATION
SYSTEM
FROM WATER SOURCE
DRAIN TO WASTE
KERACRETE
OR APPROVED EQUAL
COMPACTED SUBGRADE
GRAVEL BASE COURSE
RC STRUCTURE
AS PER ENGINEER'S DETAIL

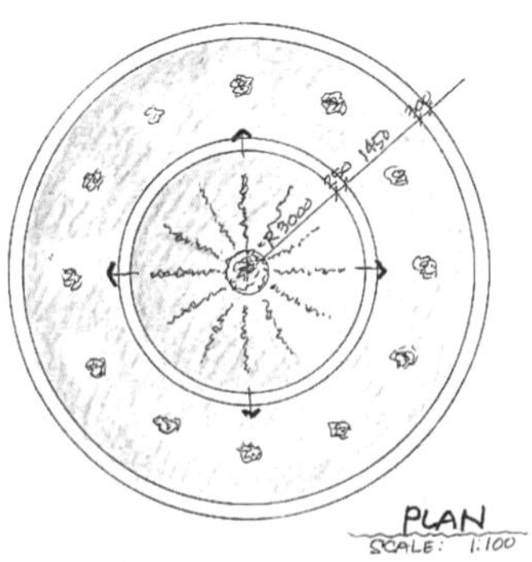

PLAN
SCALE: 1:100
DETAIL
SCALE: 1:10
STEEL DOWEL
APPROVED PAVING MATERIAL
DETAIL
SCALE: 1:20
1000 MM HIGH WATER JET AS PER SPECIALIST'S
JUMP JETS AS PER SPECIALIST
BUBBLERS AS PER SPECIALIST
NATURAL GRANITE POLISHED FINISH COLOUR: BLACK
ELEVATION/SECTION
SCALE: 1:40
FROM WATER SOURCE
TO DRAIN
TO FILTRATION
FROM WATER SOURCE
APPROVED PAVING MATERIAL
WATER FEATURE-02 DET

POND/LAKE
+0.00 WL
-0.30 BF
+0.00 TOW
POND
-2.25 WL
-1.95 BF
PLAN
SCALE: 1:50

500
1000
500
1000
500
TW/WL +0.00
2150
-2.15 FL
100 MM THK NATURAL GRANITE WEIR POLISHED FINISH COLOUR: FORESTA GREEN
NATURAL GRANITE CLADDING NATURAL CLEFT FINISH; COLOUR: FORESTA G
NATURAL GRANITE CLADDING HONED FINISH COLOUR: FORESTA GREEN
NATURAL GRANITE CLADDING POLISHED FINISH COLOUR: FORESTA GREEN
APPROVED PAVING MATERIAL
ELEVATION
SCALE 1:20
WATER WALL DETAIL
LD-

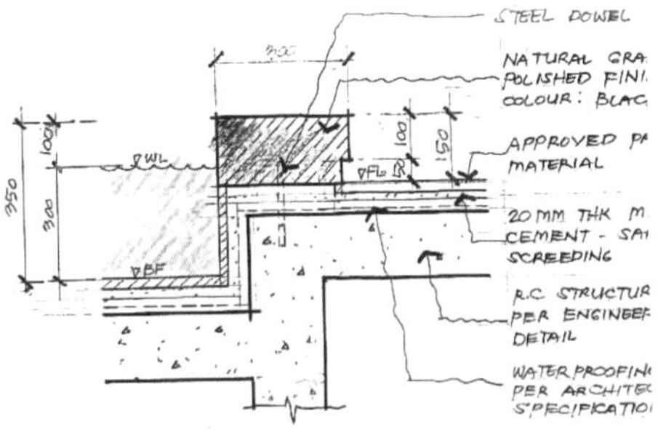

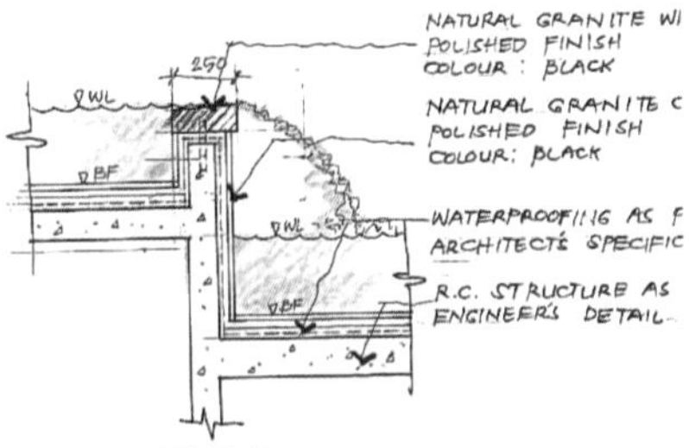

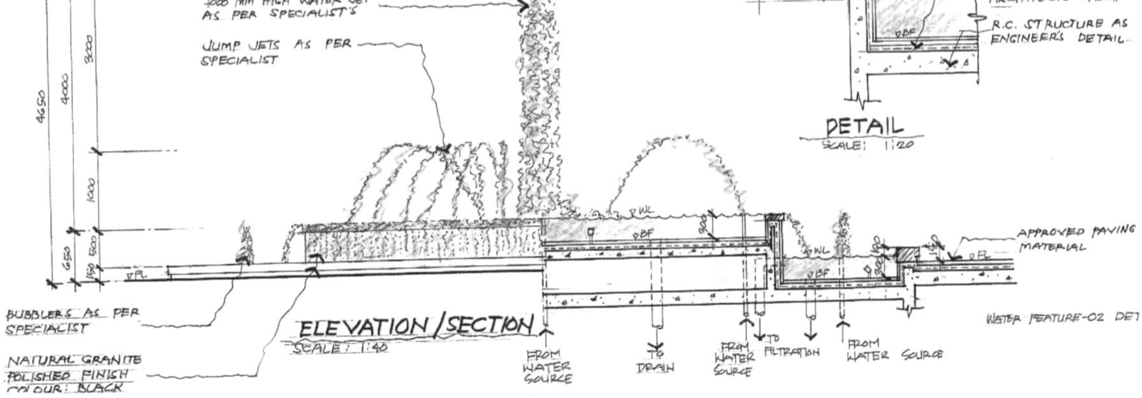

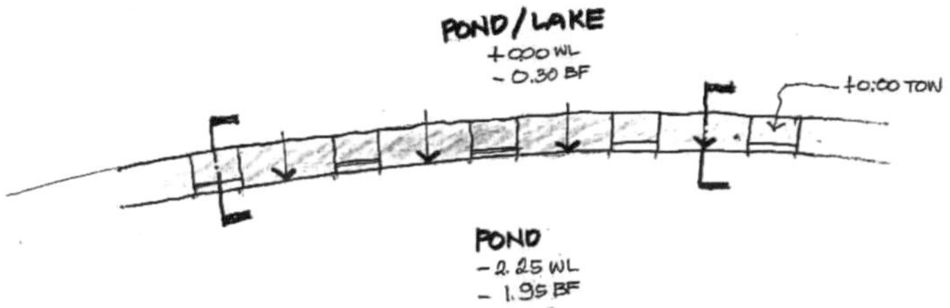

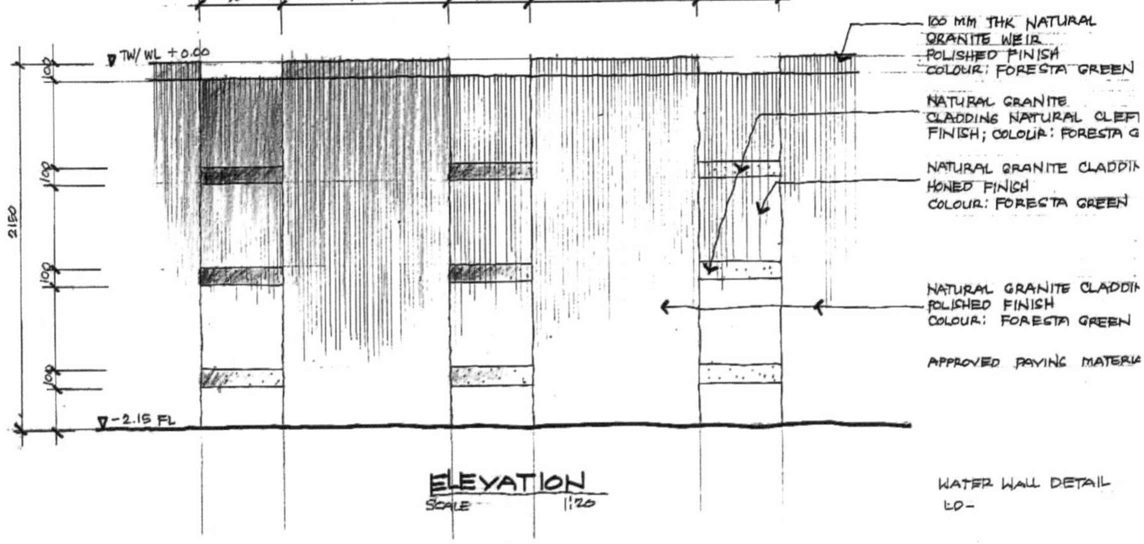

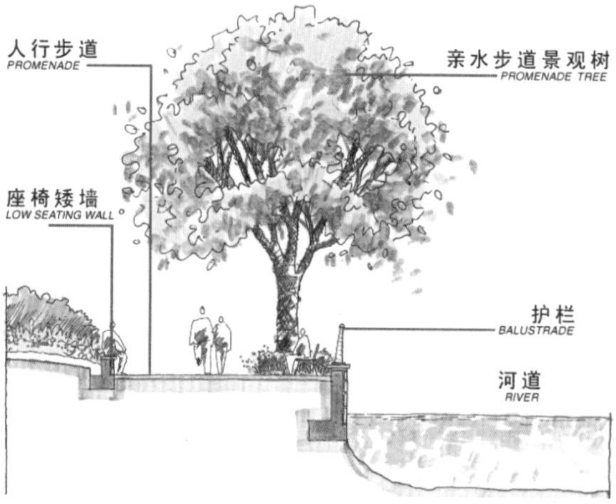
人行步道
PROMENADE
亲水步道景观树
PROMENADE TREE
座椅矮墙
LOW SEATING WALL
护栏
BALUSTRADE
河道
RIVER

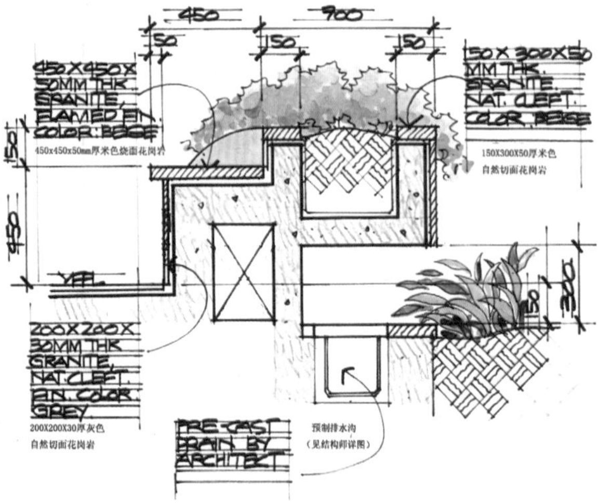
450
900
50
150
150
450X450X50MM THK. GRANITE, FLAMED FIN. COLOR: BEIGE
450x450x50mm厚米色烧面花岗岩
150X300X50 MM THK. GRANITE, NAT. CLEFT. COLOR: BEIGE
150X300X50厚米色
自然切面花岗岩
150
450
YFFL
300
200X200X30MM THK GRANITE, NAT. CLEFT. FIN. COLOR GREY
200X200X30厚灰色
自然切面花岗岩
PRE-CAST DRAIN BY ARCHITECT
预制排水沟
（见结构师详图）

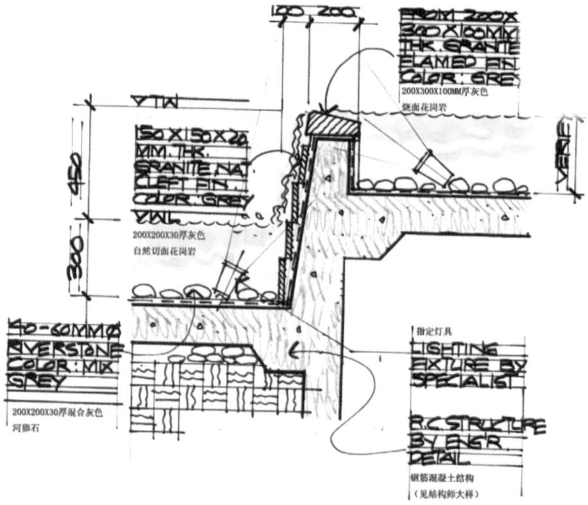
100
200
FROM 200X300X100MM THK. GRANITE FLAMED FIN COLOR: GREY
200X300X100MM厚灰色
烧面花岗岩
150X150X20 MM. THK. GRANITE NAT CLEFT FIN. COLOR: GREY
200X200X30厚灰色
自然切面花岗岩
450
300
40-60MMØ RIVERSTONE COLOR: MIX GREY
200X200X30厚混合灰色
河卵石
指定灯具
LIGHTING FIXTURE BY SPECIALIST
R.C. STRUCTURE BY ENG'R. DETAIL
钢筋混凝土结构
（见结构师大样）

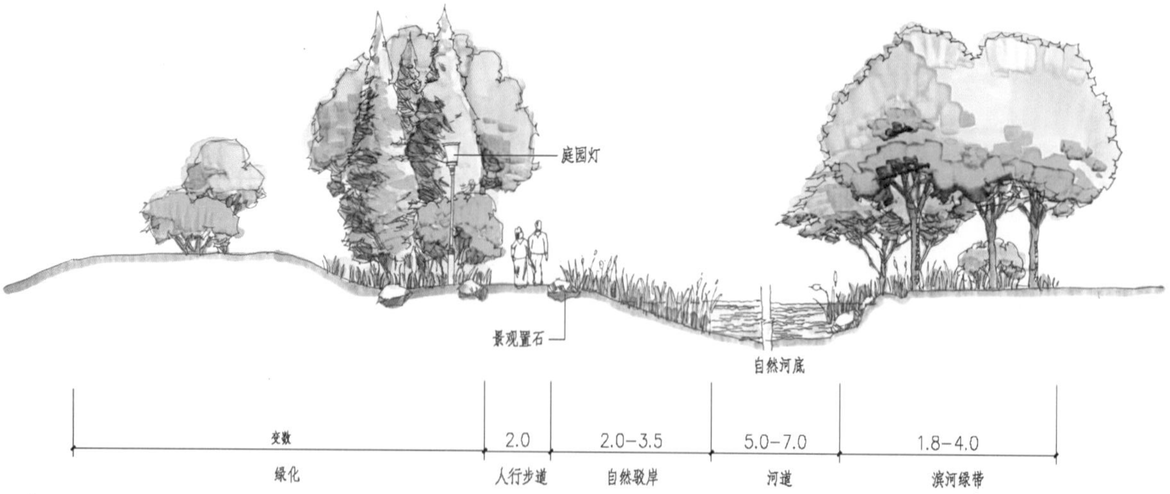
庭园灯
景观置石
自然河底
变数
2.0
2.0–3.5
5.0–7.0
1.8–4.0
绿化
人行步道
自然驳岸
河道
滨河绿带

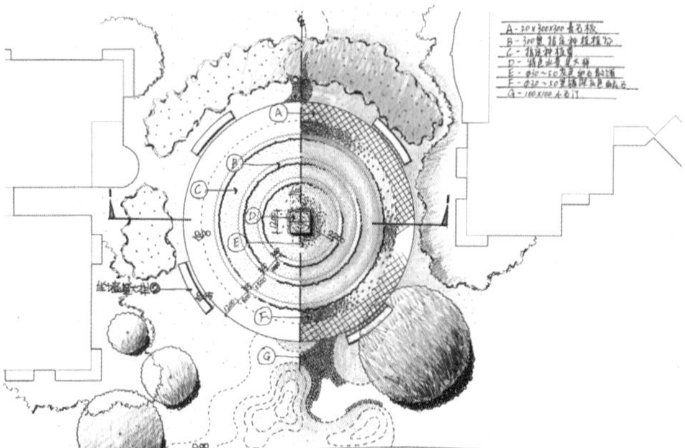

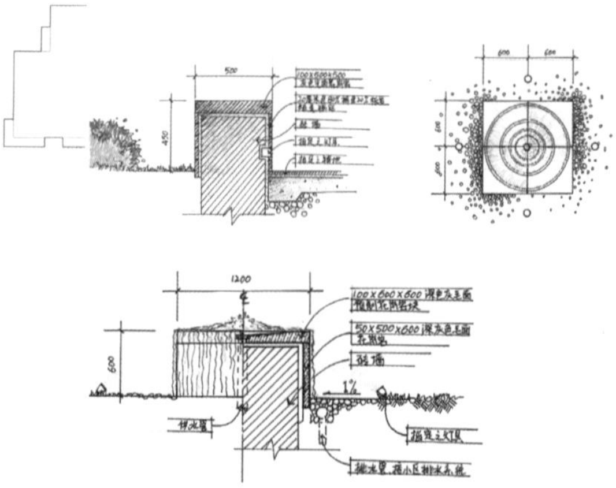

寿泉平面图 1：100
1-1剖面图 1:50

FEATURE WALL WATER SPOUT TO SPECIALIST DETAIL
FEATURE WALL STRUCTURE IN 25 MM. THK. FLAMED GRANITE CLADDING (COLOR: WHITE)
FEATURE WALL LIGHTING INSIDE TO SPECIALIST DETAIL
SELECTED WATER JETS TO SPECIALIST DETAIL
FEATURE DECORATIVE POT IN SOLID GRANITE MATERIAL (COLOR: BEIGE)
SELECTED WALL LIGHTING TO MANUFACTURER'S DETAIL

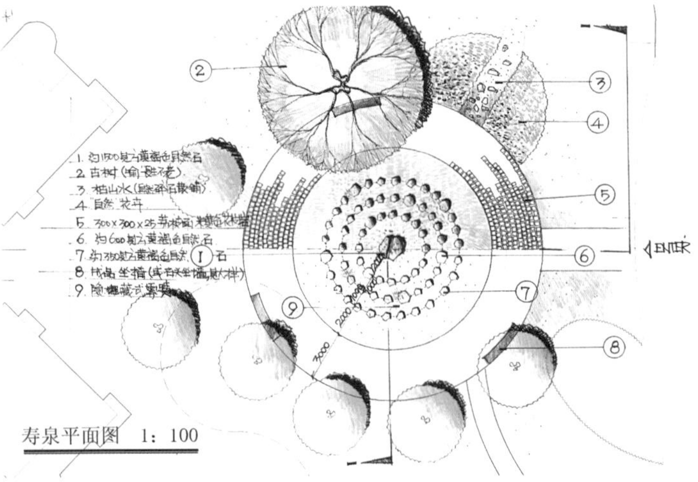

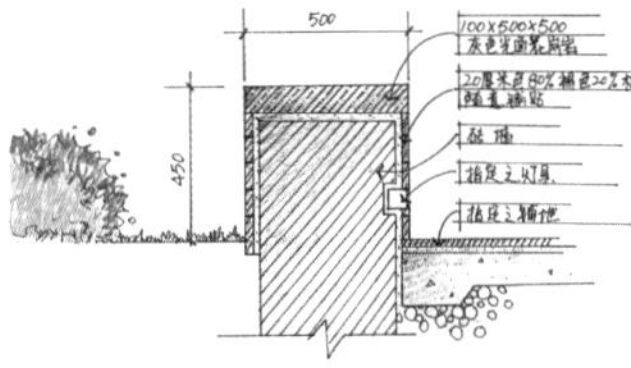

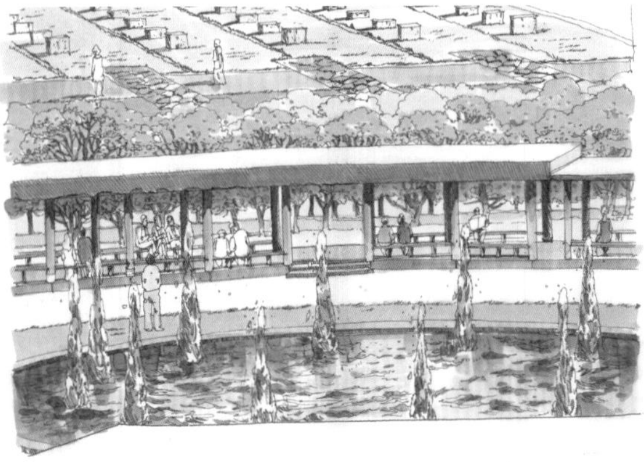

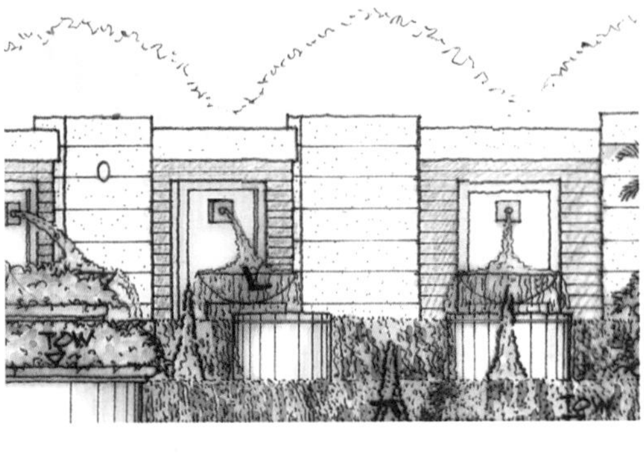

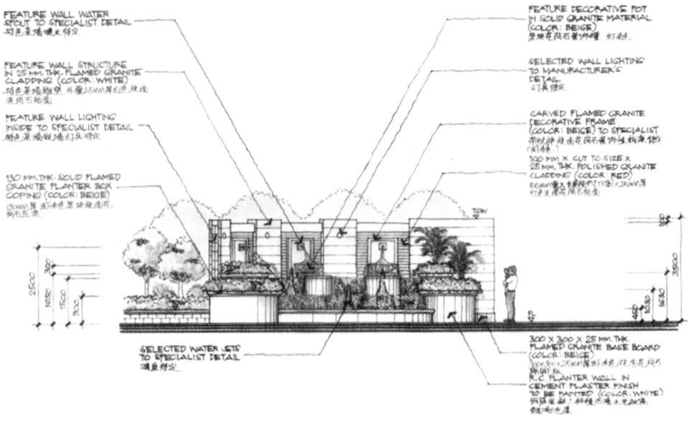

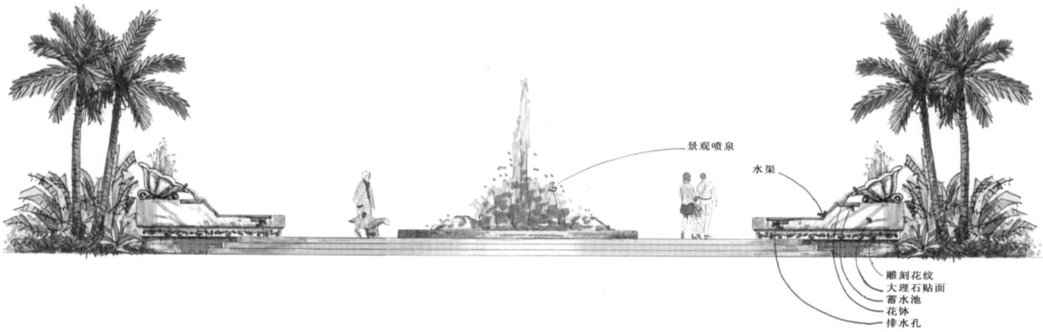

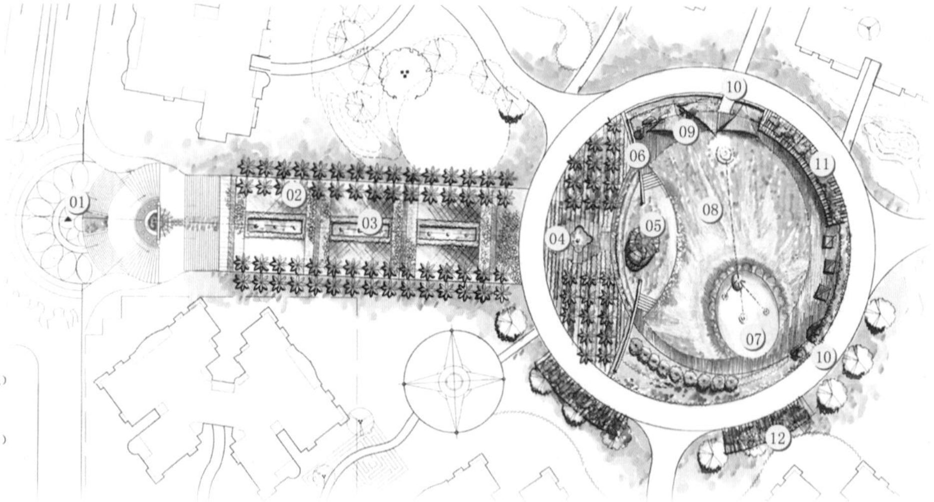

01 九龙喷泉广场
02 棕榈林
03 带状水景
04 棕榈跌水广场
05 棕榈泉（按摩池）
06 跌水景墙
07 儿童水世界
08 成人泳池
09 张拉膜（木平台）
10 出入口
11 "绿色"更衣室
12 遮荫廊架

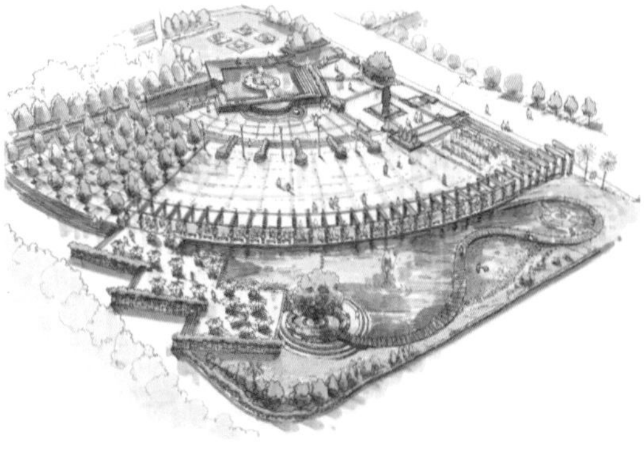

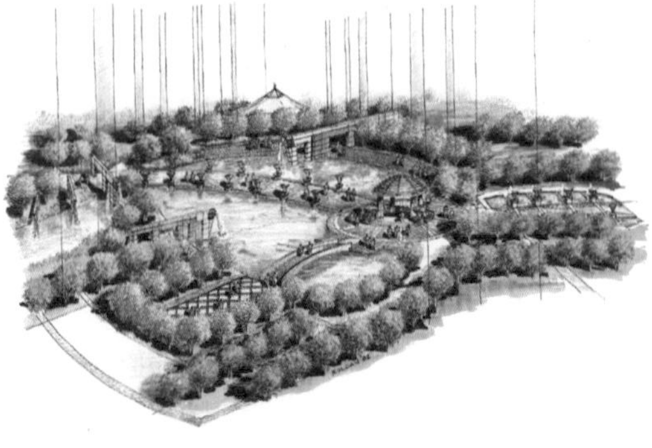

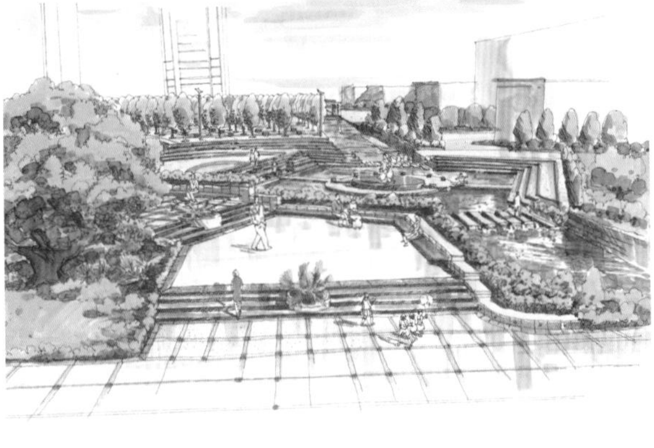

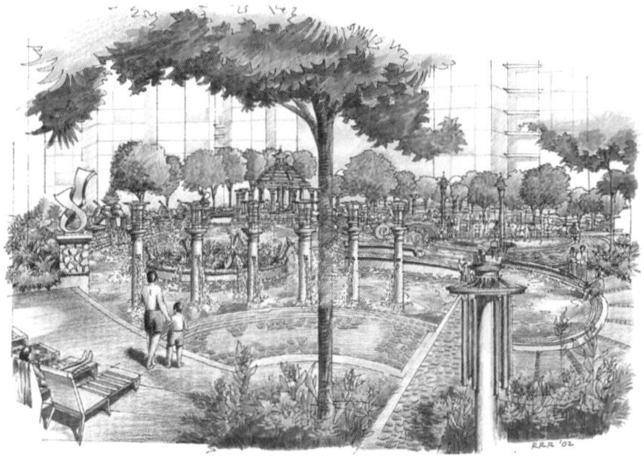

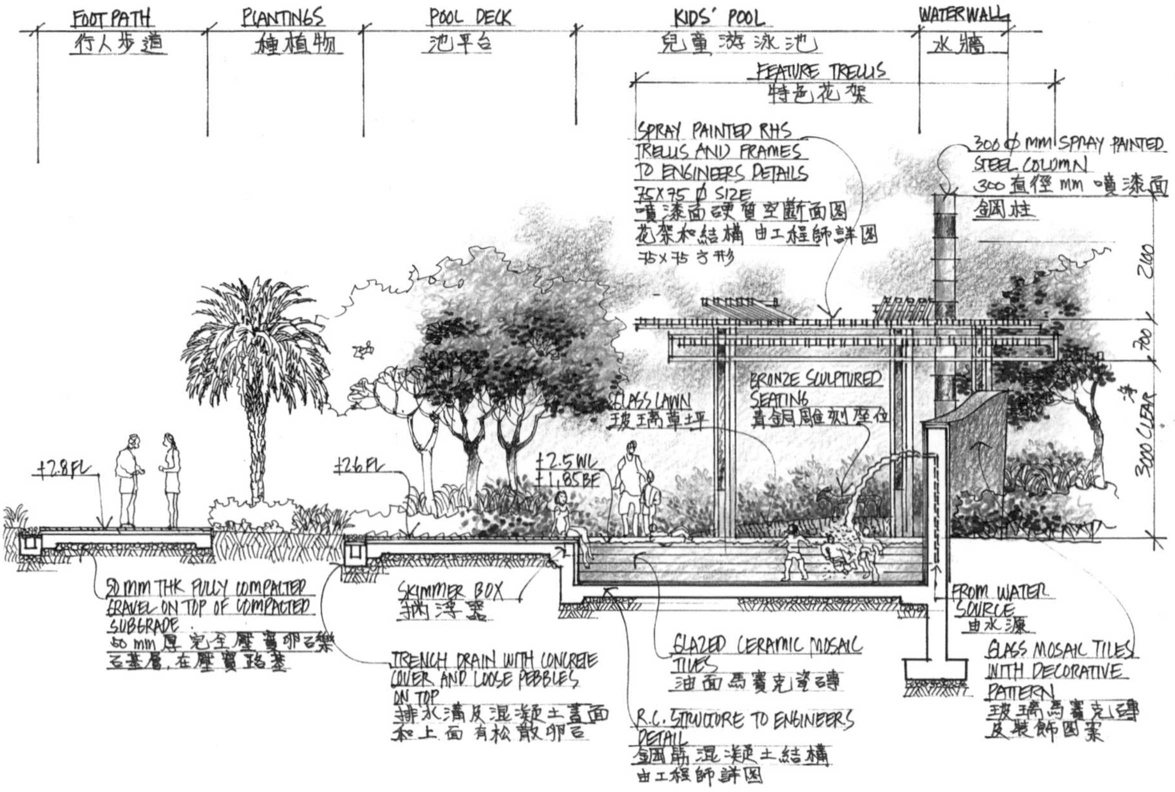
FOOT PATH
行人步道
PLANTINGS
種植物
POOL DECK
池平台
KIDS' POOL
兒童游泳池
WATERWALL
水牆
FEATURE TRELLIS
特色花架
SPRAY PAINTED RHS
TRELLIS AND FRAMES
TO ENGINEERS DETAILS
75X75 ϕ SIZE
噴漆面硬質空斷面图
花架和結構 由工程師詳图
75×75 方形
300 ϕ MM SPRAY PAINTED
STEEL COLUMN
300直徑 mm 噴漆面
鋼柱
2100
700
3000 CLEAR
GLASS LAWN
玻璃草坪
BRONZE SCULPTURED
SEATING
青銅雕刻座位
+2.8 FL
+2.6 FL
+2.5 WL
+1.85 BF
50 MM THK FULLY COMPACTED
GRAVEL ON TOP OF COMPACTED
SUBGRADE
50 mm 厚完全壓實卵石礫
石基層, 在壓實路基
SKIMMER BOX
撇浮器
TRENCH DRAIN WITH CONCRETE
COVER AND LOOSE PEBBLES
ON TOP
排水溝及混凝土蓋面
和上面有松散卵石
GLAZED CERAMIC MOSAIC
TILES
油面馬賽克瓷磚
R.C. STRUCTURE TO ENGINEERS
DETAIL
鋼筋混凝土結構
由工程師詳图
FROM WATER
SOURCE
由水源
GLASS MOSAIC TILES
WITH DECORATIVE
PATTERN
玻璃馬賽克磚
及裝飾图案

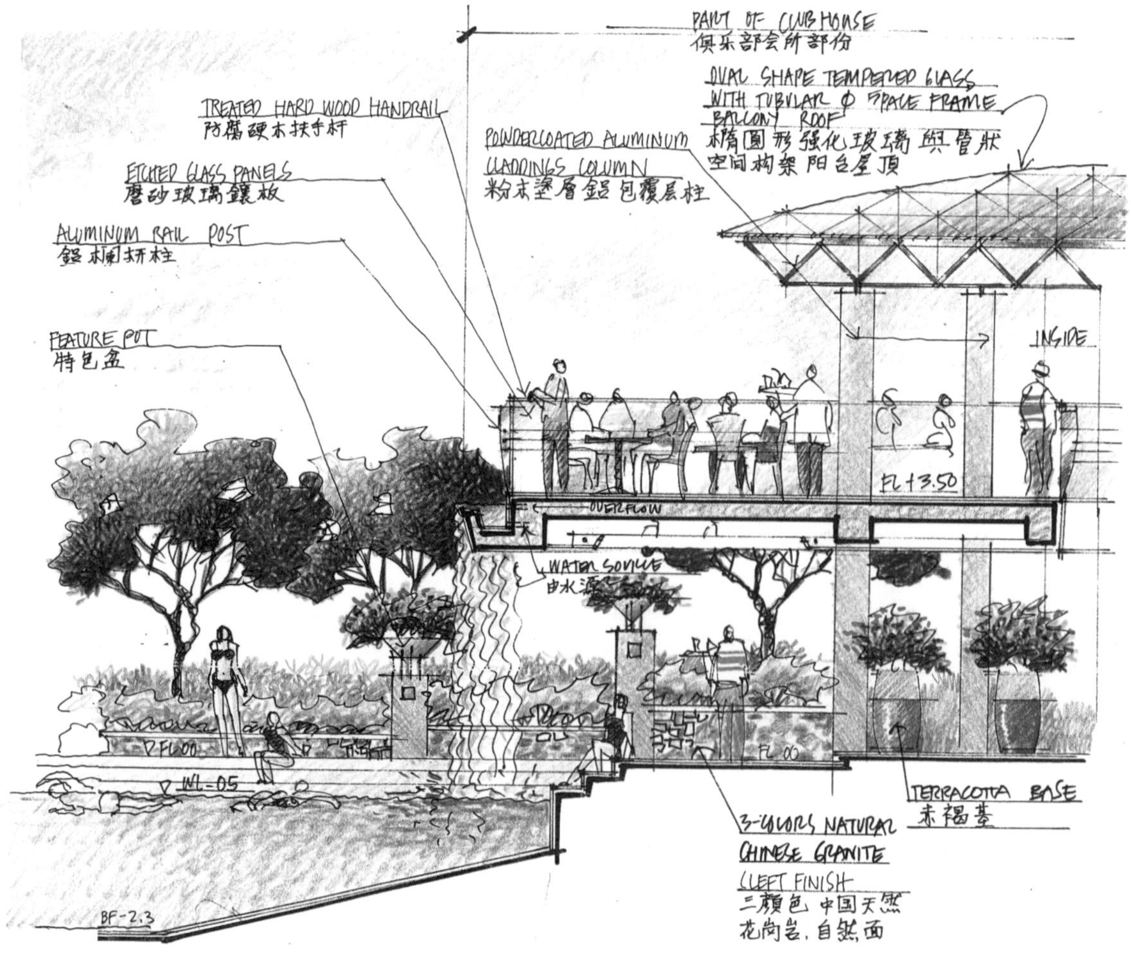
PART OF CLUB HOUSE
俱乐部会所部份
OVAL SHAPE TEMPERED GLASS
WITH TUBULAR ϕ SPACE FRAME
BALCONY ROOF
橢圓形强化玻璃與管狀
空间构架阳台屋顶
TREATED HARD WOOD HANDRAIL
防腐硬木扶手杆
POWDERCOATED ALUMINUM
CLADDINGS COLUMN
粉末塗層鋁包覆层柱
ETCHED GLASS PANELS
磨砂玻璃鑲板
ALUMINUM RAIL POST
鋁欄杆柱
FEATURE POT
特色盆
INSIDE
FL +3.50
OVERFLOW
WATER SOURCE
由水源
FL.00
WL -.05
FL.00
TERRACOTTA BASE
赤褐基
3-COLORS NATURAL
CHINESE GRANITE
CLEFT FINISH
三顏色 中国天然
花崗岩, 自然面
BF -2.3

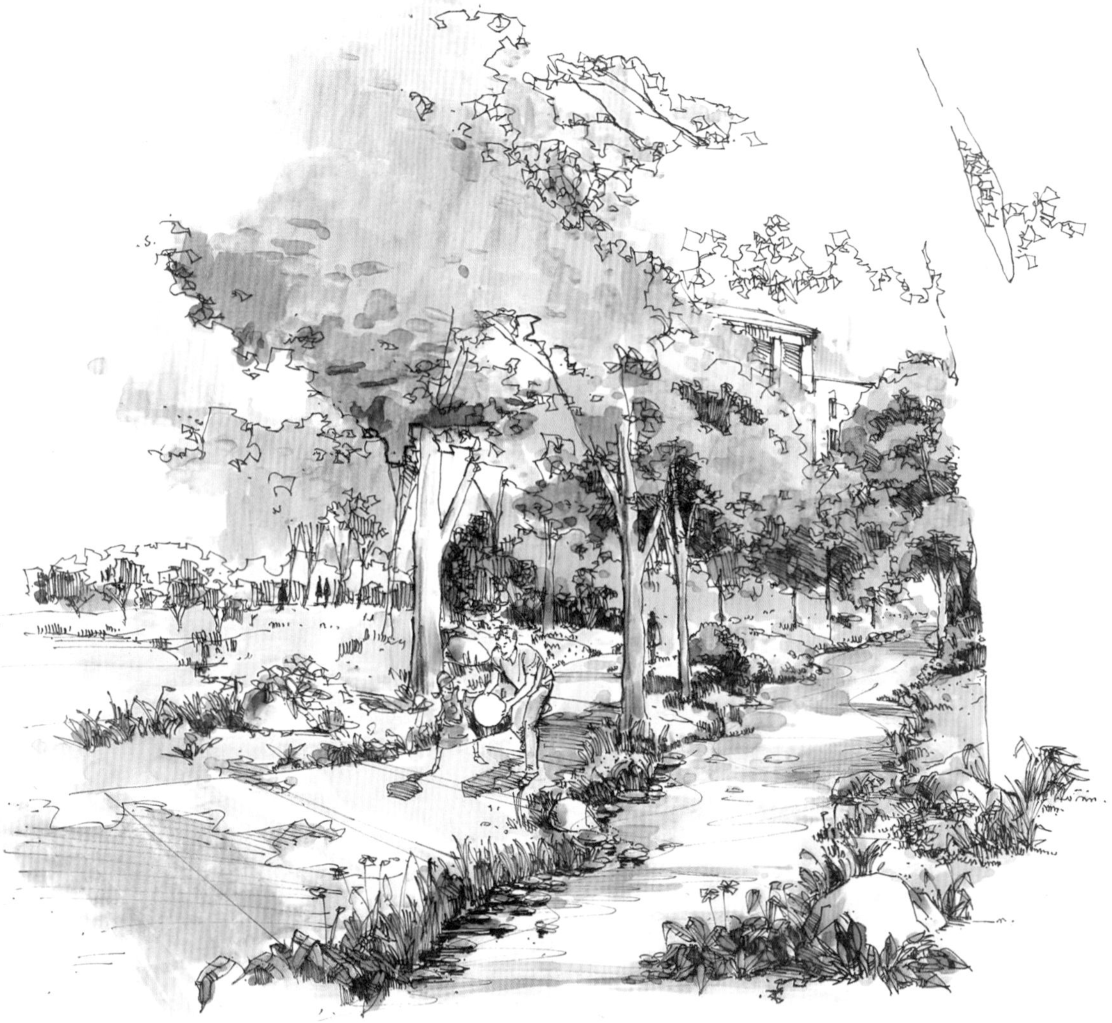

2005.

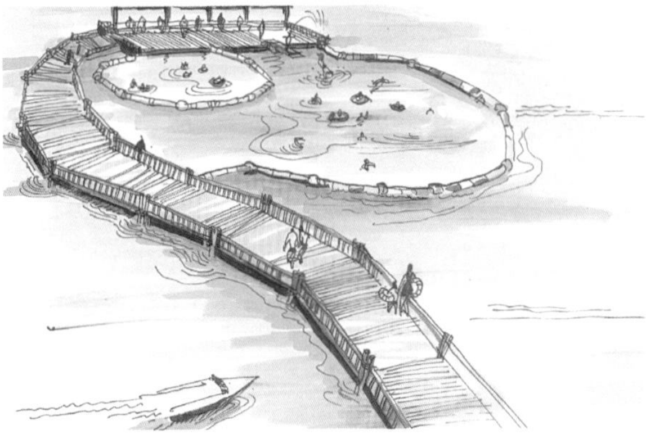

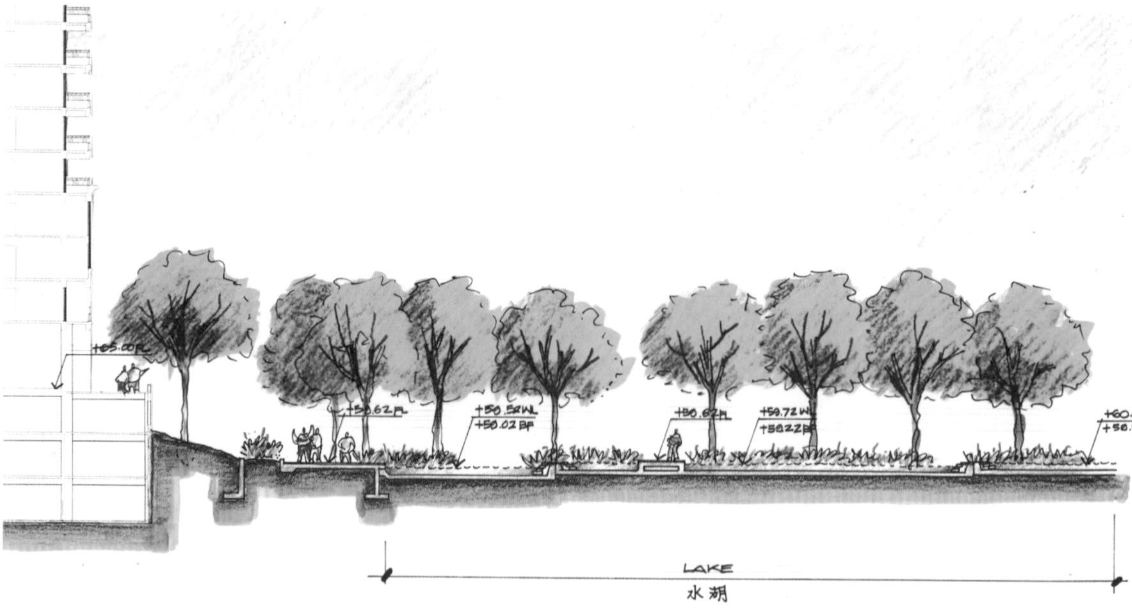
+65.00 FL
+50.02 BF
+60.00 WL
+50.50 BF
LAKE
水湖

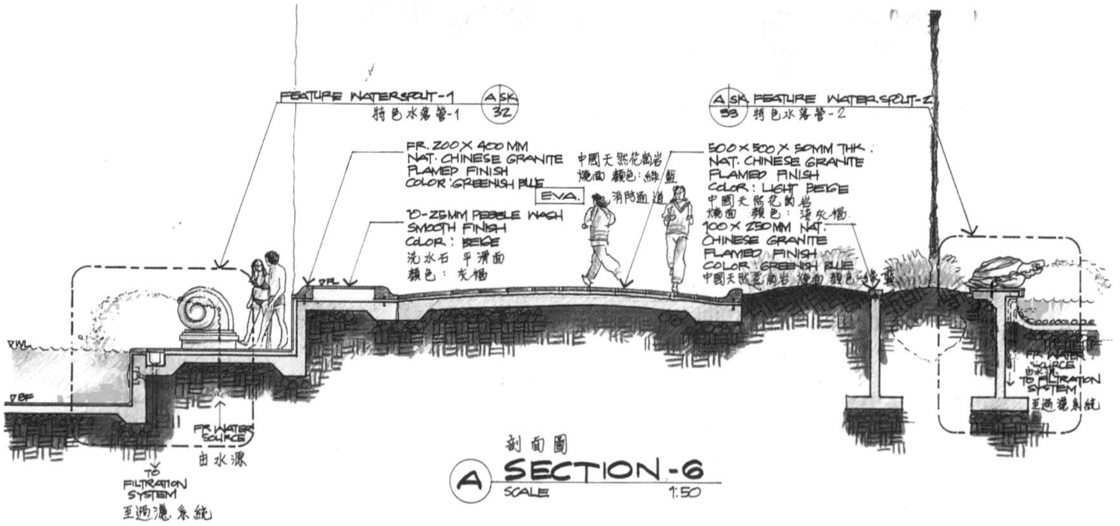
FEATURE WATERSPOUT-1
特色水落管-1
A SK 32
A SK 33
FEATURE WATERSPOUT-2
特色水落管-2
FR. 200 X 400 MM
NAT. CHINESE GRANITE
FLAMED FINISH
COLOR: GREENISH BLUE
中國天然花崗岩
燒面 顏色: 綠藍
EVA.
消防通道
10-25MM PEBBLE WASH
SMOOTH FINISH
COLOR: BEIGE
洗水石 平滑面
顏色: 灰褐
500 × 500 × 50MM THK.
NAT. CHINESE GRANITE
FLAMED FINISH
COLOR: LIGHT BEIGE
中國天然花崗岩
燒面 顏色: 淺灰褐
100 × 250MM NAT.
CHINESE GRANITE
FLAMED FINISH
COLOR: GREENISH BLUE
FR WATER SOURCE
由水源
TO FILTRATION SYSTEM
至過濾系統
剖面圖
A SECTION-6
SCALE 1:50

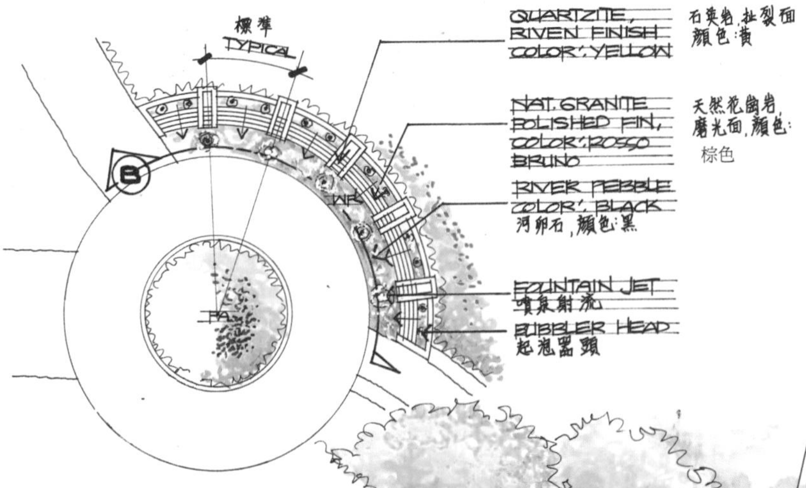
標準
TYPICAL
B
QUARTZITE, RIVEN FINISH COLOR: YELLOW
石英岩,扯裂面 顏色:黃
NAT. GRANITE POLISHED FIN, COLOR: ROSSO BRUNO
天然花崗岩 磨光面,顏色:
棕色
RIVER PEBBLE COLOR: BLACK
河卵石,顏色:黑
FOUNTAIN JET
噴泉射流
BUBBLER HEAD
起泡噐頭

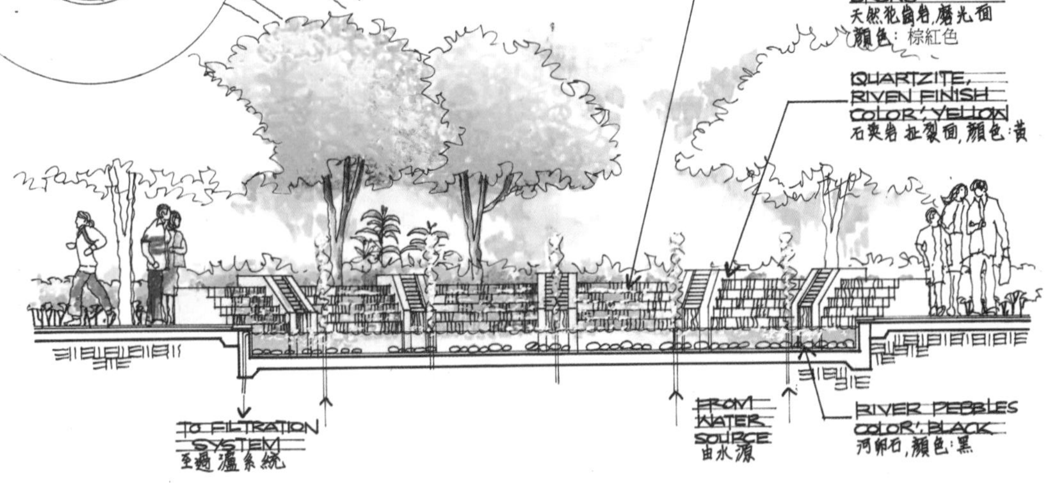
NATURAL GRANITE POLISHED FIN, COLOR: ROSSO BRUNO
天然花崗岩,磨光面
顏色:棕紅色
QUARTZITE, RIVEN FINISH COLOR: YELLOW
石英岩 扯裂面,顏色:黃
TO FILTRATION SYSTEM
至過濾系統
FROM WATER SOURCE
由水源
RIVER PEBBLES COLOR: BLACK
河卵石,顏色:黑

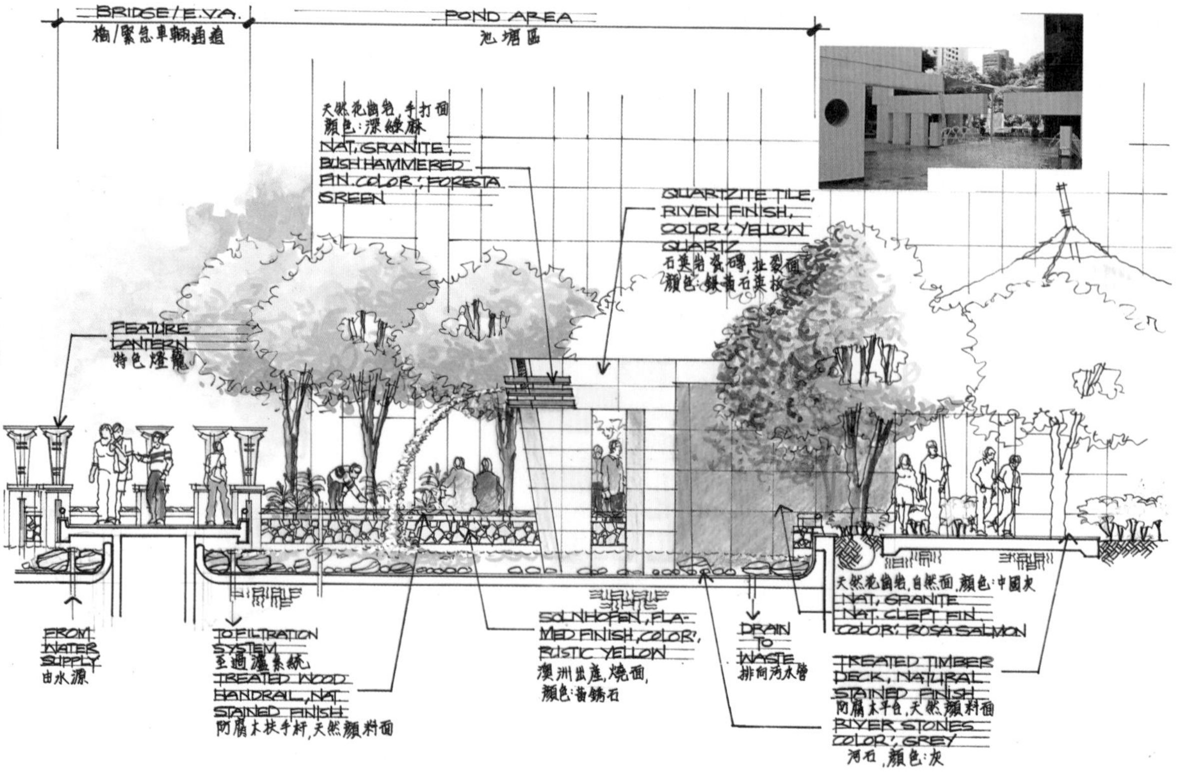
BRIDGE/E.V.A.
橋/緊急車輛通道
POND AREA
池塘區
天然花崗岩,手打面
顏色:深綠麻
NAT. GRANITE,
BUSH HAMMERED
FIN. COLOR: FORESTA
GREEN
QUARTZITE TILE,
RIVEN FINISH,
COLOR: YELLOW
QUARTZ
石英岩瓷磚,批裂面
顏色:鍍黃石英板
FEATURE
LANTERN
特色燈籠
FROM
WATER
SUPPLY
由水源
TO FILTRATION
SYSTEM
至過濾系統
TREATED WOOD
HANDRAIL, NAT.
STAINED FINISH
防腐木扶手杆,天然顏料面
SOLNHOFEN, FLA-
MED FINISH, COLOR:
RUSTIC YELLOW
澳洲出產,燒面,
顏色:黃銹石
DRAIN
TO
WASTE
排向污水管
天然花崗岩,自然面,顏色:中國灰
NAT. GRANITE
NAT. CLEFT FIN.
COLOR: ROSA SALMON
TREATED TIMBER
DECK, NATURAL
STAINED FINISH
防腐木平台,天然顏料面
RIVER STONES
COLOR: GREY
河石,顏色:灰

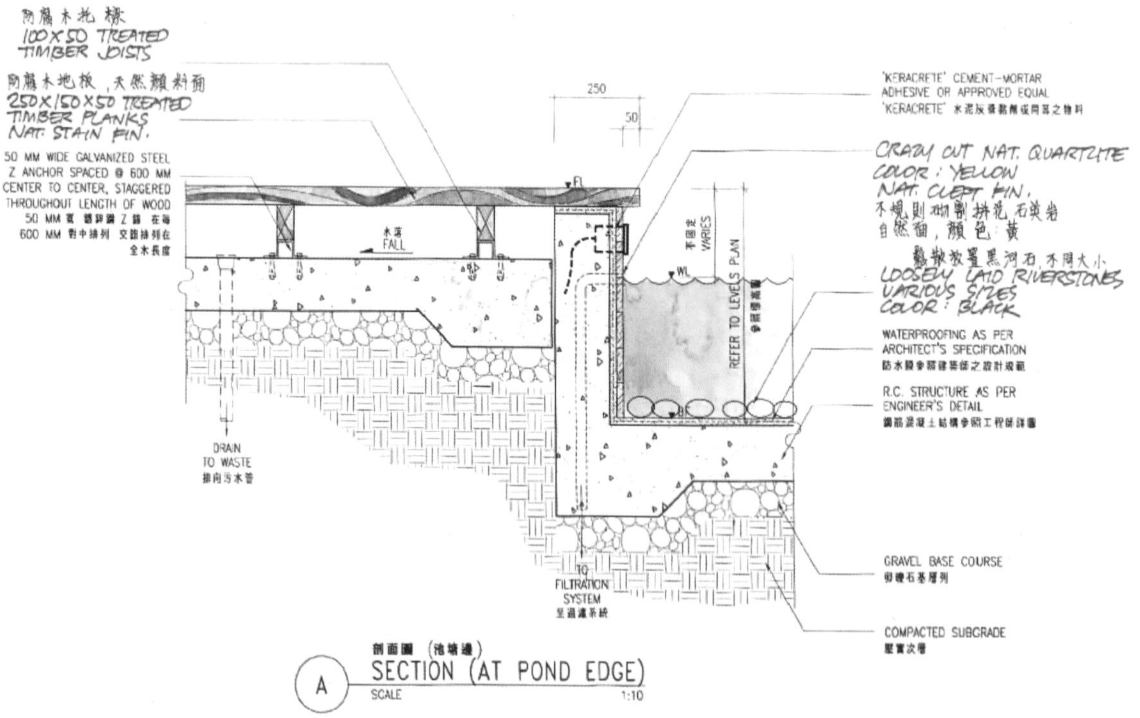
防腐木枕櫞
100X50 TREATED
TIMBER JOISTS
防腐木地板,天然顏料面
250X150X50 TREATED
TIMBER PLANKS
NAT. STAIN FIN.
50 MM WIDE GALVANIZED STEEL
Z ANCHOR SPACED @ 600 MM
CENTER TO CENTER, STAGGERED
THROUGHOUT LENGTH OF WOOD
50 MM 寬 鍍鋅鋼 Z 錨 在每
600 MM 對中排列 交錯排列在
全木長度
250
50
FL
水落
FALL
不同水
VARIES
WL
REFER TO LEVELS PLAN
DRAIN
TO WASTE
排向污水管
TO
FILTRATION
SYSTEM
至過濾系統
'KERACRETE' CEMENT-MORTAR
ADHESIVE OR APPROVED EQUAL
'KERACRETE' 水泥灰漿黏劑或同等之物料
CRAZY CUT NAT. QUARTZITE
COLOR: YELLOW
NAT. CLEFT FIN.
不規則砌割拼花石英岩
自然面,顏色:黃
鬆散放置黑河石,不同大小
LOOSELY LAID RIVERSTONES
VARIOUS SIZES
COLOR: BLACK
WATERPROOFING AS PER
ARCHITECT'S SPECIFICATION
防水層參照建築師之設計規範
R.C. STRUCTURE AS PER
ENGINEER'S DETAIL
鋼筋混凝土結構參照工程師詳圖
GRAVEL BASE COURSE
卵礫石基層列
COMPACTED SUBGRADE
壓實次層

剖面圖 (池塘邊)
A SECTION (AT POND EDGE)
SCALE 1:10

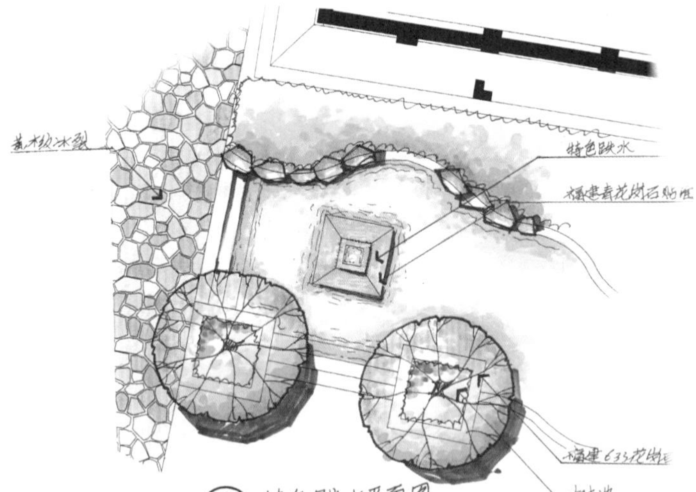

A 特色跌水平面图 SCALE: 1:75MTS.

B 特色跌水平面大样图 SCALE: 1:15MTS.

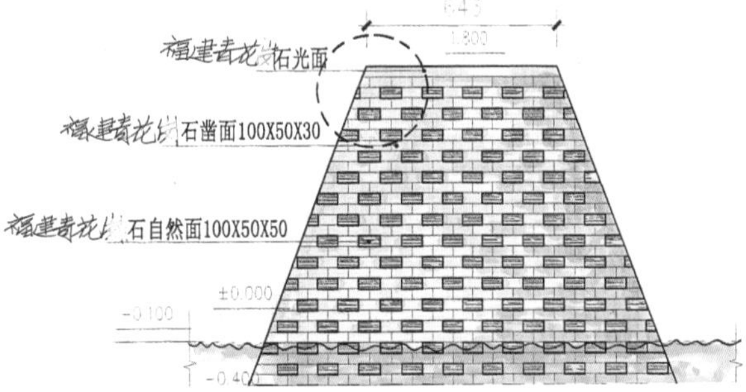

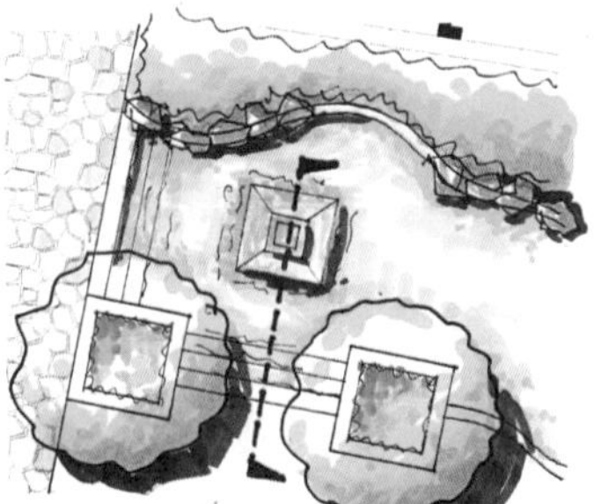

A 特色跌水平面图 SCALE: 1:100MTS

A 特色跌水剖面图 SCALE: 1:20MTS.

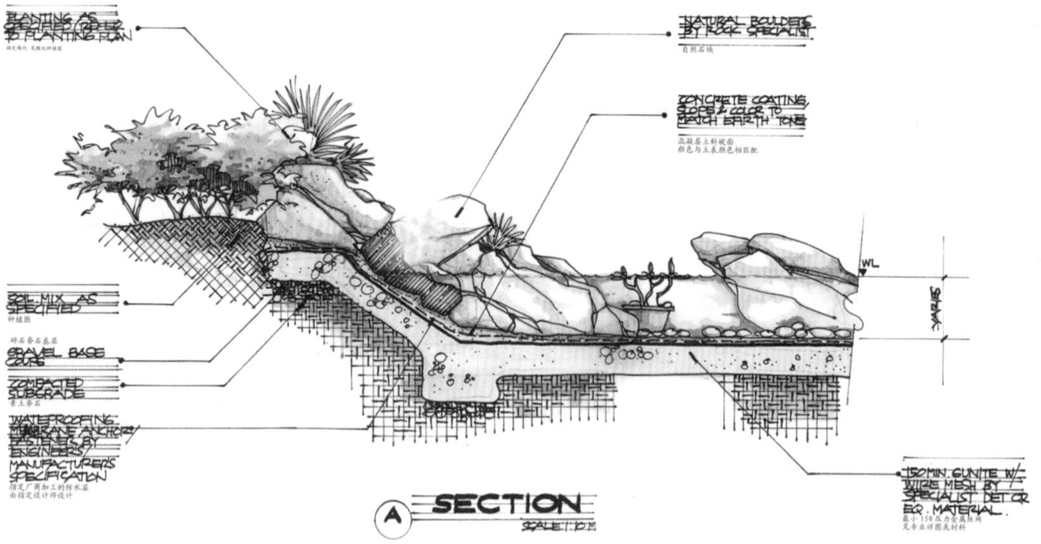
PLANTING AS SPECIFIED REFER TO PLANTING PLAN
NATURAL BOULDERS BY ROCK SPECIALIST
自然石块
CONCRETE COATING, SLOPES & COLOR TO MATCH EARTH TONE
混凝基土斜坡面
颜色与土表颜色相匹配
WL
VARIES
SOIL MIX AS SPECIFIED
种植土
碎石垫石基层
GRAVEL BASE COURSE
COMPACTED SUBGRADE
素土夯石
WATERPROOFING MEMBRANE ANCHORS FASTENER BY ENGINEERS MANUFACTURER'S SPECIFICATION
指定厂商加工的防水层
由指定设计师设计
150MIN GUNITE W/ WIRE MESH BY SPECIALIST DET OR EQ. MATERIAL
A SECTION
SCALE 1:10 M

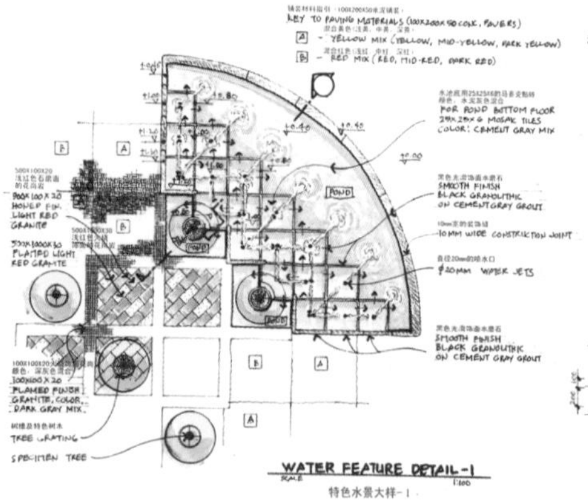
KEY TO PAVING MATERIALS (100X200X50 CONC. PAVERS)
A - YELLOW MIX (YELLOW, MID-YELLOW, DARK YELLOW)
B - RED MIX (RED, MID-RED, DARK RED)
FOR POND BOTTOM FLOOR 25X25X6 MOSAIC TILES COLOR: CEMENT GRAY MIX
SMOOTH FINISH BLACK GRANOLITHIC ON CEMENT GRAY GROUT
10MM WIDE CONSTRUCTION JOINT
φ20MM WATER JETS
SMOOTH FINISH BLACK GRANOLITHIC ON CEMENT GRAY GROUT
POND
900X100X20 HONED FIN. LIGHT RED GRANITE
500X1000X30 FLAMED LIGHT RED GRANITE
100X100X20 FLAMED FINISH GRANITE, COLOR, DARK GRAY MIX
TREE GRATING
SPECIMEN TREE
WATER FEATURE DETAIL-1
SCALE 1:100
特色水景大样-1

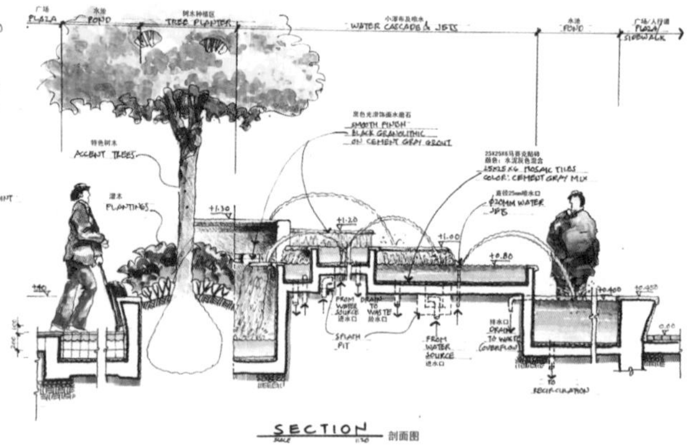
PLAZA
POND
TREE PLANTER
WATER CASCADE & JETS
POND
PLAZA/ SIDEWALK
ACCENT TREES
PLANTINGS
SMOOTH FINISH BLACK GRANOLITHIC ON CEMENT GRAY GROUT
25X25X6 MOSAIC TILES COLOR: CEMENT GRAY MIX
φ20MM WATER JETS
+1.30
+1.20
+1.00
+0.80
FROM WATER SOURCE
DRAIN TO WASTE
SPLASH PIT
FROM WATER SOURCE
DRAIN TO WASTE OVERFLOW
TO RECIRCULATION
SECTION
SCALE 1:50
剖面图

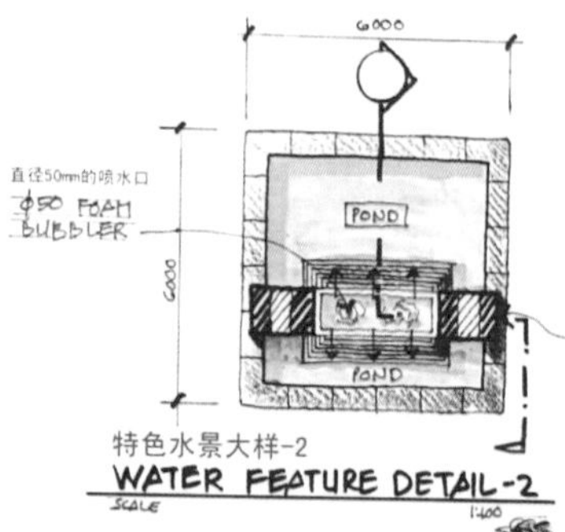
6000
直径50mm的喷水口
φ50 FOAM BUBBLER
POND
POND
柱构造
STRUCTURAL COLUMN
特色水景大样-2
WATER FEATURE DETAIL-2
SCALE 1:100

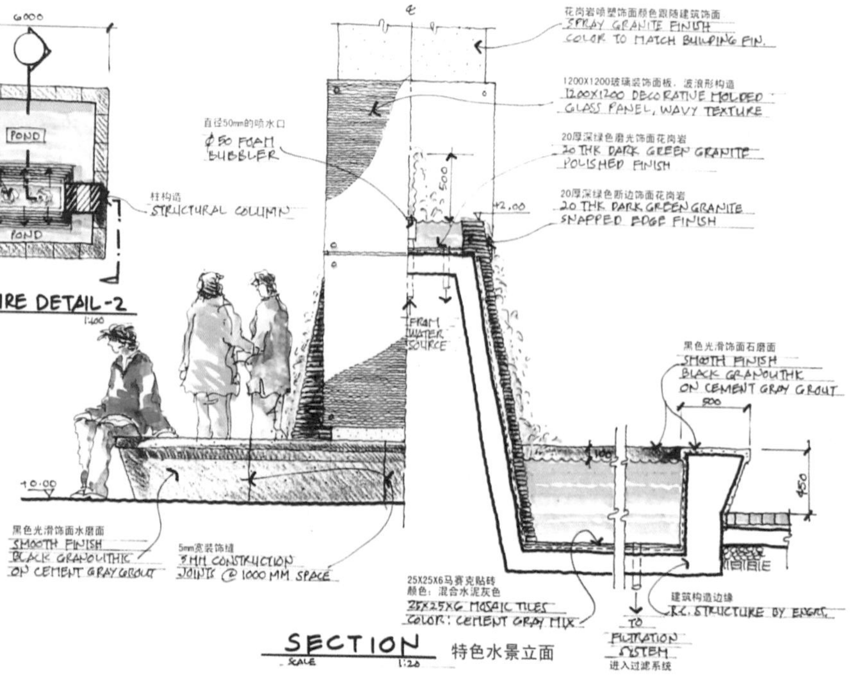
花岗岩喷塑饰面颜色跟随建筑饰面
SPRAY GRANITE FINISH COLOR TO MATCH BUILDING FIN.
1200X1200玻璃装饰面板，波浪形构造
1200X1200 DECORATIVE MOLDED GLASS PANEL, WAVY TEXTURE
直径50mm的喷水口
φ50 FOAM BUBBLER
20厚深绿色磨光饰面花岗岩
20 THK DARK GREEN GRANITE POLISHED FINISH
20厚深绿色断边饰面花岗岩
20 THK DARK GREEN GRANITE SNAPPED EDGE FINISH
+2.00
500
FROM WATER SOURCE
黑色光滑饰面石磨面
SMOOTH FINISH BLACK GRANOLITHIC ON CEMENT GRAY GROUT
500
450
±0.00
黑色光滑饰面水磨面
SMOOTH FINISH BLACK GRANOLITHIC ON CEMENT GRAY GROUT
5mm宽装饰缝
5MM CONSTRUCTION JOINTS @ 1000 MM SPACE
25X25X6马赛克贴砖
颜色：混合水泥灰色
25X25X6 MOSAIC TILES COLOR: CEMENT GRAY MIX
建筑构造边缘
R.C. STRUCTURE BY ENGRS.
TO FILTRATION SYSTEM
进入过滤系统
SECTION 特色水景立面
SCALE 1:20

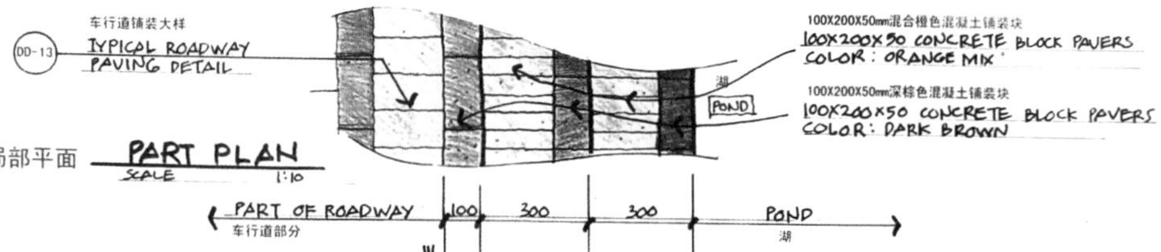
车行道铺装大样
DD-13
TYPICAL ROADWAY PAVING DETAIL
100X200X50mm混合橙色混凝土铺装块
100X200X50 CONCRETE BLOCK PAVERS
COLOR: ORANGE MIX
湖
POND
100X200X50mm深棕色混凝土铺装块
100X200X50 CONCRETE BLOCK PAVERS
COLOR: DARK BROWN
PART OF ROADWAY
车行道部分
100
300
300
POND
湖

局部平面 PART PLAN
SCALE 1:10

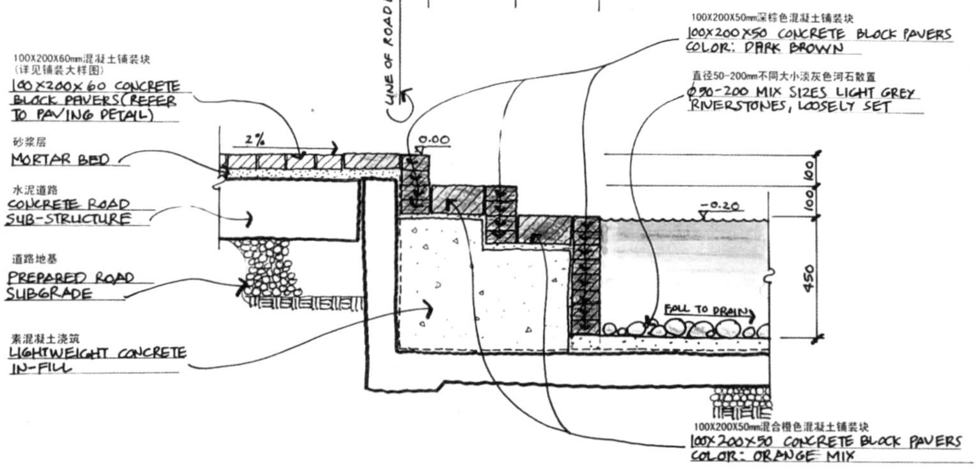
LINE OF ROAD EDGE
100X200X60mm混凝土铺装块
(详见铺装大样图)
100X200X60 CONCRETE BLOCK PAVERS (REFER TO PAVING DETAIL)
砂浆层
MORTAR BED
水泥道路
CONCRETE ROAD SUB-STRUCTURE
道路地基
PREPARED ROAD SUBGRADE
素混凝土浇筑
LIGHTWEIGHT CONCRETE IN-FILL
2%
0.00
-0.20
100
100
450
FALL TO DRAIN
100X200X50mm深棕色混凝土铺装块
100X200X50 CONCRETE BLOCK PAVERS
COLOR: DARK BROWN
直径50-200mm不同大小淡灰色河石散置
Ø50-200 MIX SIZES LIGHT GREY RIVERSTONES, LOOSELY SET
100X200X50mm混合橙色混凝土铺装块
100X200X50 CONCRETE BLOCK PAVERS
COLOR: ORANGE MIX

POND EDGE-2 湖岸设计-2
SCALE 1:10

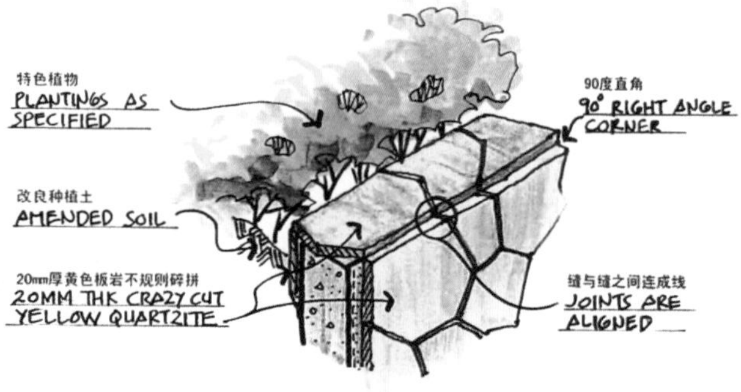
特色植物
PLANTINGS AS SPECIFIED
改良种植土
AMENDED SOIL
20mm厚黄色板岩不规则碎拼
20MM THK CRAZY CUT YELLOW QUARTZITE
90度直角
90° RIGHT ANGLE CORNER
缝与缝之间连成线
JOINTS ARE ALIGNED

DETAIL 详图
NOT TO SCALE

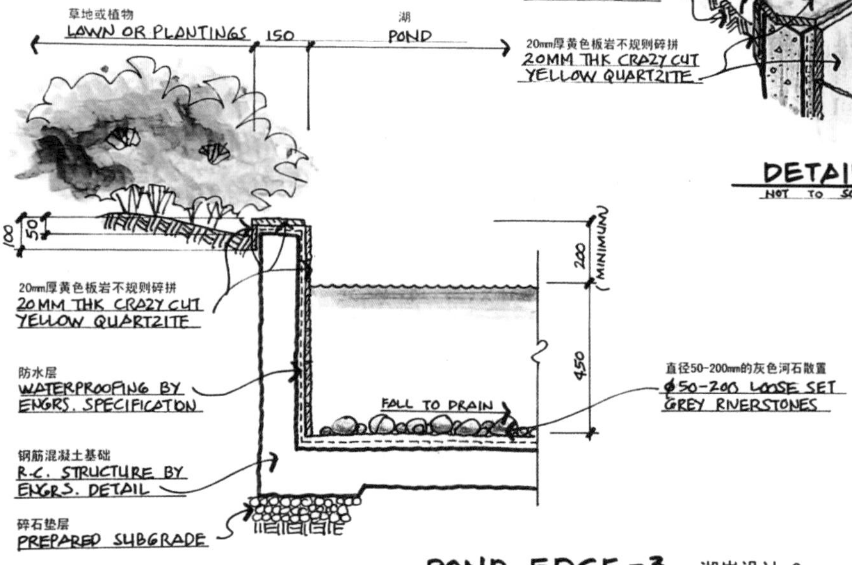
草地或植物
LAWN OR PLANTINGS
150
湖
POND
100
50
200 (MINIMUM)
450
20mm厚黄色板岩不规则碎拼
20MM THK CRAZY CUT YELLOW QUARTZITE
防水层
WATERPROOFING BY ENGRS. SPECIFICATION
钢筋混凝土基础
R.C. STRUCTURE BY ENGRS. DETAIL
碎石垫层
PREPARED SUBGRADE
FALL TO DRAIN
直径50-200mm的灰色河石散置
Ø50-200 LOOSE SET GREY RIVERSTONES

POND EDGE-3 湖岸设计-3
SCALE 1:10

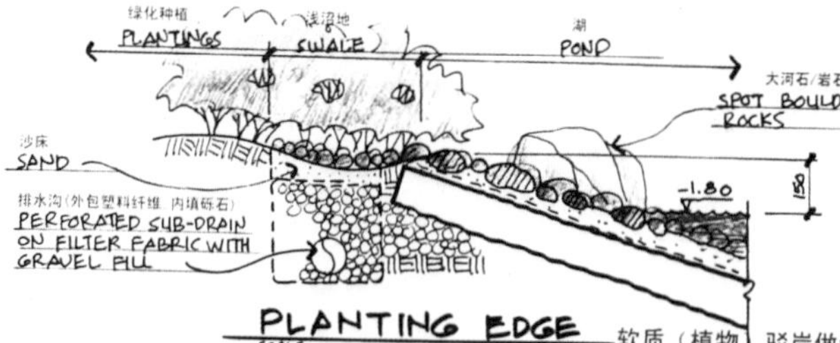

PLANTING EDGE SCALE 1:10 软质（植物）驳岸做法

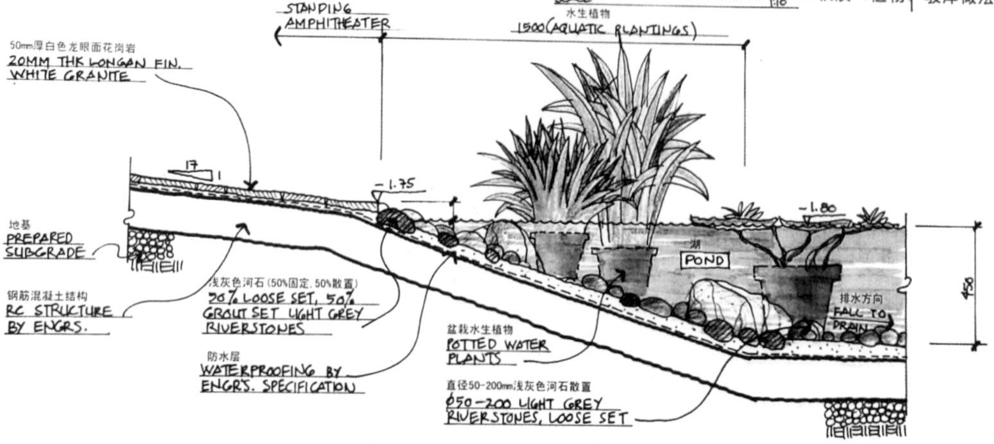

POND EDGE-5 SCALE 1:10 湖岸设计-5

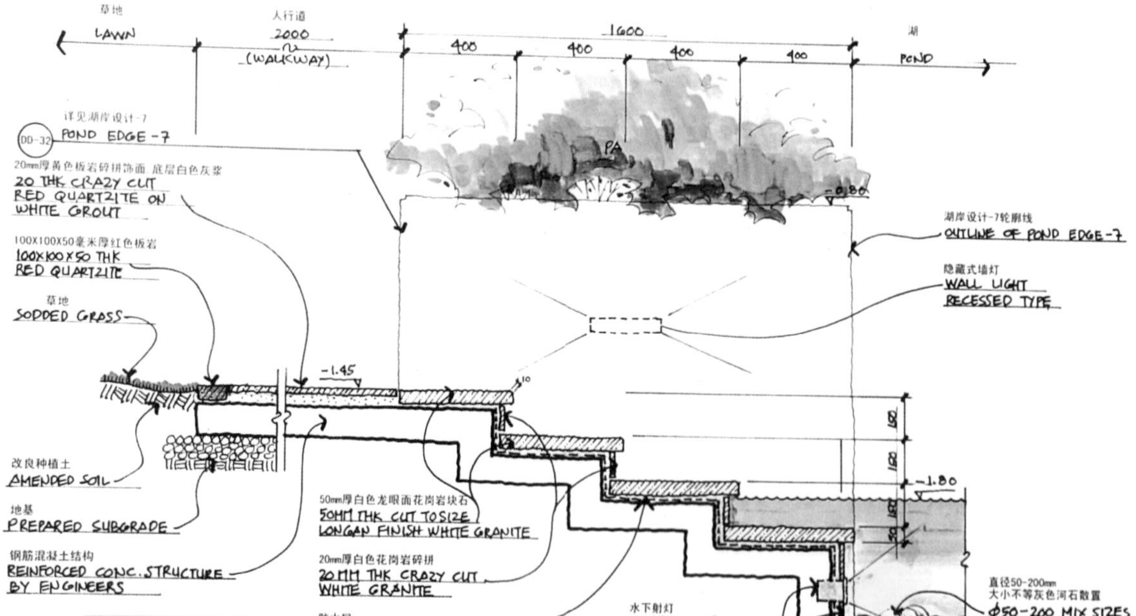

POND EDGE-6 SCALE 1:10
湖岸设计-6

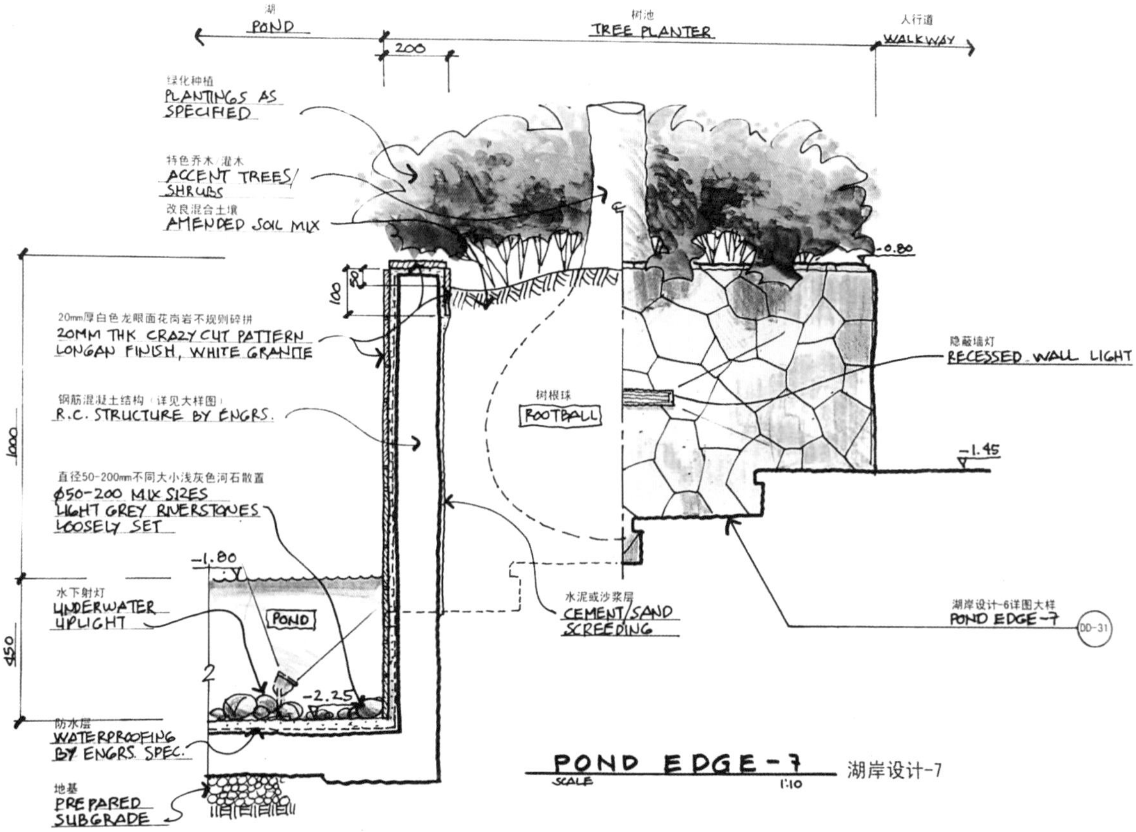
湖
POND
树池
TREE PLANTER
人行道
WALKWAY
200
绿化种植
PLANTINGS AS SPECIFIED
特色乔木 灌木
ACCENT TREES/ SHRUBS
改良混合土壤
AMENDED SOIL MIX
-0.80
100
80
20mm厚白色龙眼面花岗岩不规则碎拼
20MM THK CRAZY CUT PATTERN LONGAN FINISH, WHITE GRANITE
隐蔽墙灯
RECESSED WALL LIGHT
钢筋混凝土结构（详见大样图）
R.C. STRUCTURE BY ENGRS.
树根球
ROOTBALL
1000
-1.45
直径50-200mm不同大小浅灰色河石散置
φ50-200 MIX SIZES LIGHT GREY RIVERSTONES LOOSELY SET
-1.80
水下射灯
UNDERWATER UPLIGHT
POND
水泥或沙浆层
CEMENT/SAND SCREEDING
湖岸设计-6详图大样
POND EDGE-7
DD-31
450
-2.25
防水层
WATERPROOFING BY ENGRS. SPEC.
地基
PREPARED SUBGRADE

POND EDGE-7 湖岸设计-7

SCALE 1:10

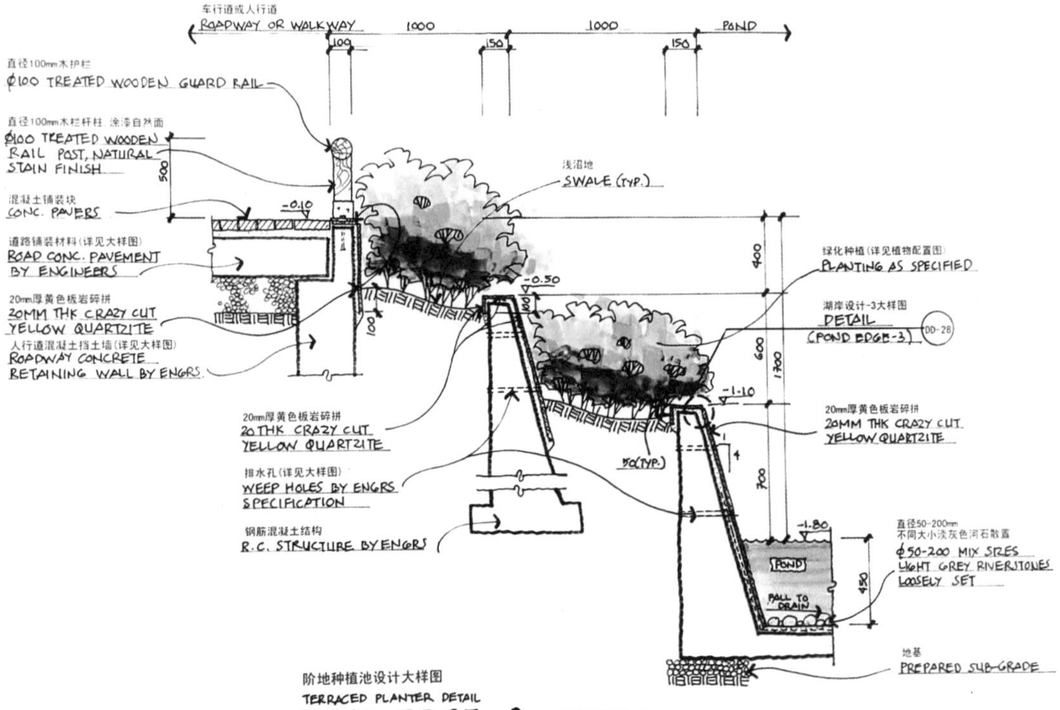
车行道或人行道
ROADWAY OR WALKWAY
1000
1000
POND
100
150
150
直径100mm木护栏
φ100 TREATED WOODEN GUARD RAIL
直径100mm木栏杆柱，油漆自然面
φ100 TREATED WOODEN RAIL POST, NATURAL STAIN FINISH
500
浅沼地
SWALE (TYP.)
混凝土铺装块
CONC. PAVERS
-0.10
道路铺装材料（详见大样图）
ROAD CONC. PAVEMENT BY ENGINEERS
400
绿化种植（详见植物配置图）
PLANTING AS SPECIFIED
-0.50
20mm厚黄色板岩碎拼
20MM THK CRAZY CUT YELLOW QUARTZITE
100
湖岸设计-3大样图
DETAIL (POND EDGE-3)
DD-28
人行道混凝土挡土墙（详见大样图）
ROADWAY CONCRETE RETAINING WALL BY ENGRS.
600
1700
-1.10
20mm厚黄色板岩碎拼
20 THK CRAZY CUT YELLOW QUARTZITE
20mm厚黄色板岩碎拼
20MM THK CRAZY CUT YELLOW QUARTZITE
50(TYP.)
排水孔（详见大样图）
WEEP HOLES BY ENGRS SPECIFICATION
700
-1.80
钢筋混凝土结构
R.C. STRUCTURE BY ENGRS
直径50-200mm不同大小浅灰色河石散置
φ50-200 MIX SIZES LIGHT GREY RIVERSTONES LOOSELY SET
POND
FALL TO DRAIN
450
地基
PREPARED SUB-GRADE

阶地种植池设计大样图

TERRACED PLANTER DETAIL

POND EDGE-8 湖岸设计-8

SCALE 1:15

PART A LANDSCAPE ANALYSIS (III)
A區景觀分析圖(三)

PART A LANDSCAPE ANALYSIS (V)
A區景觀分析圖(五)

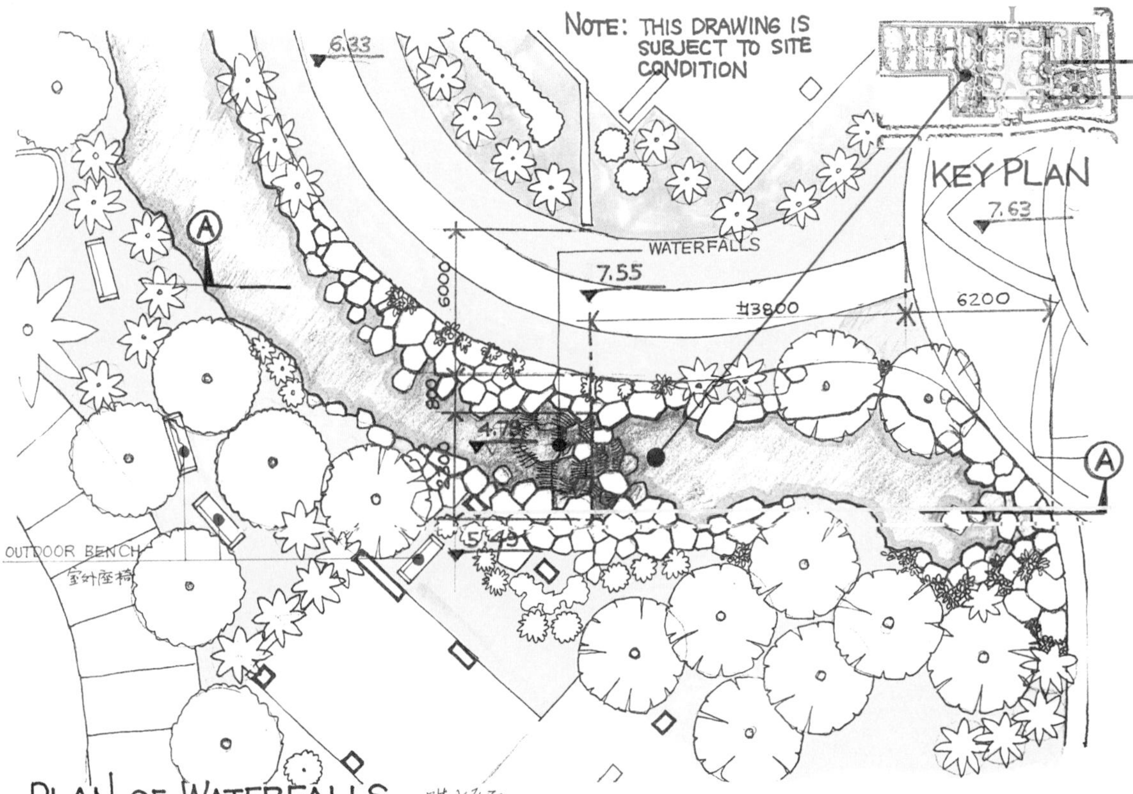
NOTE: THIS DRAWING IS SUBJECT TO SITE CONDITION
KEY PLAN
6.33
7.63
WATERFALLS
7.55
6000
±3800
6200
4.75
5.45
A
OUTDOOR BENCH
室外座椅
PLAN OF WATERFALLS
跌水平面
SCALE 1:125

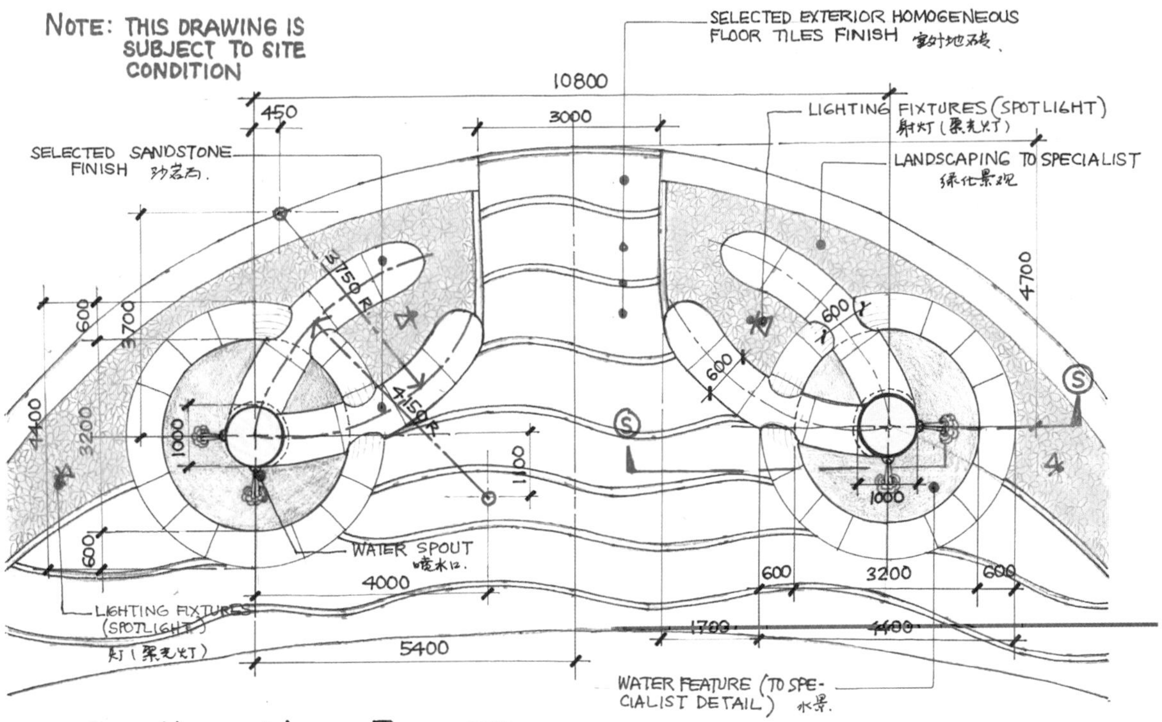
NOTE: THIS DRAWING IS SUBJECT TO SITE CONDITION
SELECTED EXTERIOR HOMOGENEOUS FLOOR TILES FINISH 室外地砖
10800
450
3000
LIGHTING FIXTURES (SPOTLIGHT)
射灯(聚光灯)
SELECTED SANDSTONE FINISH 砂岩石
LANDSCAPING TO SPECIALIST
绿化景观
4700
3700
600
3200
4400
1000
1100
WATER SPOUT
喷水口
4000
600
3200
600
1700
1400
5400
LIGHTING FIXTURES (SPOTLIGHT)
灯(聚光灯)
WATER FEATURE (TO SPECIALIST DETAIL) 水景
DETAIL OF WATER FEATURE
SCALE 1:50

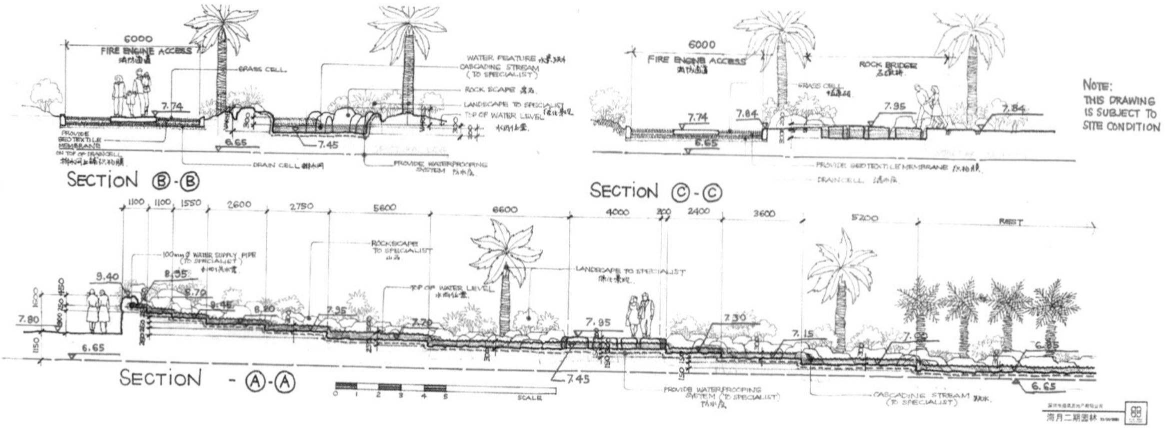
FIRE ENGINE ACCESS
SECTION Ⓑ-Ⓑ
SECTION Ⓒ-Ⓒ
SECTION -Ⓐ-Ⓐ
NOTE:
THIS DRAWING IS SUBJECT TO SITE CONDITION

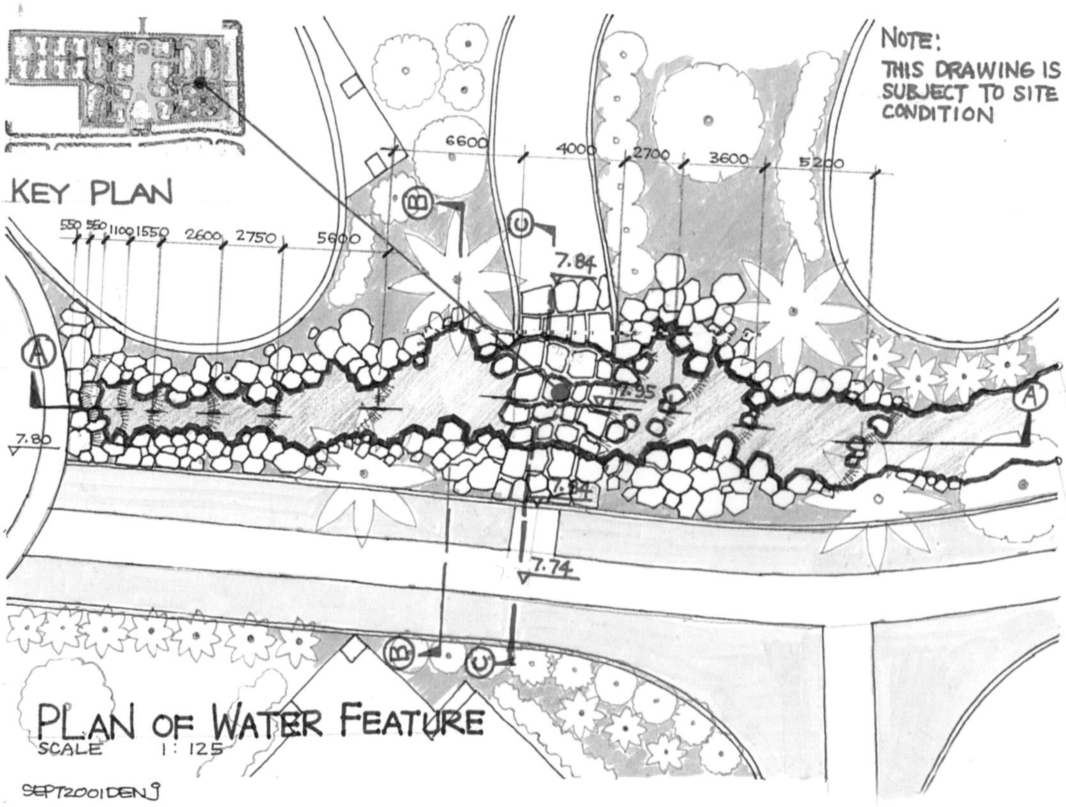
NOTE:
THIS DRAWING IS SUBJECT TO SITE CONDITION
KEY PLAN
PLAN OF WATER FEATURE
SCALE 1: 125
SEPT2001DENG

2 景墙

在园林小品中，景墙具有隔断、导游、衬景、装饰、保护等作用。景墙的形式也是多种多样，一般根据材料、断面的不同，有高矮、曲直、虚实、光洁、粗糙、有椽无椽等形式。景墙既要美观，又要坚固耐久。常用材料有转、混凝土、花格围墙、石墙、铁花格围墙等。景观常将这些墙巧妙地组合与变化，并结合树、石、建筑、花木等其他因素，以及墙上的漏窗、门洞的巧妙处理，形成空间有序、富有层次、虚实相间、明暗变化的静观效果。

一、景墙定义

园内划分空间、组织景色、安排导游而布置的围墙，能够反映文化，兼有美观、隔断、通透的作用的景观墙体。

二、景墙功能

景墙不仅在于营造公园内的景点，而且是在改善市容市貌及城市文化建设的重要手段。而“文化墙”这一概念更是把景墙在城市文化建设中的特殊作用做了概念性总结。园墙在园林中起划分内外范围、分隔内部空间和遮挡劣景的作用。精巧的园墙还可装饰园景。

三、中国传统园林的景墙

传统园林景墙按材料和构造可分为版筑墙、乱石墙、磨砖墙、白粉墙等。分隔院落空间多用白粉墙，墙头配以青瓦。用白粉墙衬托山石、花木，犹如在白纸上绘制山水花卉，意境尤佳。园墙与假山之间可即可离，各有其妙。园墙与水面之间宜有道路、石峰、花木点缀，景物映于墙面和水中，可增加意趣。产竹地区常就地取材，用竹编园墙，既经济又富有地方色彩，但不够坚固耐久，不宜作永久性园墙。

园林中的墙有分隔空间、组织导游、衬托景物、装饰美化或遮蔽视线的作用，是园林空间构图的一个重要因素。

墙的形式有云墙（波形墙）、梯形墙、漏明墙、白粉墙、钢筋混凝土花格墙、虎皮石墙、竹篱笆墙等。其建造材料丰富，施工简便。《园冶》中说，“宜石宜砖，宜漏宜磨，各有所制。”

我国江南古典园林中的墙多是白粉墙。白粉墙面不仅能与灰黑色瓦顶、栗褐色门窗有着鲜明的色彩对比，而且能衬托出山石、竹丛和花木藤萝的多姿多彩。在阳光照射下，墙面上水光树影变幻莫测，形成一幅幅美丽的画面。墙上又常设漏窗、空窗和洞门，形成虚实、明暗对比，使墙面的变化更加丰富多彩。

墙上的漏窗又名透花窗，可用以分隔景区，使空间似隔非隔、景物若隐若现，富于层次。

通过漏窗看到的各种对景可以使人目不遐接而又不致一览无遗，能收到虚中有实、实中有虚、隔而不断的艺术效果。漏窗本身的图案在不同的光线照射下可产生各种富有变化的阴影，使平直呆板的墙面显得活泼生动。

漏窗的窗框有方形、长方形、圆形、六角形、八角形、扇形以及其他各种不规则的形状。漏窗的花纹图案灵活多样，从构图上看，有几何形体的与自然形体的两种。几何形体的图案有万字、菱花、橄榄、冰纹、鱼鳞、秋叶、海棠、葵花、如意、波纹等。几何形体图案多用砖、木、瓦片（筒瓦、板瓦）制做。自然形体图案有以花卉为题材的，如松柏、牡丹、梅、竹、兰、芭蕉、荷花等等；也有以鸟兽为题材的，如鹤、鹿、凤凰、孔雀、蝙蝠等。

园林的墙上还常有不装窗扇的窗孔，称空窗。空窗除能采光外，还常作为取景框，使游人在游览过程中不断地获得新的画面。空窗后常置石峰、竹丛、芭蕉之类，形成一幅幅小品图画。空窗还能使空间相互渗透，可产生增加景深、扩大空间的效果。空窗的形式也有方形、长方形、六角形、圆形、扇形、葫芦形、秋叶行、瓶形等各种。空窗的高度多以人的视点高度为标准来决定，以便于眺望。

在江南古典园林中，空窗的边框常用青灰色方砖镶砌，周围刨出挺秀的线脚，经打磨光滑之后与白粉墙配合成朴素明净的色调对比。现代，则常用斩假石、水刷石等材料制做边框。

园林中的墙还可与山石、竹丛、灯具、雕塑、花池、花坛、花架等组合独立成景。园墙的位置选择除考虑其功能之外，还应考虑造景的要求。

园墙的设置多与地形结合，平坦的地形多建成平墙，坡地或山地则就势建成阶梯形，为了避免单调，有的建成波浪形的云墙。划分内外范围的园墙内侧常用土山、花台、山石、树丛、游廊等把墙隐蔽起来，使有限空间产生无限景观的效果。

国外常用木质的或金属的通透栅栏作园墙，园内景色能透出园外。英国自然风景园常用干沟式的“隐垣”作为边界，远处看不见园墙，园景与周围的田野连成一片。园内空间分隔常用高 2 米以上的高绿篱。

新建公园绿地的园墙，在传统作法的基础上广泛使用新材料、新技术。多采用较低矮和较通透的形式，普遍应用预制混凝土和金属的花格、栏栅。混凝土花格可以整体预制或用预制块拼砌，经久耐用；金属花格栏栅轻巧精致，遮挡最小，施工方便，小型公园应用最多。

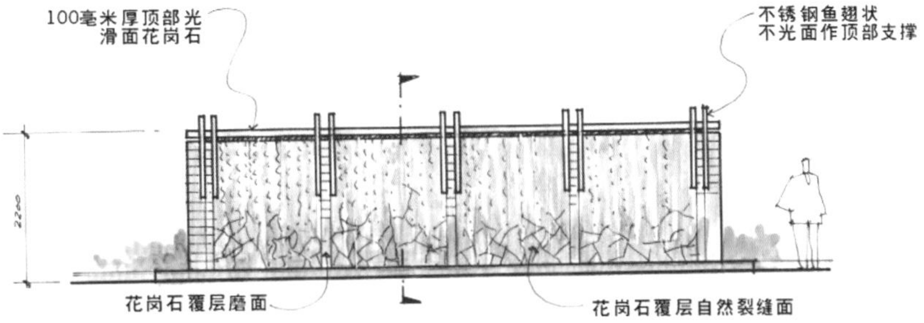
100毫米厚顶部光
滑面花岗石
不锈钢鱼翅状
不光面作顶部支撑
2200
花岗石覆层磨面
花岗石覆层自然裂缝面

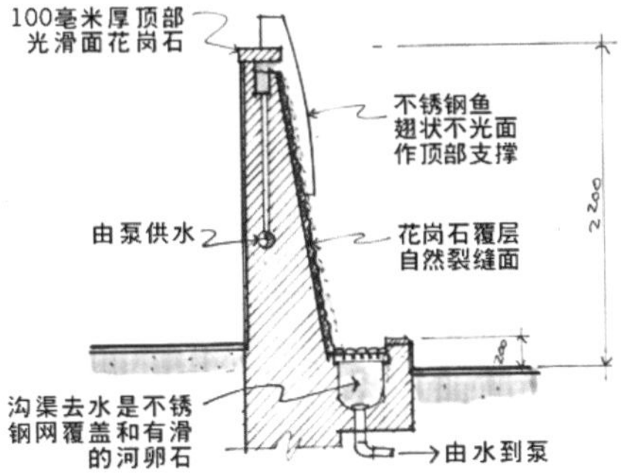
100毫米厚顶部
光滑面花岗石
不锈钢鱼
翅状不光面
作顶部支撑
由泵供水
花岗石覆层
自然裂缝面
2200
沟渠去水是不锈
钢网覆盖和有滑
的河卵石
由水到泵

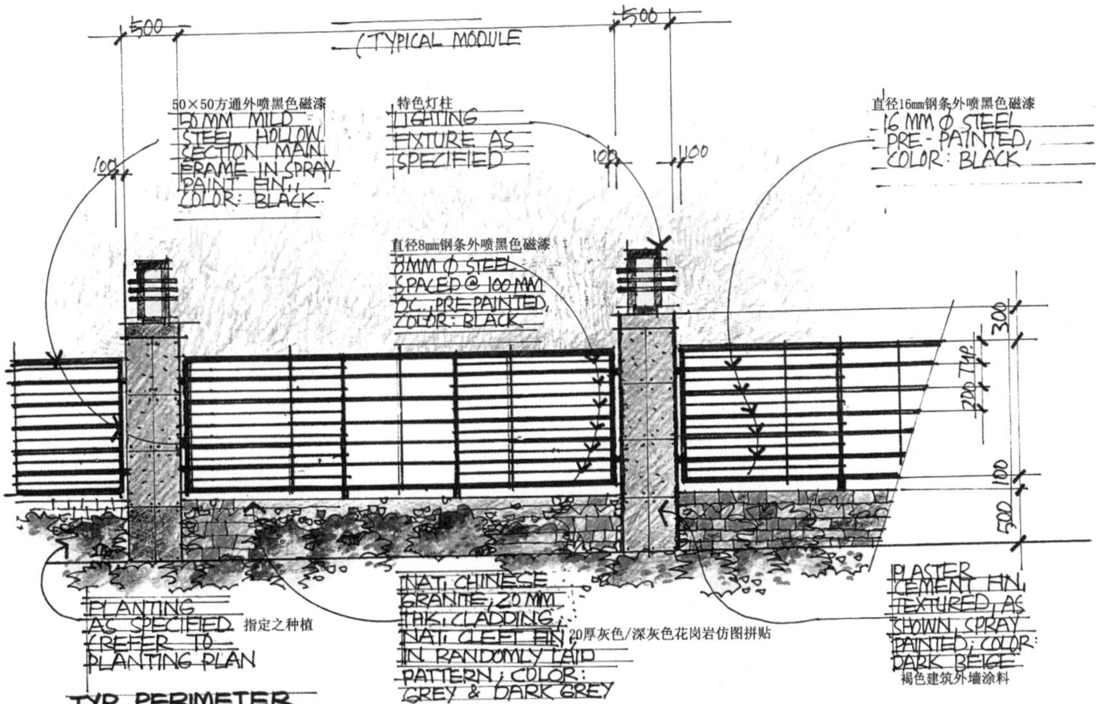
500
500
(TYPICAL MODULE
50×50方通外喷黑色磁漆
50MM MILD
STEEL HOLLOW
SECTION MAIN
FRAME IN SPRAY
PAINT FIN.,
COLOR: BLACK
特色灯柱
LIGHTING
FIXTURE AS
SPECIFIED
直径16mm钢条外喷黑色磁漆
16 MM Ø STEEL
PRE-PAINTED,
COLOR: BLACK
100
100
1100
直径8mm钢条外喷黑色磁漆
8MM Ø STEEL
SPACED @ 100 MM
O.C. PRE-PAINTED,
COLOR: BLACK
300
200 TYP.
100
500
PLANTING
AS SPECIFIED
(REFER TO
PLANTING PLAN
指定之种植
NAT. CHINESE
GRANITE, 20 MM
THK. CLADDING,
NAT. CLEFT FIN.
IN RANDOMLY LAID
PATTERN; COLOR:
GREY & DARK GREY
20厚灰色/深灰色花岗岩仿图拼贴
PLASTER
CEMENT FIN.
TEXTURED, AS
SHOWN, SPRAY
PAINTED; COLOR:
DARK BEIGE
褐色建筑外墙涂料

A TYP. PERIMETER FENCE ELEV.
SCALE 1:30
围墙标准段立面图

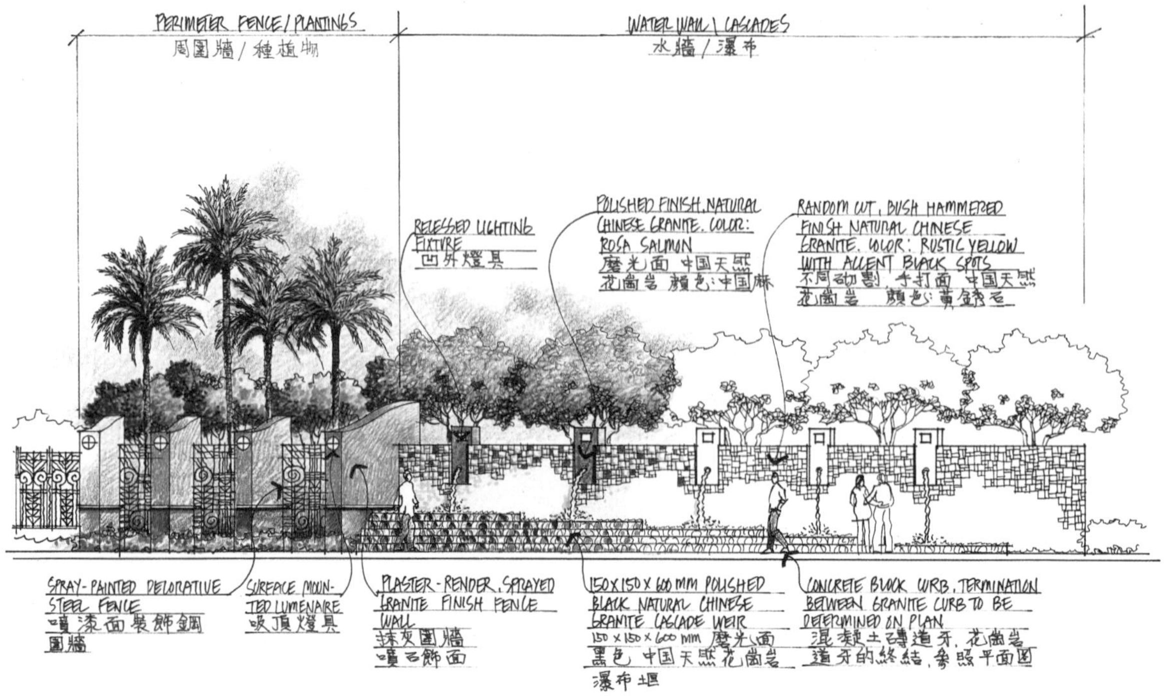
PERIMETER FENCE / PLANTINGS
周圍墙 / 種植物
WATER WALL / CASCADES
水墙 / 瀑布
RECESSED LIGHTING FIXTURE
凹外燈具
POLISHED FINISH, NATURAL CHINESE GRANITE, COLOR: ROSA SALMON
磨光面 中国天然花崗岩 顏色:中国麻
RANDOM CUT, BUSH HAMMERED FINISH NATURAL CHINESE GRANITE, COLOR: RUSTIC YELLOW WITH ACCENT BLACK SPOTS
不同切割, 手打面 中国天然花崗岩 顏色:黄銹石
SPRAY-PAINTED DECORATIVE STEEL FENCE
噴漆面裝飾鋼圍墙
SURFACE MOUNTED LUMINAIRE
吸頂燈具
PLASTER-RENDER, SPRAYED GRANITE FINISH FENCE WALL
抹灰圍墙 噴石飾面
150X150X600 MM POLISHED BLACK NATURAL CHINESE GRANITE CASCADE WEIR
150×150×600 mm 磨光面 黑色 中国天然花崗岩 瀑布堰
CONCRETE BLOCK CURB, TERMINATION BETWEEN GRANITE CURB TO BE DETERMINED ON PLAN
混凝土磚道牙, 花崗岩道牙的終結, 參照平面图

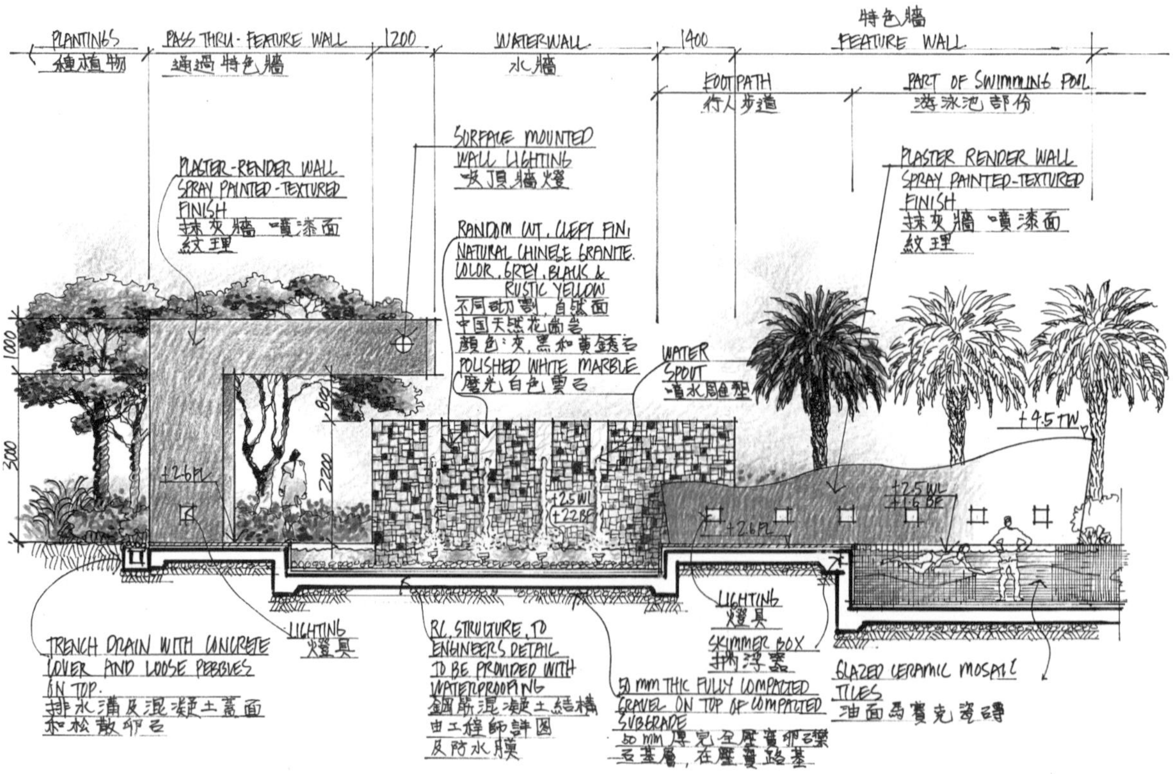
特色墙
FEATURE WALL
PLANTINGS
種植物
PASS THRU-FEATURE WALL
通過特色墙
1200
WATERWALL
水墙
1400
FOOTPATH
行人步道
PART OF SWIMMING POOL
游泳池部份
SURFACE MOUNTED WALL LIGHTING
吸頂墙燈
PLASTER-RENDER WALL SPRAY PAINTED-TEXTURED FINISH
抹灰墙 噴漆面 紋理
RANDOM CUT, CLEFT FIN. NATURAL CHINESE GRANITE. COLOR, GREY, BLACK & RUSTIC YELLOW
不同切割, 自然面 中国天然花崗岩 顏色:灰, 黑和黄銹石
POLISHED WHITE MARBLE
磨光白色雲石
WATER SPOUT
噴水雕塑
PLASTER RENDER WALL SPRAY PAINTED-TEXTURED FINISH
抹灰墙 噴漆面 紋理
1000
800
3000
2200
+2.6FL
+2.5WL
+2.2BF
+2.6FL
+2.5WL
+1.6BF
+4.5TW
TRENCH DRAIN WITH CONCRETE COVER AND LOOSE PEBBLES ON TOP.
排水溝及混凝土蓋面和松散卵石
LIGHTING
燈具
RC. STRUCTURE, TO ENGINEER'S DETAIL TO BE PROVIDED WITH WATERPROOFING
鋼筋混凝土結構由工程師詳图及防水膜
LIGHTING
燈具
SKIMMER BOX
撇浮器
50 MM THIC FULLY COMPACTED GRAVEL ON TOP OF COMPACTED SUBGRADE
50 mm 厚完全壓實卵礫石基層, 在壓實路基
GLAZED CERAMIC MOSAIC TILES
油面馬賽克瓷磚

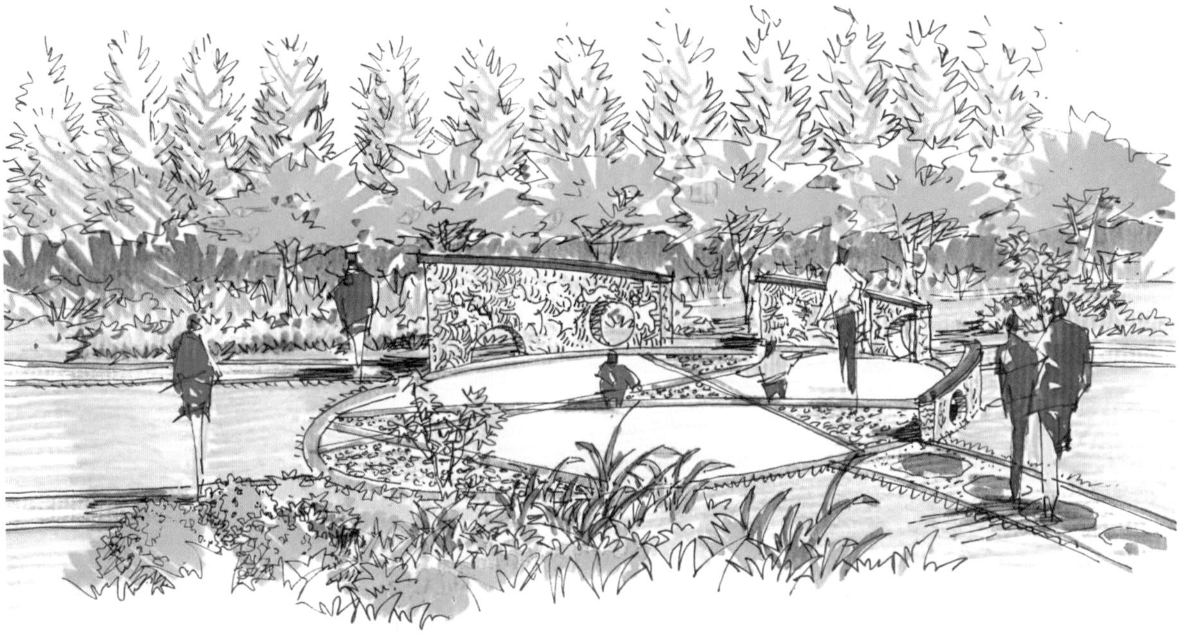

1区块透视图1

1区块透视图2

入口景墙平面

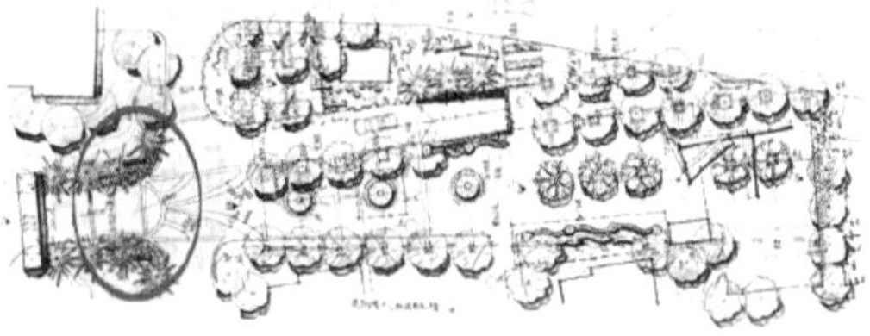

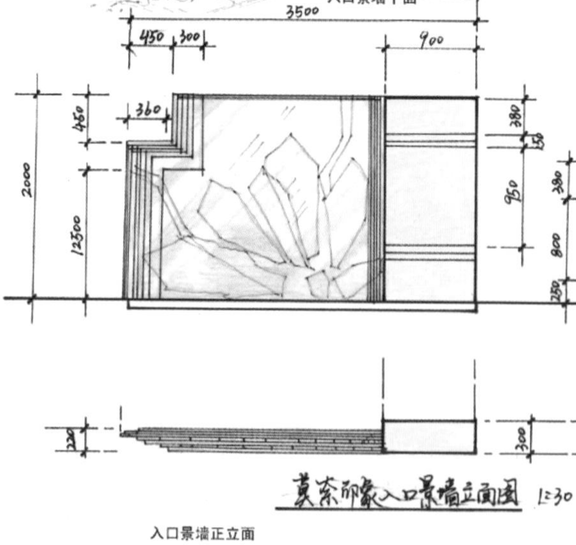

入口景墙正立面

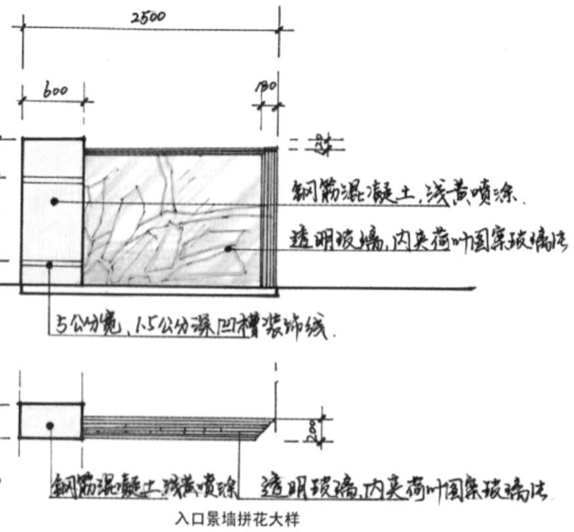

入口景墙拼花大样

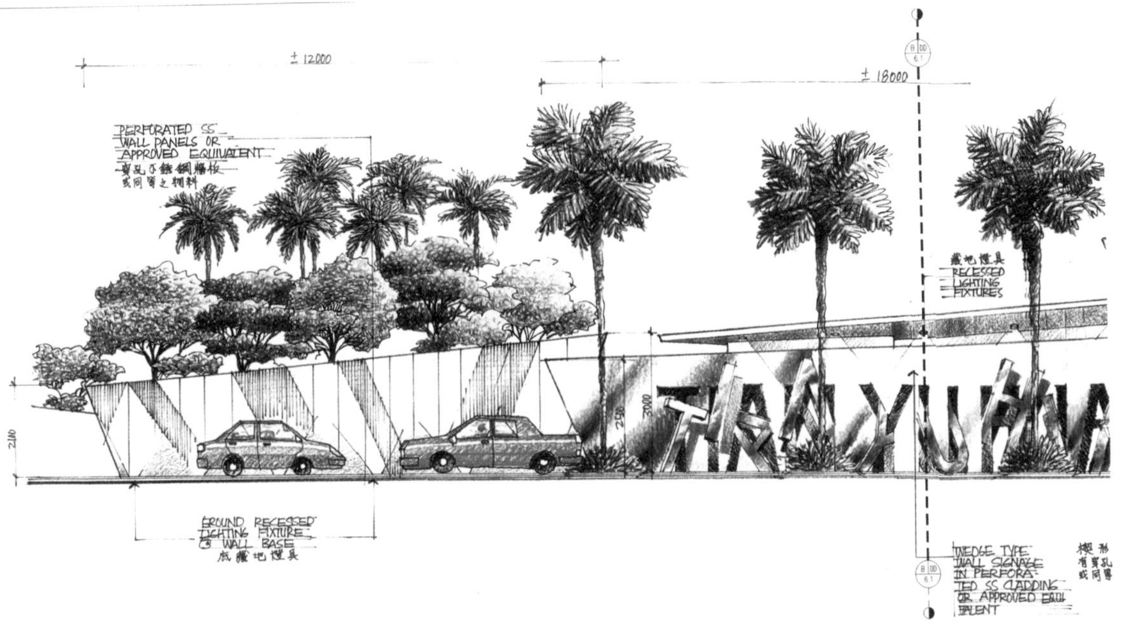
± 12000
± 18000
PERFORATED SS WALL PANELS OR APPROVED EQUIVALENT
RECESSED LIGHTING FIXTURES
GROUND RECESSED LIGHTING FIXTURE @ WALL BASE
WEDGE TYPE WALL SIGNAGE IN PERFORATED SS CLADDING OR APPROVED EQUIVALENT

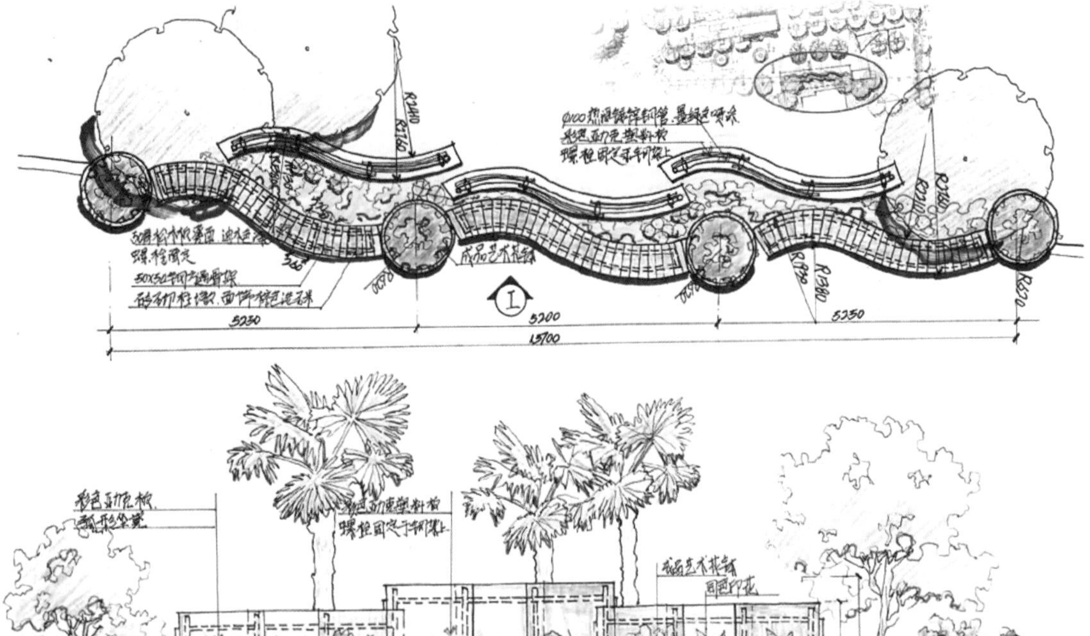
5230
5200
5250
13700

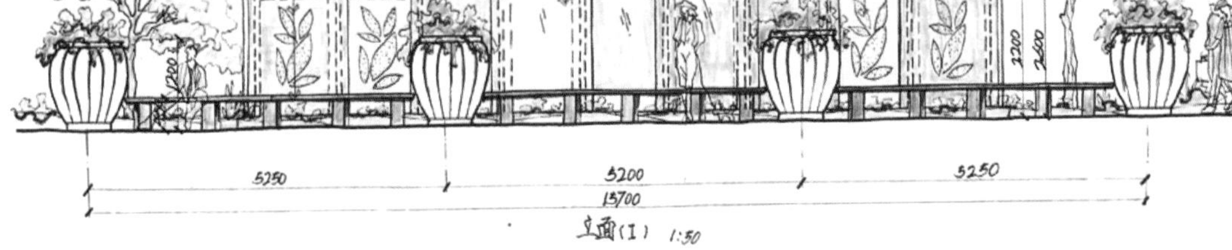
5250
5200
5250
13700

立面(I) 1:50

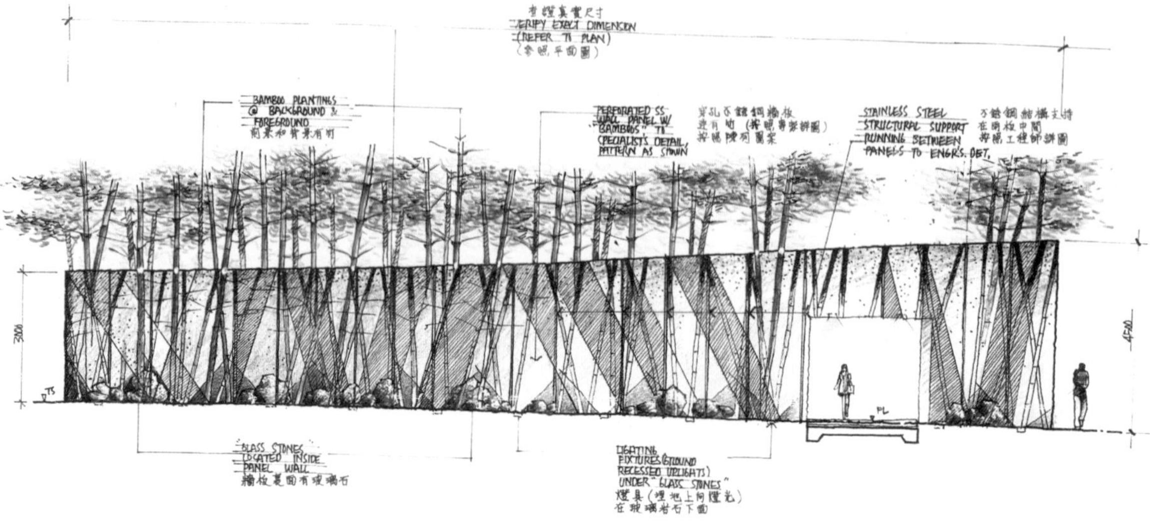
VERIFY EXACT DIMENSION
(REFER TO PLAN)
(參照平面圖)
BAMBOO PLANTINGS @ BACKGROUND & FOREGROUND
前景和背景有竹
PERFORATED SS WALL PANEL W/ "BAMBOOS" TO SPECIALIST'S DETAIL, PATTERN AS SHOWN
STAINLESS STEEL STRUCTURAL SUPPORT RUNNING BETWEEN PANELS TO ENGR'S. DET.
GLASS STONES LOCATED INSIDE PANEL WALL
LIGHTING FIXTURES (GROUND RECESSED UPLIGHTS) UNDER "GLASS STONES"
3000
4500

±12500
3000
RECESSED LIGHTING FIXTURES
GATE WITH CHINESE TEXTS TO SPECIALIST'S DETAIL
閘門連有中文
按照專家詳圖
WEDGE TYPE WALL SIGNAGE IN PERFORATED SS CLADDING OR APPROVED EQUIVALENT
楔形牆壁標誌，
有穿孔不銹鋼牆身飾面
或同等之物料
RC WALL IN PLASTER CEMENT FIN, TEXTURED, COLOR: EARTH RED
水泥抹灰面的
鋼筋混凝土牆
紋理面 顏色：礦材紅

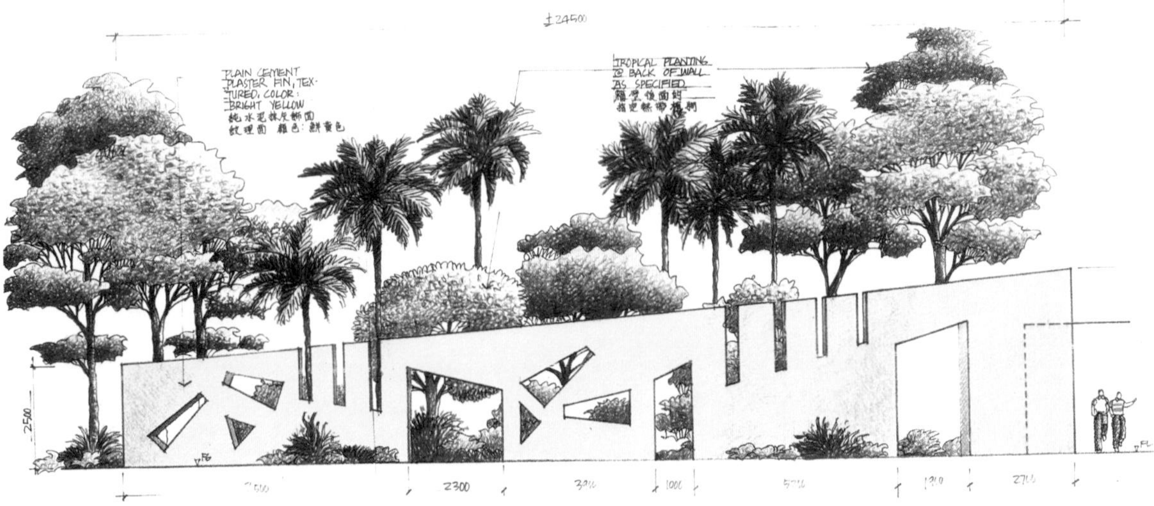

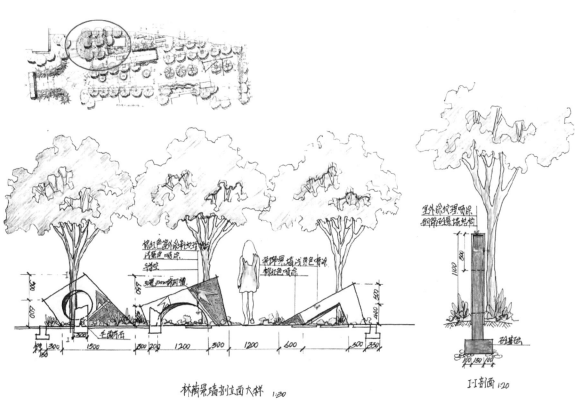

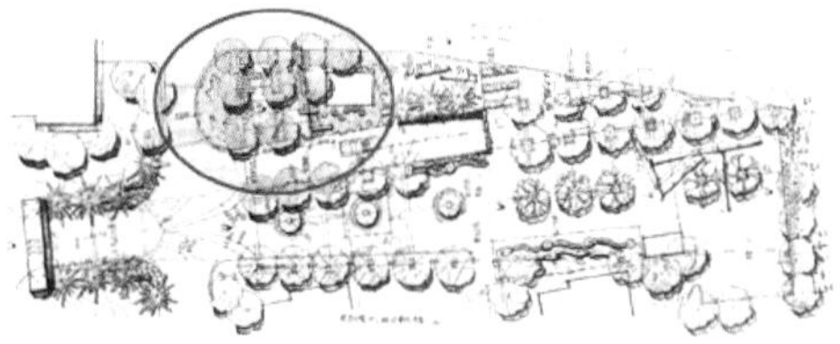

林荫景墙剖立面大样 1:30

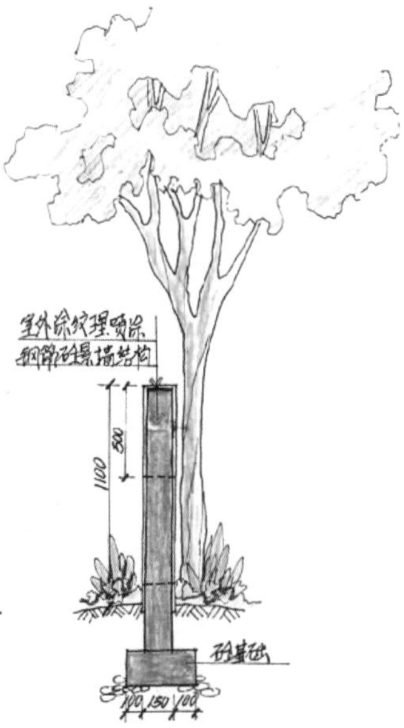

I-I剖面 1:20

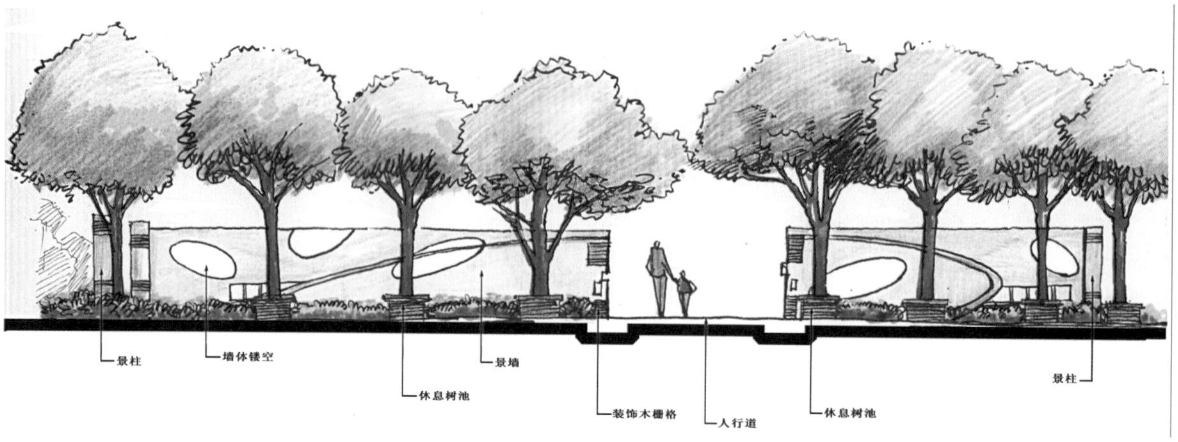

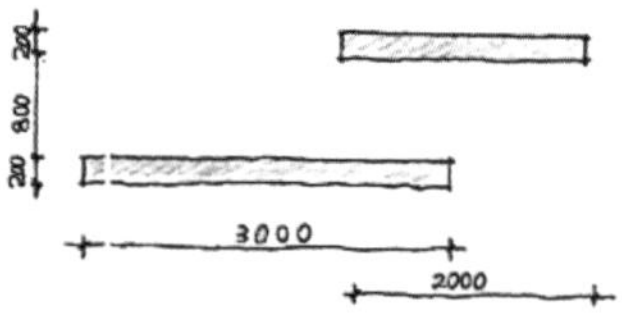

装饰景墙平面

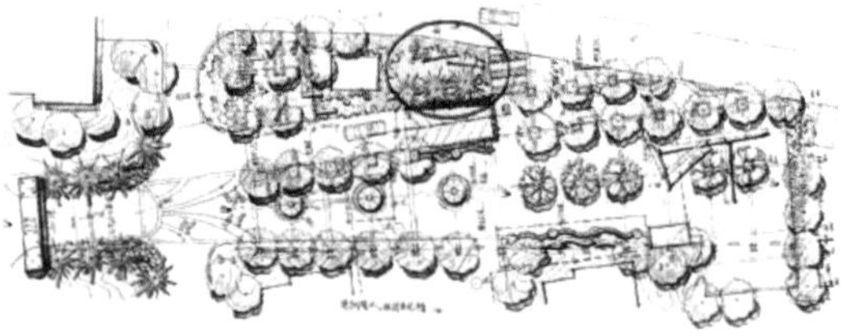

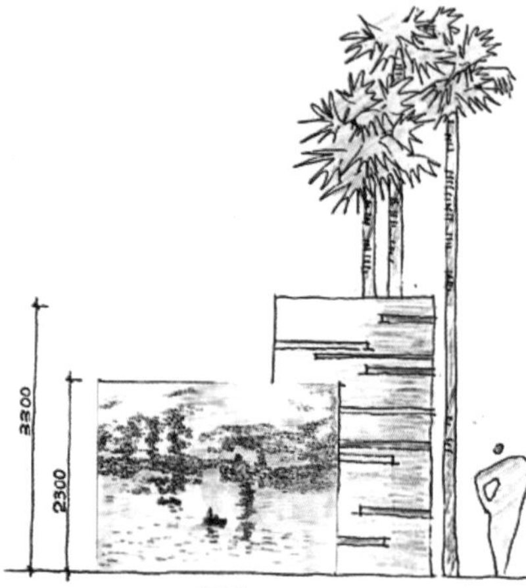

比例 1:500

装饰景墙正立面

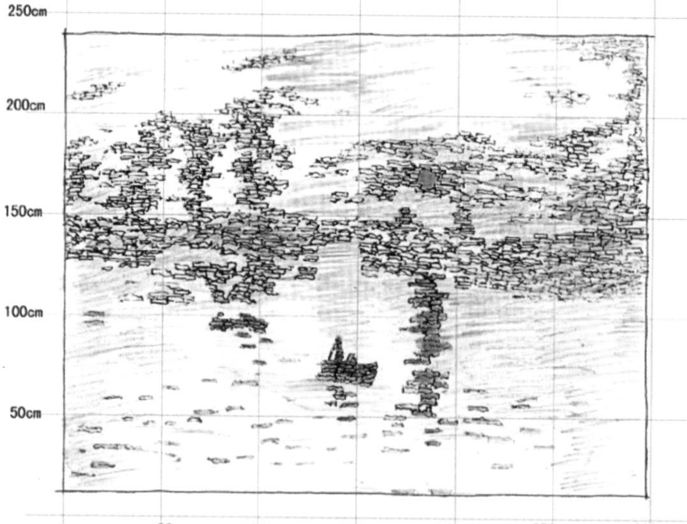

装饰景墙拼花大样

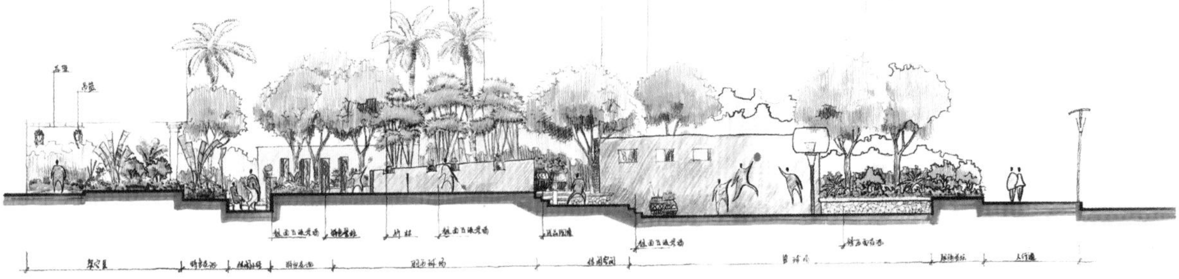

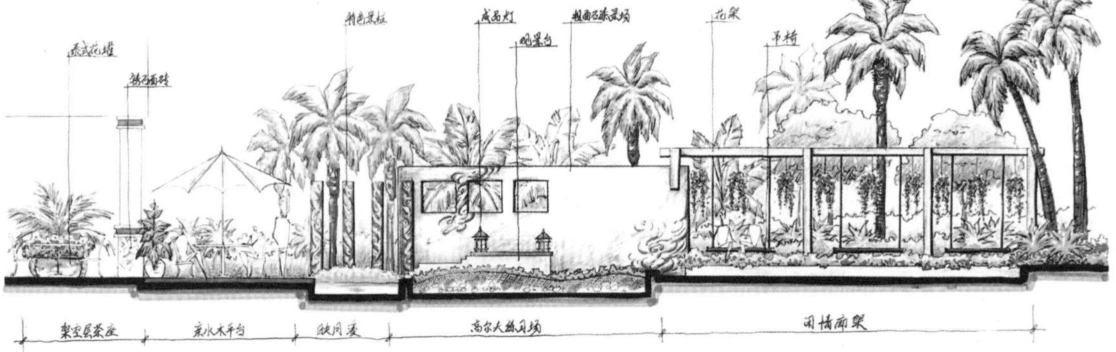

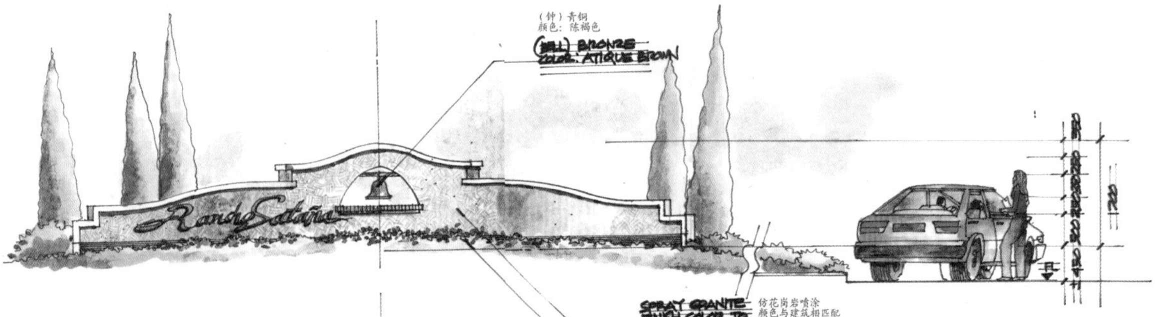

B ELEVATION SCALE 1:40M

立面图

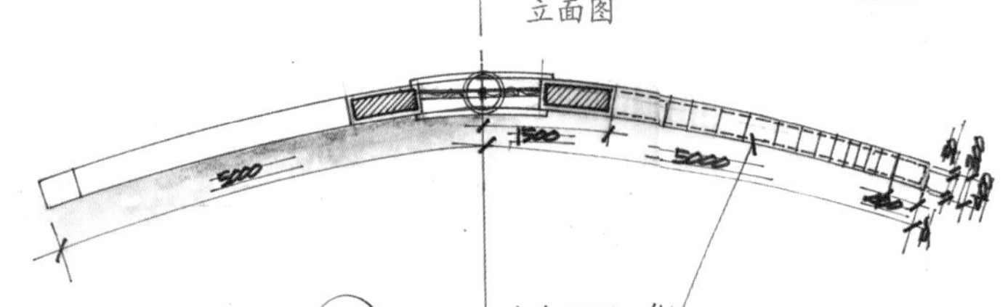

A PLAN SCALE 1:40M

平面图

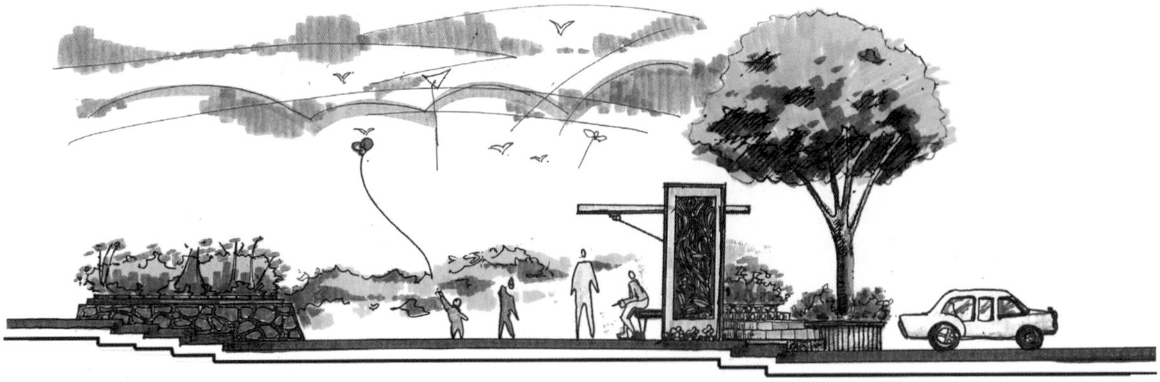

林荫道
特色弧形景墙
草坪

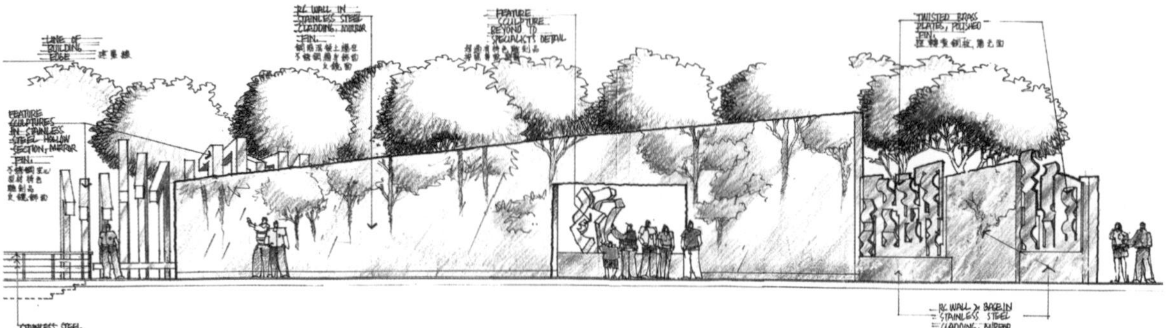
LINE OF BUILDING EDGE
FEATURE SCULPTURES IN STAINLESS STEEL HOLLOW SECTION, MIRROR FIN.
RC WALL IN STAINLESS STEEL CLADDING, MIRROR FIN.
FEATURE SCULPTURE BEYOND TO SPECIALIST'S DETAIL
TWISTED BRASS PLATES, POLISHED FIN.
STAINLESS STEEL RAILING W/ SS GUY WIRES
RC WALL W/ BAGE IN STAINLESS STEEL CLADDING, MIRROR FIN.

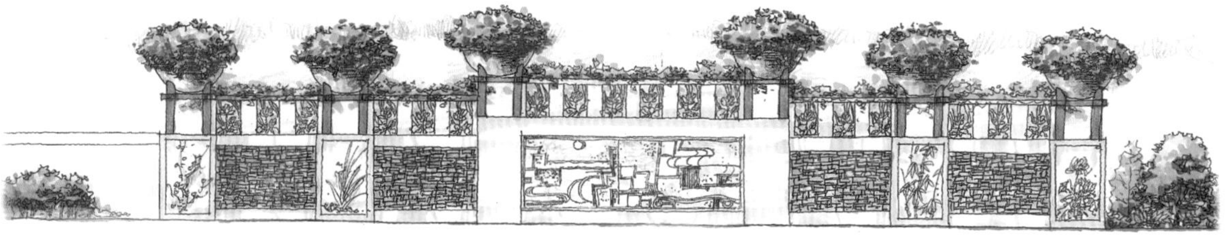

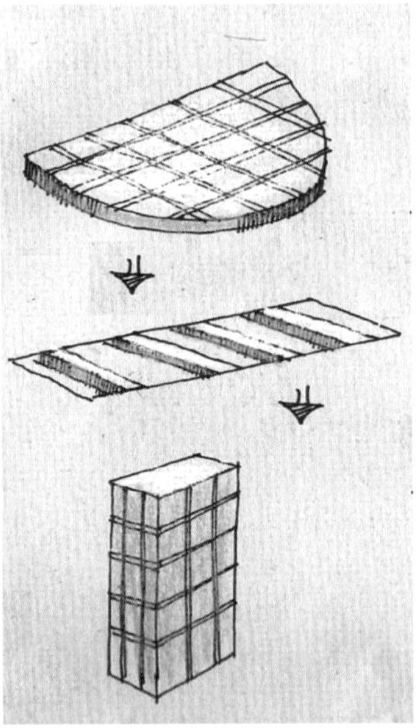

竹子景墙立面示意

玻璃金属构筑物示意

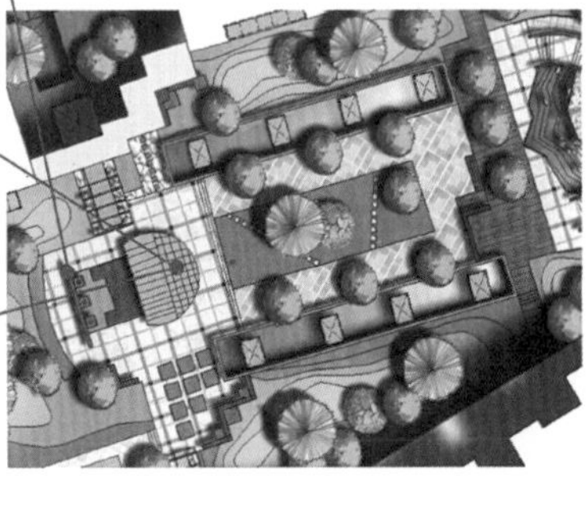

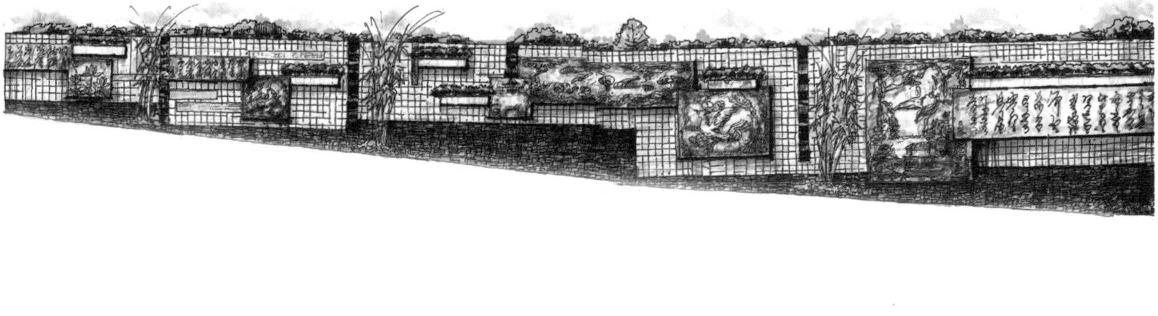

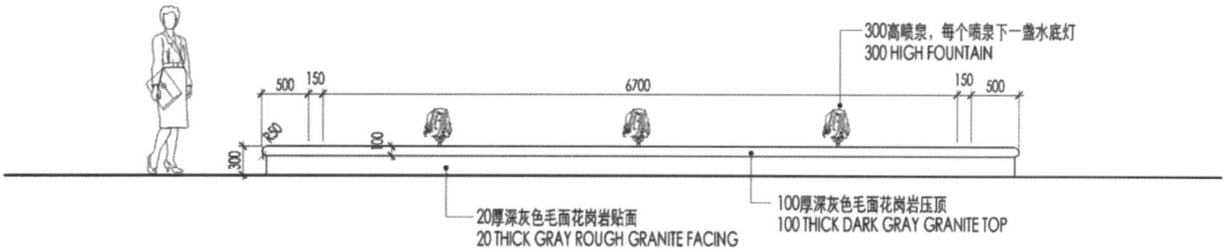

ZANDT.

300高喷泉，每个喷泉下一盏水底灯
300 HIGH FOUNTAIN
500
150
6700
150
500
250
100
300
20厚深灰色毛面花岗岩贴面
20 THICK GRAY ROUGH GRANITE FACING
100厚深灰色毛面花岗岩压顶
100 THICK DARK GRAY GRANITE TOP

1
水 景 4 景 墙 立 面 图
ELEVATION OF WATER FEATURE3
SCALE BAR:
0
500
2500

100厚深灰色毛面花岗岩压顶
100 THICK DARK GRAY GRANITE TOP
300高喷泉，每个喷泉下一盏水底灯
300 HIGH FOUNTAIN
500
150
6700
150
500
300
500
□30-50黑色鹅卵石排水沟
□30-50 BLACK PEBBLE
300x300x20深灰色毛面花岗岩贴池底池壁
300x300x20 DARK GRAY GRANITE

2
水 景 4 景 墙 剖 面 图
ELEVATION OF WATER FEATURE4
SCALE BAR:
0
500
2500

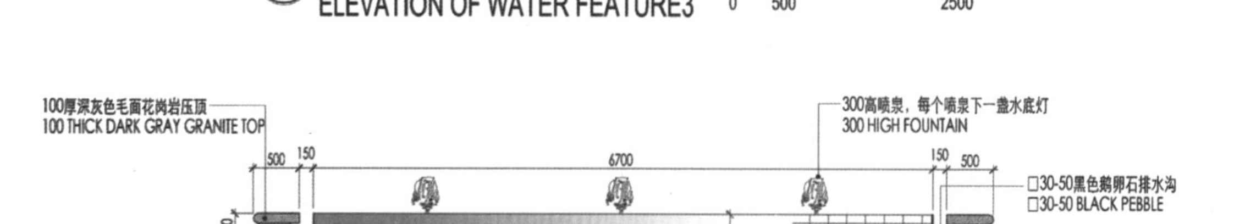

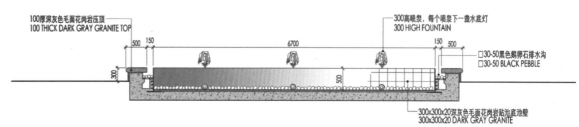

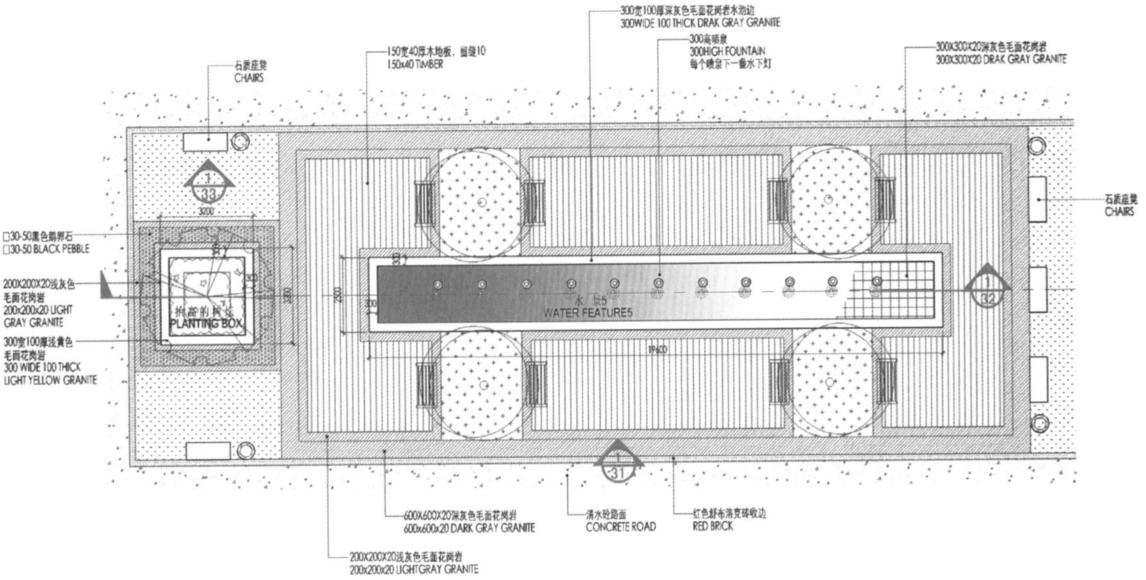

300宽100厚深灰色毛面花岗岩水池边
300WIDE 100 THICK DRAK GRAY GRANITE
300高喷泉
300HIGH FOUNTAIN
每个喷泉下一盏水下灯
300X300X20深灰色毛面花岗岩
300X300X20 DRAK GRAY GRANITE
150宽40厚木地板，留缝10
150x40 TIMBER
石质座凳
CHAIRS
30-50黑色鹅卵石
30-50 BLACK PEBBLE
200X200X20浅灰色毛面花岗岩
200x200x20 LIGHT GRAY GRANITE
300宽100厚浅黄色毛面花岗岩
300 WIDE 100 THICK LIGHT YELLOW GRANITE
PLANTING BOX
WATER FEATURES5
石质座凳
CHAIRS
600X600X20深灰色毛面花岗岩
600x600x20 DARK GRAY GRANITE
清水砼路面
CONCRETE ROAD
红色砂布清克砖收边
RED BRICK
200X200X20浅灰色毛面花岗岩
200x200x20 LIGHTGRAY GRANITE

1 水 景5 区 景 墙 平 面 图
FEATURE WALL PLAN OF WATER FEATURE5
SCALE BAR:
0 1M 5M

TYP. SHED/TRELLIS DETAIL
大样图
LANTERN
特色灯
SEATWALL
座墙
3 STEPS UP
RAILING
栏杆
50x300MM TREATED TIMBER PLANKS
50X300已防腐处理木地板
PART PLAN: ENJOYING TO FISH
钓鱼台平面图
1:150
砂岩壁画
SANDSTONE MURAL
壁画墙(描述水下生物)
MURAL WALL (DEPICTING UNDERWATER)
CARVED GRANITE POTS
带雕刻花岗岩花盆
100X200MM TREATED TIMBER RAILING
100X200木栏杆
EXISTING FINISH
现有的完成面
ELEVATION
立面图
1:75
TYPICAL SHED/TRELLIS DETAIL
典型凉棚大样图
WILLOW TREES
柳树
带雕刻花岗岩花盆
GRANITE CURB
花岗岩路缘
特色灯
LANTERN
TREATED WOOD
已处理木板
SANDSTONE MURAL
砂岩壁画
壁画墙(描述水下生物)
SECTION
剖面图
1:50

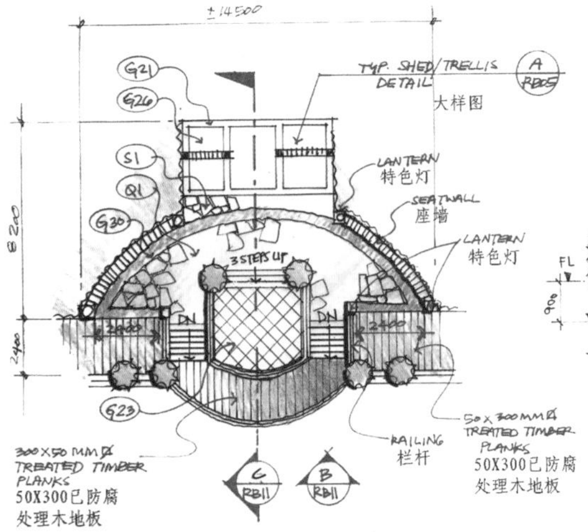

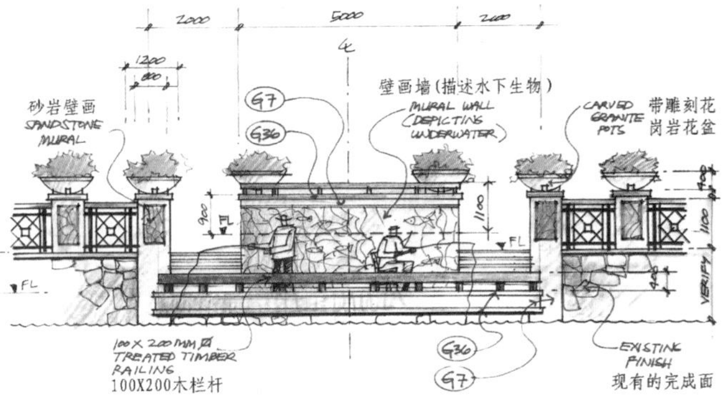

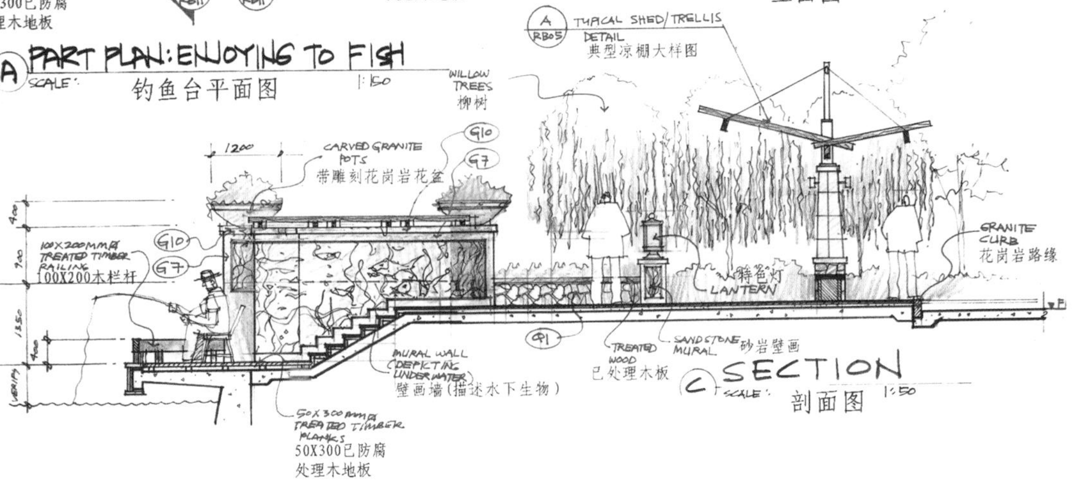

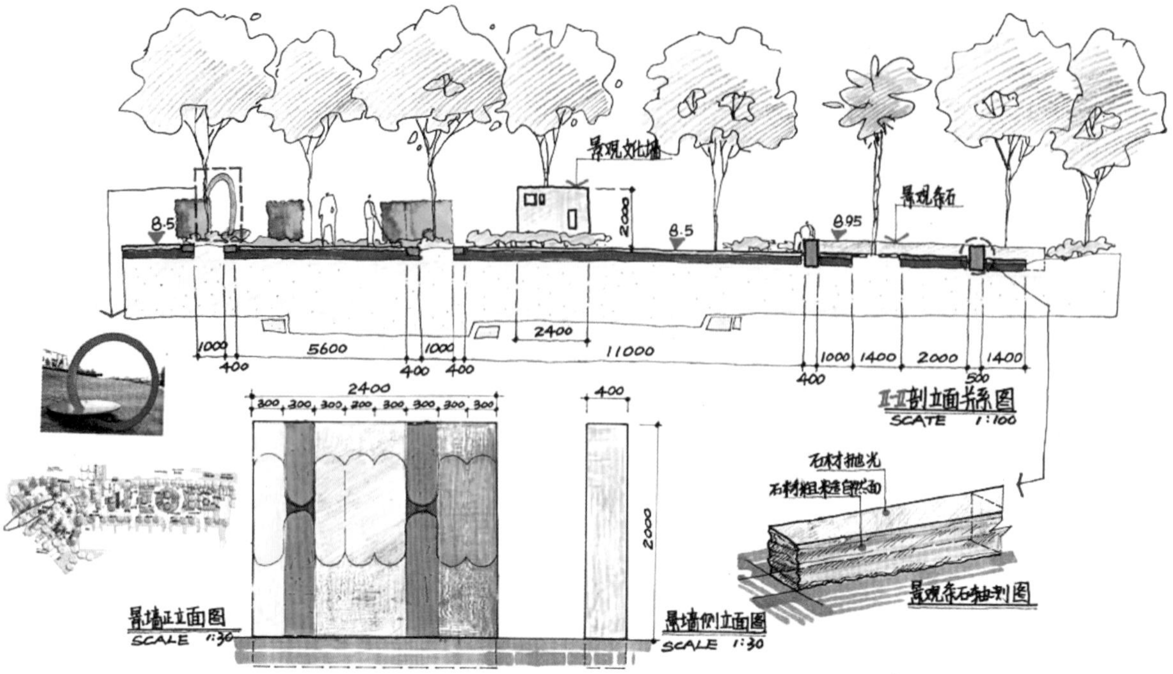
景观文化墙
景观条石
8.5
8.95
2000
1000
5600
400
2400
11000
1400
500
I-I剖立面关系图
石材抛光
景观条石轴测图
300
400
景墙正立面图
SCALE 1:30
景墙侧立面图
SCALE 1:30

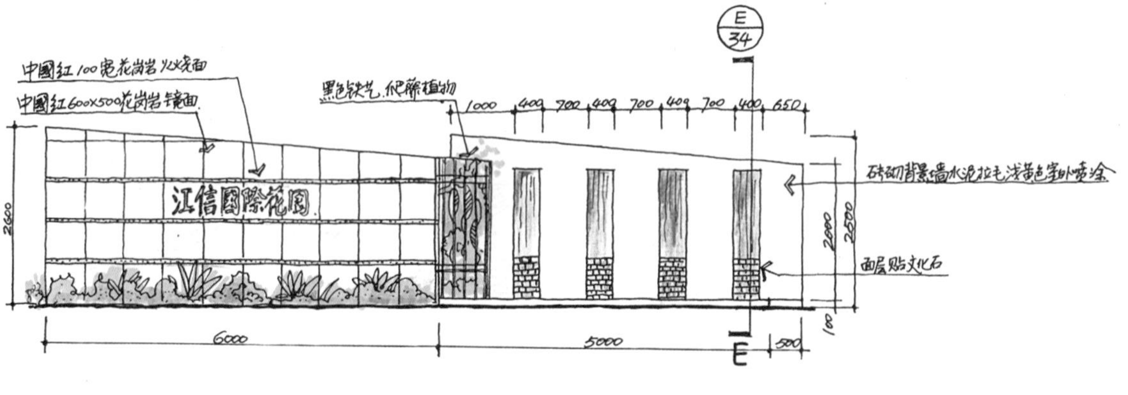
中國红600×500花岗岩镜面
黑色铁艺,爬藤植物
江信国际花园
面层贴文化石
E
34
1000
400
700
650
2600
2000
2500
6000
5000
500

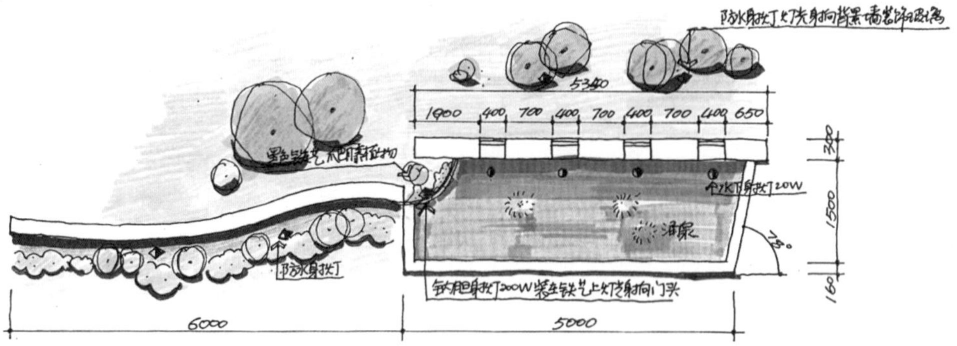
5340
1000
400
700
650
300
1500
160
水下射灯20W
喷泉
防水射灯
6000
5000

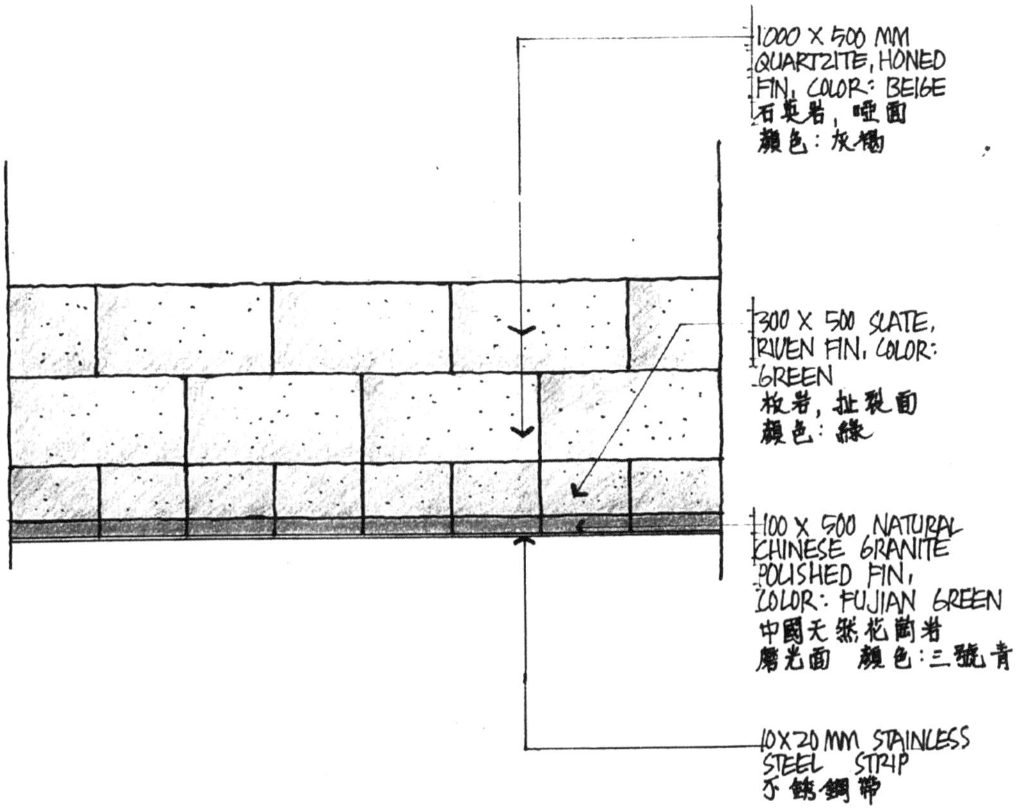
1000 X 500 MM QUARTZITE, HONED FIN, COLOR: BEIGE
石英岩，哑面
颜色：灰褐
300 X 500 SLATE, RIVEN FIN, COLOR: GREEN
板岩，扯裂面
颜色：绿
100 X 500 NATURAL CHINESE GRANITE POLISHED FIN, COLOR: FUJIAN GREEN
中國天然花崗岩
磨光面 顏色：三號青
10X20 MM STAINLESS STEEL STRIP
不銹鋼帶

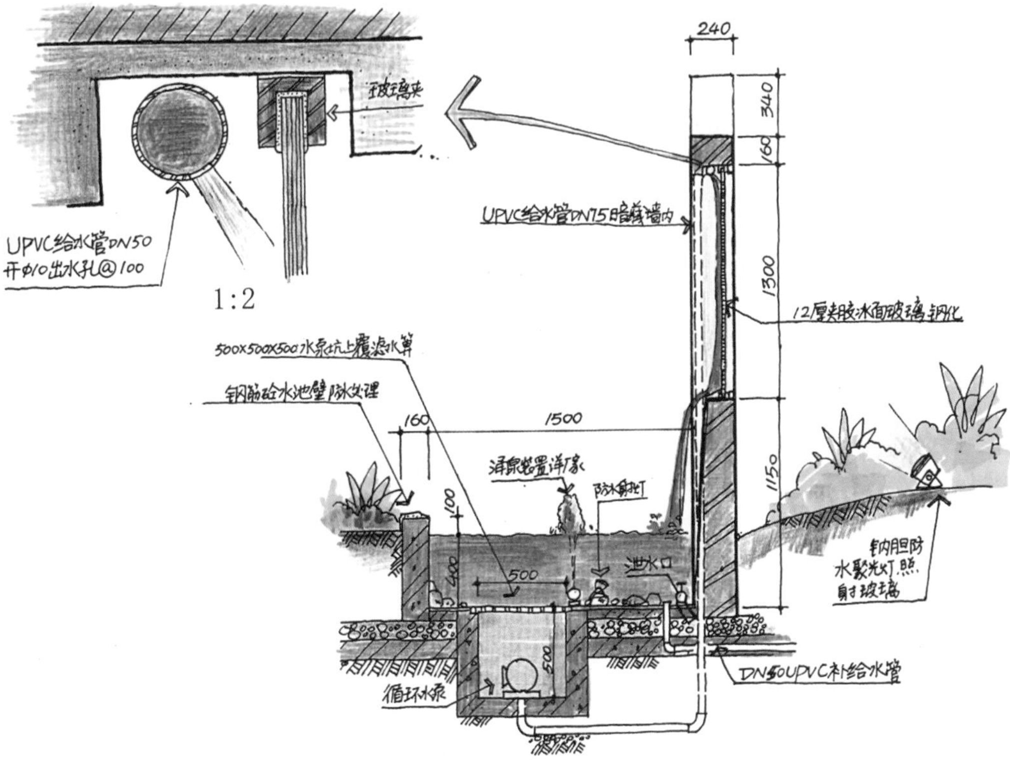
玻璃夹
UPVC给水管DN50
开φ10出水孔@100
1:2
240
340
160
1300
1150
UPVC给水管DN75暗藏墙内
12厚夹胶冰面玻璃 钢化
500X500X500水泵坑上覆滤水箅
钢筋砼水池壁防水处理
160
1500
涌泉装置详图
防水射灯
泄水口
100
400
500
500
循环水泵
DN50UPVC补给水管
钢胆防水聚光灯照射玻璃

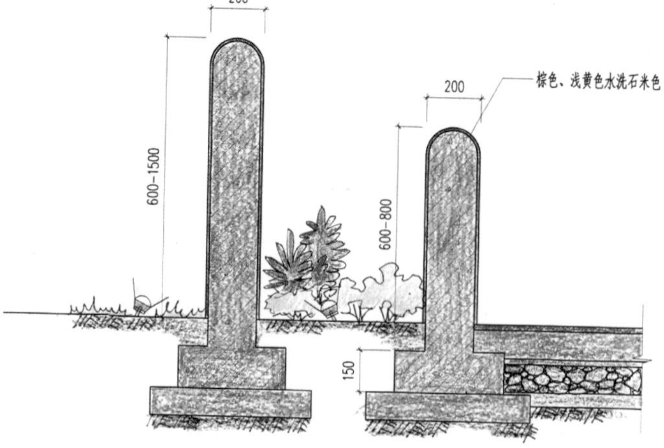
200
200
棕色、浅黄色水洗石米色
600-1500
600-800
150

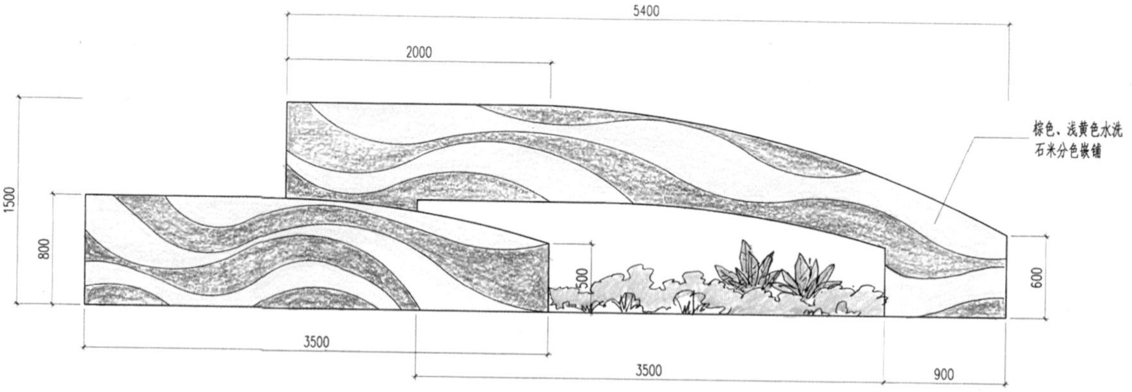
5400
2000
1500
800
500
600
棕色、浅黄色水洗
石米分色嵌铺
3500
3500
900

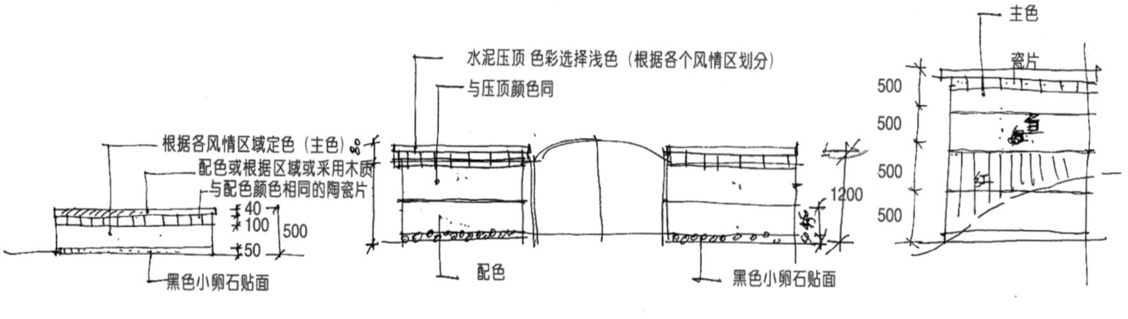
根据各风情区域定色（主色）
配色或根据区域或采用木质
与配色颜色相同的陶瓷片
40
100
500
50
黑色小卵石贴面
水泥压顶 色彩选择浅色（根据各个风情区划分）
与压顶颜色同
80
1200
配色
黑色小卵石贴面
主色
瓷片
500
500
500
500

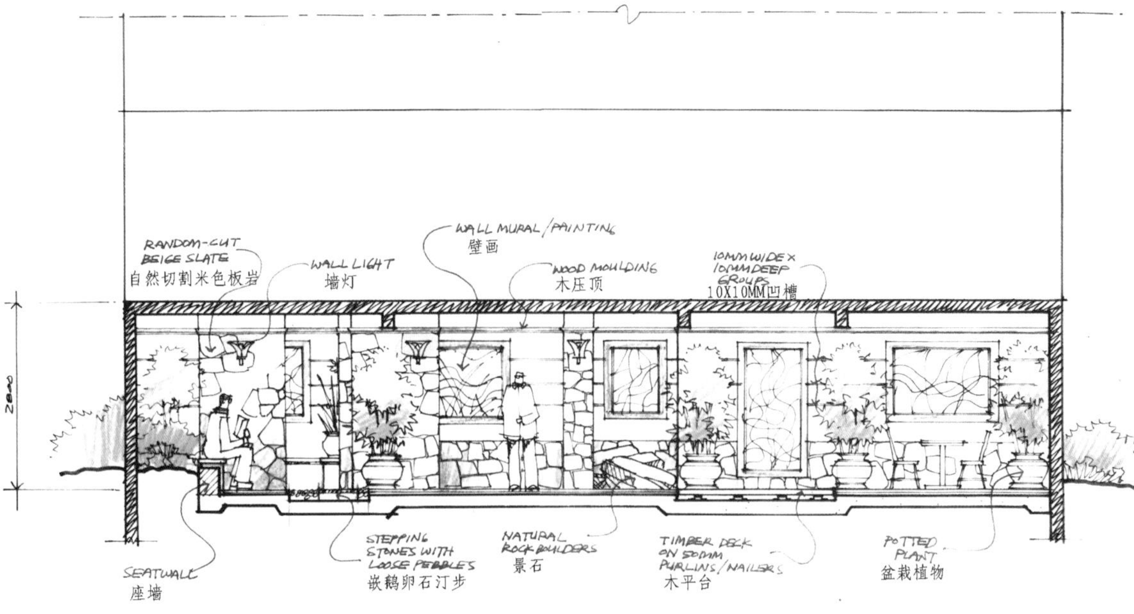

A SECTION THRU' A
SCALE: 1:50
剖面图A

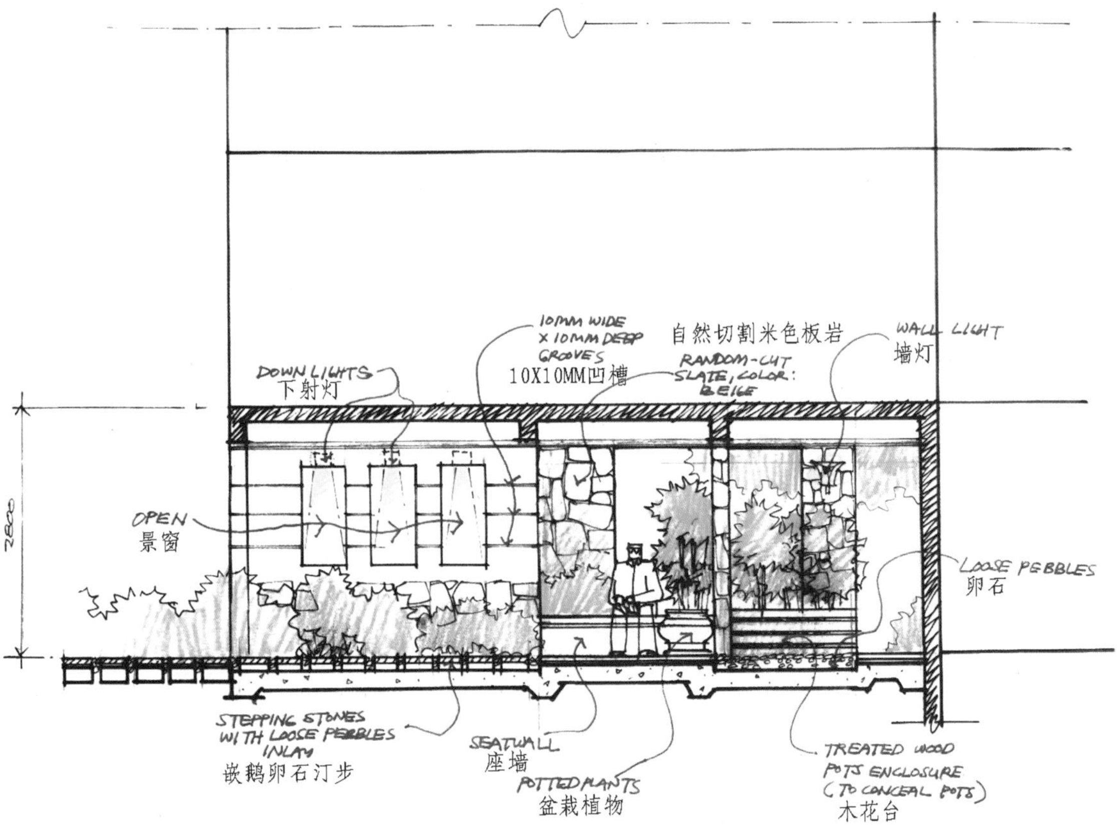

B SECTION THRU' B
SCALE: 1:50
剖面图B

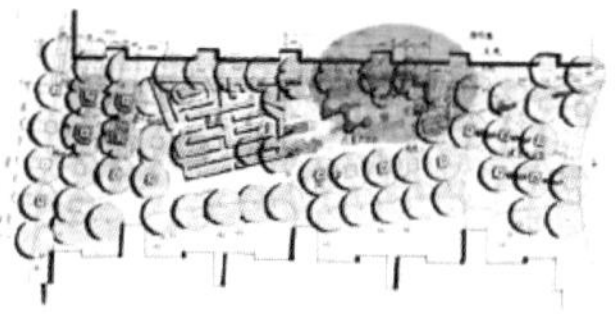

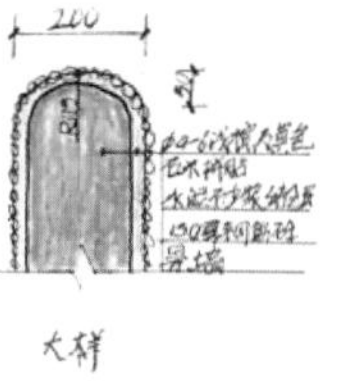

大样

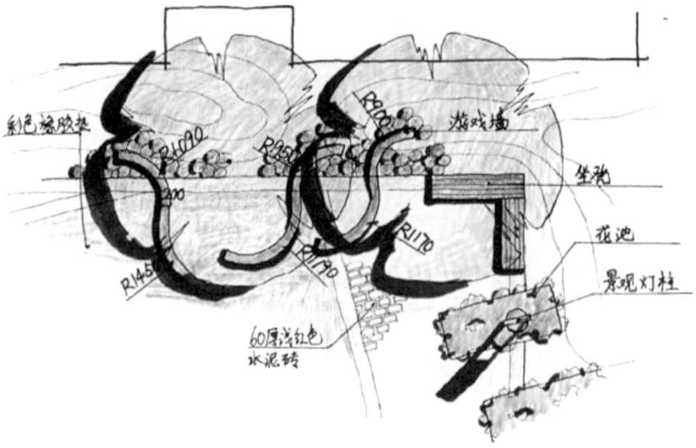

儿童游戏墙平面大样 1:75

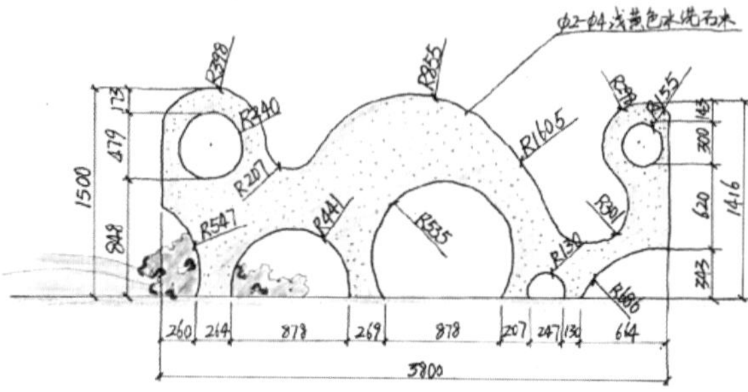

儿童游戏墙立面大样 1:30

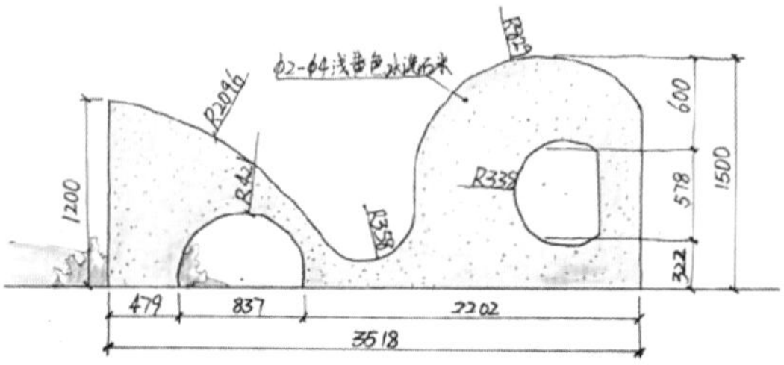

儿童游戏墙立面大样 1:30

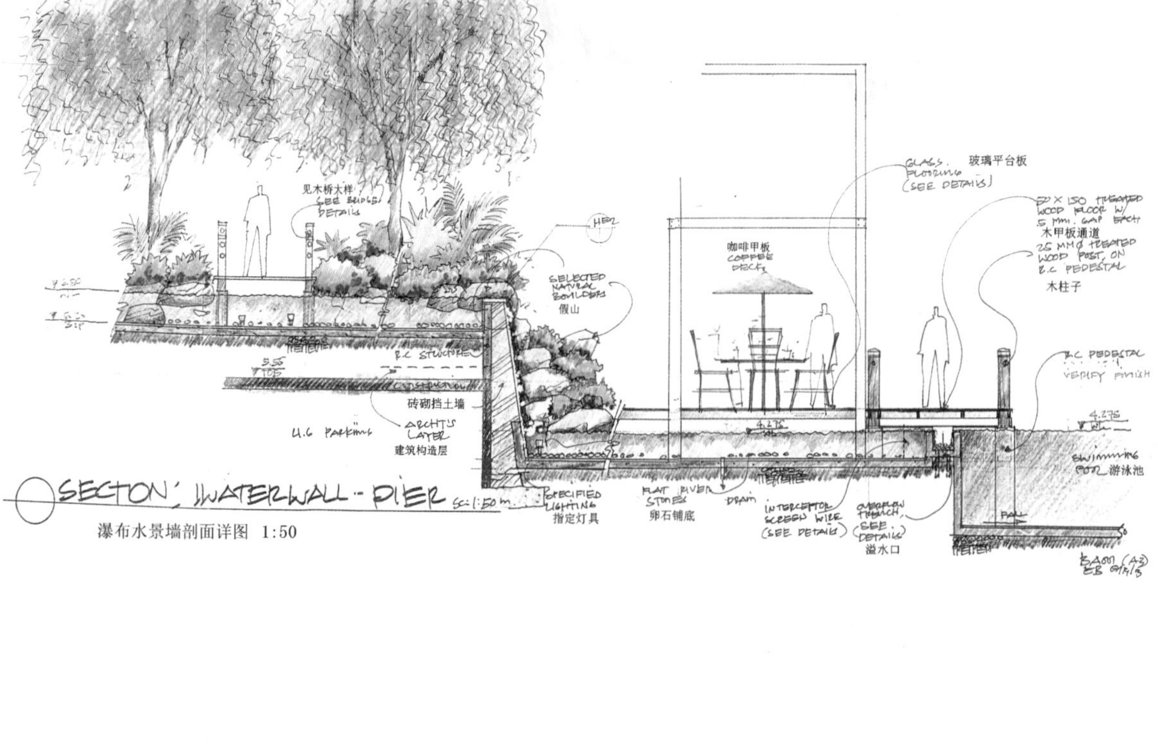

瀑布水景墙剖面详图 1:50

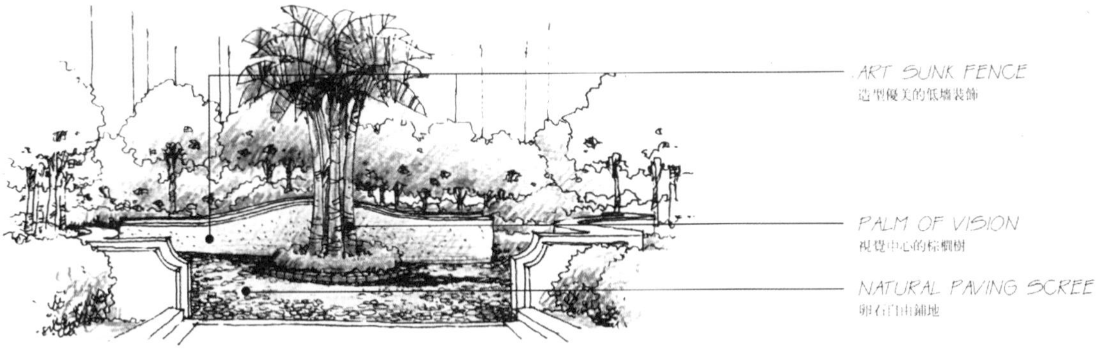
ART SUNK FENCE
造型優美的低牆裝飾
PALM OF VISION
視覺中心的棕櫚樹
NATURAL PAVING SCREE
卵石自由鋪地

FOUNDATION OF BORDER
花盆基座，一側爲擋土牆
GRC FLOWER POT
GRC花盆
STEP TRANSITION FROM
STREET TO GARDEN
由道路過渡到花園的臺階
WALKWAY TILE WITH COLOR

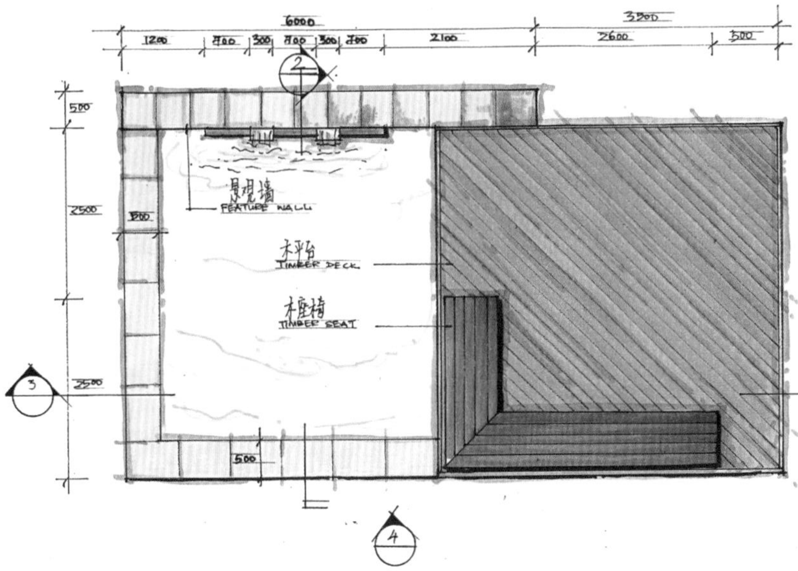
6000
3500
1200
2100
2600
500
景观墙
FEATURE WALL
木平台
TIMBER DECK
木座椅
TIMBER SEAT

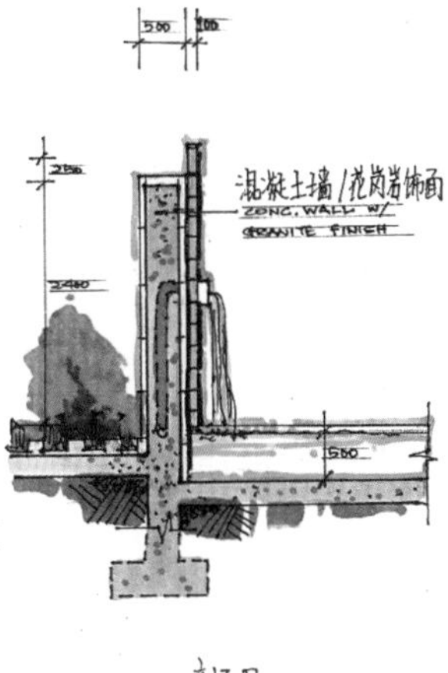
混凝土墙/花岗岩饰面
CONC. WALL W/
GRANITE FINISH
500

剖面图
2 SECTION
SCALE 1:50

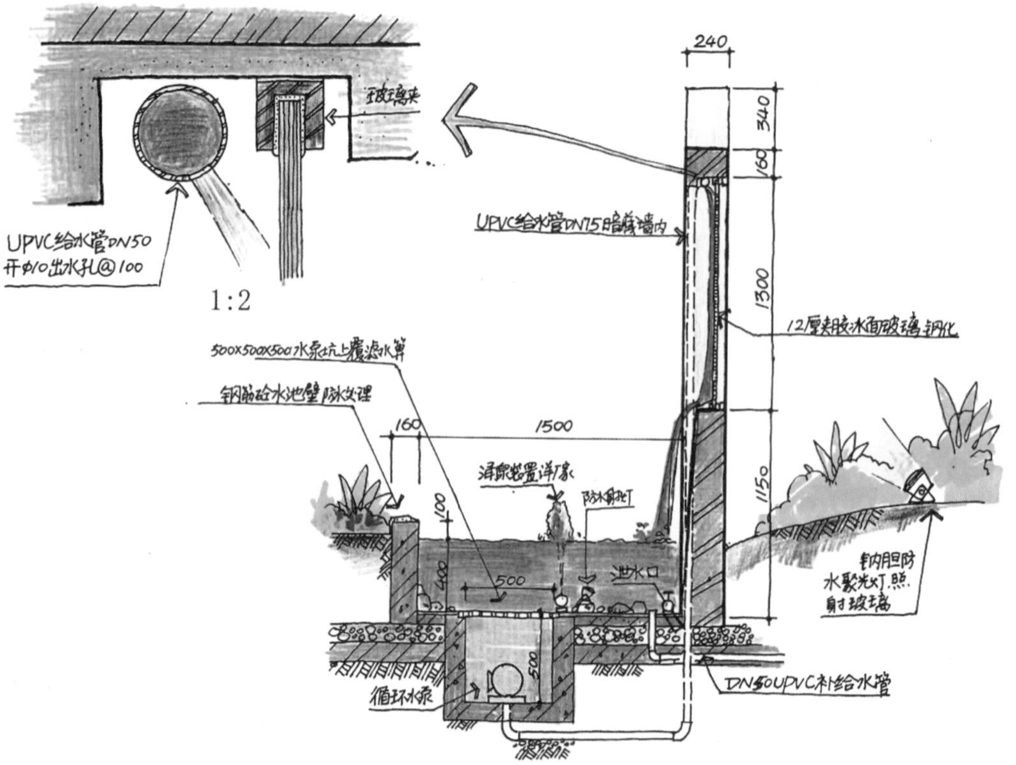

E 背景墙剖面大样　1:20

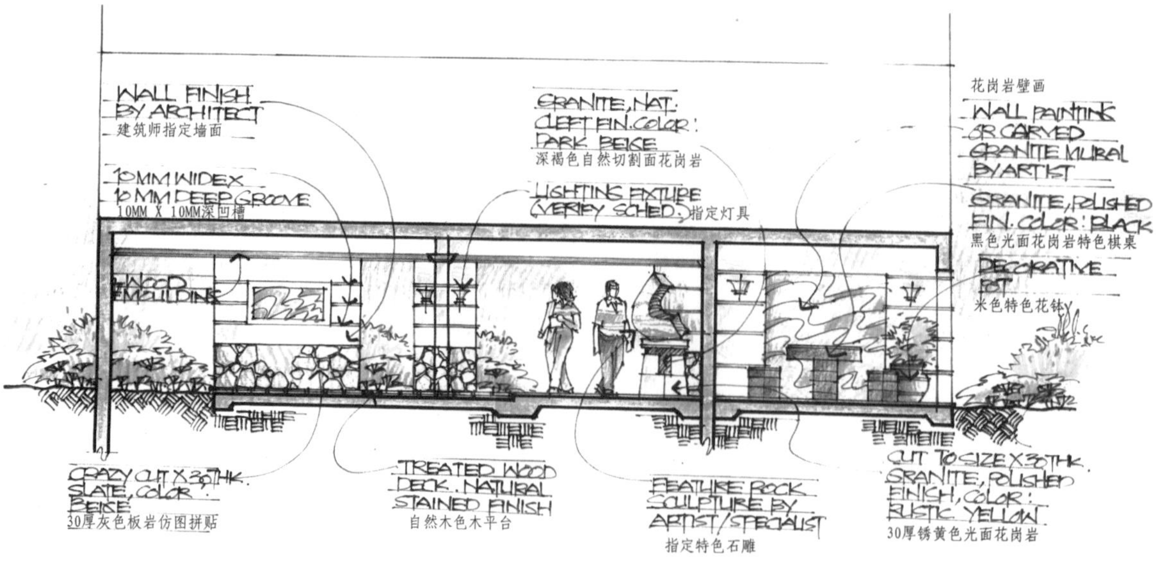

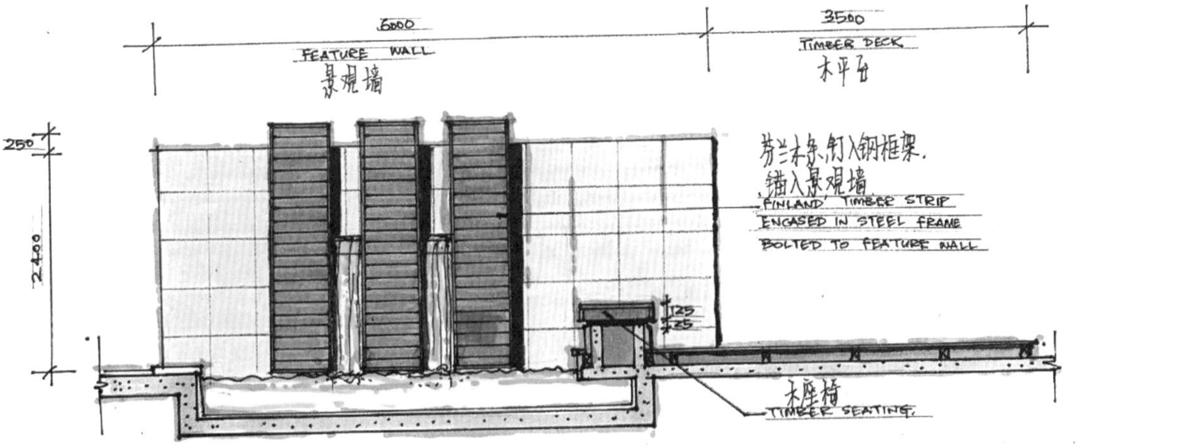

6000
3500
FEATURE WALL
景观墙
TIMBER DECK
木平台
250
2400
芬兰木条钉入钢框架，
锚入景观墙
FINLAND TIMBER STRIP
ENCASED IN STEEL FRAME
BOLTED TO FEATURE WALL
125
25
木座椅
TIMBER SEATING

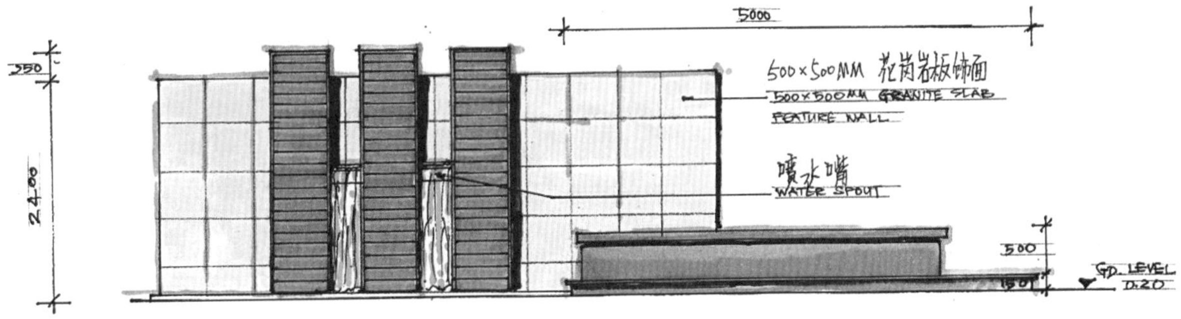

5000
250
2400
500×500MM 花岗岩板饰面
500×500MM GRANITE SLAB
FEATURE WALL
喷水嘴
WATER SPOUT
500
150
GD. LEVEL
0.20

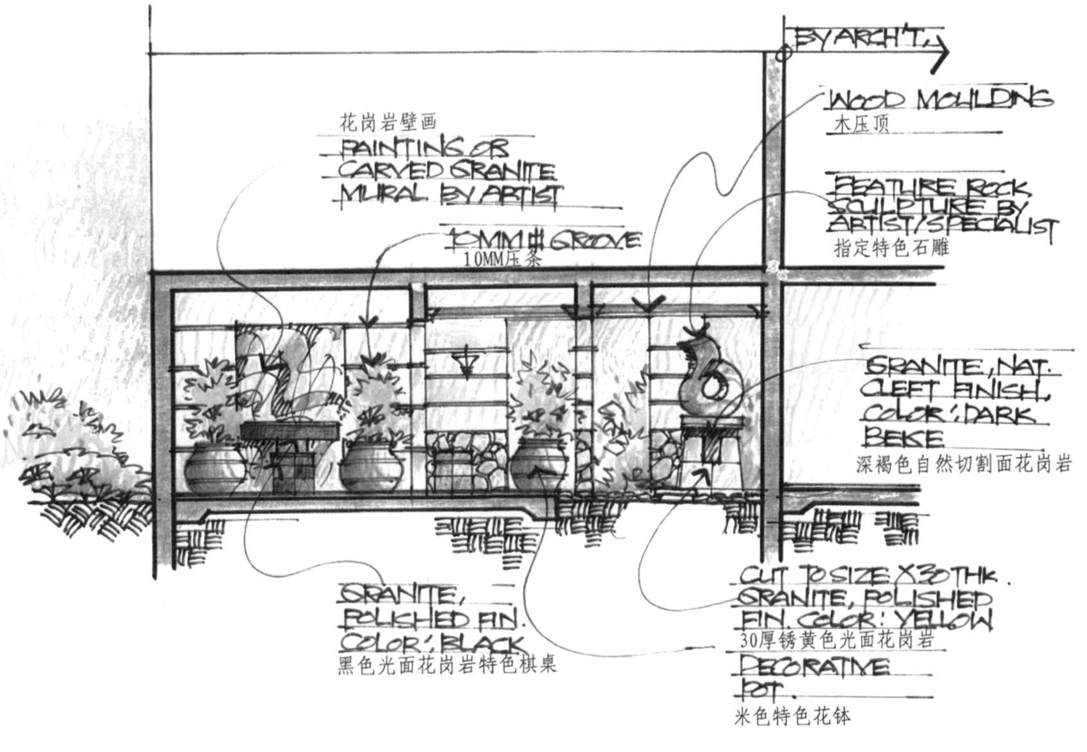

BY ARCH'T.
WOOD MOULDING
木压顶
花岗岩壁画
PAINTING OR
CARVED GRANITE
MURAL BY ARTIST
FEATURE ROCK
SCULPTURE BY
ARTIST/SPECIALIST
指定特色石雕
10MM # GROOVE
10MM压条
GRANITE, NAT.
CLEFT FINISH,
COLOR: DARK
BEIGE
深褐色自然切割面花岗岩
GRANITE,
POLISHED FIN.
COLOR: BLACK
黑色光面花岗岩特色棋桌
CUT TO SIZE X30 THK.
GRANITE, POLISHED
FIN. COLOR: YELLOW
30厚锈黄色光面花岗岩
DECORATIVE
POT.
米色特色花钵

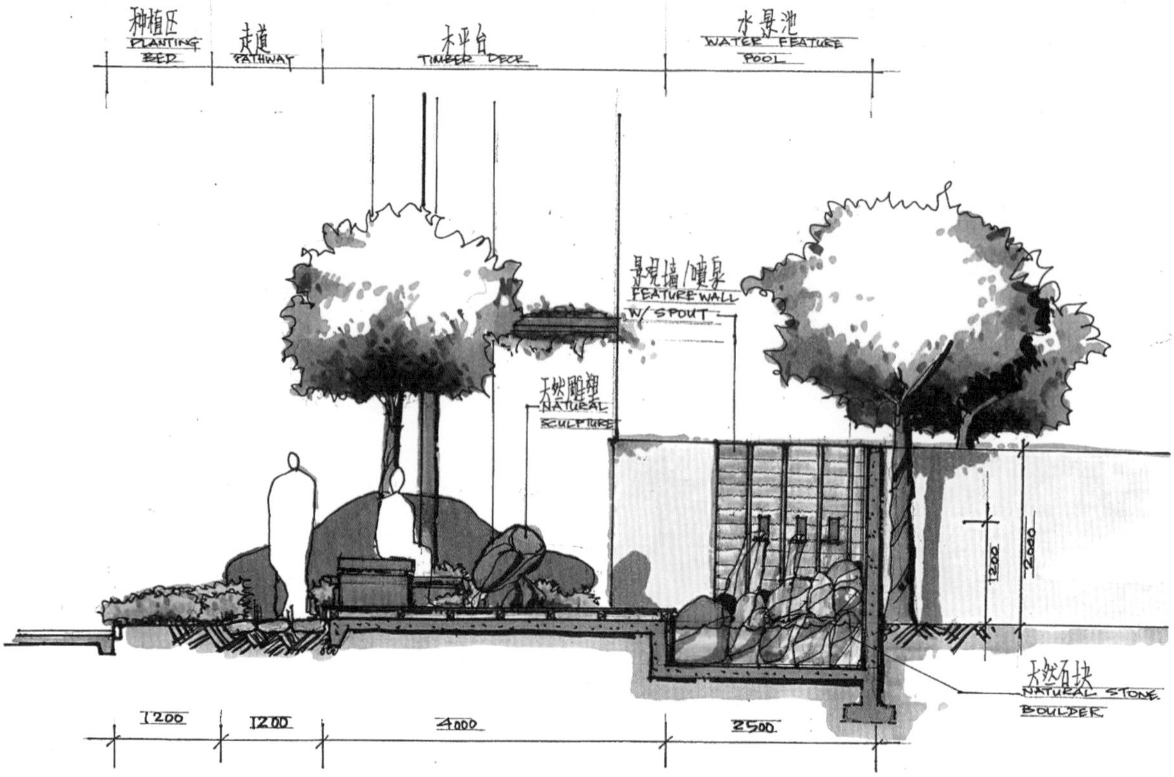
种植区
PLANTING BED
走道
PATHWAY
木平台
TIMBER DECK
水景池
WATER FEATURE POOL
景观墙/喷泉
FEATURE WALL W/ SPOUT
天然雕塑
NATURAL SCULPTURE
天然石块
NATURAL STONE BOULDER
1200
2000
1200
1200
4000
2500

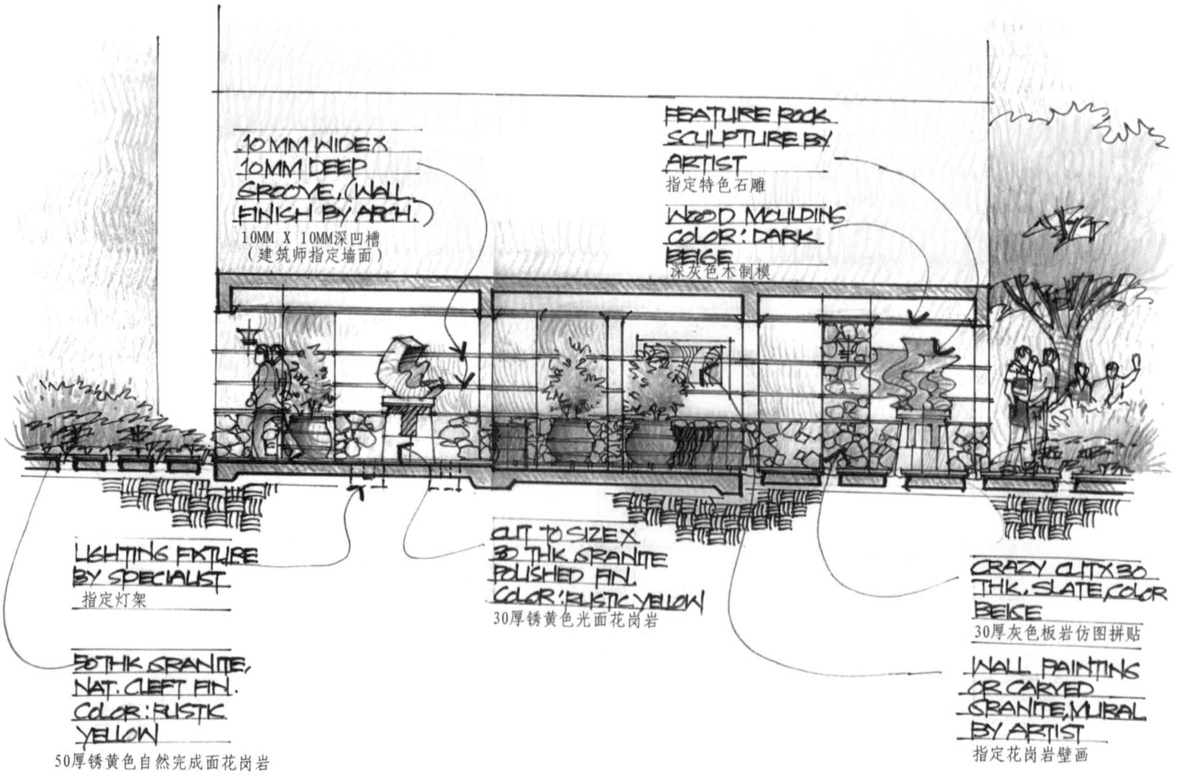
10MM WIDE X 10MM DEEP GROOVE, (WALL FINISH BY ARCH.)
10MM X 10MM深凹槽
（建筑师指定墙面）
FEATURE ROCK SCULPTURE BY ARTIST
指定特色石雕
WOOD MOULDING COLOR: DARK BEIGE
深灰色木制模
LIGHTING FIXTURE BY SPECIALIST
指定灯架
CUT TO SIZE X 30 THK GRANITE POLISHED FIN. COLOR: RUSTIC YELLOW
30厚锈黄色光面花岗岩
CRAZY CUT X 30 THK. SLATE, COLOR BEIGE
30厚灰色板岩仿图拼贴
50 THK GRANITE, NAT. CLEFT FIN. COLOR: RUSTIC YELLOW
50厚锈黄色自然完成面花岗岩
WALL PAINTING OR CARVED GRANITE MURAL BY ARTIST
指定花岗岩壁画

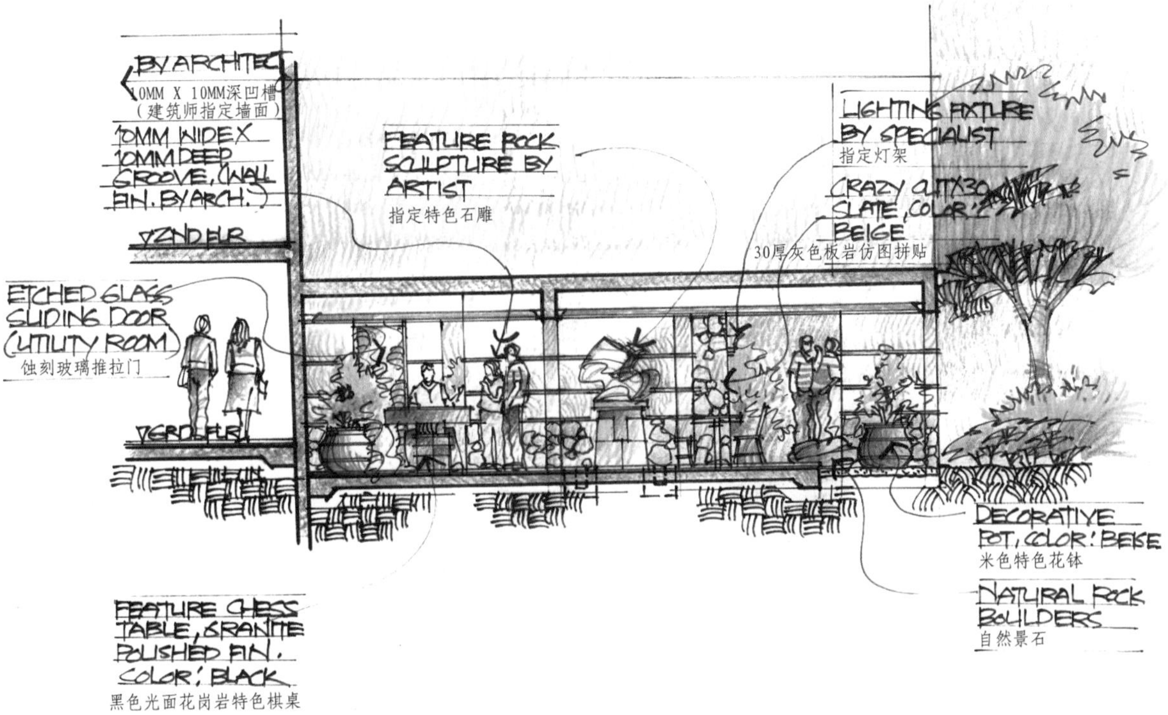
BY ARCHITEC
10MM X 10MM深凹槽
（建筑师指定墙面）
10MM WIDE X 10MM DEEP GROOVE. (WALL FIN. BY ARCH.)
FEATURE ROCK SCULPTURE BY ARTIST
指定特色石雕
LIGHTING FIXTURE BY SPECIALIST
指定灯架
CRAZY CUT X30 SLATE, COLOR: BEIGE
30厚灰色板岩仿图拼贴
ETCHED GLASS SLIDING DOOR (UTILITY ROOM)
蚀刻玻璃推拉门
DECORATIVE POT, COLOR: BEIGE
米色特色花钵
NATURAL ROCK BOULDERS
自然景石
FEATURE CHESS TABLE, GRANITE POLISHED FIN. COLOR: BLACK
黑色光面花岗岩特色棋桌

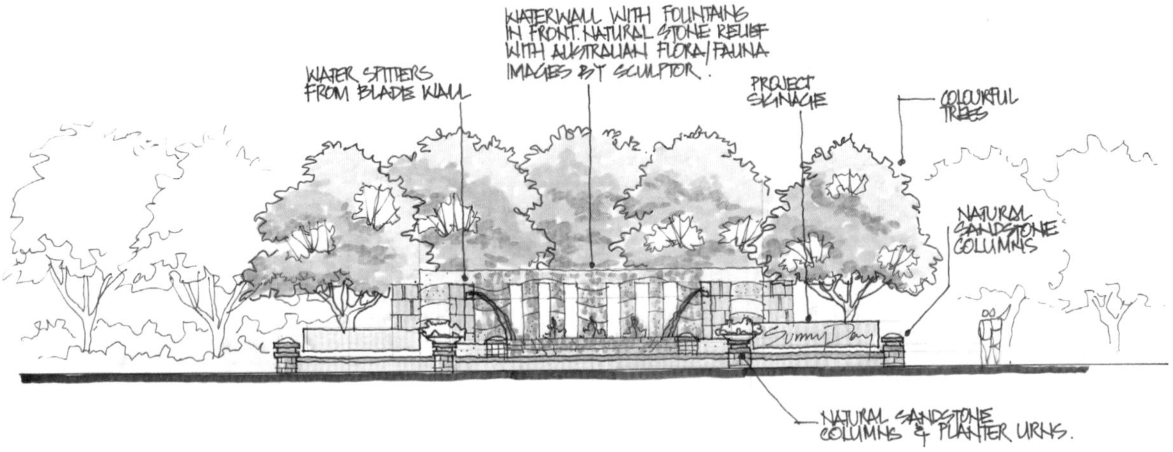
WATERWALL WITH FOUNTAINS IN FRONT. NATURAL STONE RELIEF WITH AUSTRALIAN FLORA/FAUNA IMAGES BY SCULPTOR.
WATER SPITTERS FROM BLADE WALL
PROJECT SIGNAGE
COLOURFUL TREES
NATURAL SANDSTONE COLUMNS
NATURAL SANDSTONE COLUMNS & PLANTER URNS.

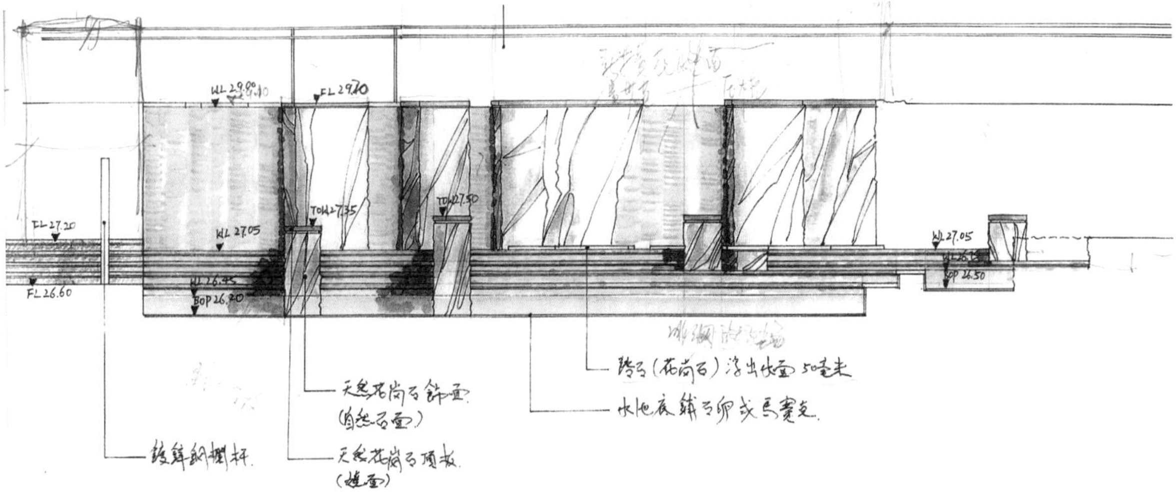
WL 29.80
FL 29.70
FL 27.20
WL 27.05
TOW 27.35
TOW 27.50
WL 26.45
BOP 26.20
FL 26.60
WL 27.05
BOP 26.50
踏石（花岗石）浮出水面50毫米
天然花岗石饰面（自然石面）
水池底铺石卵或马赛克
镀锌钢栏杆
天然花岗石顶板（烧面）

WATER FEATURE

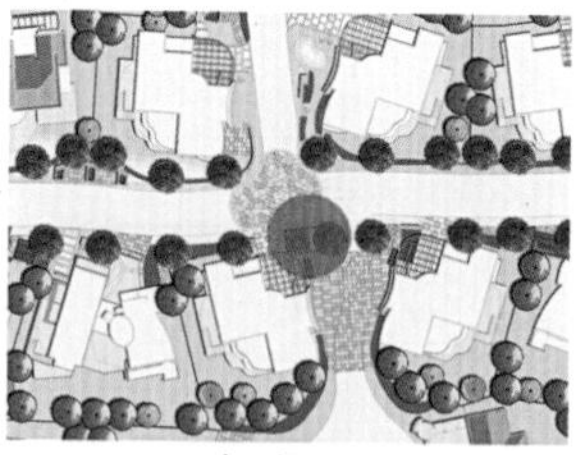

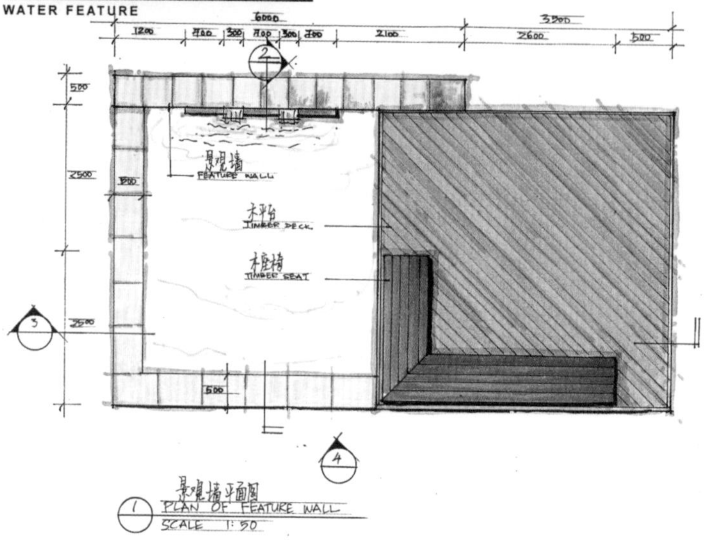
6000
3500
1200
2100
2600
500
2500
景观墙
FEATURE WALL
木平台
TIMBER DECK
木座椅
TIMBER SEAT
景观墙平面图
PLAN OF FEATURE WALL
SCALE 1:50

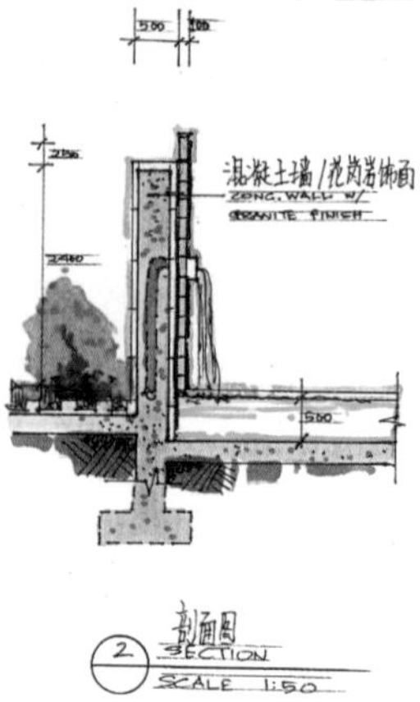
混凝土墙/花岗岩饰面
CONC. WALL W/
GRANITE FINISH
剖面图
SECTION
SCALE 1:50

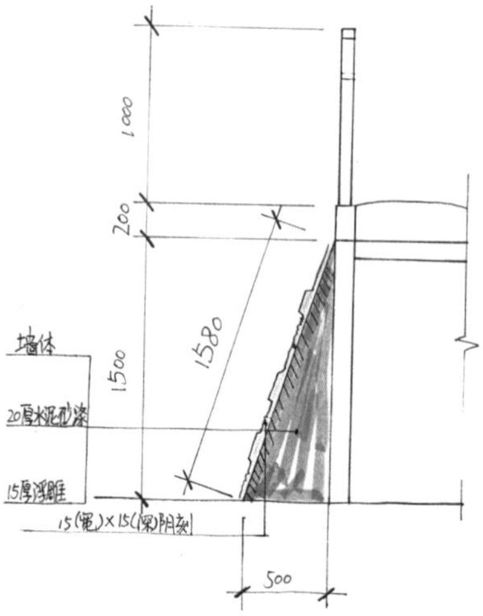
1000
200
1500
1580
墙体
20厚水泥砂浆
15厚浮雕
15(宽)×15(深)阴刻
500

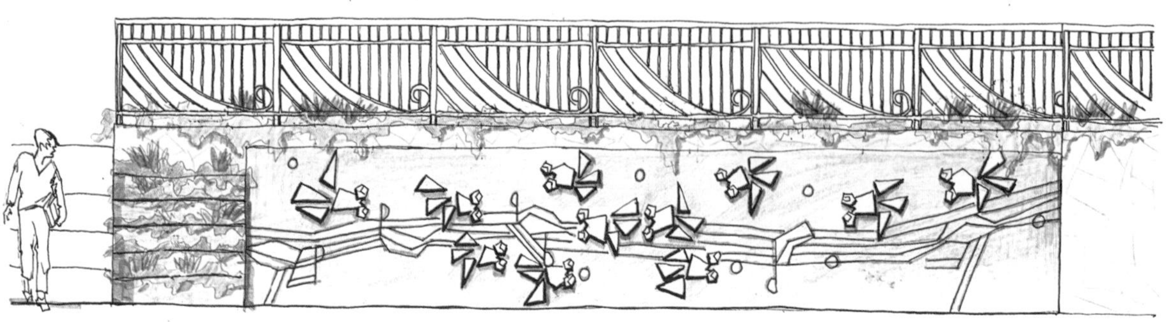

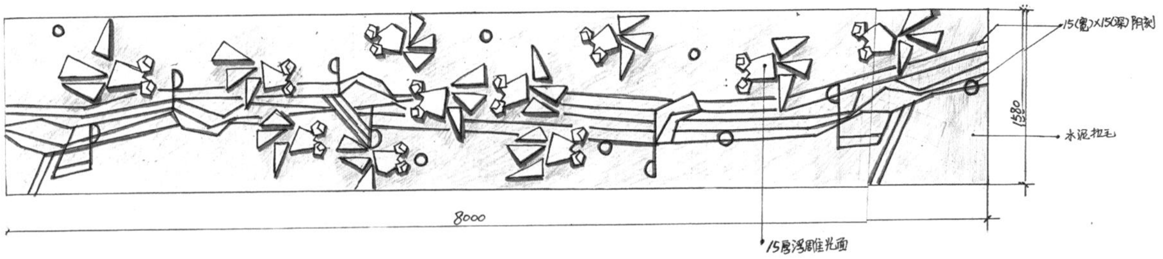
15(宽)×150(深)阴刻
1580
水泥拉毛
8000

架空层“禄”景墙大样图

比例：1：20

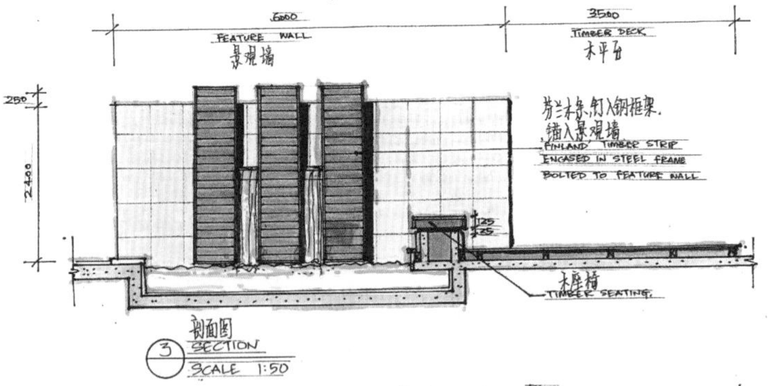
6000
FEATURE WALL
景观墙
3500
TIMBER DECK
木平台
250
2400
芬兰木条,钉入钢框架,
锚入景观墙
'FINLAND' TIMBER STRIP
ENCASED IN STEEL FRAME
BOLTED TO FEATURE WALL
25
25
木座椅
TIMBER SEATING
剖面图
3
SECTION
SCALE 1:50

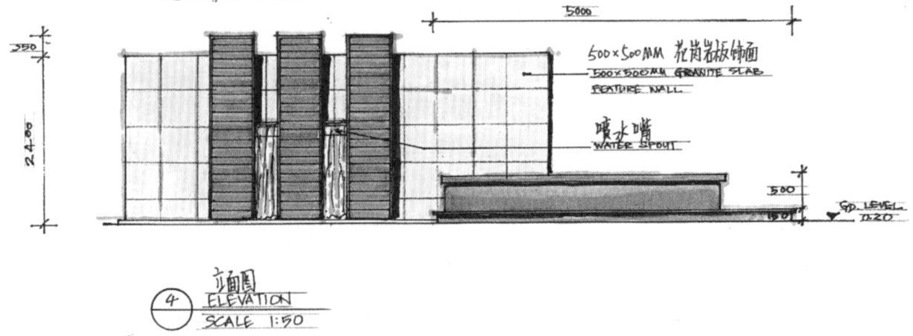
5000
350
2400
500×500MM 花岗岩板饰面
500×500MM GRANITE SLAB
FEATURE WALL
喷水嘴
WATER SPOUT
500
150
GD. LEVEL
D.20
立面图
4
ELEVATION
SCALE 1:50

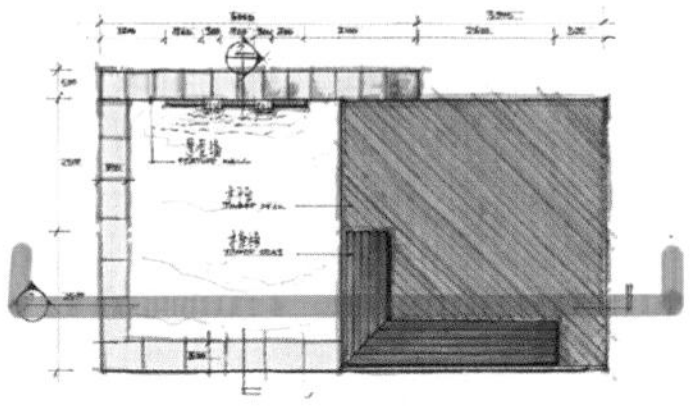

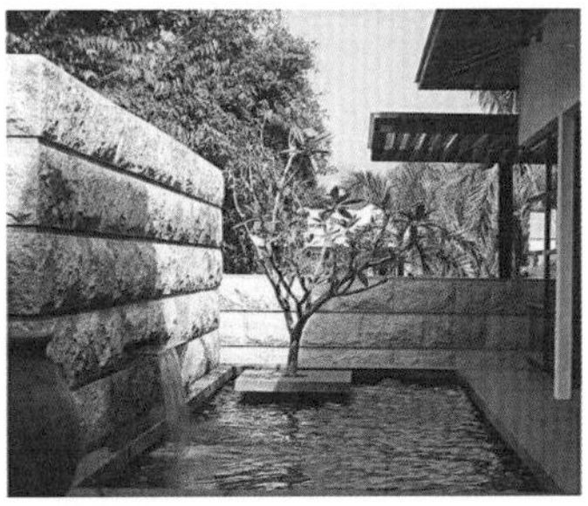

WATER FEATURE

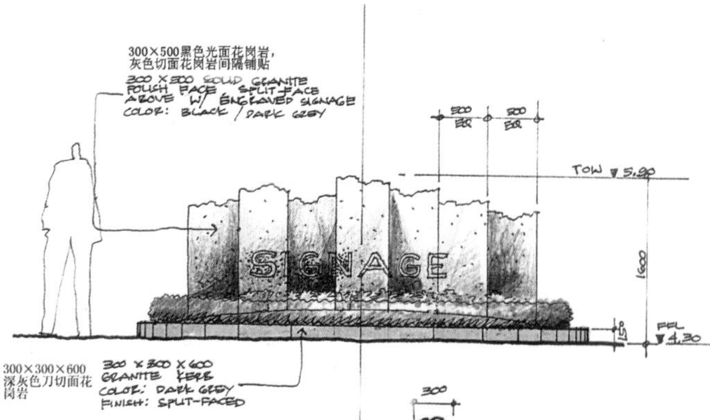

SG2 ELEVATION: SIGNAGE SC: 1:30 M.
标识牌立面

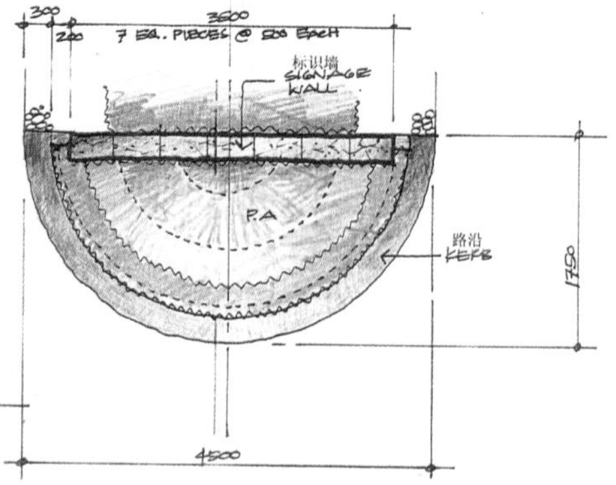

SG1 PLAN: SIGNAGE SC: 1:40 M.
标识牌平面

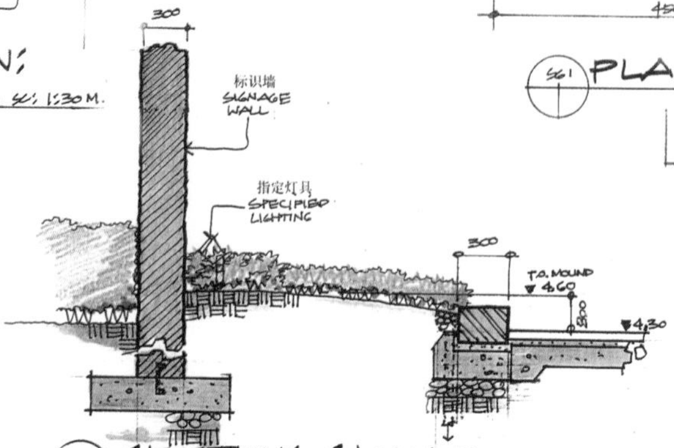

SG3 SECTION: SIGNAGE SC: 1:20 M.
标识牌剖面

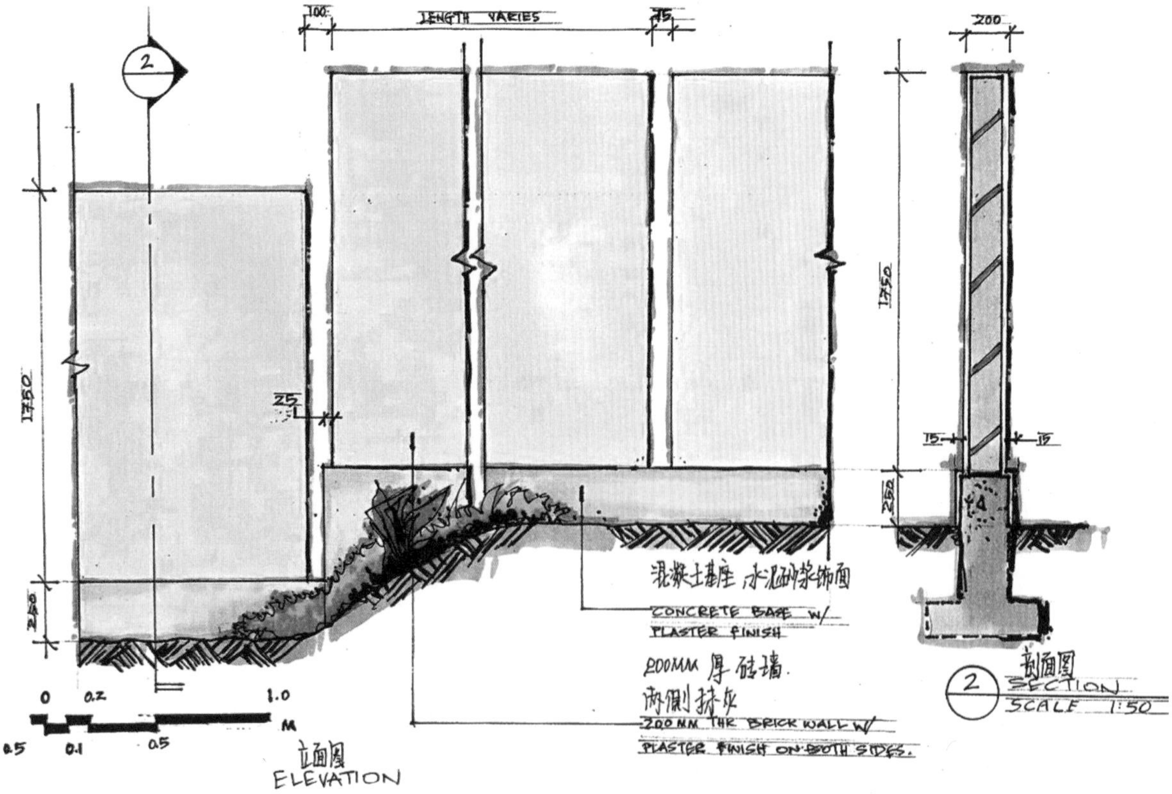

1 墙详图
WALL DETAILS
SCALE 1:50

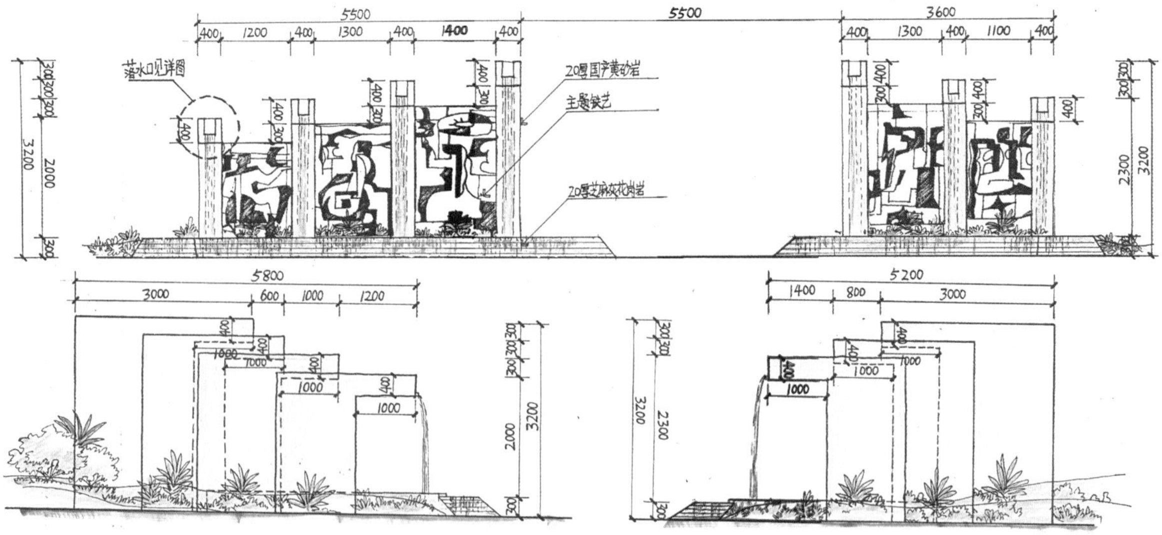
落水口见详图
20厚国产黄砂岩
主题铁艺
20厚芝麻灰花岗岩

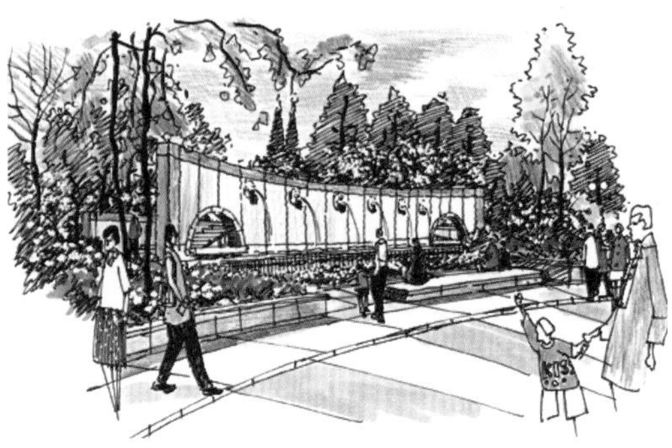

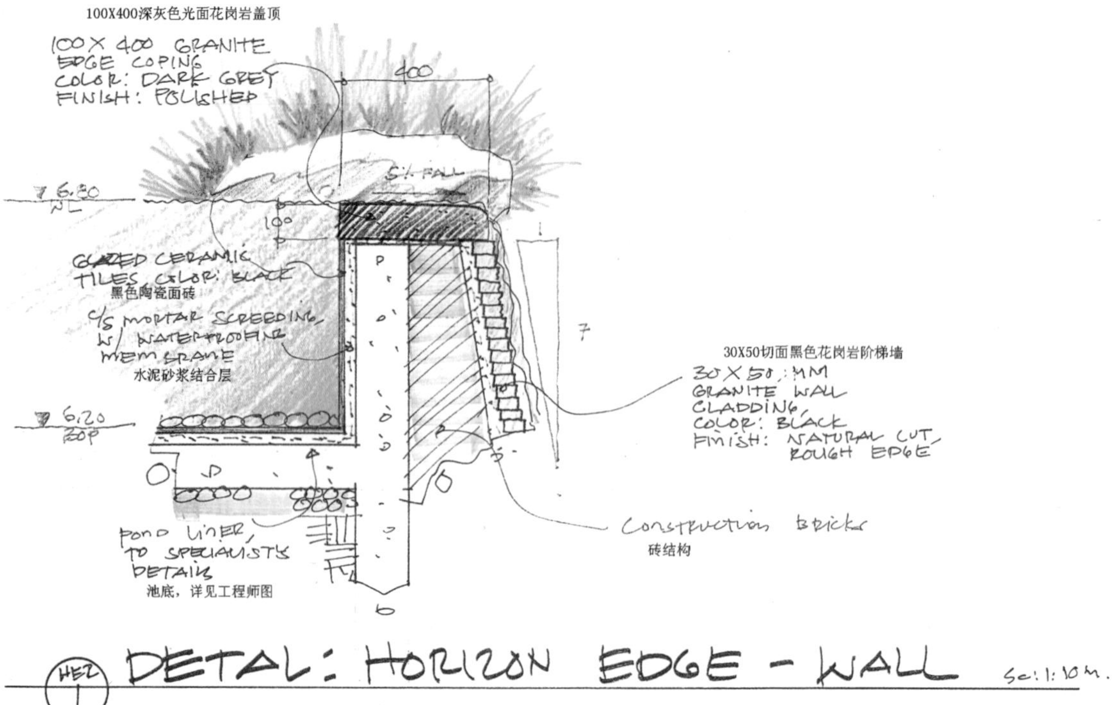
100X400深灰色光面花岗岩盖顶
100 X 400 GRANITE EDGE COPING COLOR: DARK GREY FINISH: POLISHED
400
5% FALL
6.80 WL
100
GLAZED CERAMIC TILES COLOR: BLACK
黑色陶瓷面砖
C/S MORTAR SCREEDING W/ WATERPROOFING MEMBRANE
水泥砂浆结合层
6.20 BOP
30X50切面黑色花岗岩阶梯墙
30 X 50 MM GRANITE WALL CLADDING COLOR: BLACK FINISH: NATURAL CUT, ROUGH EDGE
CONSTRUCTION BRICKS
砖结构
POND LINER, TO SPECIALIST'S DETAILS
池底，详见工程师图
HEZ
DETAL: HORIZON EDGE - WALL

瀑布水墙详图 1:10

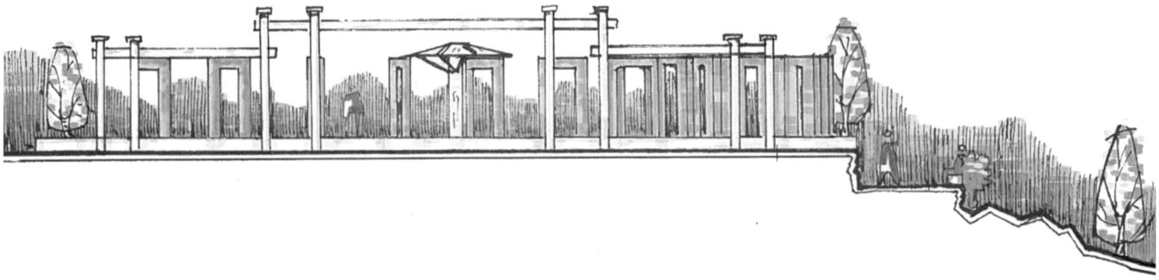

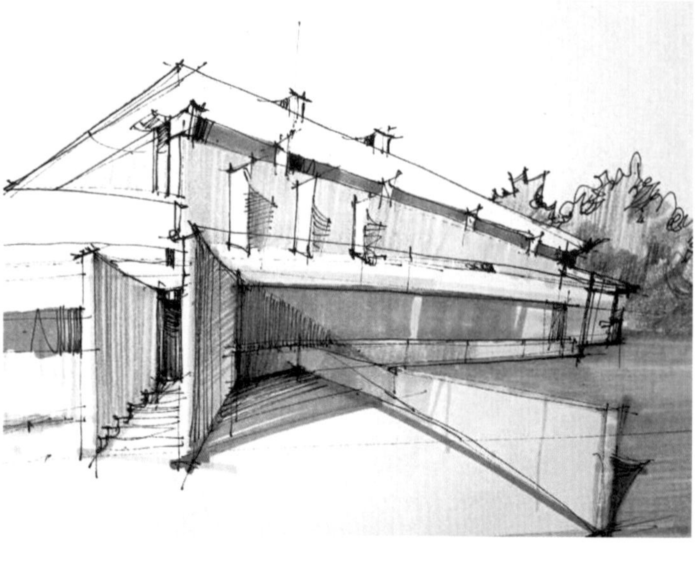

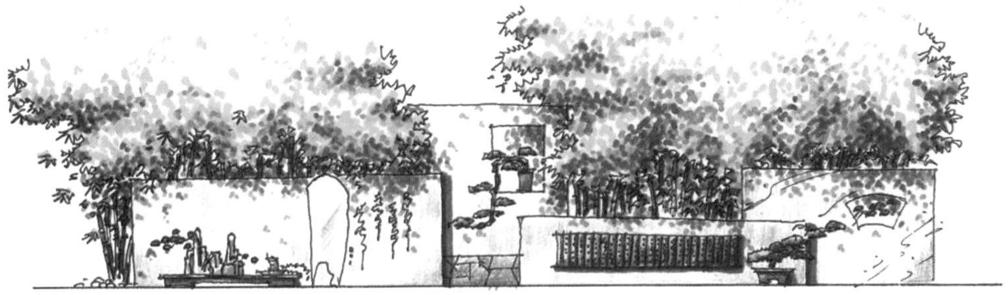

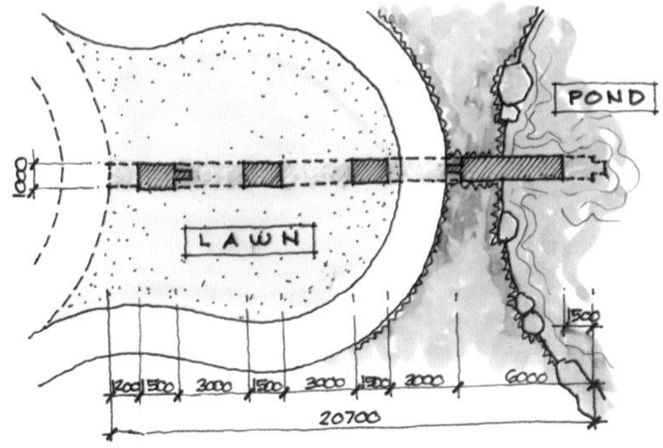
POND
1000
LAWN
1500
1200
1500
3000
1500
3000
1500
3000
6000
20700

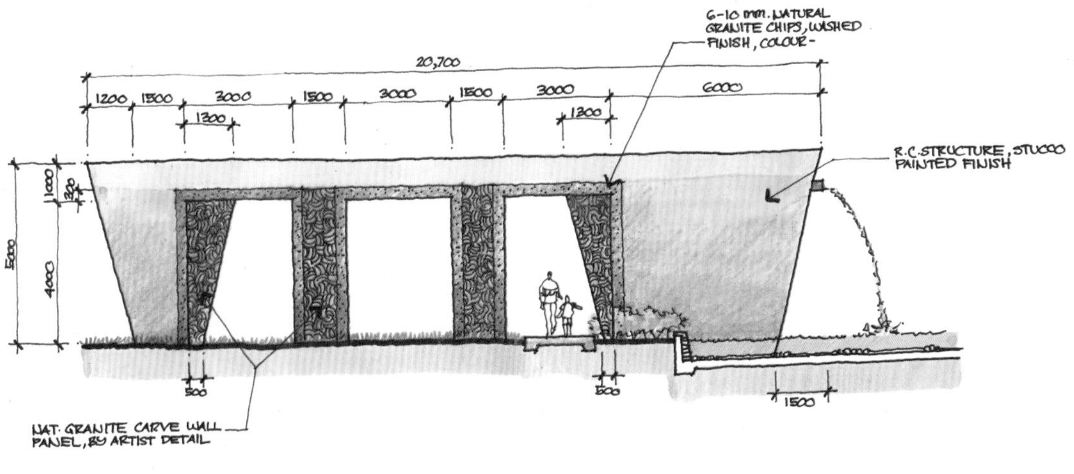
6-10 mm. NATURAL
GRANITE CHIPS, WASHED
FINISH, COLOUR-
20,700
1200
1500
3000
1500
3000
1500
3000
6000
1300
1300
R.C.STRUCTURE, STUCCO
PAINTED FINISH
1000
300
5000
4000
500
500
1500
NAT. GRANITE CARVE WALL
PANEL, BY ARTIST DETAIL

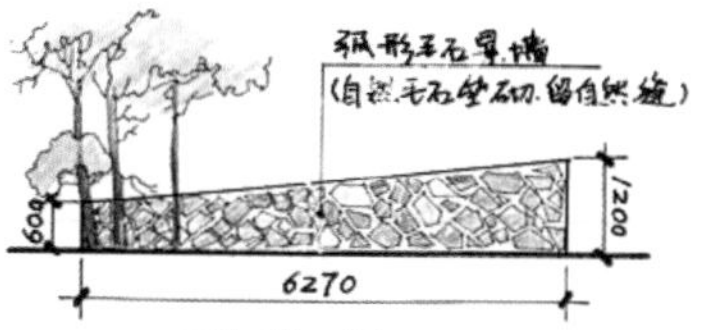

3C立面展开图 1:75

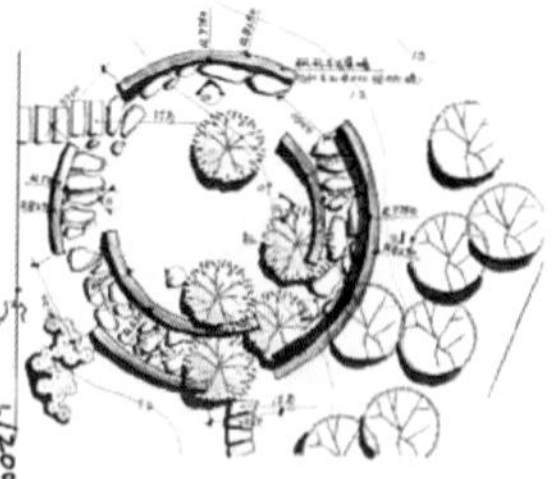

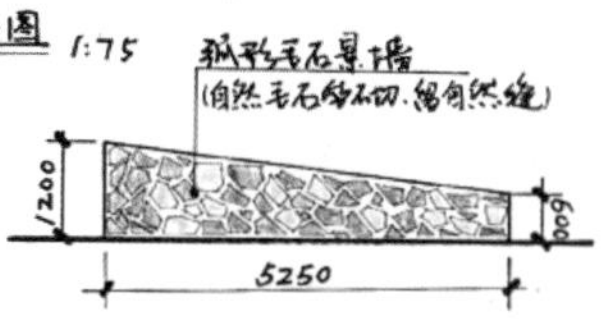

3D立面展开图 1:75

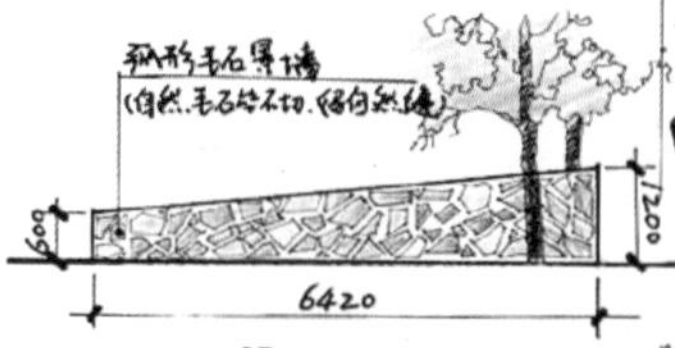

3E立面展开图 1:75

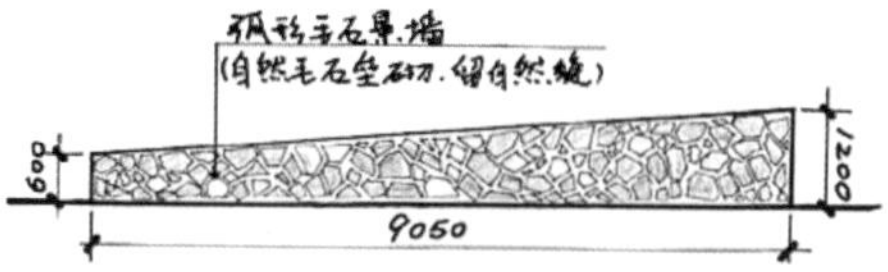

3A立面展开图 1:75

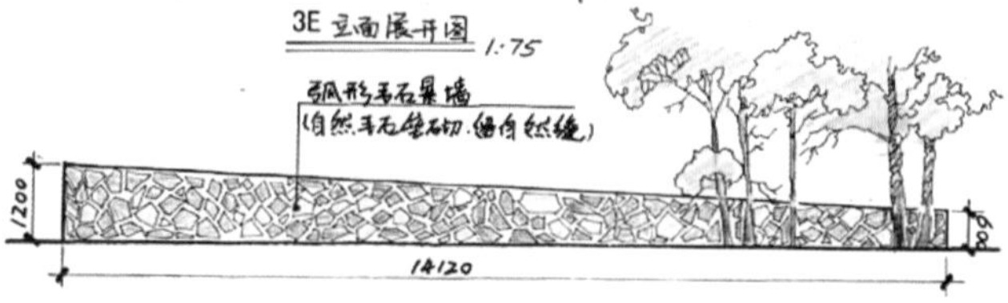

3B立面展开图 1:75

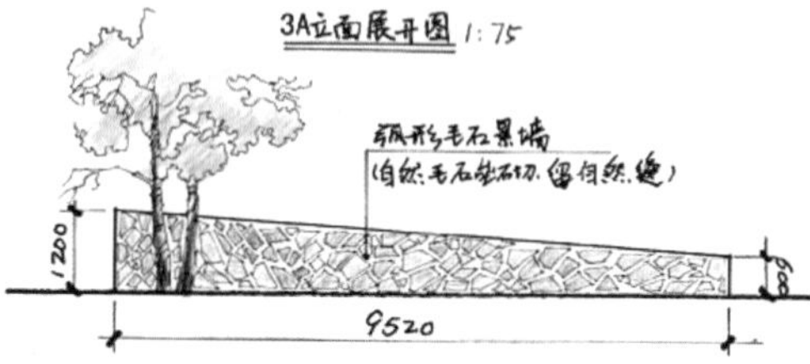

3F立面展开图 1:75

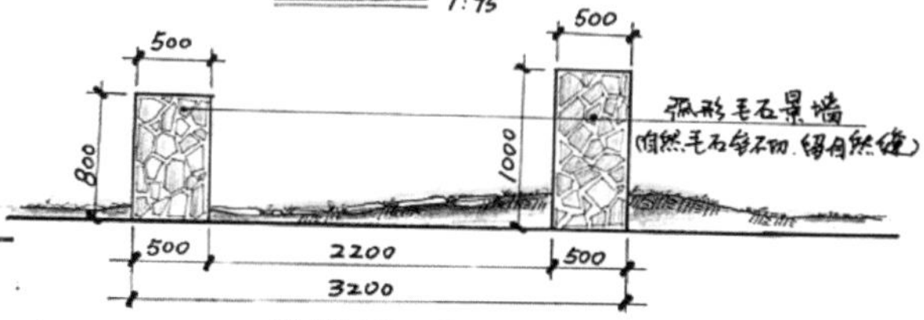

3G-3G剖面图 1:30

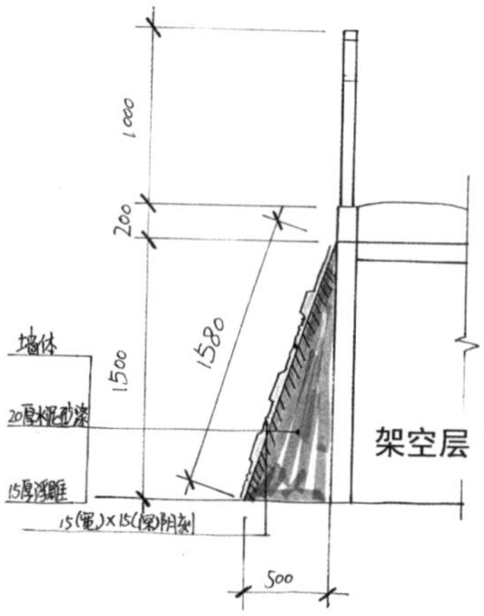

架空层“禄”景墙剖立面

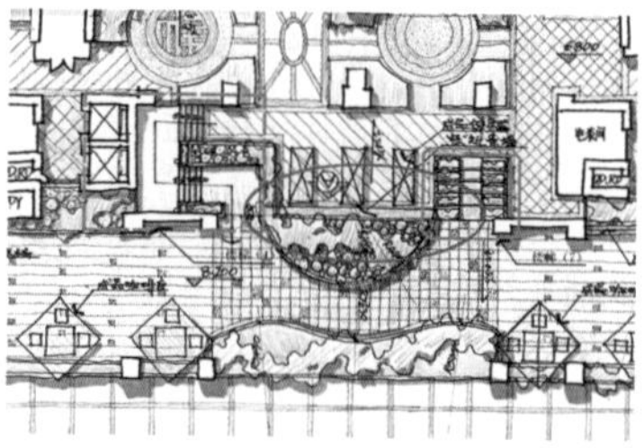

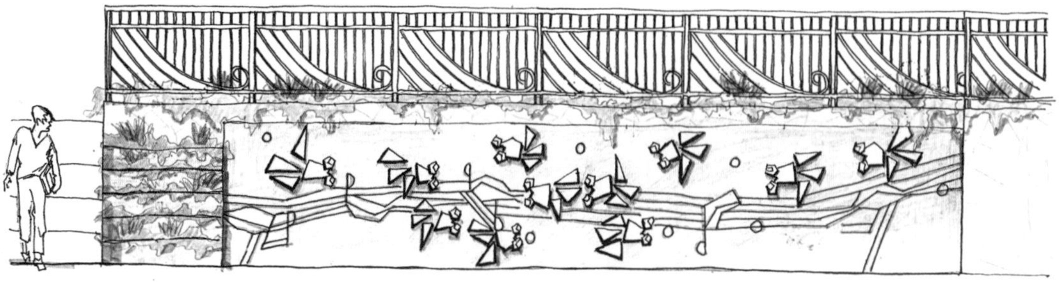

架空层“禄”A立面

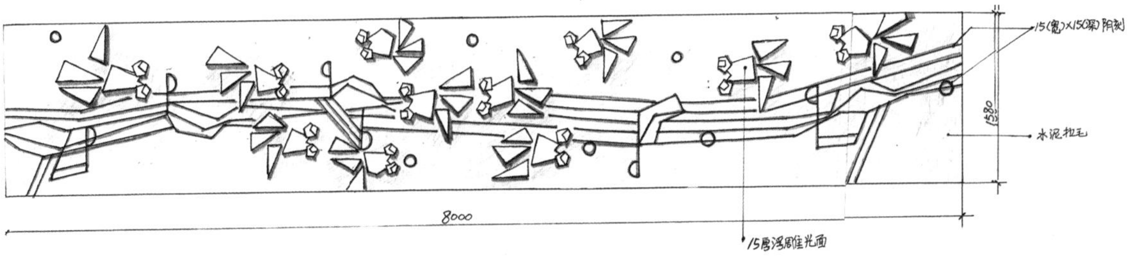

架空层“禄”景墙大样图

比例：1：20

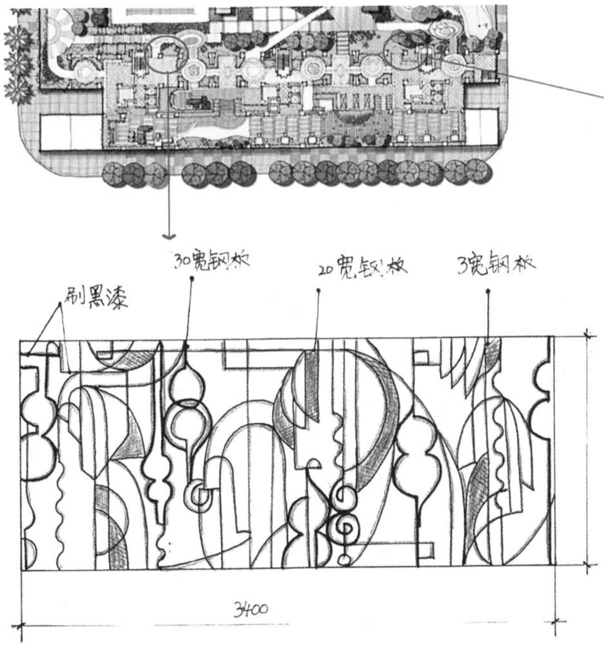

“寿”铁艺大样

比例：1：20

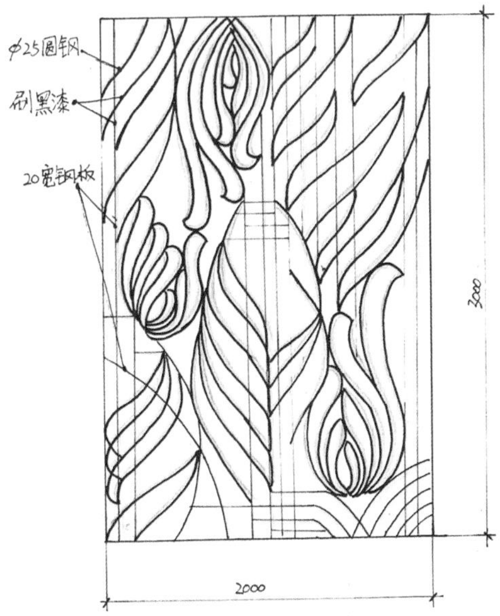

“福”铁艺大样

比例：1：20

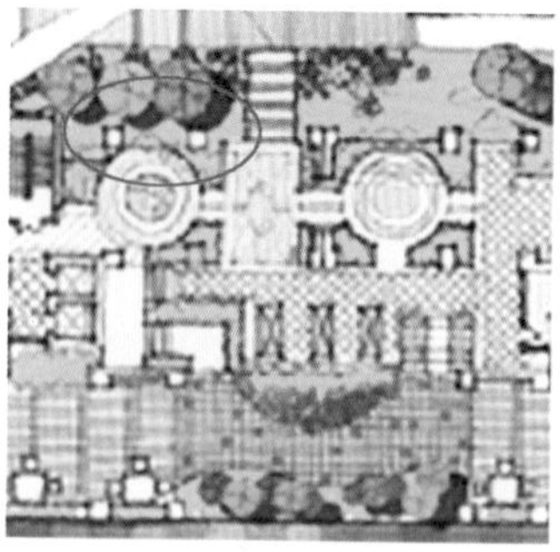

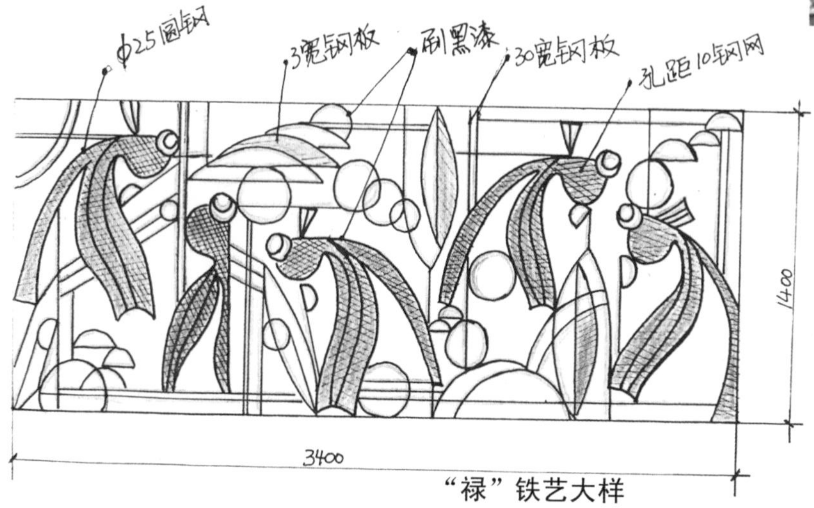

"禄"铁艺大样

比例：1：20

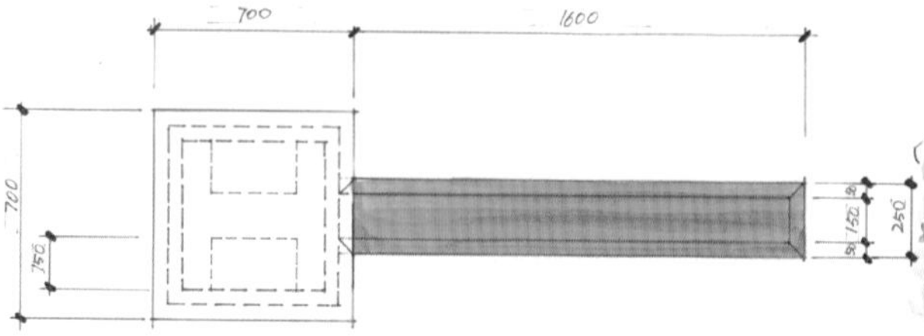

平面图

比例 1:15

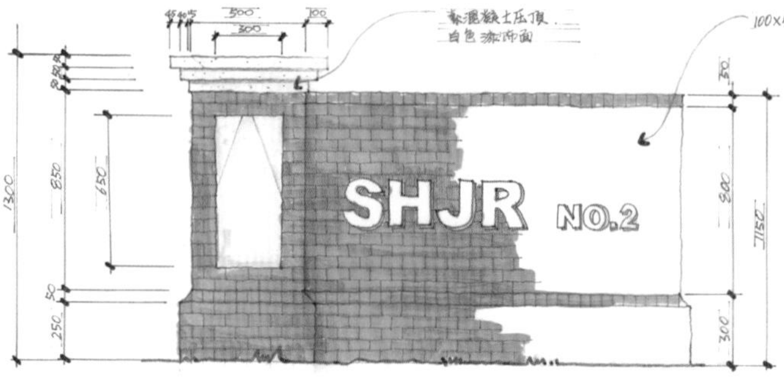

正立面图

比例 1:15

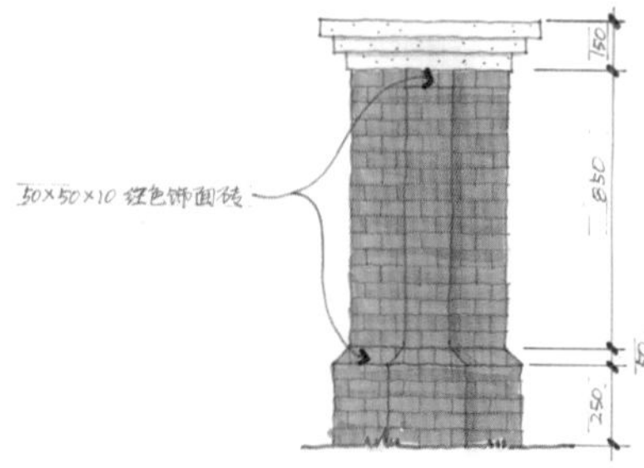

侧立面图

比例 1:15

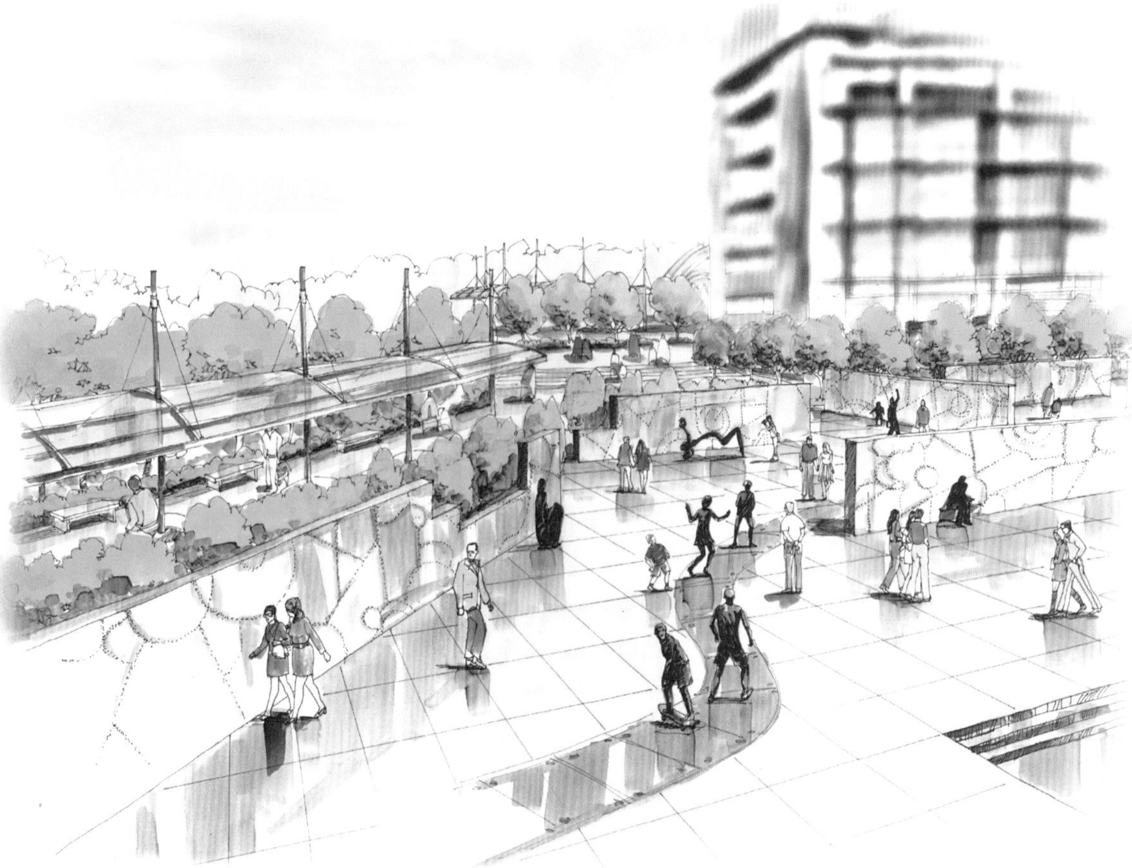

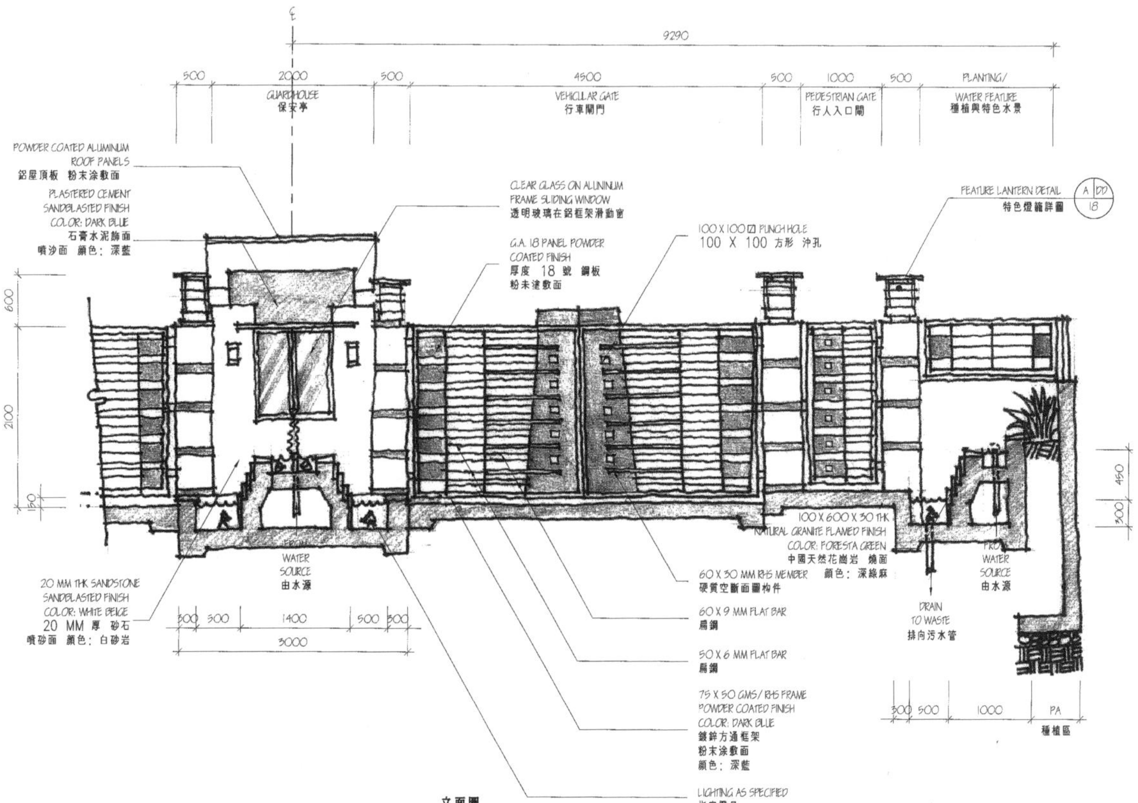
9290
500
2000
500
4500
500
1000
500
GUARDHOUSE
保安亭
VEHICULAR GATE
行車閘門
PEDESTRIAN GATE
行人入口閘
PLANTING/
WATER FEATURE
種植與特色水景
POWDER COATED ALUMINUM
ROOF PANELS
鋁屋頂板 粉末涂敷面
PLASTERED CEMENT
SANDBLASTED FINISH
COLOR: DARK BLUE
石膏水泥飾面
噴沙面 顏色：深藍
CLEAR GLASS ON ALUMINUM
FRAME SLIDING WINDOW
透明玻璃在鋁框架滑動窗
G.A. 18 PANEL POWDER
COATED FINISH
厚度 18 號 鋼板
粉未塗敷面
100 X 100 ☑ PUNCH HOLE
100 X 100 方形 沖孔
FEATURE LANTERN DETAIL
特色燈籠詳圖
A
DD
18
600
2100
150
450
300
100 X 600 X 30 THK
NATURAL GRANITE FLAMED FINISH
COLOR: FORESTA GREEN
中國天然花崗岩 燒面
顏色：深綠麻
60 X 30 MM RHS MEMBER
硬質空斷面圖構件
WATER
SOURCE
由水源
20 MM THK SANDSTONE
SANDBLASTED FINISH
COLOR: WHITE BEIGE
20 MM 厚 砂石
噴砂面 顏色：白砂岩
300
300
1400
500
300
3000
60 X 9 MM FLAT BAR
扁鋼
50 X 6 MM FLAT BAR
扁鋼
75 X 50 GMS/RHS FRAME
POWDER COATED FINISH
COLOR: DARK BLUE
鍍鋅方通框架
粉末涂敷面
顏色：深藍
LIGHTING AS SPECIFIED
指定燈具
DRAIN
TO WASTE
排向污水管
WATER
SOURCE
由水源
300
500
1000
PA
種植區

A
立面圖
ELEVATION
SCALE 1:40

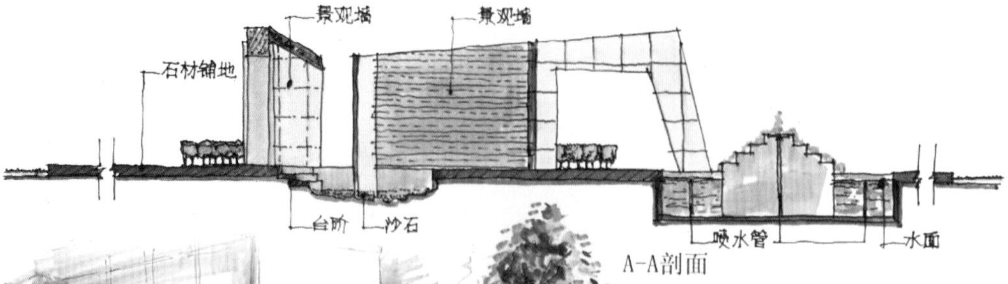
景观墙
景观墙
石材铺地
台阶
沙石
喷水管
水面
A-A剖面

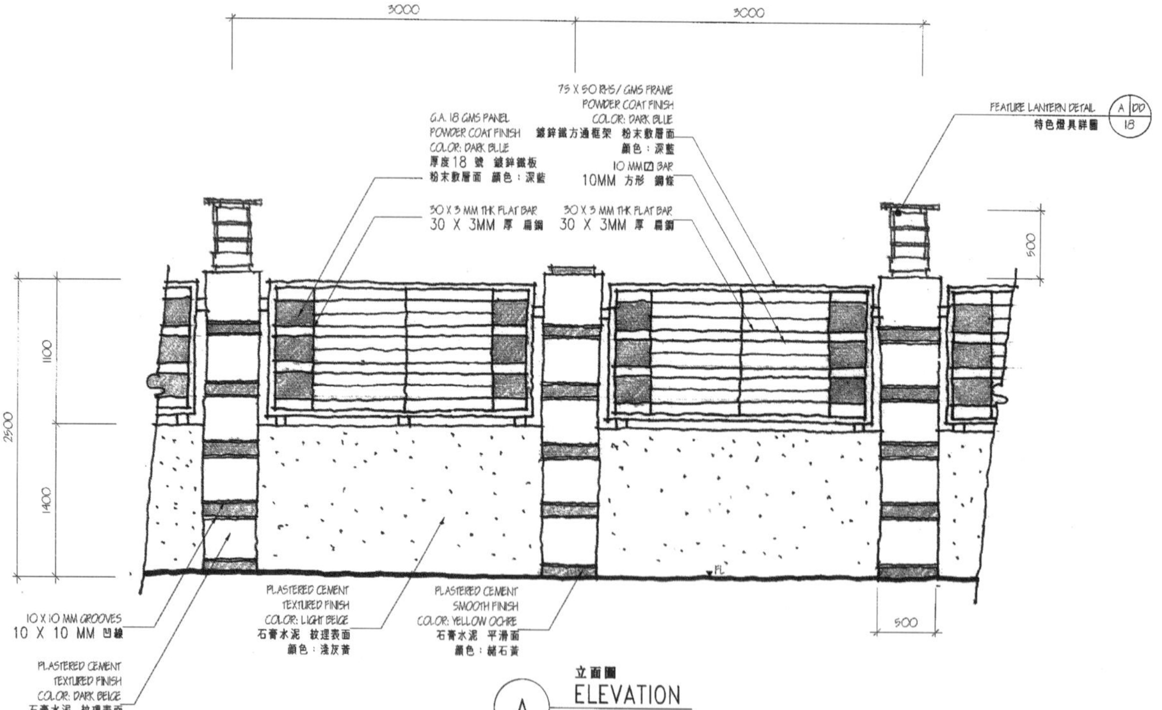
3000
3000
75 X 50 RHS / GMS FRAME
POWDER COAT FINISH
COLOR: DARK BLUE
鍍鋅鐵方通框架 粉末敷層面
顏色：深藍
G.A. 18 GMS PANEL
POWDER COAT FINISH
COLOR: DARK BLUE
厚度18 號 鍍鋅鐵板
粉末敷層面 顏色：深藍
10 MM ☐ BAR
10MM 方形 鋼條
30 X 3 MM THK FLAT BAR
30 X 3MM 厚 扁鋼
30 X 3 MM THK FLAT BAR
30 X 3MM 厚 扁鋼
FEATURE LANTERN DETAIL
特色燈具詳圖
A DD 18
500
2500
1100
1400
FL
500
10 X 10 MM GROOVES
10 X 10 MM 凹線
PLASTERED CEMENT
TEXTURED FINISH
COLOR: DARK BEIGE
石膏水泥 紋理表面
顏色：深灰黃
PLASTERED CEMENT
TEXTURED FINISH
COLOR: LIGHT BEIGE
石膏水泥 紋理表面
顏色：淺灰黃
PLASTERED CEMENT
SMOOTH FINISH
COLOR: YELLOW OCHRE
石膏水泥 平滑面
顏色：赭石黃

立面圖
A ELEVATION
SCALE 1:30

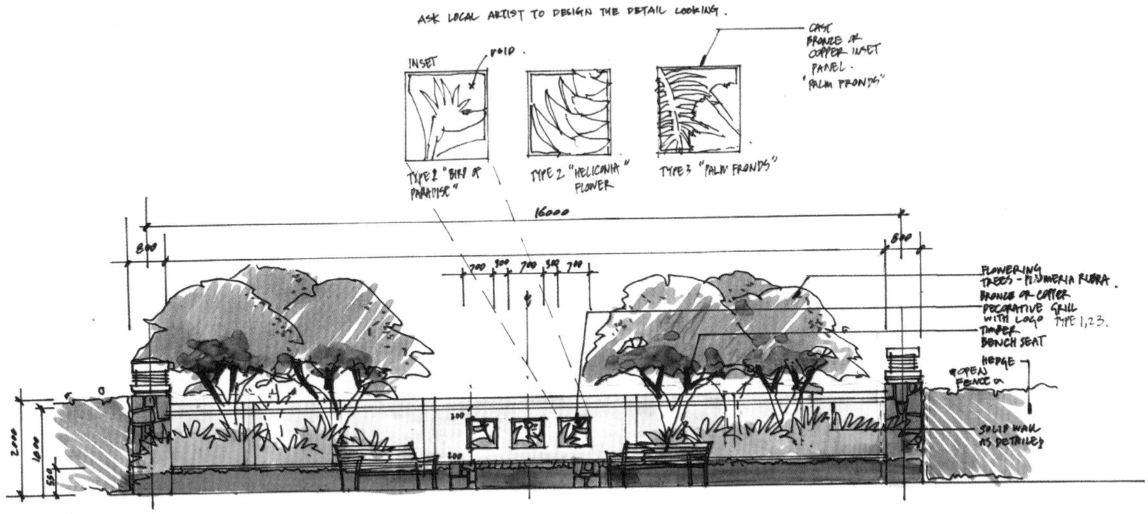
ASK LOCAL ARTIST TO DESIGN THE DETAIL LOOKING.
INSET
VOID.
CAST BRONZE OR COPPER INSET PANEL. "PALM FRONDS"
TYPE 1 "BIRD OF PARADISE"
TYPE 2 "HELICONIA" FLOWER
TYPE 3 "PALM FRONDS"
16000
800
800
700 300 700 300 700
FLOWERING TREES - PLUMERIA RUBRA.
BRONZE OR COPPER DECORATIVE GRILL WITH LOGO TYPE 1,2,3.
TIMBER BENCH SEAT
HEDGE
OPEN FENCE
SOLID WALL AS DETAILED
2200
1000
500
200
200

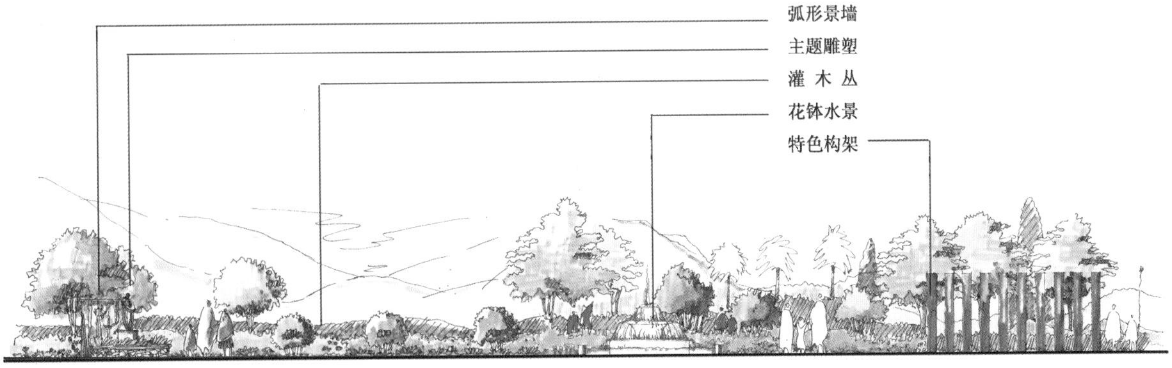
弧形景墙
主题雕塑
灌 木 丛
花钵水景
特色构架

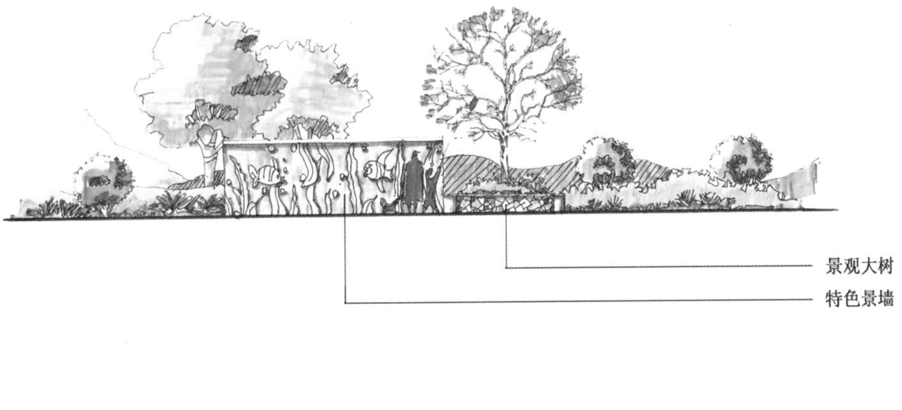
景观大树
特色景墙

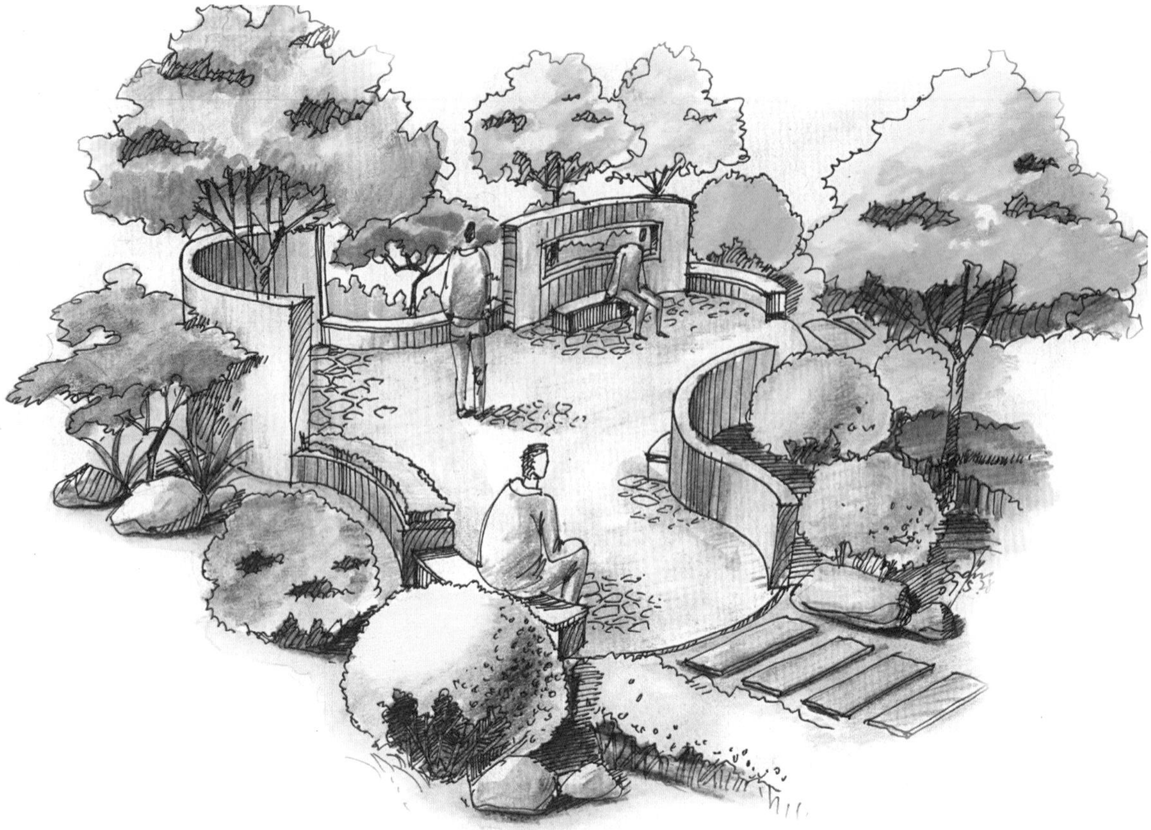

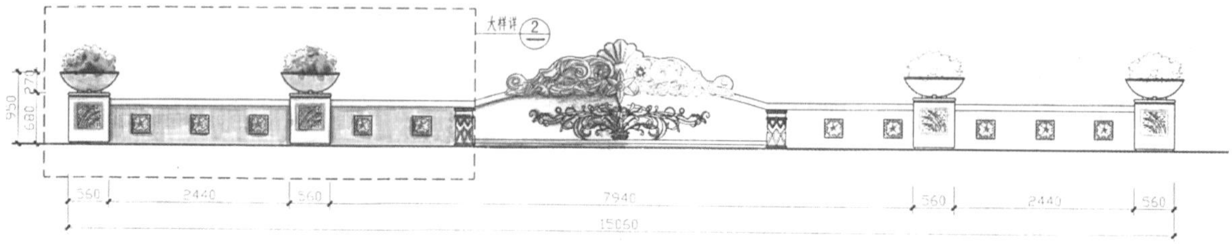

① 景观矮墙立面图 1:50

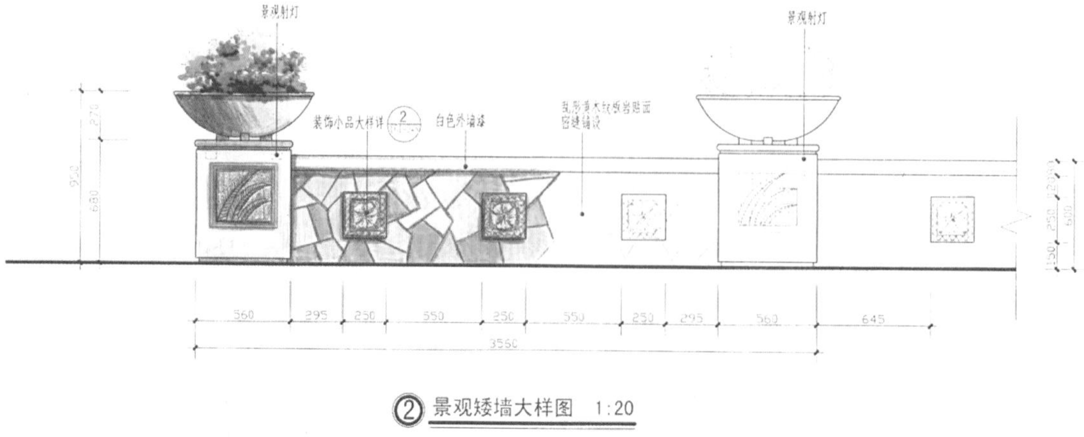

② 景观矮墙大样图 1:20

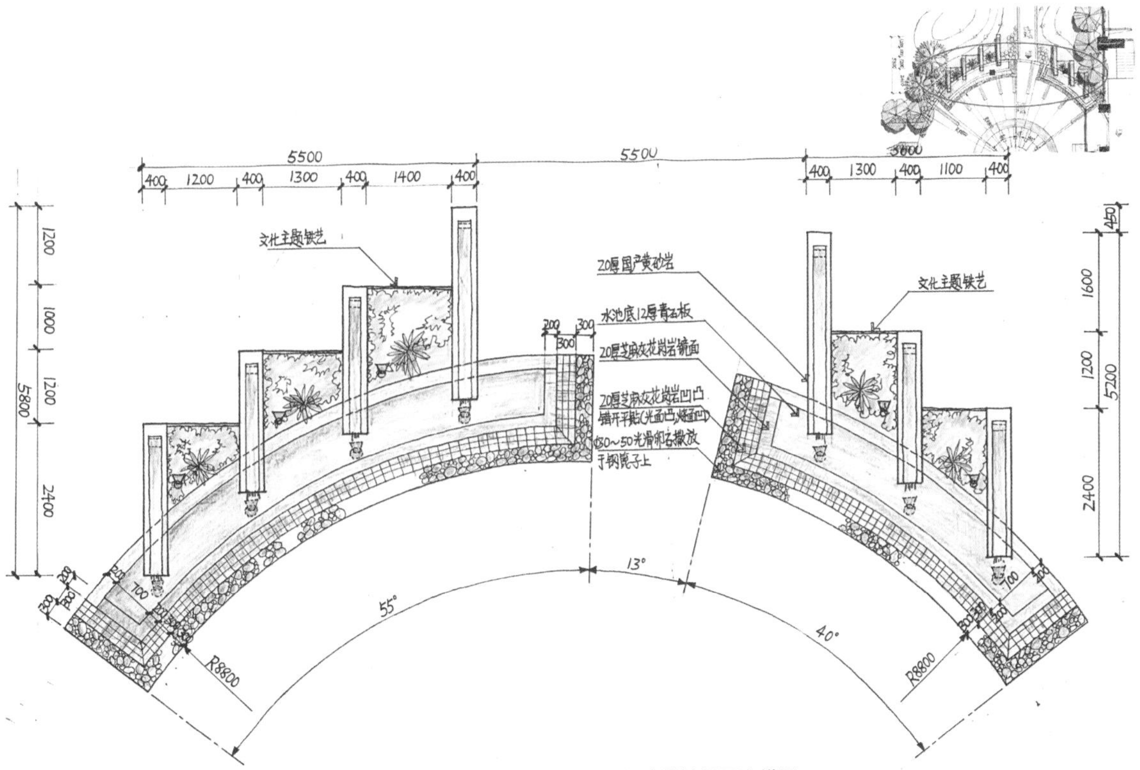

主入口内广场水景墙平面大样图　1：50

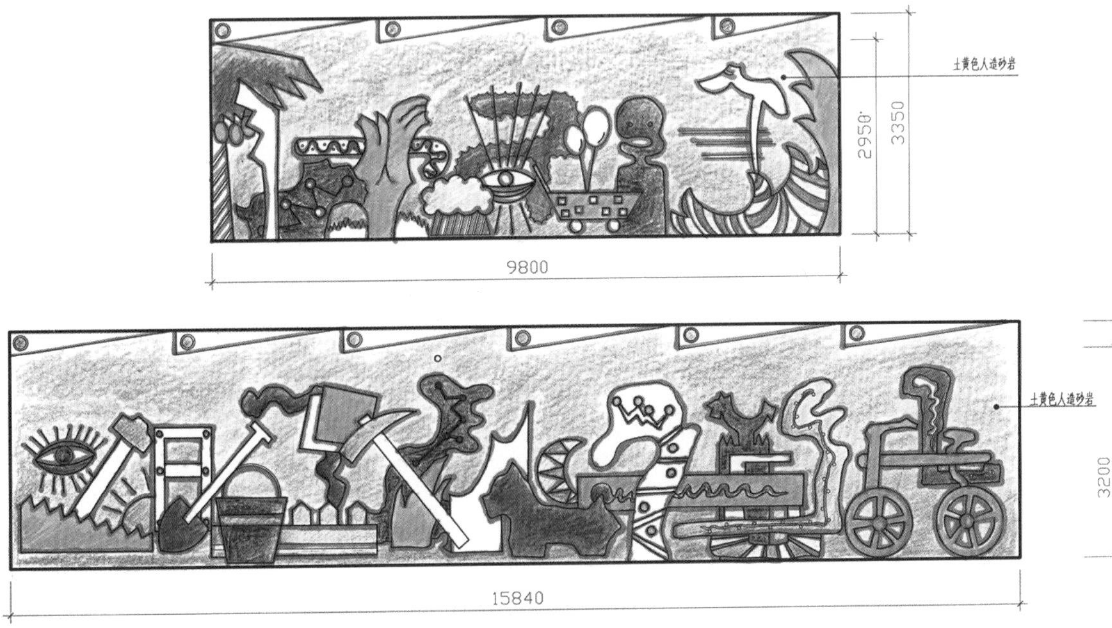

① 入口广场挡墙立面装饰大样图 1:50

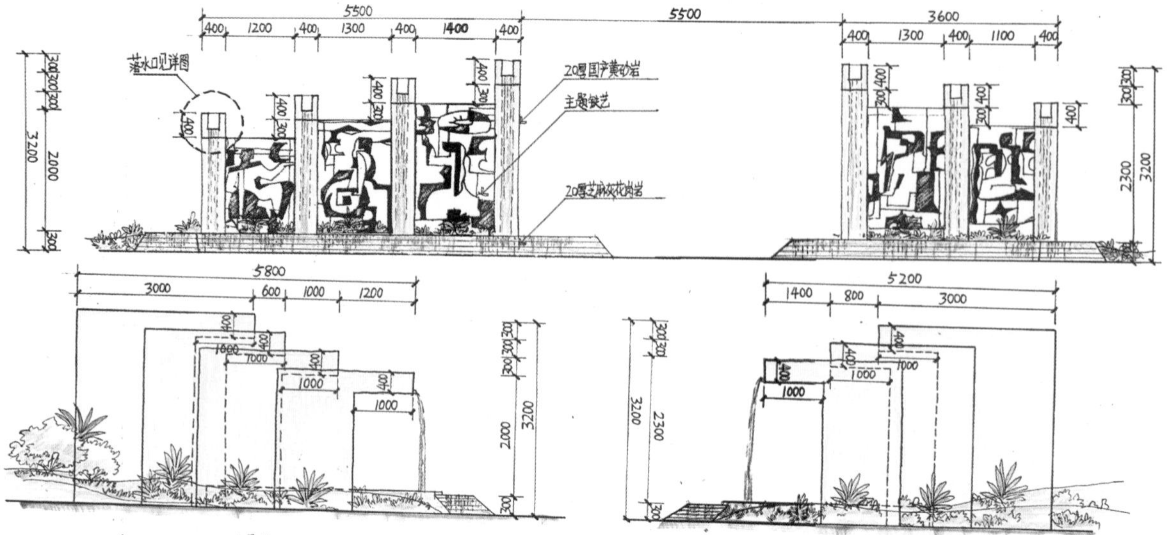

水景墙立面 1:50

水景墙立面 1:50

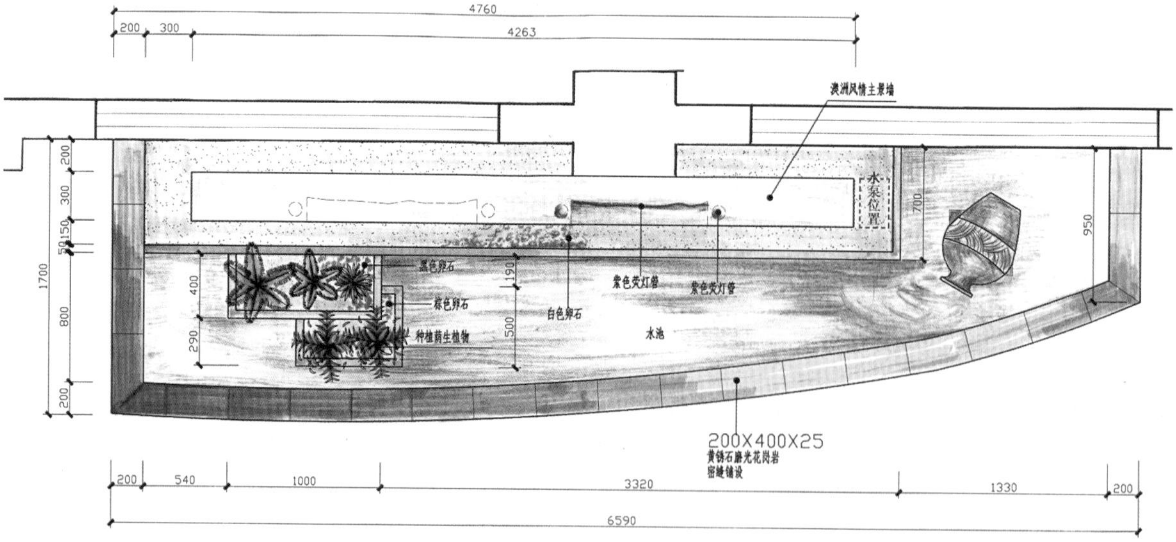

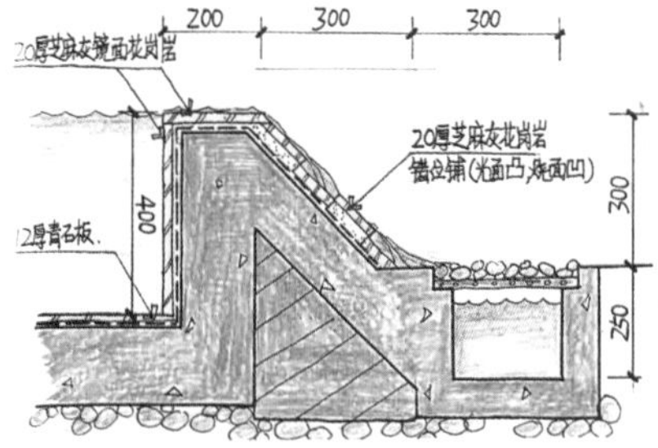

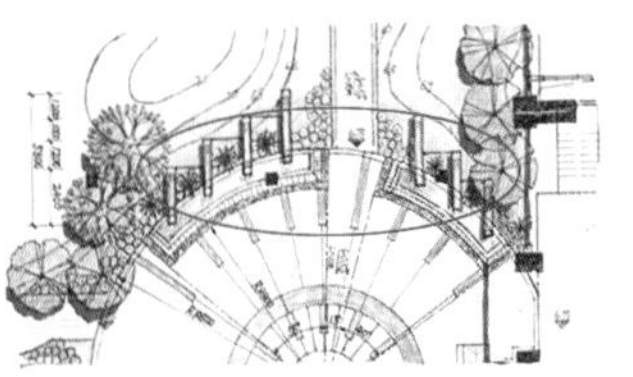

局部详图 1:10

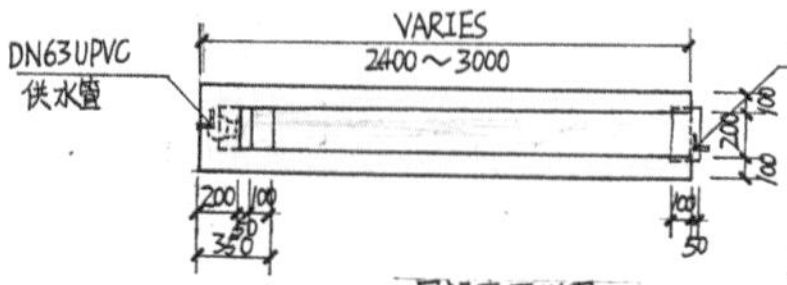

局部平面详图 1:30

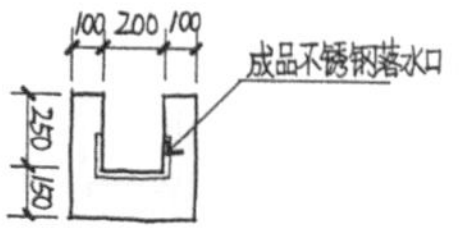

落水口大样图 1:20

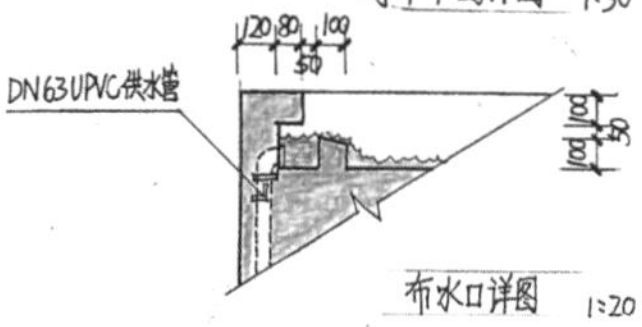

布水口详图 1:20

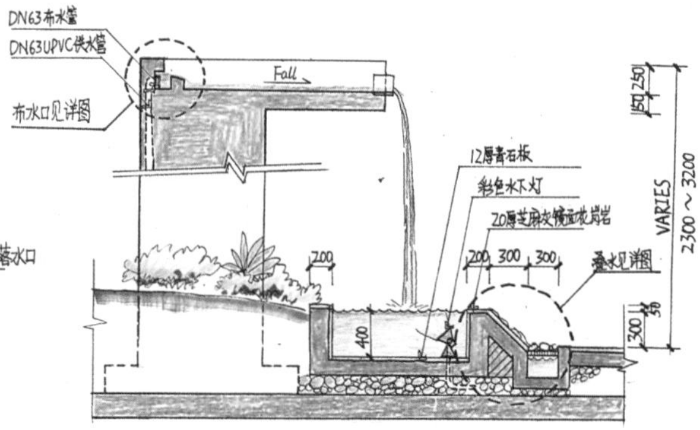

水景墙剖立面 1:30

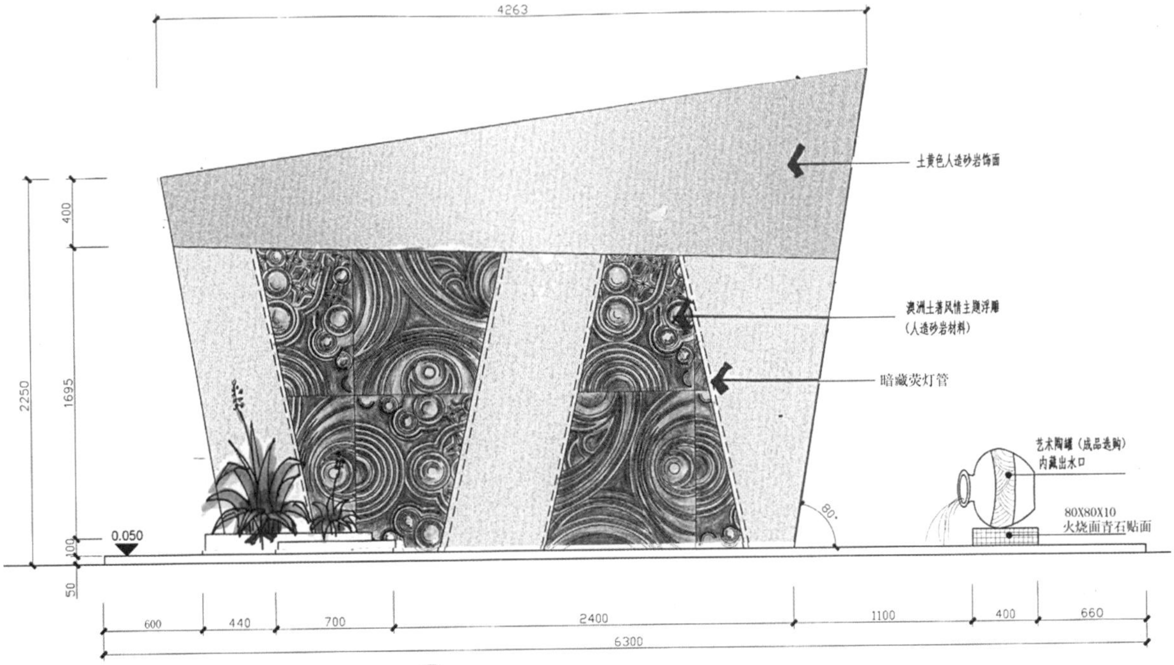

② 咖啡吧景墙立面图 1:20

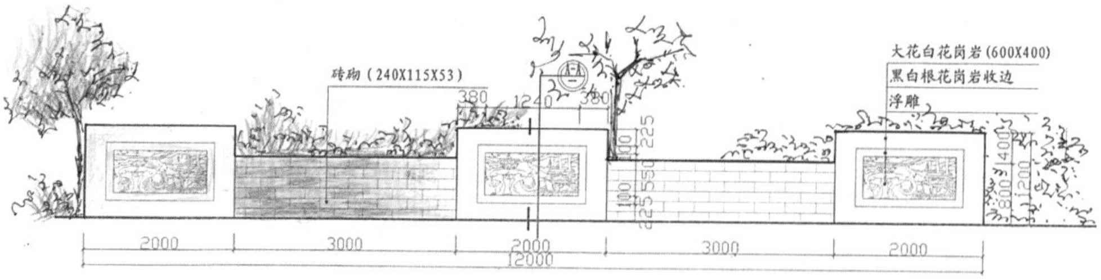

① 景墙立面展开图 1:50

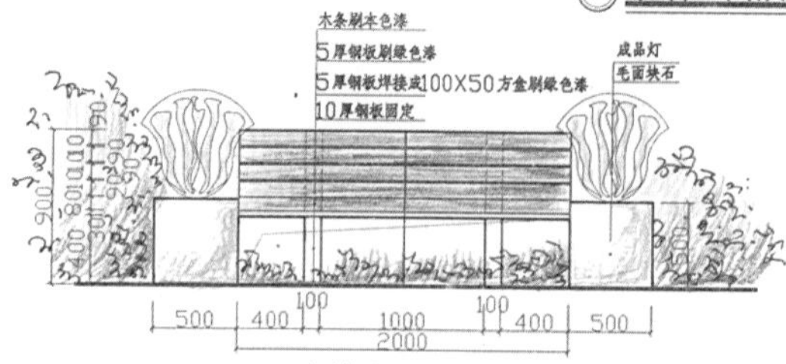

③ 坐椅立面图 1:30

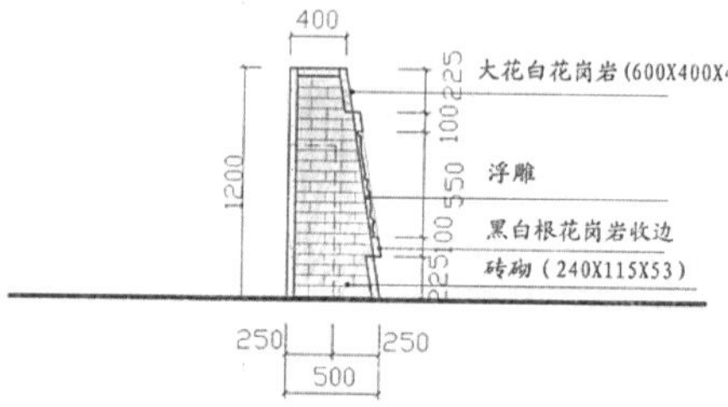

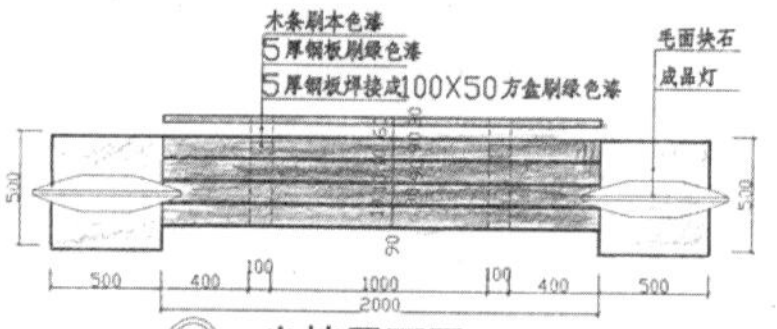

④ 坐椅平面图 1:30

② A-A剖面图 1:30

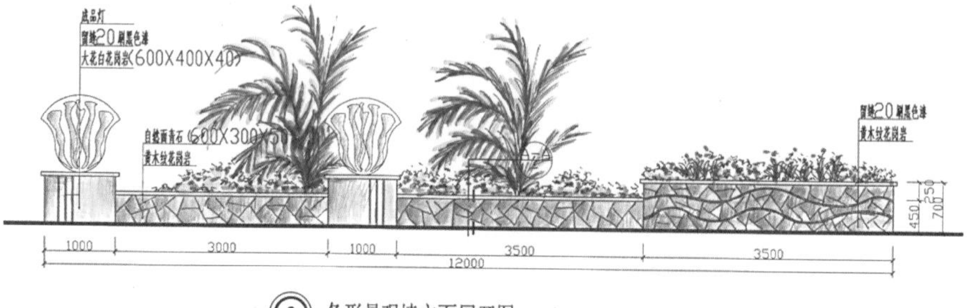

⑥ 条形景观墙立面展开图 1:50

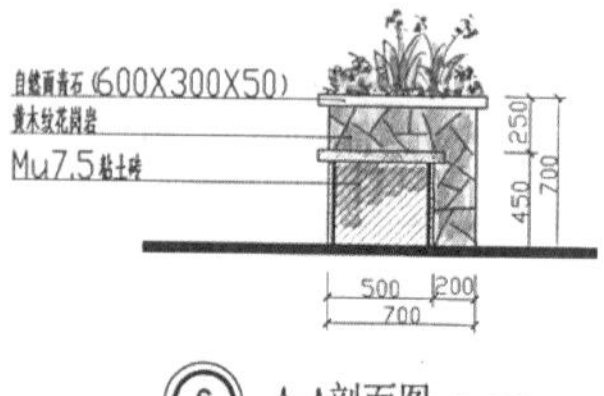

⑥ A-A剖面图 1:50

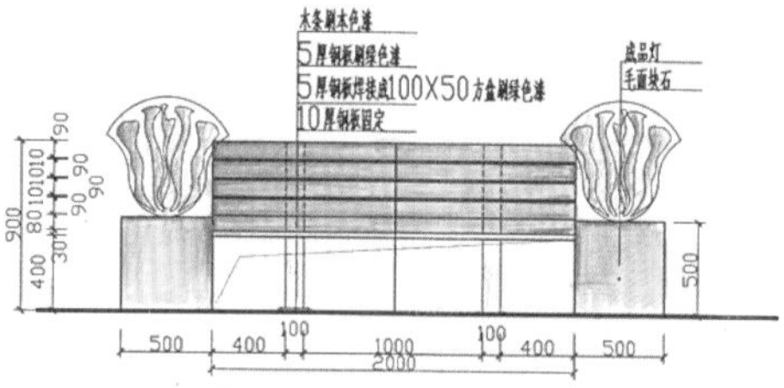

⑦ 坐椅立面图 1:30

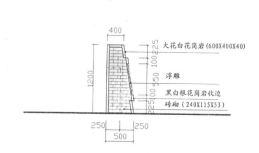

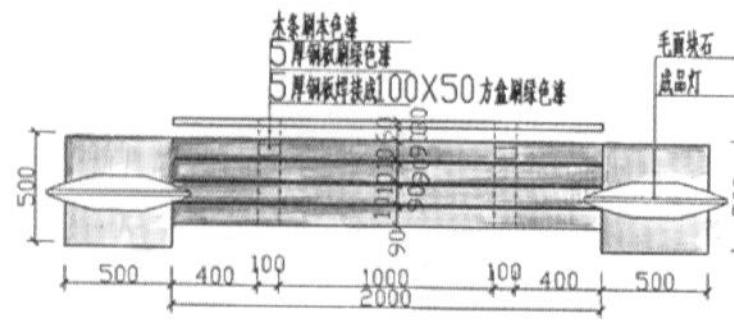

⑦ 坐椅平面图 1:30

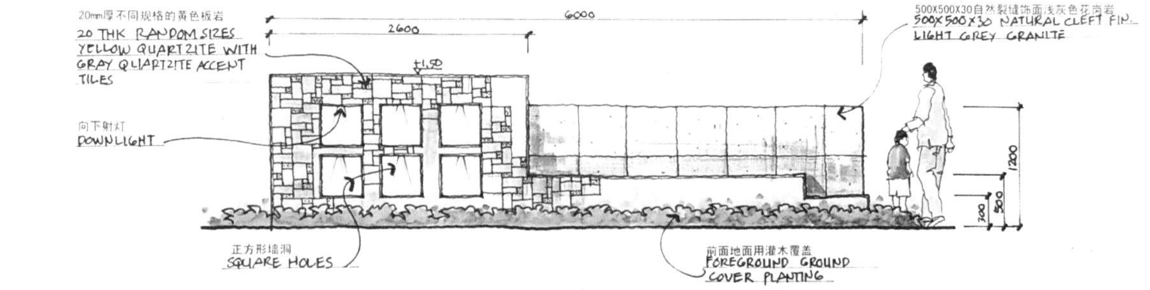
20mm厚不同规格的黄色板岩
20 THK RANDOM SIZES YELLOW QUARTZITE WITH GRAY QUARTZITE ACCENT TILES
向下射灯
DOWNLIGHT
正方形墙洞
SQUARE HOLES
6000
2600
+1.50
500X500X30自然裂墙饰面浅灰色花岗岩
500X500X30 NATURAL CLEFT FIN. LIGHT GREY GRANITE
1200
500
300
前面地面用灌木覆盖
FOREGROUND GROUND COVER PLANTING

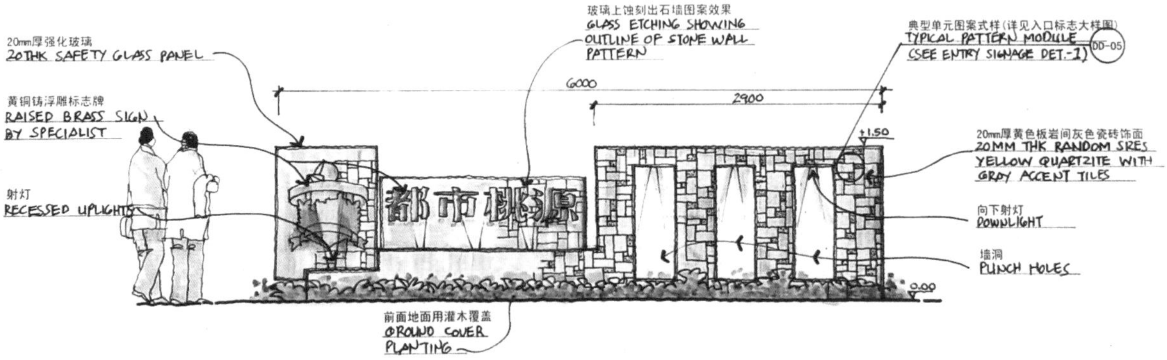
20mm厚强化玻璃
20THK SAFETY GLASS PANEL
黄铜铸浮雕标志牌
RAISED BRASS SIGN BY SPECIALIST
射灯
RECESSED UPLIGHTS
玻璃上蚀刻出石墙图案效果
GLASS ETCHING SHOWING OUTLINE OF STONE WALL PATTERN
6000
2900
都市桃源
典型单元图案式样(详见入口标志大样图)
TYPICAL PATTERN MODULE (SEE ENTRY SIGNAGE DET.-1)
DD-05
+1.50
20mm厚黄色板岩间灰色瓷砖饰面
20MM THK RANDOM SIZES YELLOW QUARTZITE WITH GRAY ACCENT TILES
向下射灯
DOWNLIGHT
墙洞
PUNCH HOLES
0.00
前面地面用灌木覆盖
GROUND COVER PLANTING

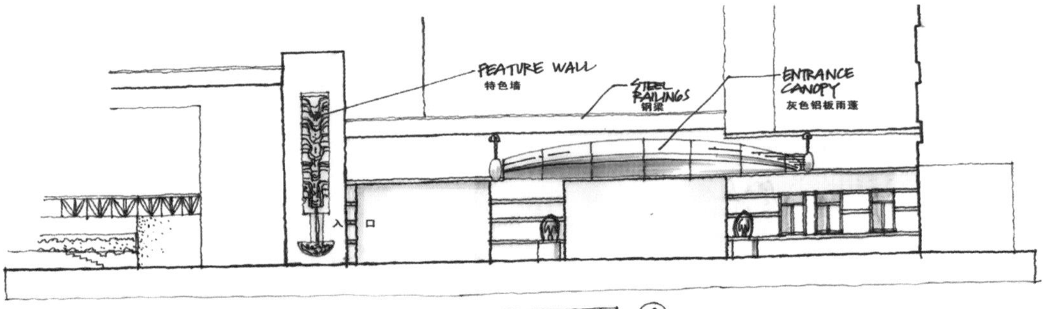
FEATURE WALL
特色墙
STEEL RAILINGS
钢梁
ENTRANCE CANOPY
灰色铝板雨篷
入口

ELEVATION ①
立面图

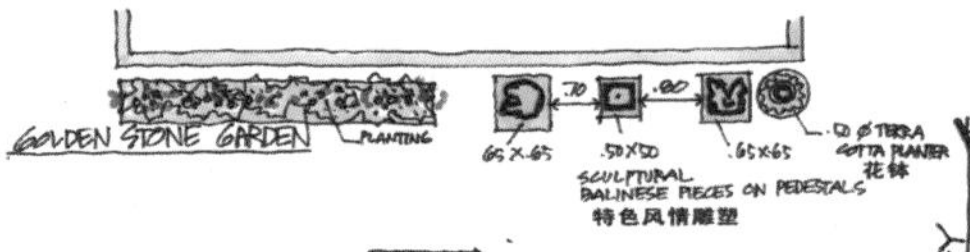
GOLDEN STONE GARDEN
PLANTING
65×65
.50×50
.65×65
SCULPTURAL BALINESE PIECES ON PEDESTALS
特色风情雕塑
TERRA COTTA PLANTER
花钵

PLAN
平面图

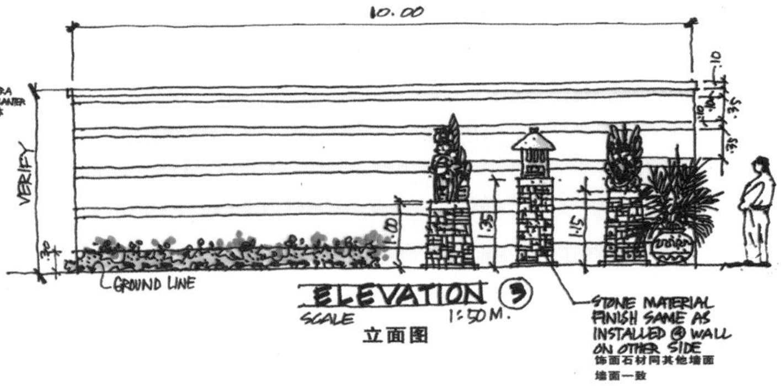
10.00
VERIFY
GROUND LINE
STONE MATERIAL FINISH SAME AS INSTALLED @ WALL ON OTHER SIDE
饰面石材同其他墙面墙面一致

ELEVATION ③
SCALE 1:50M.
立面图

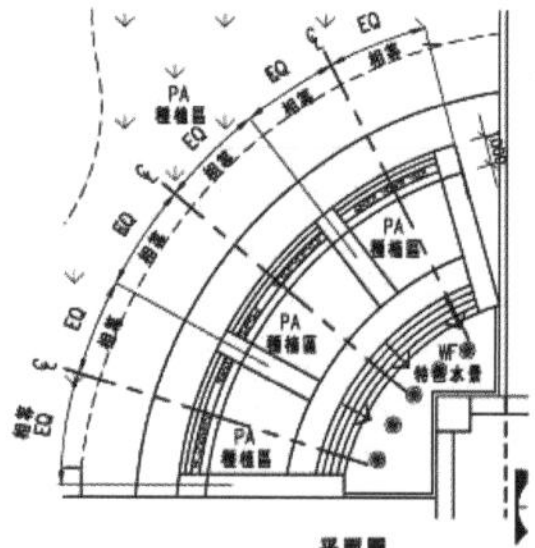

A PLAN
SCALE 1:200

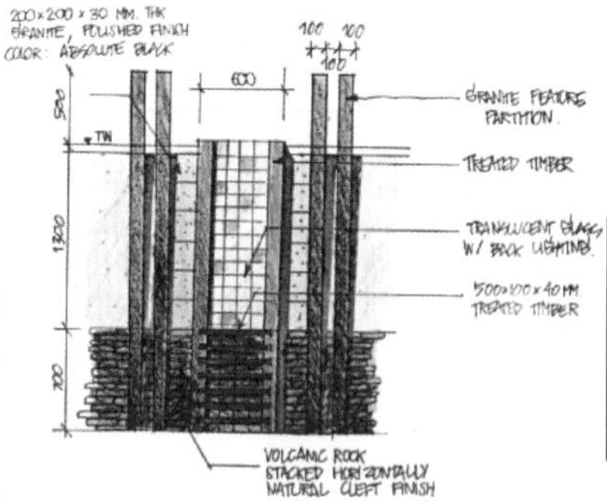

B ELEVATION
SCALE 1:30

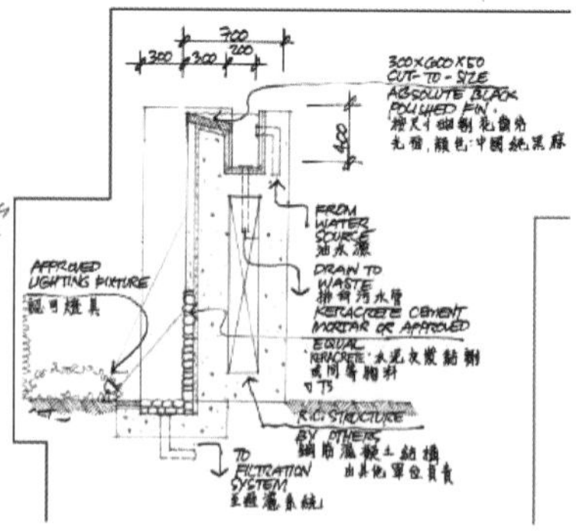

C SECTION
SCALE 1:30

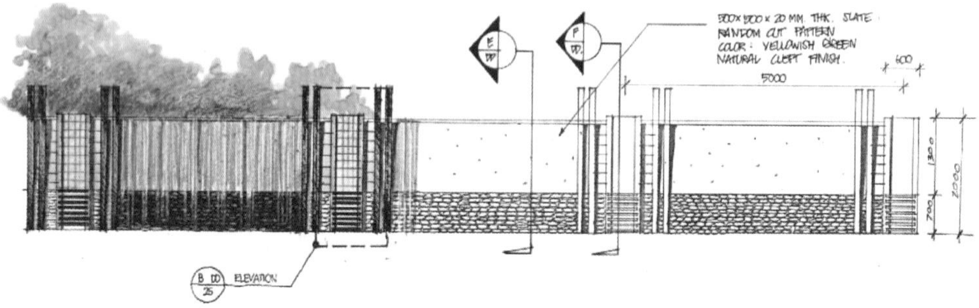

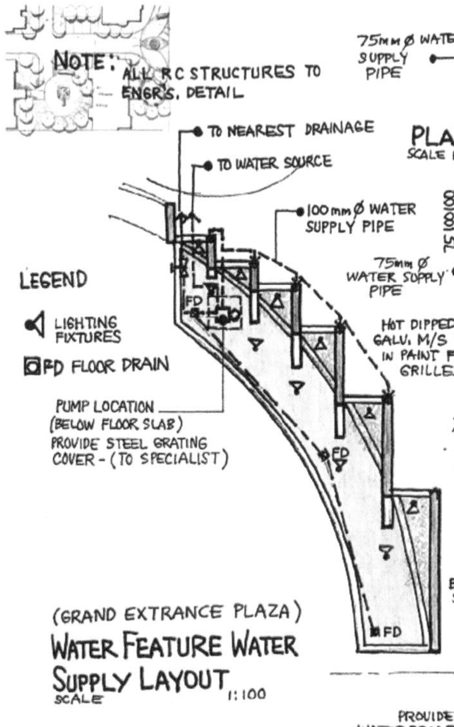

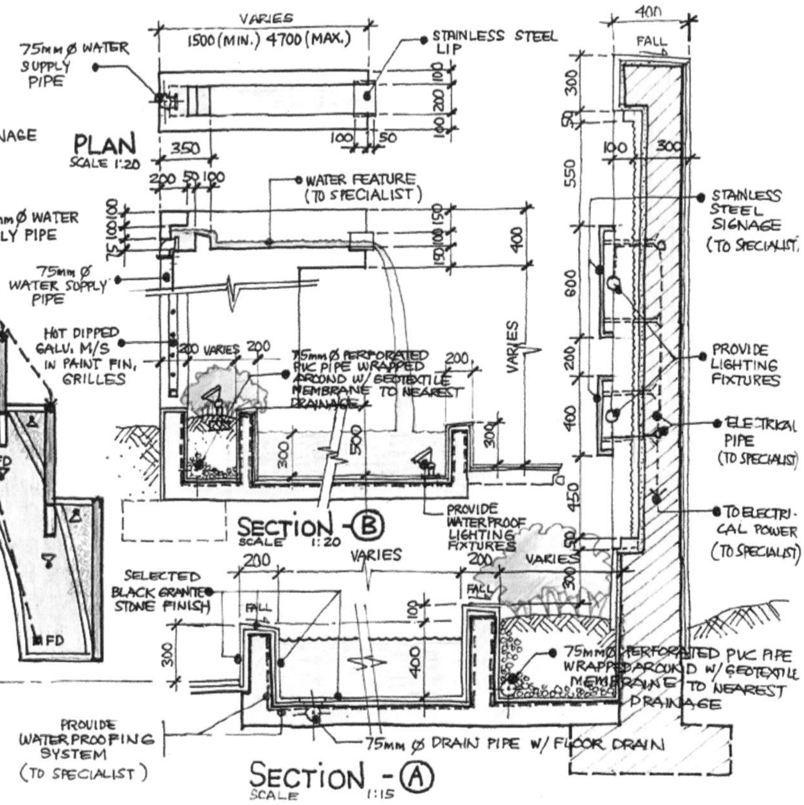

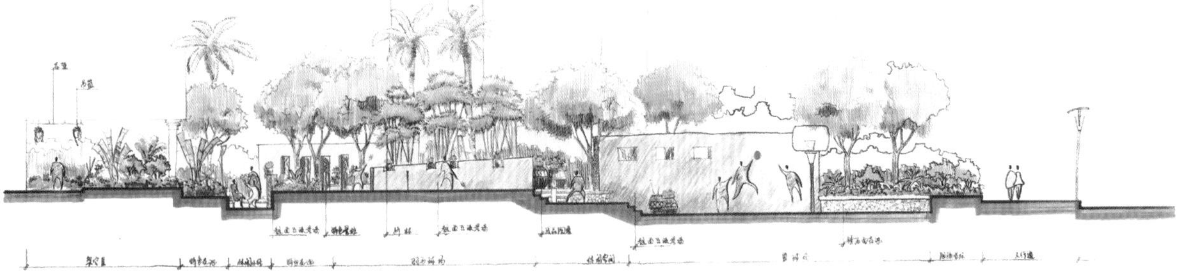

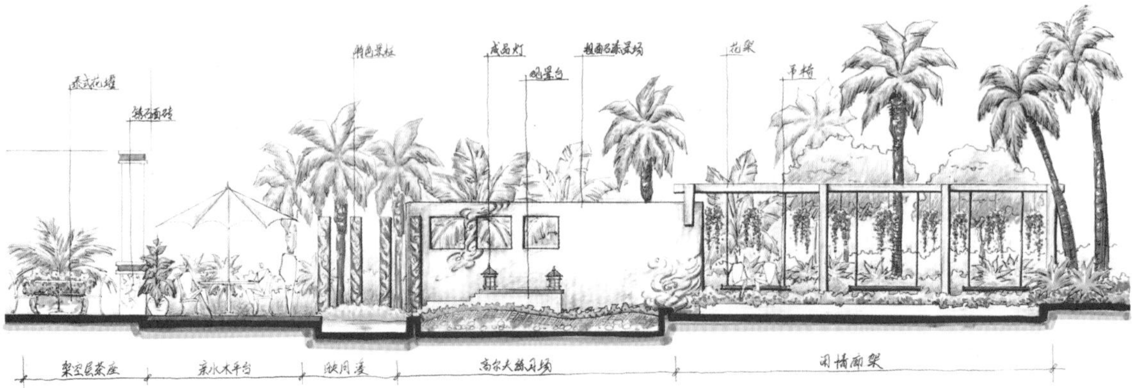
泰式花罐
特色景柱
成品灯
观景台
花架
吊椅
架空层茶座
亲水木平台
高尔夫练习场
闲情廊架

FEATURE POT
POLE TOP LIGHTING FIXTURES
(VERTICAL SCULPTURAL WALL)
RANDOM CUT, CLEFT FINISH, DARK AND
LIGHT GRAY WITH RUSTIC YELLOW SPOTS
(VERTICAL SCULPTURAL WALL)

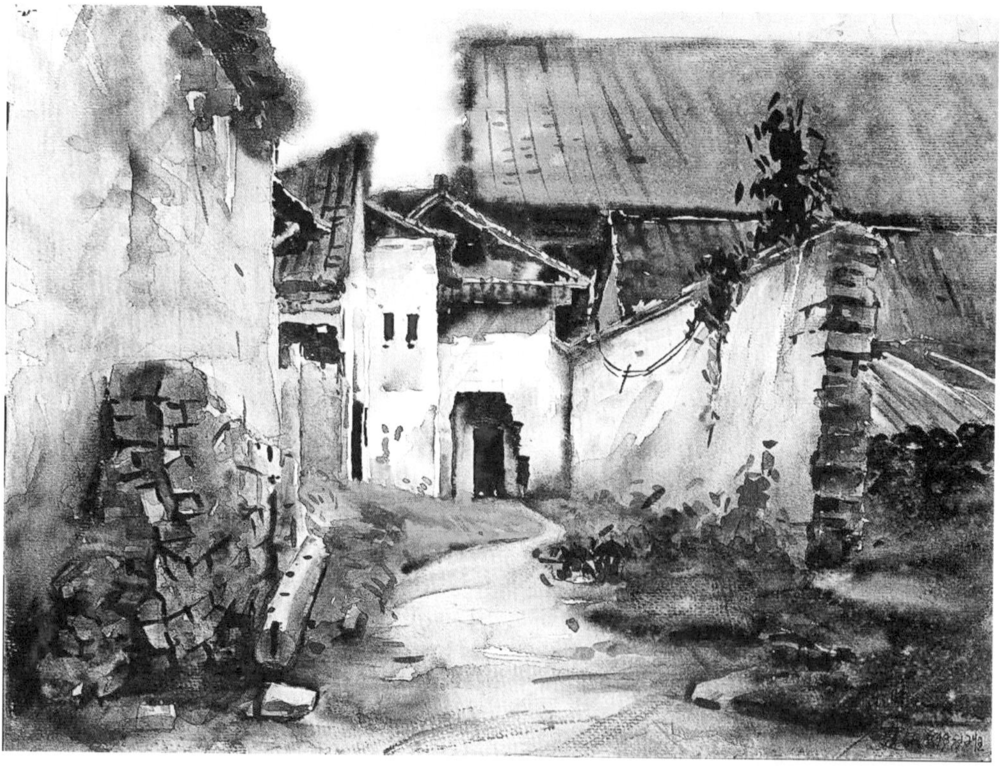

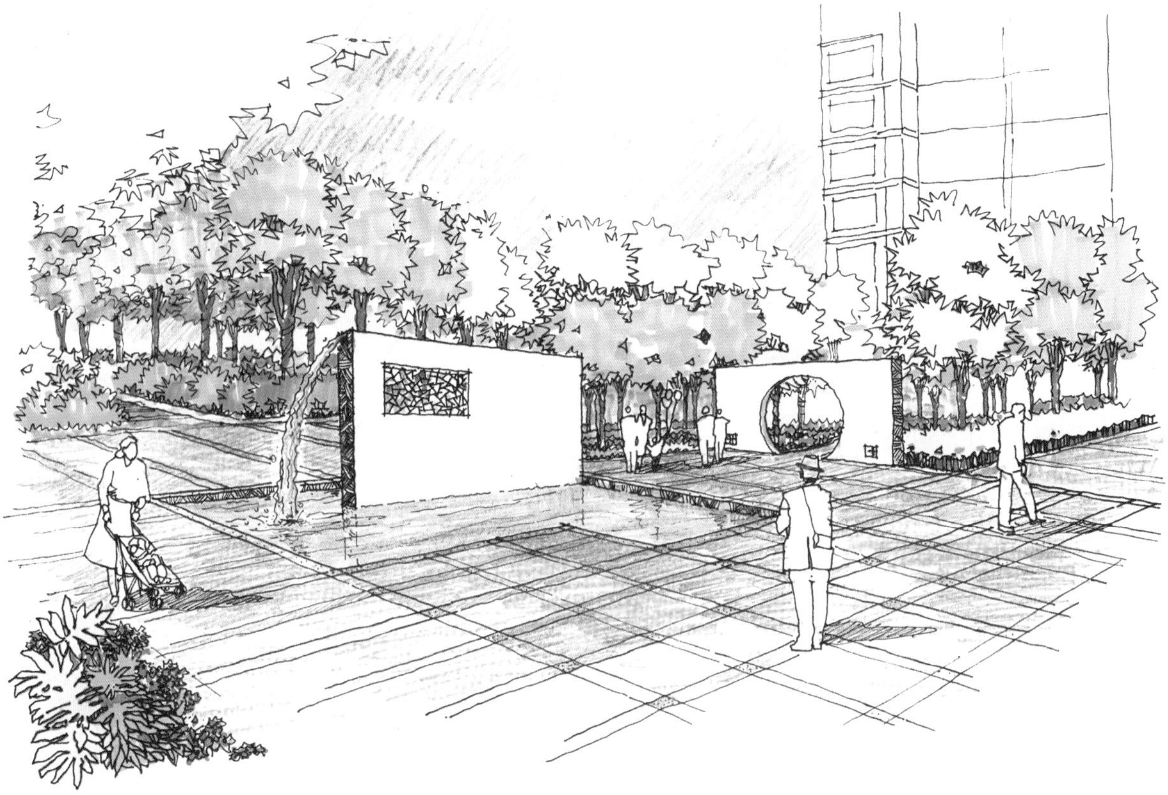

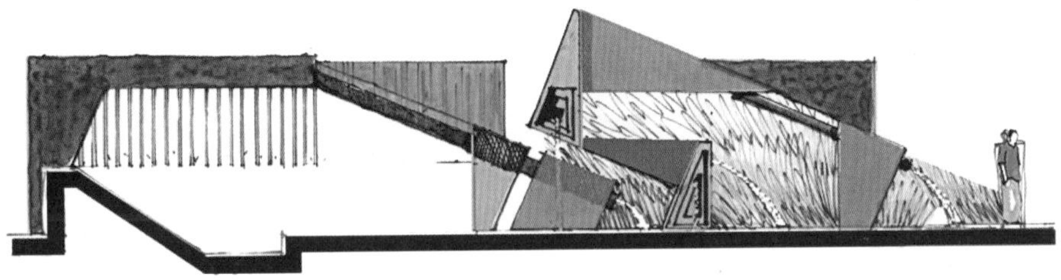

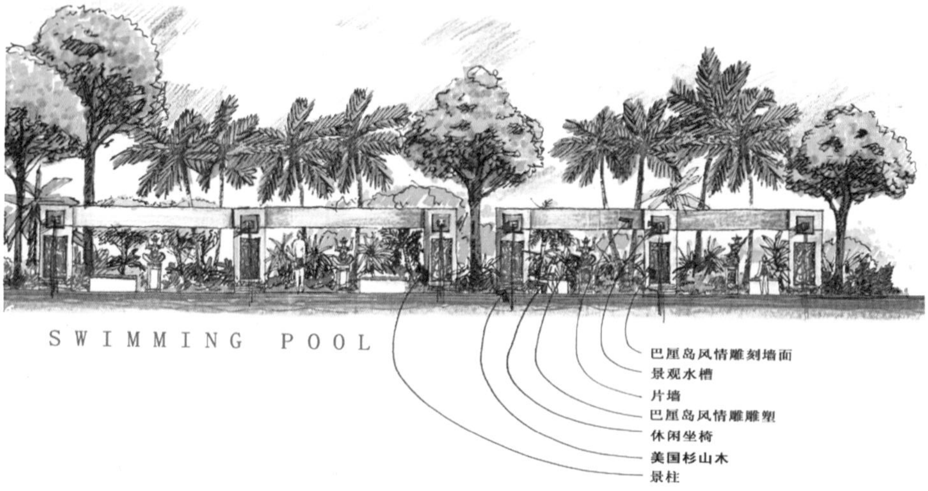
SWIMMING POOL
巴厘岛风情雕刻墙面
景观水槽
片墙
巴厘岛风情雕雕塑
休闲坐椅
美国杉山木
景柱

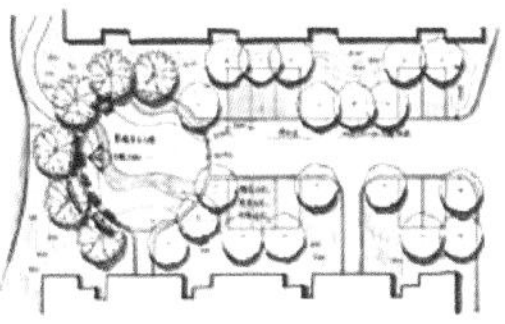

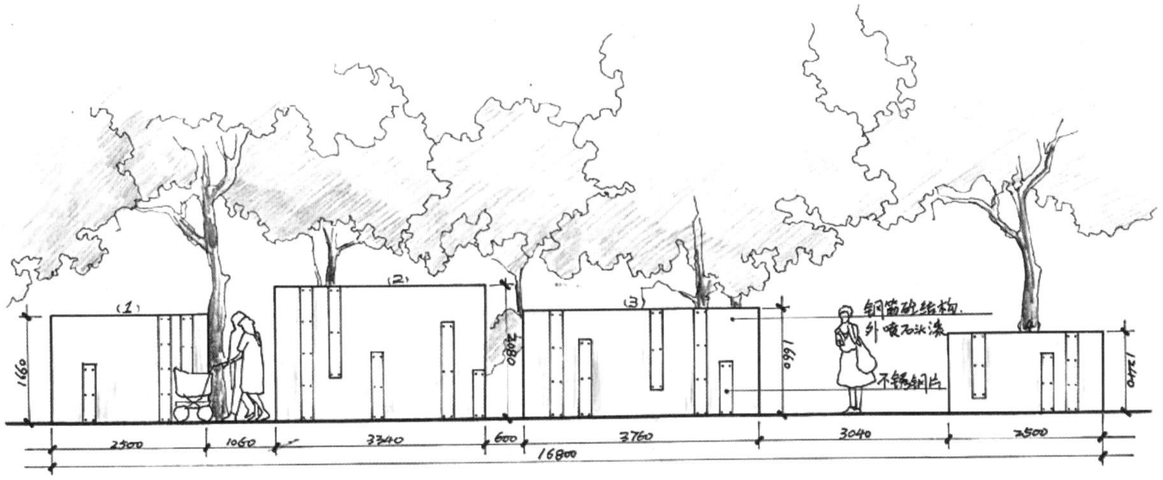

(1)
(2)
(3)
钢筋砼结构
不锈钢片
1660
2080
1660
1470
2500
1060
3340
600
3760
3040
2500
16800

A立面详图 1:50

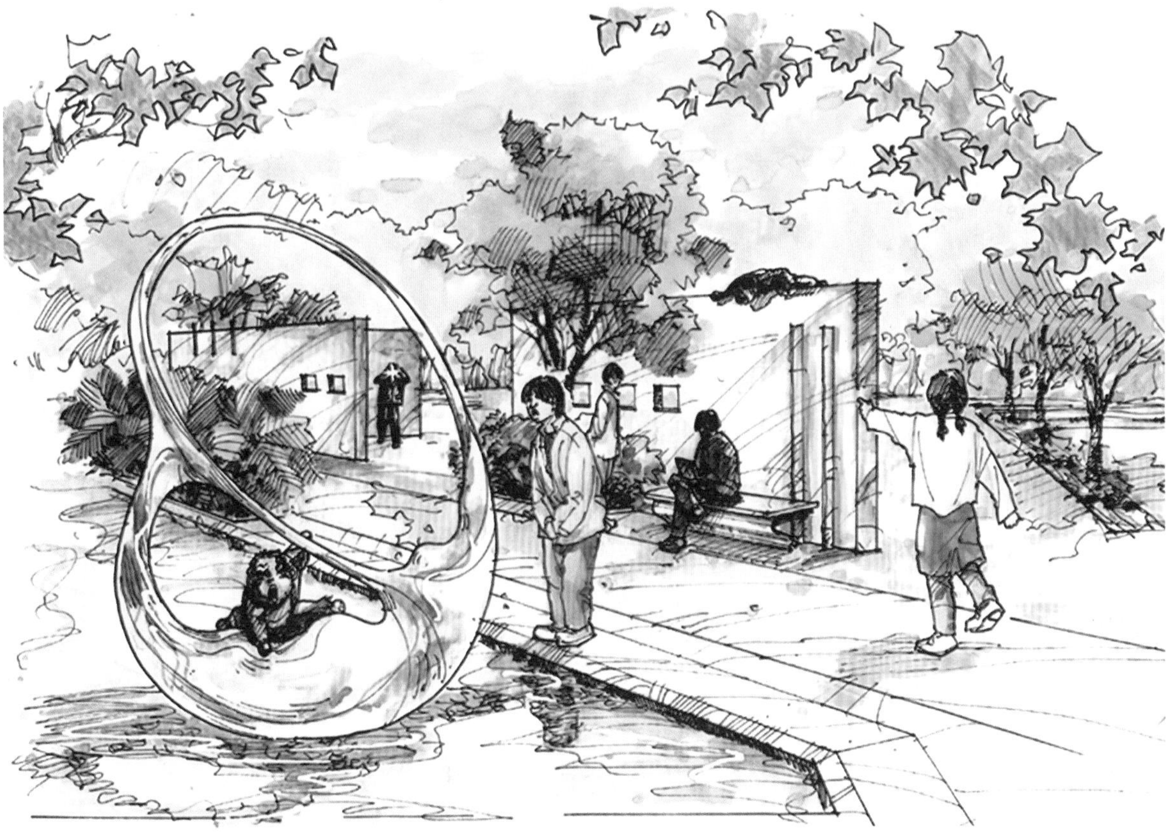

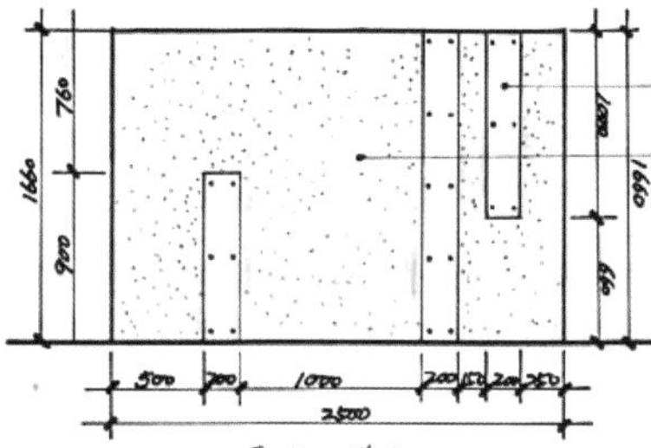

景墙(1)详图
1:30

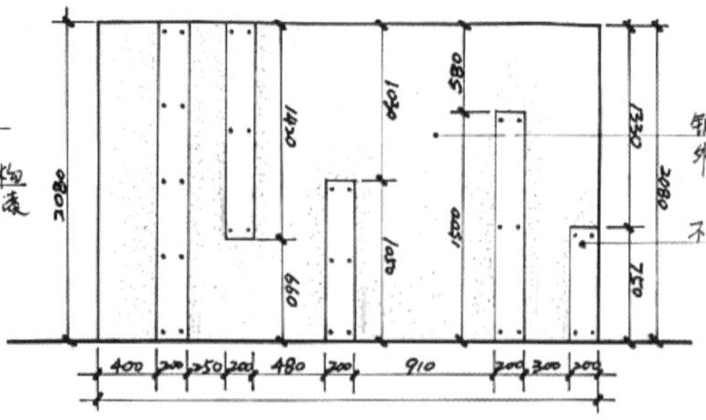

景墙(2)详图
1:30

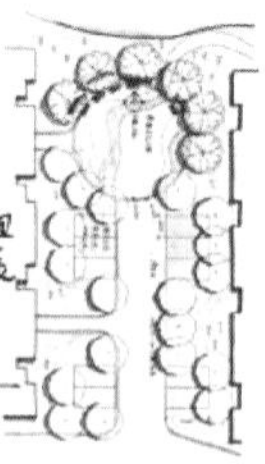

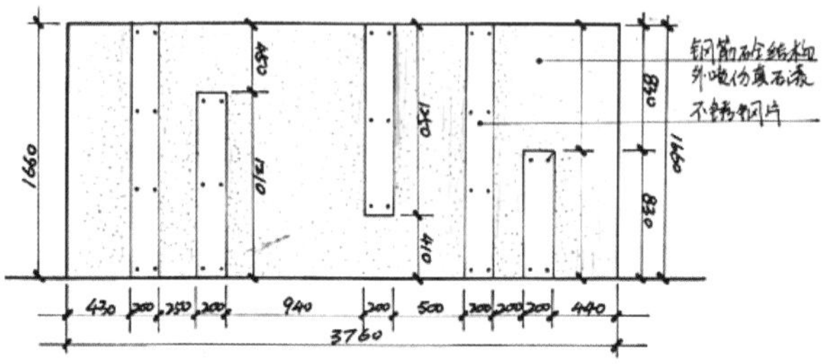

景墙(3)详图
1:30

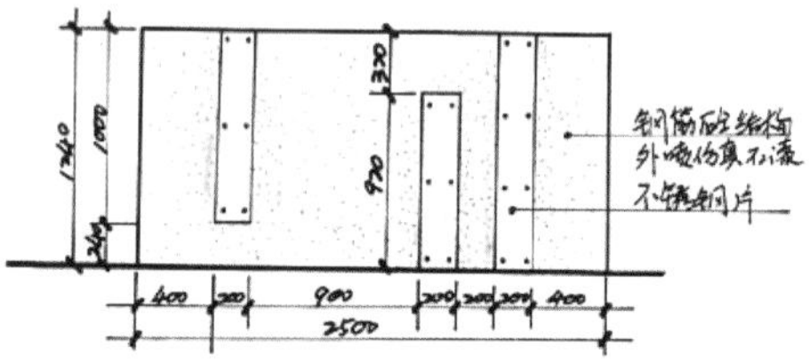

景墙(4)详图
1:30

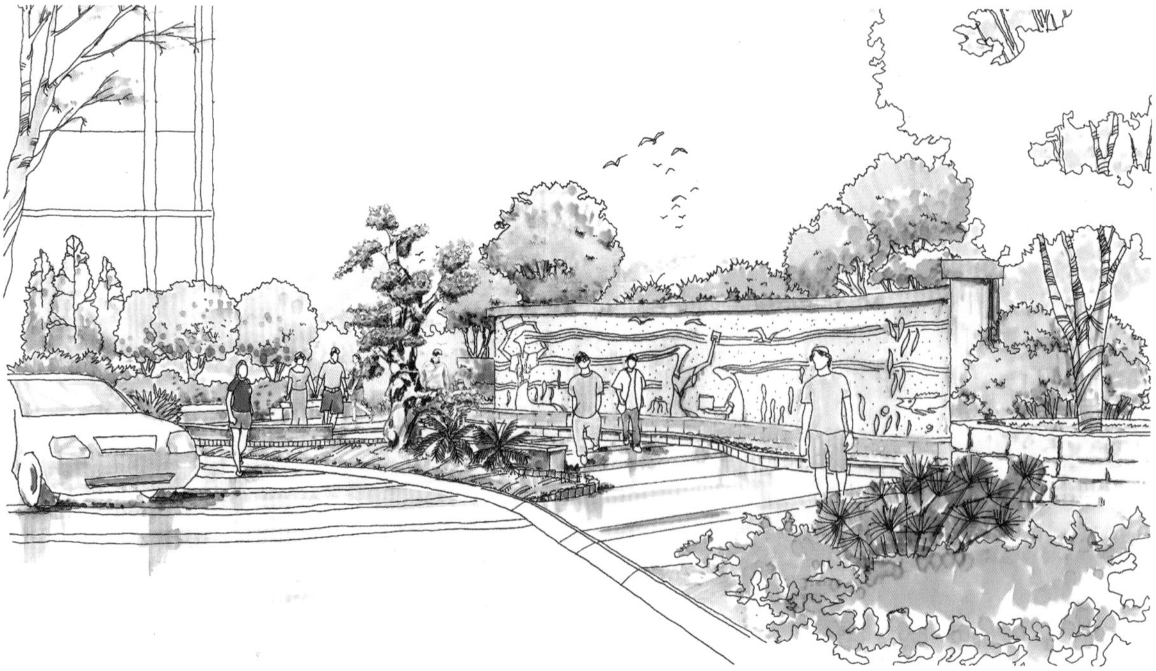

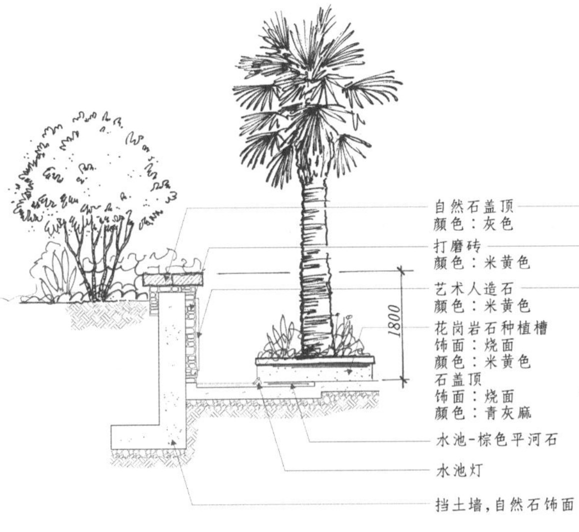

剖面图-A

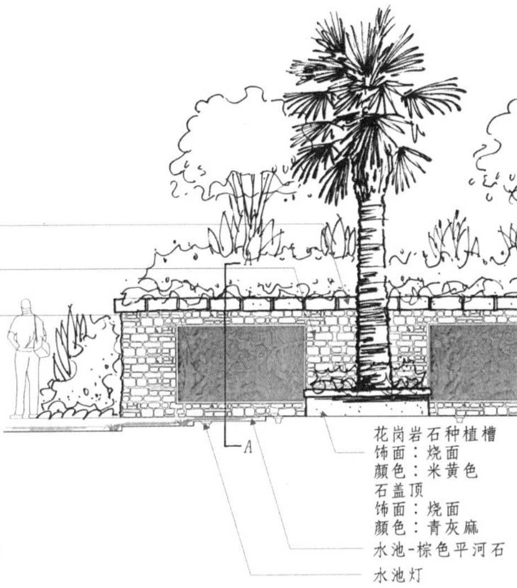

剖面图-B

挡土墙及标准自然水池剖面图 比例：1:50

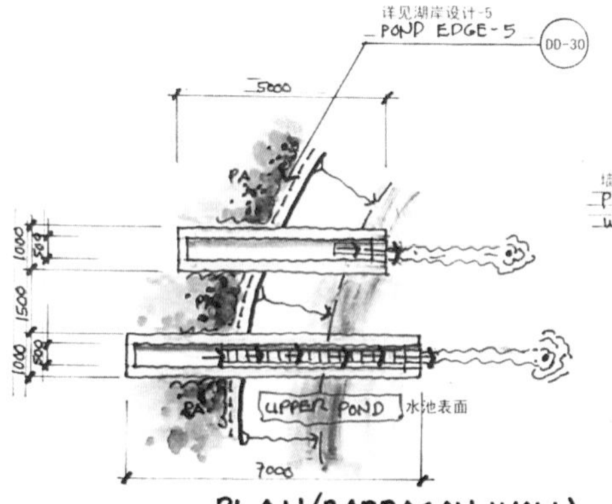

PLAN(BARRAGAN WALL)
SCALE 喷水景墙平面图 1:400

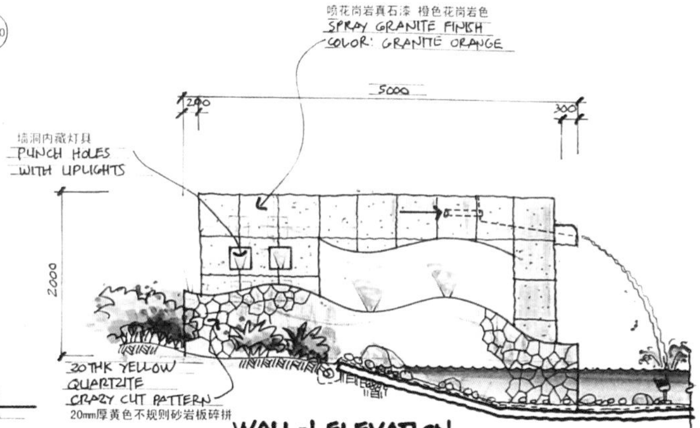

WALL-1 ELEVATION
SCALE 1:40 喷水景墙立面图-1

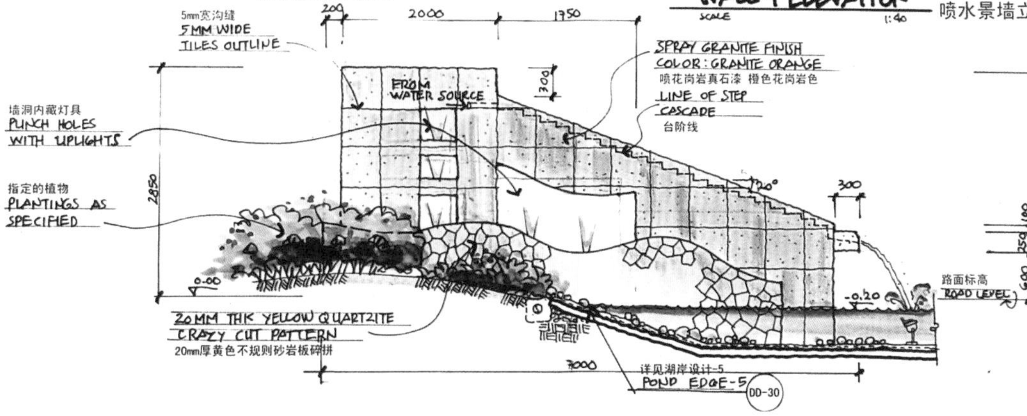

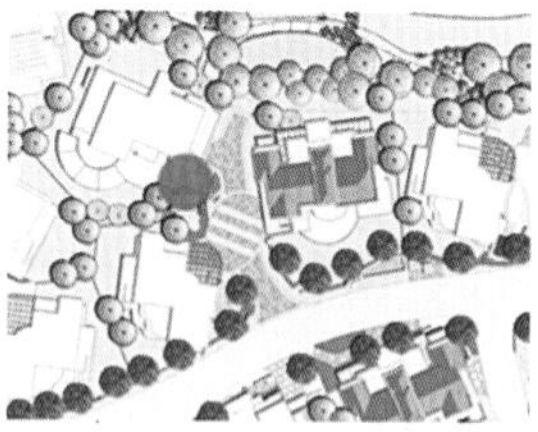

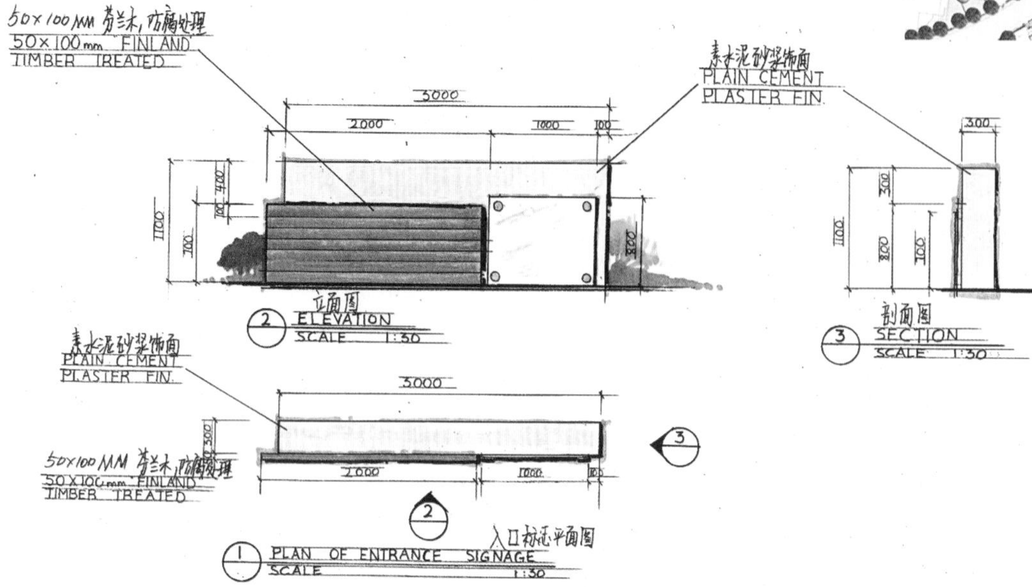
50×100MM 芬兰木,防腐处理
50×100mm FINLAND
TIMBER TREATED
素水泥砂浆饰面
PLAIN CEMENT
PLASTER FIN.
3000
2000
1000
100
1100
400
100
700
800
300
立面图
2 ELEVATION
SCALE 1:50
剖面图
3 SECTION
SCALE 1:30
素水泥砂浆饰面
PLAIN CEMENT
PLASTER FIN.
3000
300
50×100MM 芬兰木,防腐处理
50×100mm FINLAND
TIMBER TREATED
2000
1000
100
入口标志平面图
1 PLAN OF ENTRANCE SIGNAGE
SCALE 1:30

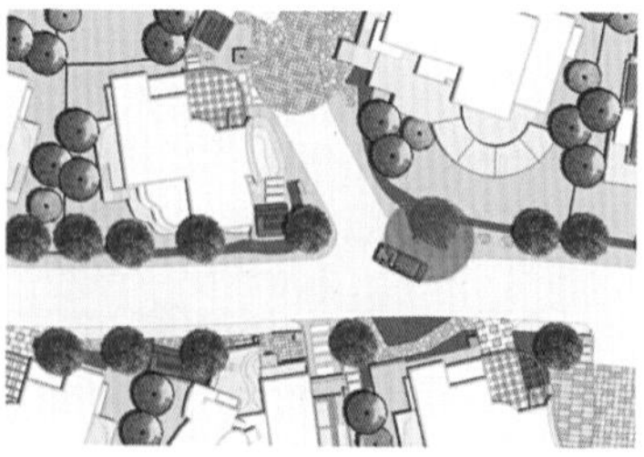

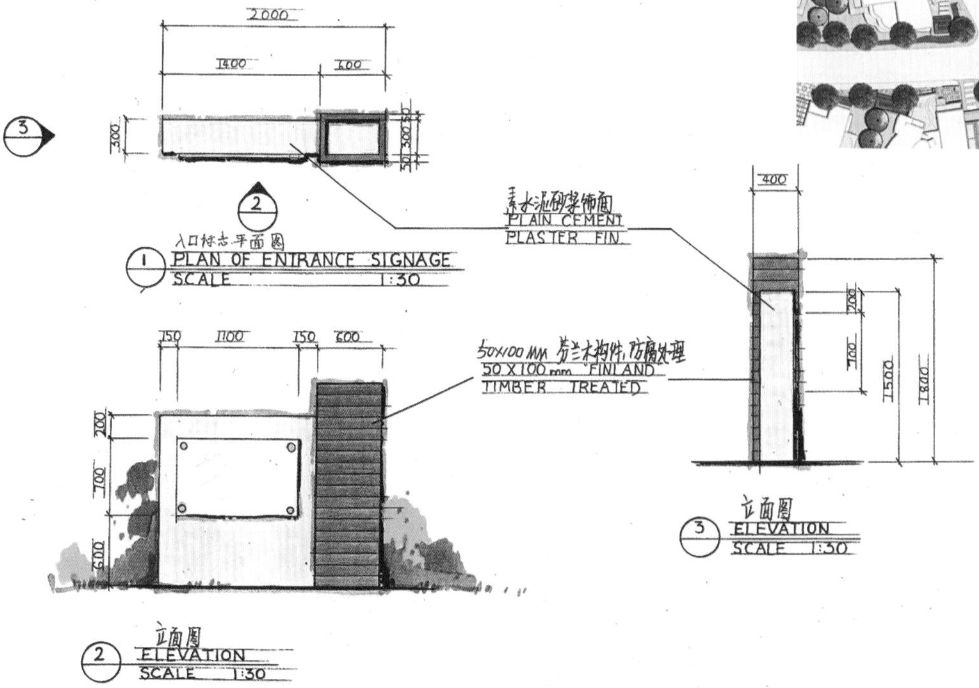
2000
1400
600
300
入口标志平面图
1 PLAN OF ENTRANCE SIGNAGE
SCALE 1:30
素水泥砂浆饰面
PLAIN CEMENT
PLASTER FIN.
400
150
1100
150
600
50×100MM 芬兰木构件,防腐处理
50×100mm FINLAND
TIMBER TREATED
200
700
600
1500
1800
立面图
2 ELEVATION
SCALE 1:30
立面图
3 ELEVATION
SCALE 1:30

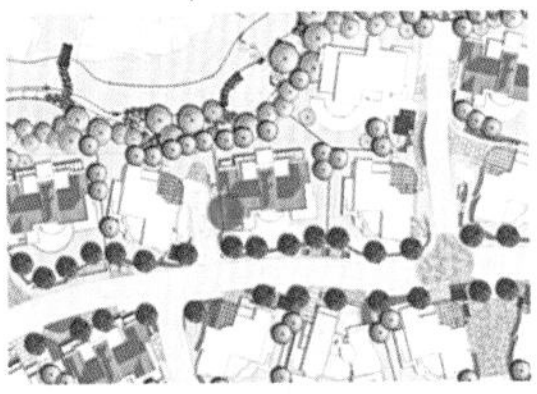

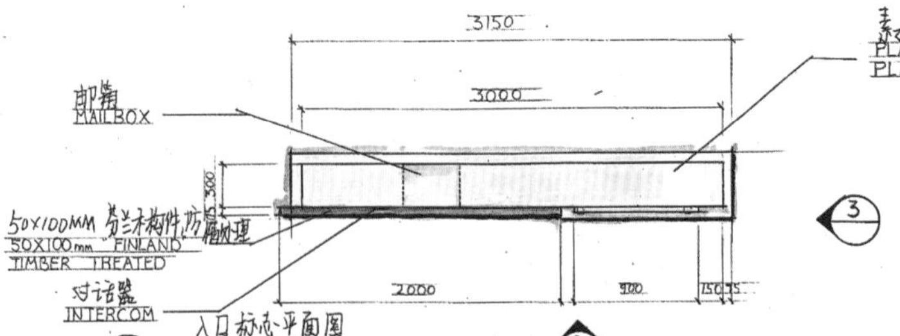

入口标志平面图
1 PLAN OF ENTRANCE SIGNAGE
SCALE 1:30

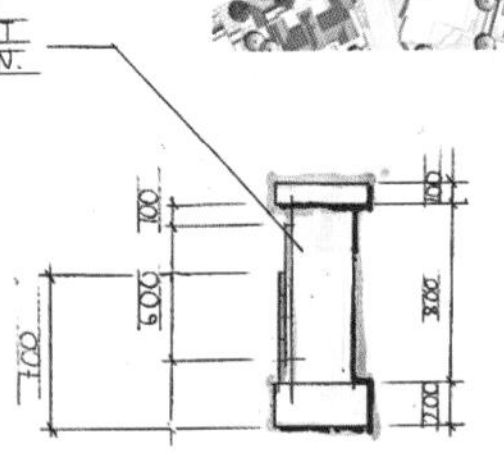

立面图
3 ELEVATION
SCALE 1:30

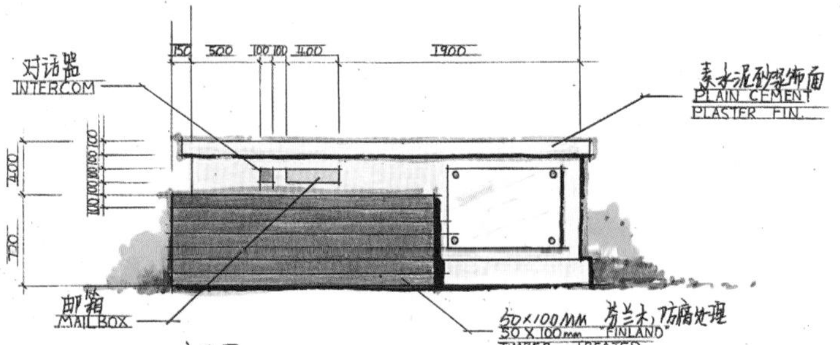

立面图
2 ELEVATION
SCALE 1:30

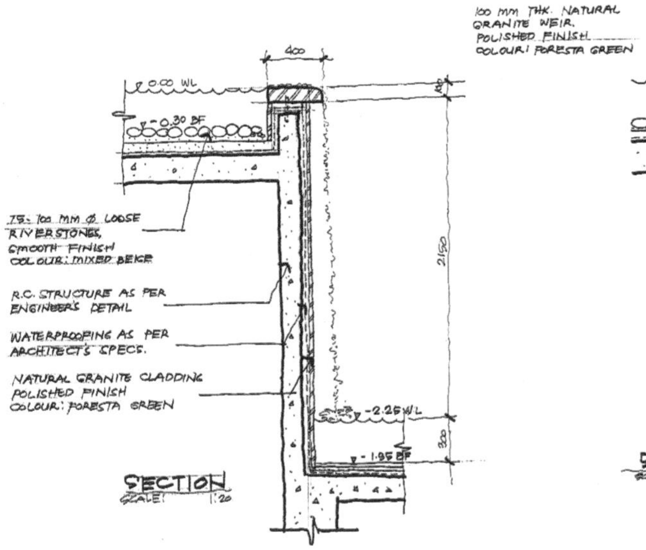

SECTION
SCALE 1:20

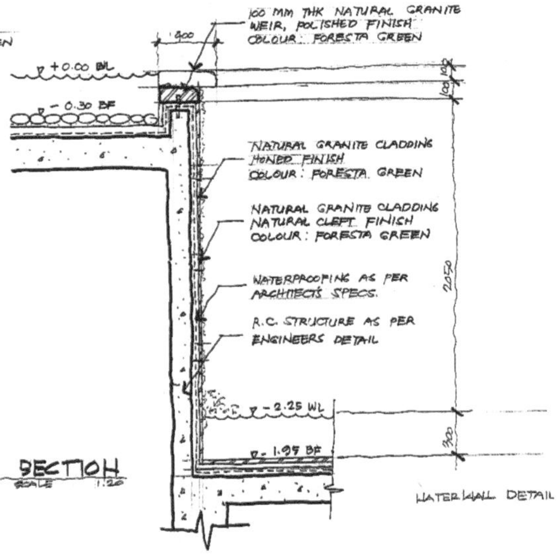

SECTION
SCALE 1:20

WATERWALL DETAIL

FOREST PLANTING ABOVE
ARCADE KIOSK RETAIL GLASS SHOP FRONT
BRIDGE
STREAM
AQUATIC PLANTING
GRANITE COLONADE WITH TIMBER TRELLIS
LIGHTS
STEPS TO LAWN

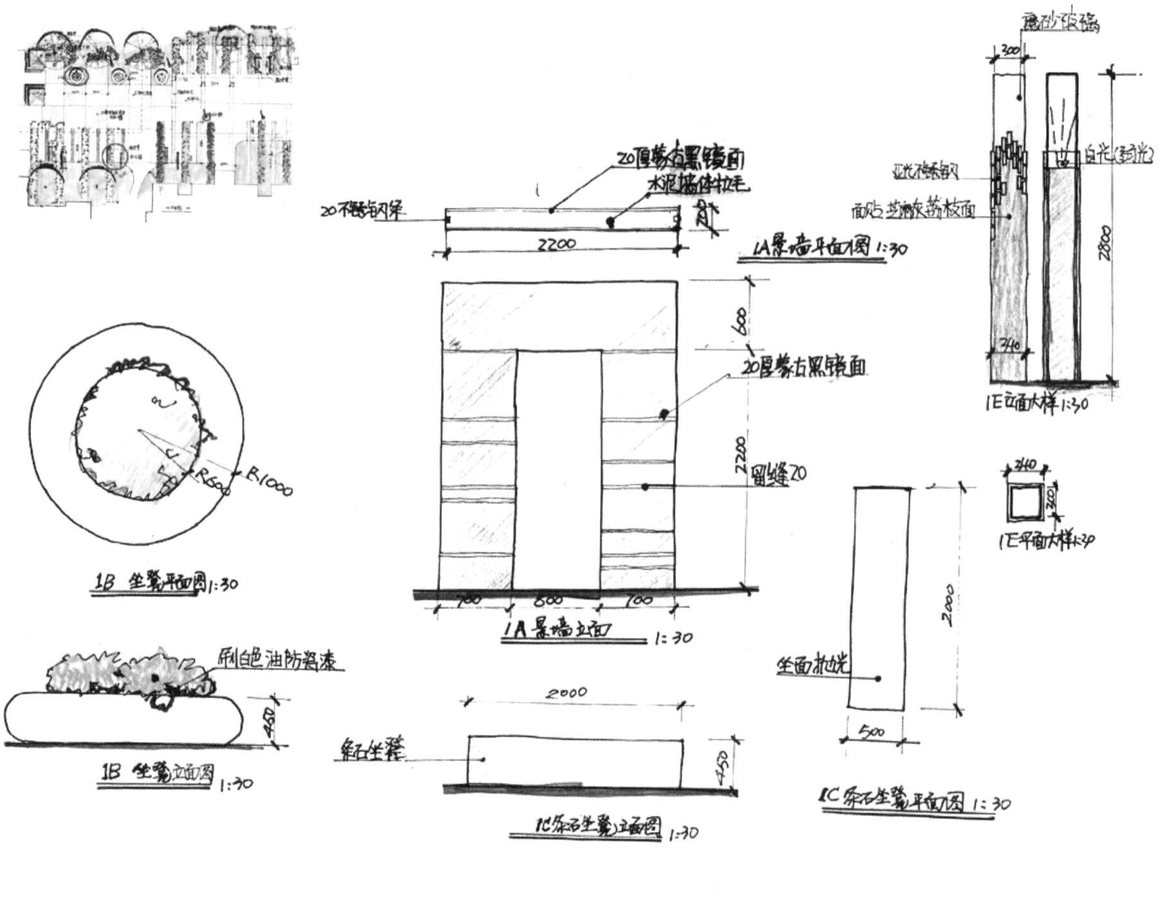
20厚蒙古黑镜面
水泥墙体拉毛
2200
1A景墙平面图 1:30
600
20厚蒙古黑镜面
2200
留缝20
700
800
700
1A景墙立面 1:30
R600
R1000
1B 坐凳平面图 1:30
刷白色油防潮漆
450
1B 坐凳立面图 1:30
2000
450
1C条石坐凳立面图 1:30
300
白光(射光)
2800
240
1E立面大样 1:30
240
300
1E平面大样 1:30
2000
500
1C条石坐凳平面图 1:30

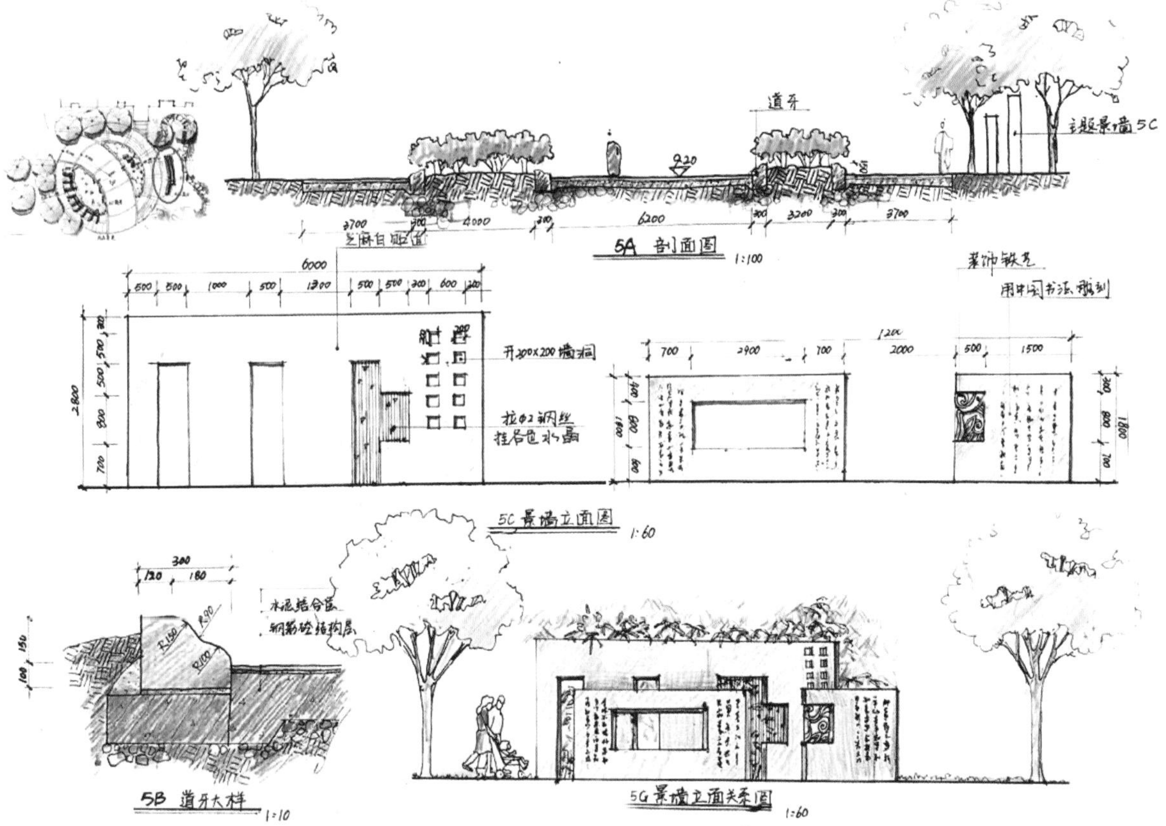
道牙
主题景墙5C
5A 剖面图
1:100
装饰铁艺
用中国书法雕刻
开200x200墙洞
拉Φ2钢丝
挂各色水晶
5C 景墙立面图
1:60
水泥结合层
钢筋砼结构层
5B 道牙大样
1:10
5G 景墙立面关系图
1:60

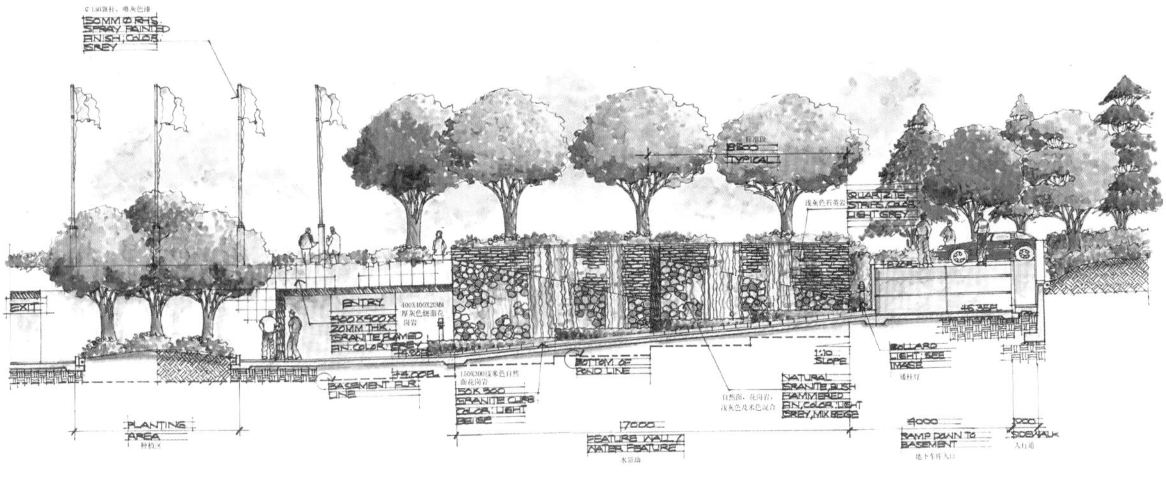
ENTRY
EXIT
BOTTOM OF POND LINE
BASEMENT FLR. LINE
1:10 SLOPE
BOLLARD LIGHT, SEE IMAGE
PLANTING AREA
FEATURE WALL / WATER FEATURE
RAMP DOWN TO BASEMENT
SIDEWALK

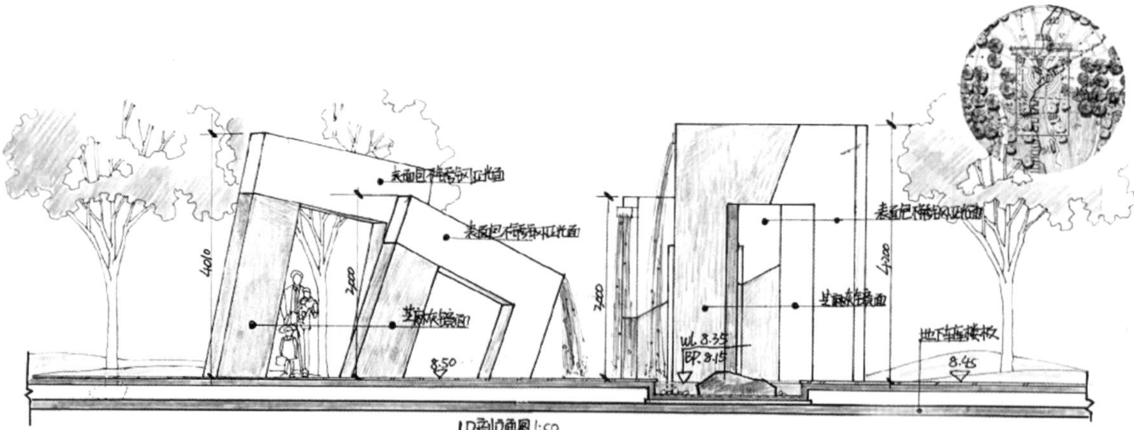

1D剖面图 1:50

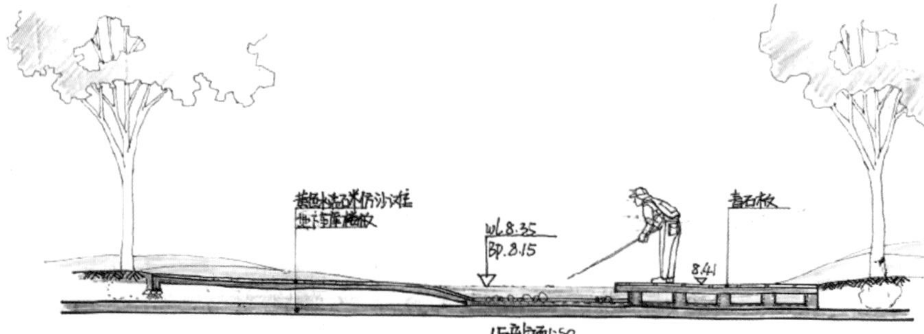

1E剖面 1:50

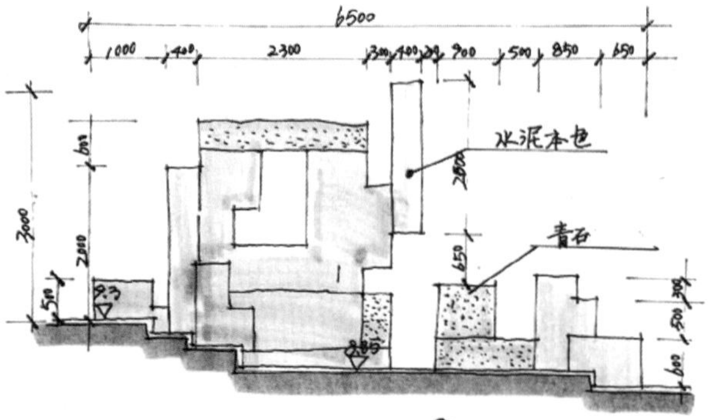

①

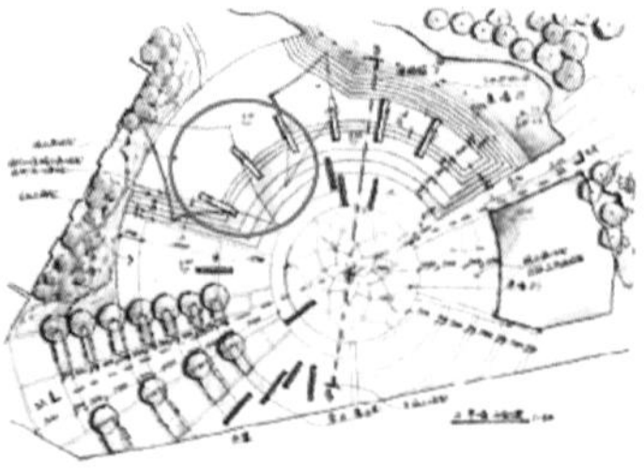

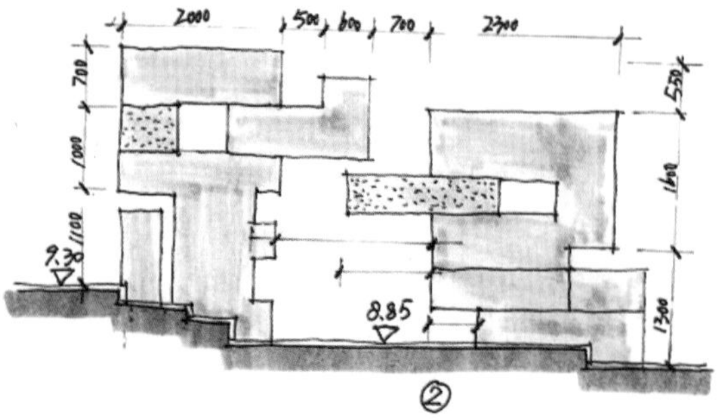

②

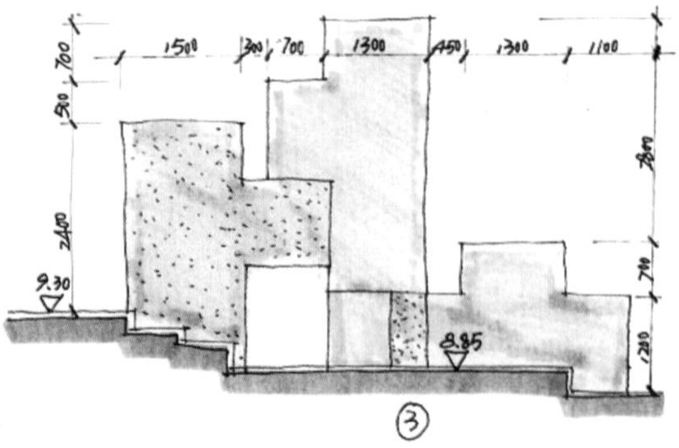

③

2F 景墙大样图 1:50

特色景墙
欧式构架
花钵喷泉
特色小品
植物群落

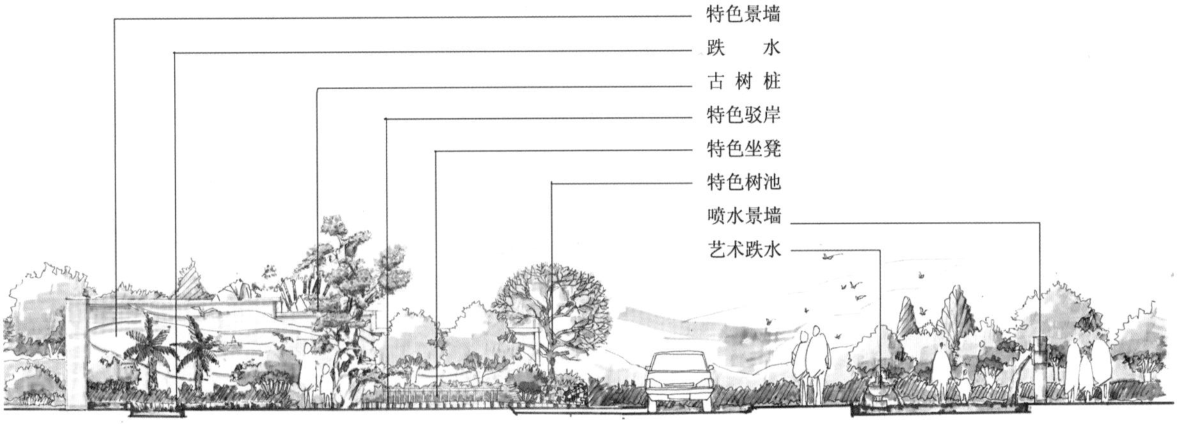
特色景墙
跌 水
古树桩
特色驳岸
特色坐凳
特色树池
喷水景墙
艺术跌水

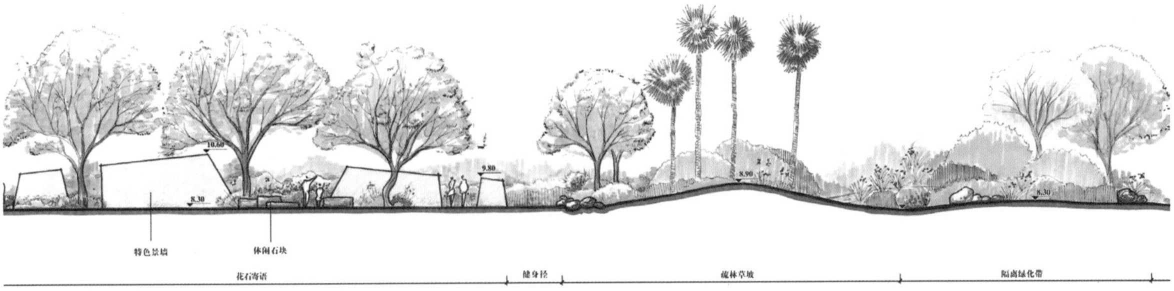
10.60
9.80
8.30
8.90
8.30
特色景墙
休闲石块
花石寄语
健身径
疏林草坡
隔离绿化带

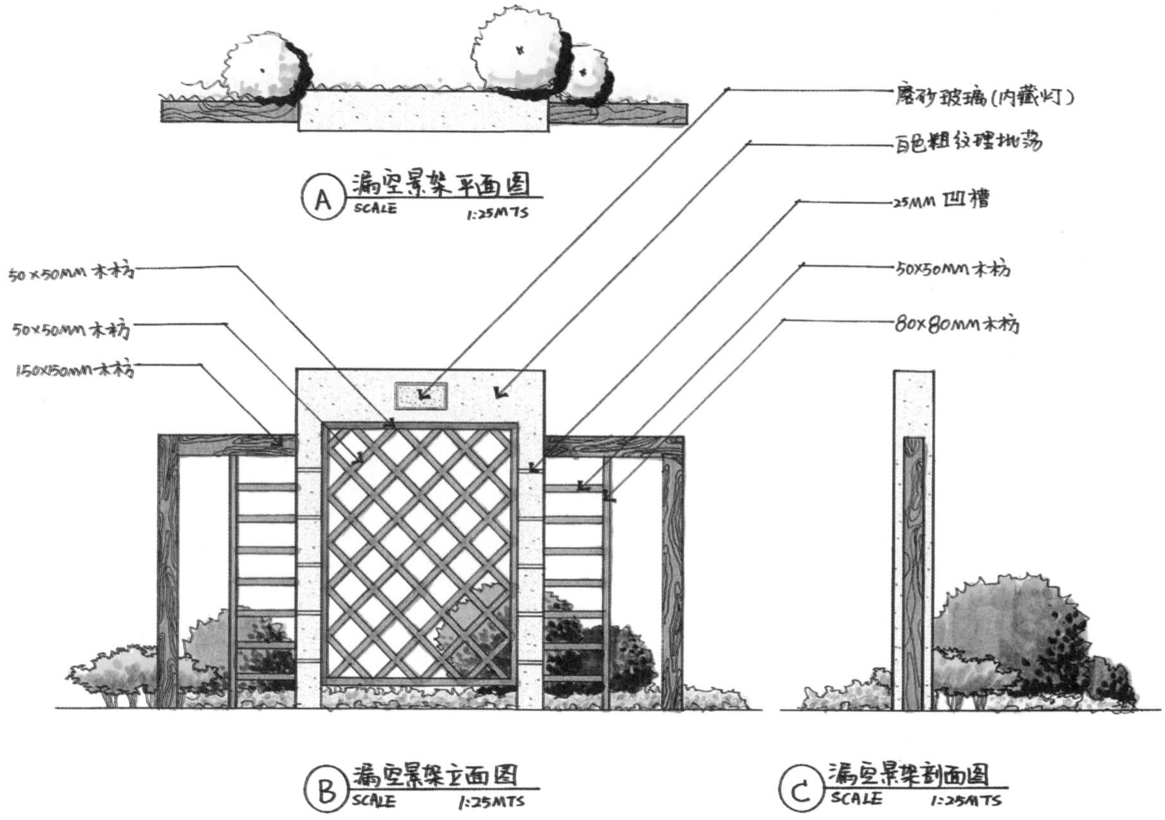
磨砂玻璃(内藏灯)
白色粗纹理批荡
25MM 凹槽
50x50MM 木枋
80x80MM 木枋
50×50MM 木枋
50×50MM 木枋
150×150MM 木枋
A 漏空景架平面图 SCALE 1:25MTS
B 漏空景架立面图 SCALE 1:25MTS
C 漏空景架剖面图 SCALE 1:25MTS

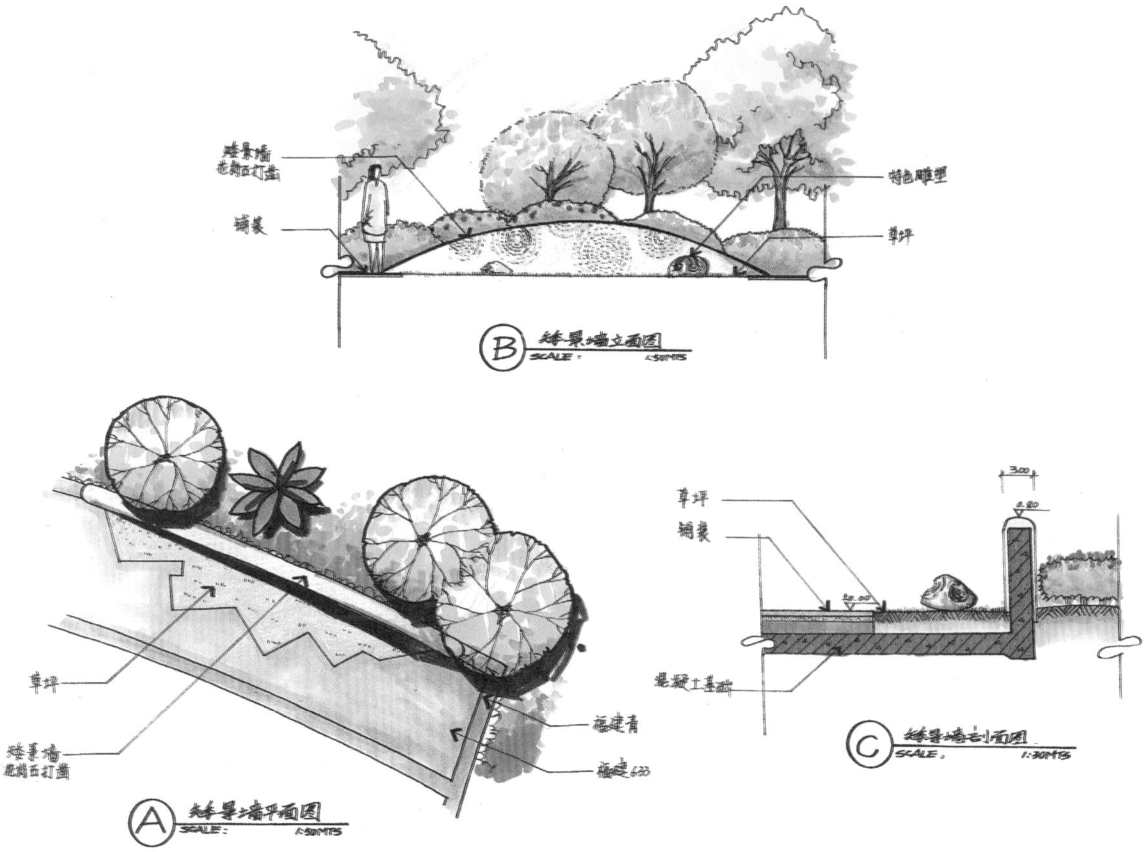
矮景墙
花岗石打凿
铺装
特色雕塑
草坪
B 矮景墙立面图
草坪
矮景墙
花岗石打凿
福建青
福建633
A 矮景墙平面图 SCALE 1:50MTS
草坪
铺装
混凝土基础
C 矮景墙剖面图 SCALE 1:30MTS

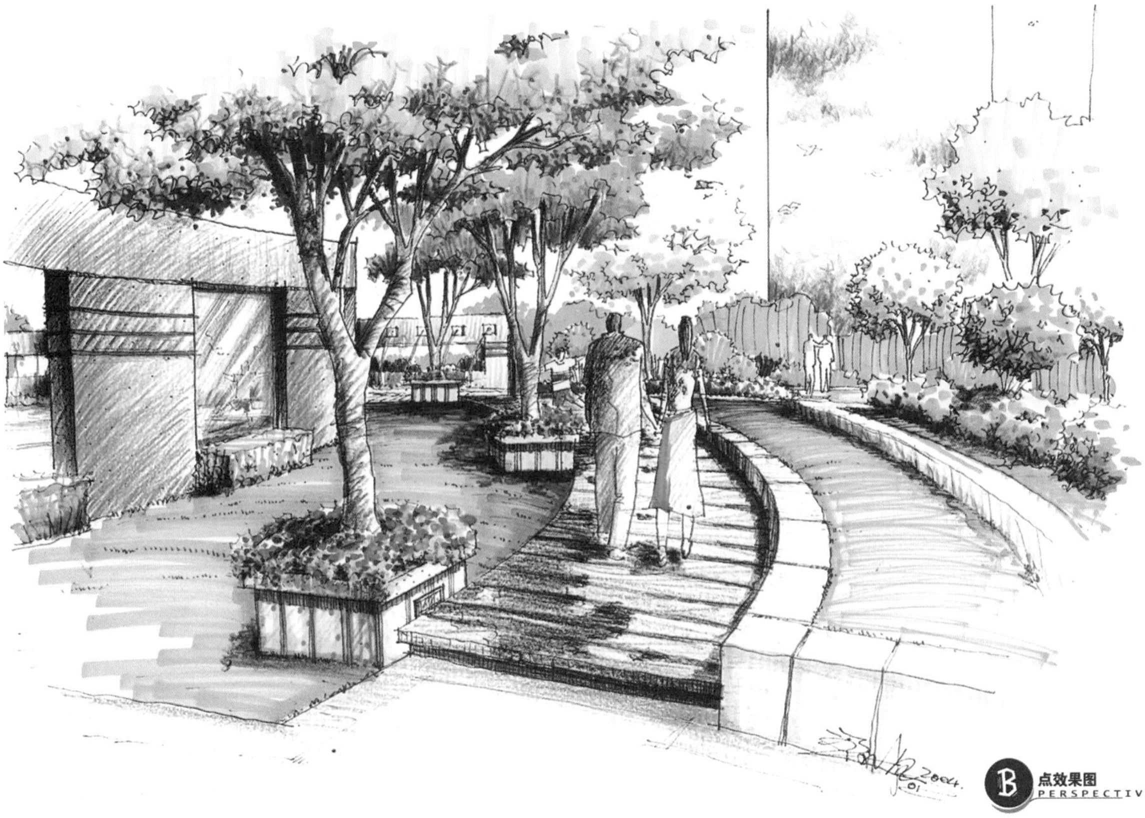

B 点效果图 PERSPECTIVE

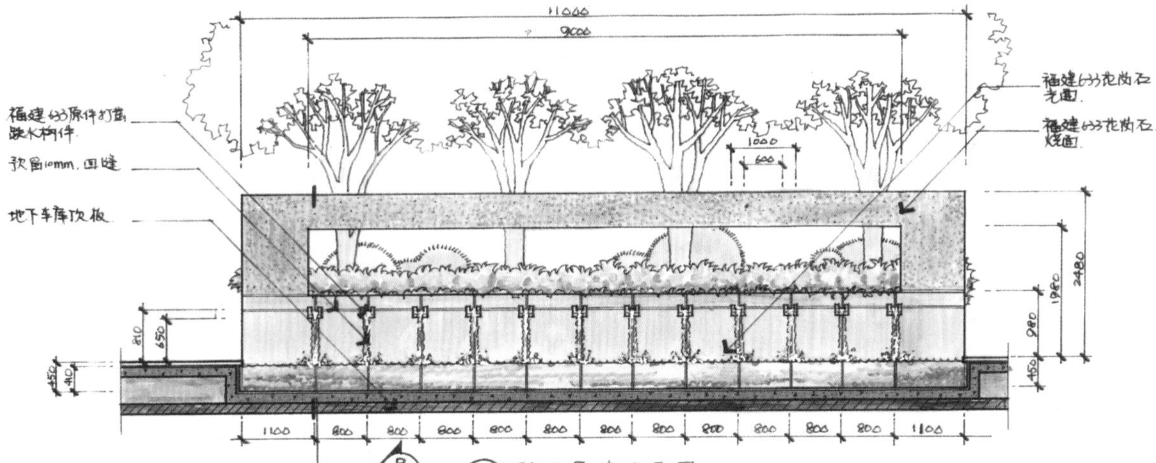

A 跌水景墙立面图
SCALE 1:50 HTS

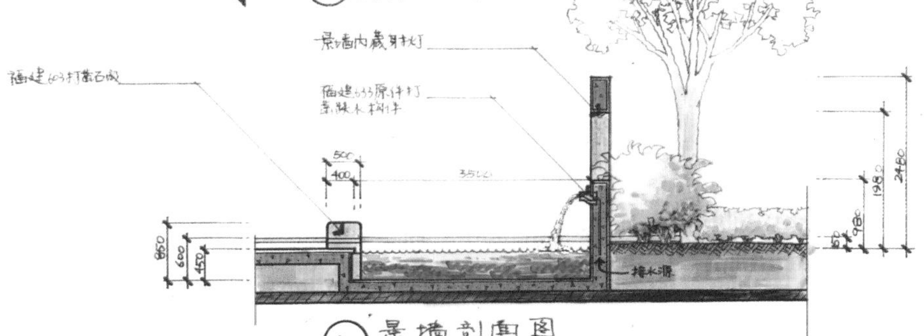

B 景墙剖面图
SCALE 1:50 MTS

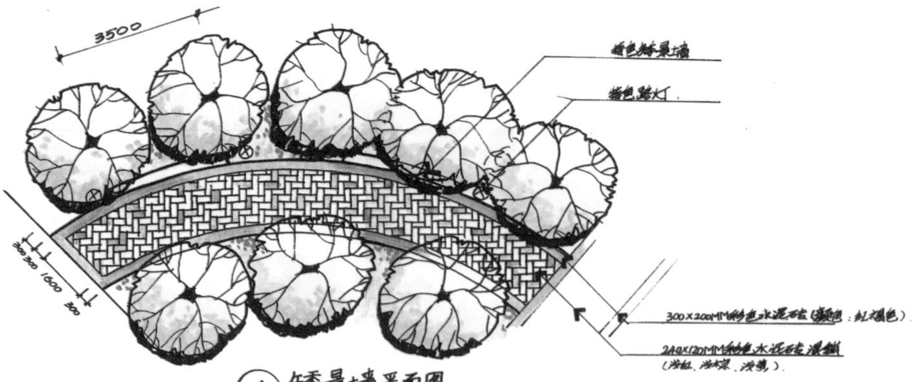

A 矮景墙平面图
SCALE 1:100MTS

B 矮景墙立面图
SCALE 1:50MTS

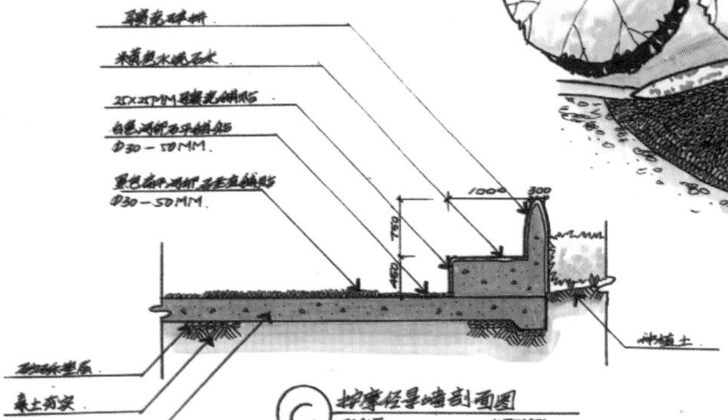

C 按摩径景墙剖面图
SCALE: 1:75MTS

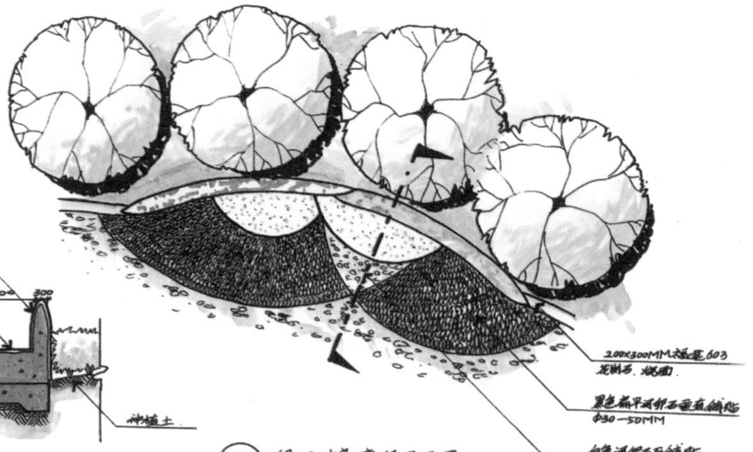

A 卵石按摩径平面图
SCALE: 1:75MTS

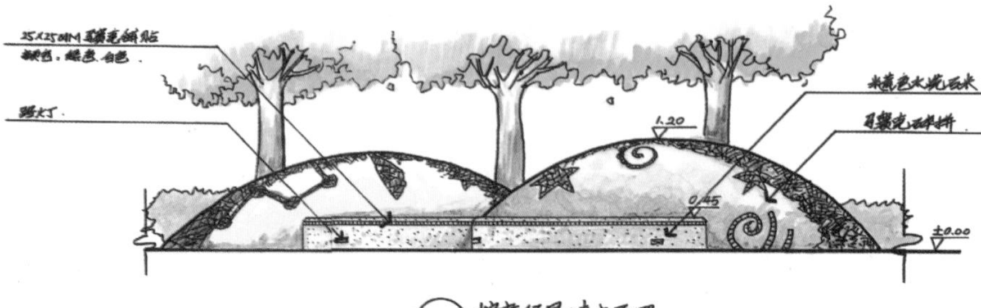

B 按摩径景墙立面图.
SCALE: 1:50MTS.

水景墙立面图

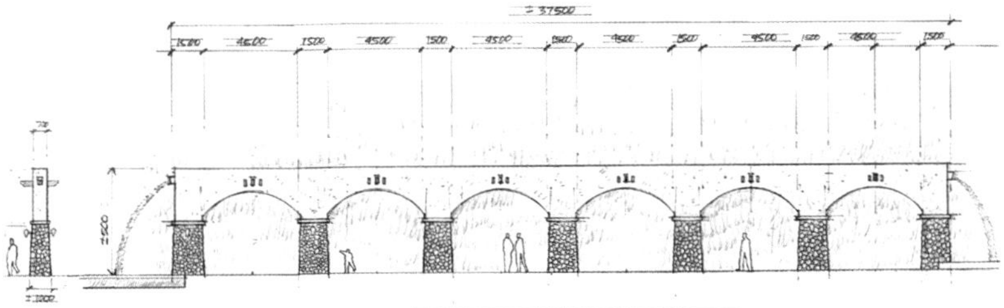

AQUADUCT ELEV PROFILE
SCALE 1:75 MTS
水景墙立面图

深圳·中航"阳光新苑"环境扩初设计

SZ002/DD23 L&A

水景墙水池剖面图

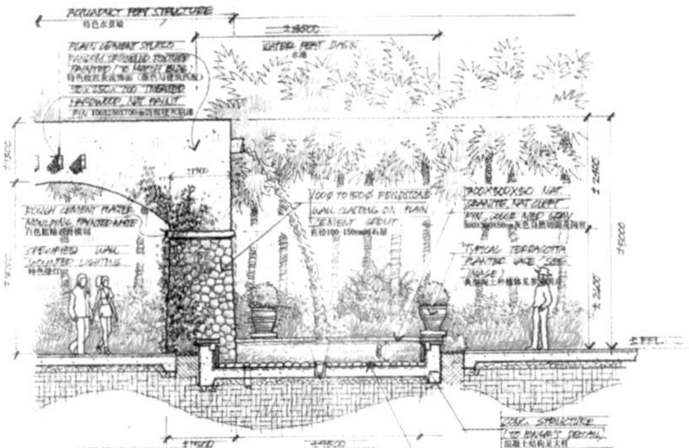

水景墙水池剖面图
PART ELEV AQUADUCT & WATER BASIN SECTION
SCALE 1:50 MTS

深圳·中航"阳光新苑"环境扩初设计

SZ002/DD24 L&A

太仓·上海假日 景观扩初设计

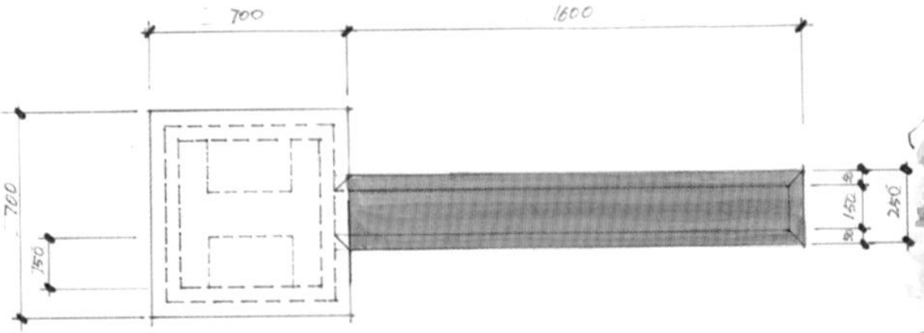

平面图
比例 1:15

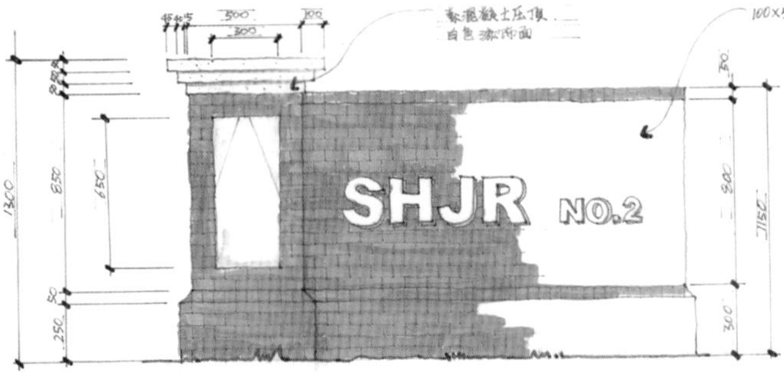

正立面图
比例 1:15

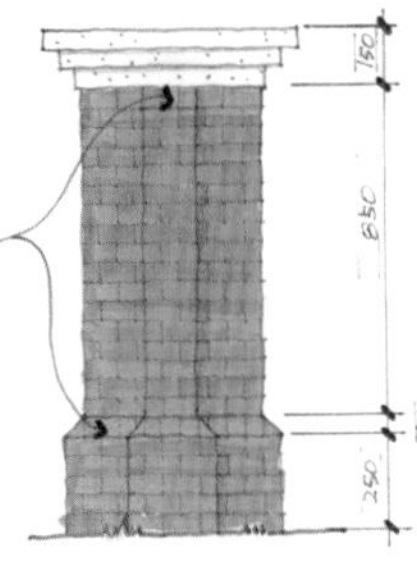

侧立面图
比例 1:15

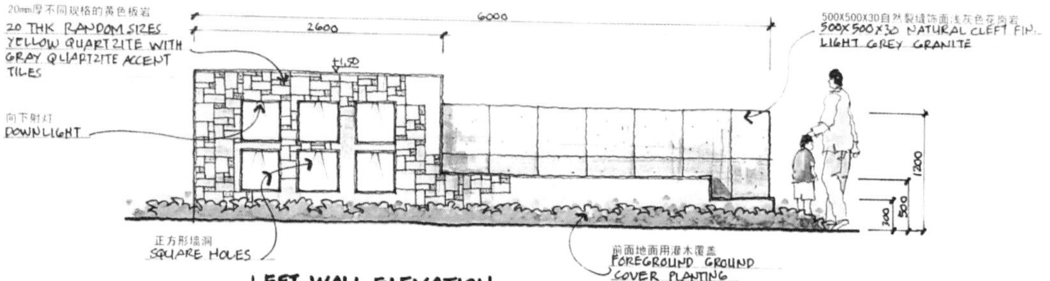

LEFT WALL ELEVATION
SCALE 1:30
景墙左立面图

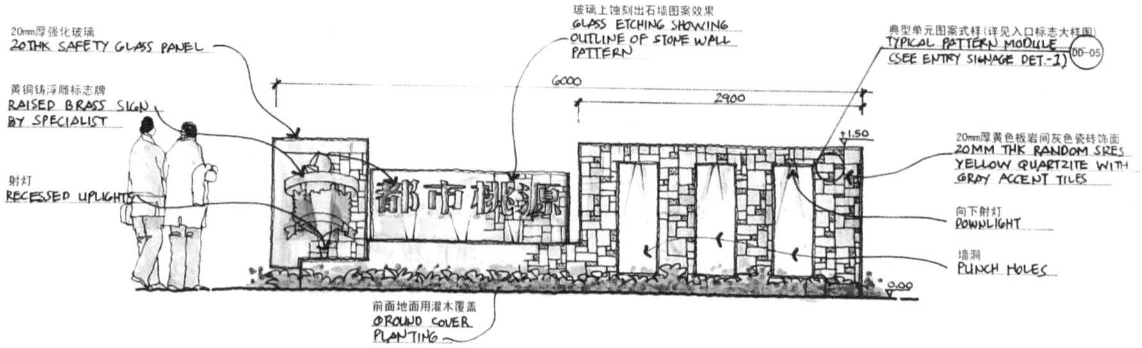

RIGHT WALL ELEVATION
SCALE 1:30
景墙右立面图

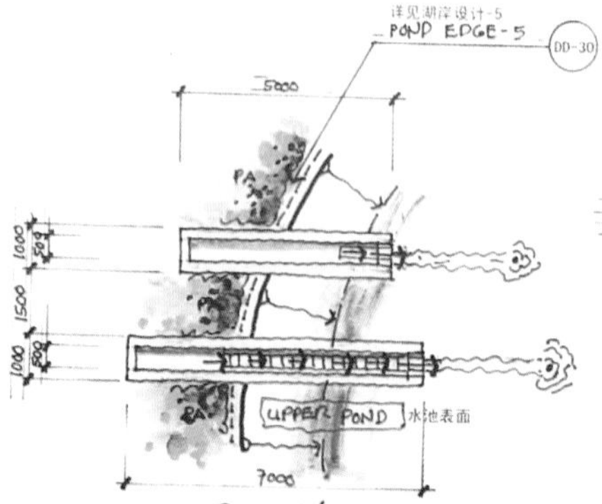

PLAN(BARRAGAN WALL)
SCALE 1:100
喷水景墙平面图

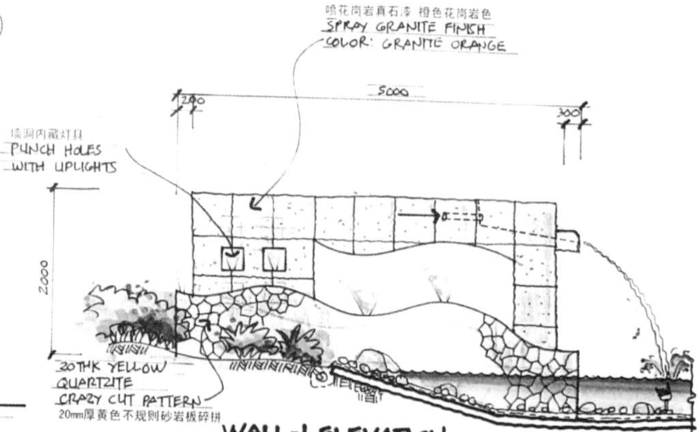

WALL-1 ELEVATION
SCALE 1:40
喷水景墙立面图-1

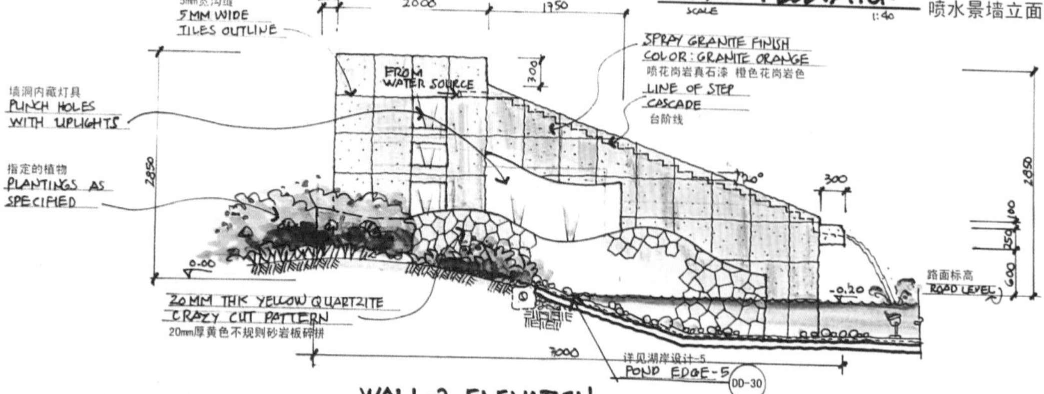

WALL-2 ELEVATION
SCALE 1:40
喷水景墙立面图-2

3 栏杆、围墙

栏杆中国古称阑干，也称勾阑，是桥梁和建筑上的安全设施。栏杆在使用中起分隔、导向的作用，使被分割区域边界明确清晰，设计好的栏杆，很具装饰意义。周代礼器座上有类似栏杆的构件。汉代以卧棂式栏杆为最多。六朝盛行钩片勾阑。栏杆转角立望柱或寻杖绞口造者，均可见于云冈石窟、敦煌壁画。元明清的木栏杆比较纤细，而石栏杆逐渐脱离木制栏杆的形制，趋向厚重。清末以后，西方古典比例、尺度和装饰的栏杆形式进入中国。现代栏杆的材料和造型更为多样。

一、栏杆简介

从形式上看，栏杆可分为节间式与连续式两种。前者由立柱，扶手及横挡组成，扶手支撑于立柱上；后者具有连续的扶手，由扶手，栏杆柱及底座组成。常见种类有：木制栏杆、石栏杆、不锈钢栏杆、铸铁栏杆、铸造石栏杆、水泥栏杆、组合式栏杆。

一般低栏高 0.2 ~ 0.3 米，中栏 0.8 ~ 0.9 米，高栏 1.1 ~ 1.3 米。栏杆柱的间矩一般为 0.5 ~ 2 米。

1. 形式 有漏空和实体两类。漏空的由立杆、扶手组成，有的加设

有横档或花饰。实体的是由栏板、扶手构成，也有局部漏空的。栏杆还可做成坐凳或靠背式的。栏杆的设计，应考虑安全、适用、美观、节省空间和施工方便等。

2. 构造 建造栏杆的材料有木、石、混凝土、砖、瓦、竹、金属、有机玻璃和塑料等。栏杆的高度主要取决于使用对象和场所，一般高 900 毫米；幼儿园、小学楼梯栏杆还可建成双道扶手形式，分别供成人和儿童使用；在高险处可酌情加高。楼梯宽度超过 1.4 米时，应设双面栏杆扶手（靠墙一面设置靠墙扶手），大于 2.4 米时，须在中间加一道栏杆扶手。居住建筑中，栏杆不宜有过大空档或可攀登的横档。

二、栏杆款式

1. 铁栏杆 栏杆和基座相连接，有以下几种形式：①插入式：将开脚扁铁、倒刺铁件等插入基座预留的孔穴中，用水泥砂浆或细石混凝土浆填实固结；②焊接式：把栏杆立柱（或立杆）焊于基座中预埋的钢板、套管等铁件上；③螺栓结合式：可用预埋螺丝母套接，或用板底螺帽栓紧贯穿基板的立杆。上述方法也适用于侧向斜撑式铁栏杆。

钢筋混凝土栏杆多用预制立杆，下端同基座插筋焊接或预埋铁件相连，上端同混凝土扶手中的钢筋相接，浇筑而成。

2. 木栏杆 以榫接为主。若为望柱，则应将柱底卯入楼梯斜梁，扶
手再与望柱榫接。

3. 栏板式栏杆 可采用现浇或预制的钢筋混凝土板和钢丝网水泥板，
也可用砖砌。室内的还可考虑使用钢化玻璃和有机玻璃等。扶手多为木制的，常以木螺丝固定于立杆顶端的通长扁铁条上（木立杆时为榫接）。也可用金属焊接和螺钉固接或以金属作骨衬，饰以木质和塑料面层，或为混凝土浇注、水磨石抹面等。断面形式和尺寸应根据功能需要。

4. 石栏杆 采用大理石或花岗岩制作，上有石扶手，中间是石栏杆，下方的底板根据需要，可要可不要，如果是楼梯的开头一端还有一根较大将军柱。拼接处主要是用铁条和专用云石胶连接。由于其由天然石材经物理加工制作，所以抗老化能力较强，外观较厚重，具有现代气息。室外多用花岗岩材质为主，室内则多用大理石材质。

5. 夹胶玻璃护栏 使用不锈钢立柱及 A3 钢立柱，立柱配件锁住玻璃，
玻璃多采用 6+6 安全干夹玻璃，不绣钢圆管扶手，特色是玻璃为主要配件，款式比较现代，适合通透式楼房及高档商场等公共场所。

三、安全要求

1. 在低窗台附加栏杆，重外观效果更得重安全

常见的低窗台距地 0.5 米左右，如果紧贴内墙增加 0.4 米栏杆或栅栏肯定达到规范要求的防护措施。但由于美观要求和利用窗台的需求，很多人喜欢将栏杆设在紧贴窗扇的位置，如果窗台台面太大，如凸窗等，小孩经常站在窗台上眺望，而且使用者也必须站到窗台上开启窗户，这时，附加在窗台上的栏杆本身高度应达到 0.9 米，如果窗台太低，住户往往会无意识攀登到窗台上，不宜简单附加低栏杆，否则危险是没有充分杜绝的。

2. 以固定窗作为低窗台的防护措施时，仍有危险存在

近年来，采用低窗台或落地窗的住宅越来越多，有若干情况存在安全问题。如固定窗框强度不够，使用者轻趴在窗框上会导致玻璃破裂；落地窗仅用固定玻璃，没有必要的防护，儿童玩耍、椅子翻倒等正常活动会碰破玻璃，造成险情；在高层住宅的高层套型中采用落地窗时，如果没有必要的防护设施，老年人普遍反映外眺时眩晕。类似情况只要引起投诉，设计人员就要承担

一定责任。同时，根据国外相关规范和我国部分地区标准，2001 年 5 月 1 日起实施的中国工程建设标准化协会标准《斜屋顶下可居住空间技术规程》提出：“当斜屋顶窗的单块玻璃面积大于 1.5 平方米时，应采用安全玻璃”。因此，设计固定扇落地玻璃窗时，务必采取确实可行的安全防护措施。

3. 封闭阳台的栏杆，不可采用窗台的高度

关于阳台，《住宅设计规范》（GB50096-2011）要求“住宅的阳台栏板或栏杆净高，六层及六层以下的不应低于 1.05m；七层及七层以上的不应低于 1.10m。（注：窗外有阳台或平台时可不受此限制。窗台的净高或防护栏杆的高度均应从可踏面起算，保证净高达到 0.90m。）”。比窗台护栏要求 0.9 米高些。规范编制的初衷是，阳台往往三面临空，是全家向外眺望活动比较集中的地方，对栏杆的防护要求应该高些。近年来阳台封闭现象比较普遍，一些工程在设计阶段就按照封闭阳台设计，并认为封闭阳台的栏杆高度可按窗台要求降低。但在施工图审查或工程监理中经常引起争议。质检和监理部门明确认定阳台是阳台，窗户是窗户，指出如果将阳台当窗户，工程图中出现许多不能自圆其说的矛盾，比如面积计算、日照间距、窗地比指标等。一些工程设计虽然按封闭阳台设计，实施时仍然交给住户自行处理，引起事故或纠纷使设计者十分被动。所以应考虑到，封闭阳台并没有改变阳台三面临空，是全家向外眺望活动比较集中的地方等性质，并且阳台是否封闭应是住户自己的选择，目前封闭的阳台日后敞开的可能性完全存在，必要的安全防护措施不能减少或降低标准，因此封闭阳台的栏杆决不可采用窗台的高度。

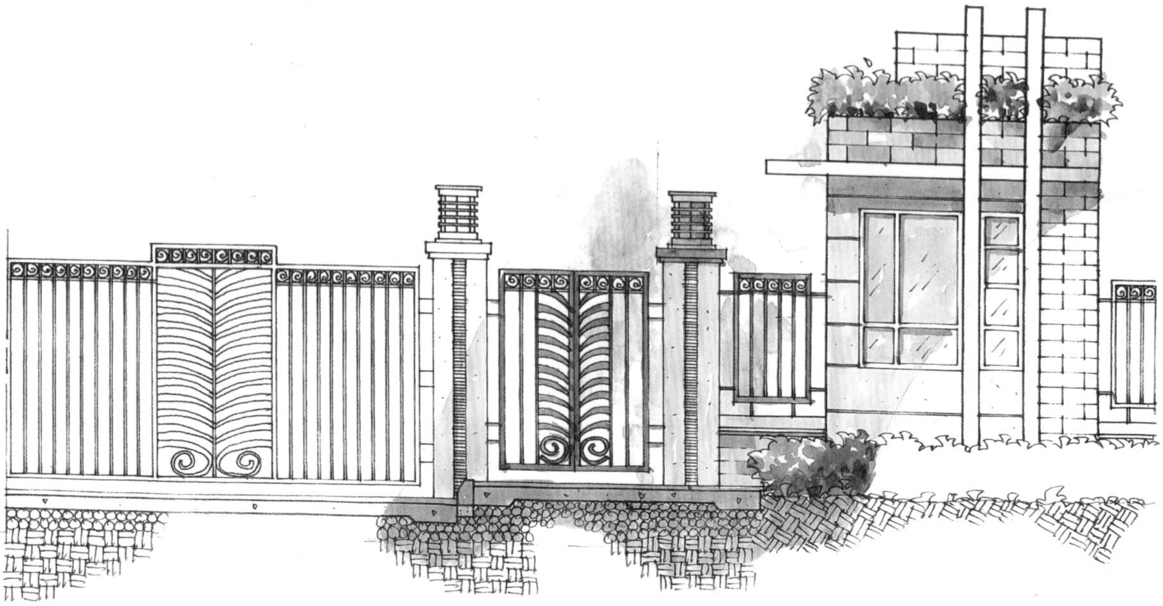

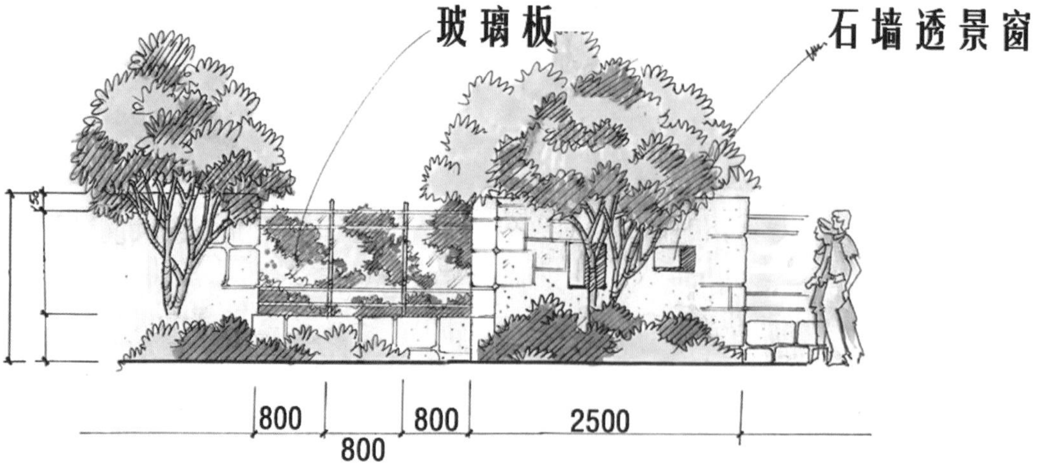
玻璃板
石墙透景窗
800
800
800
2500

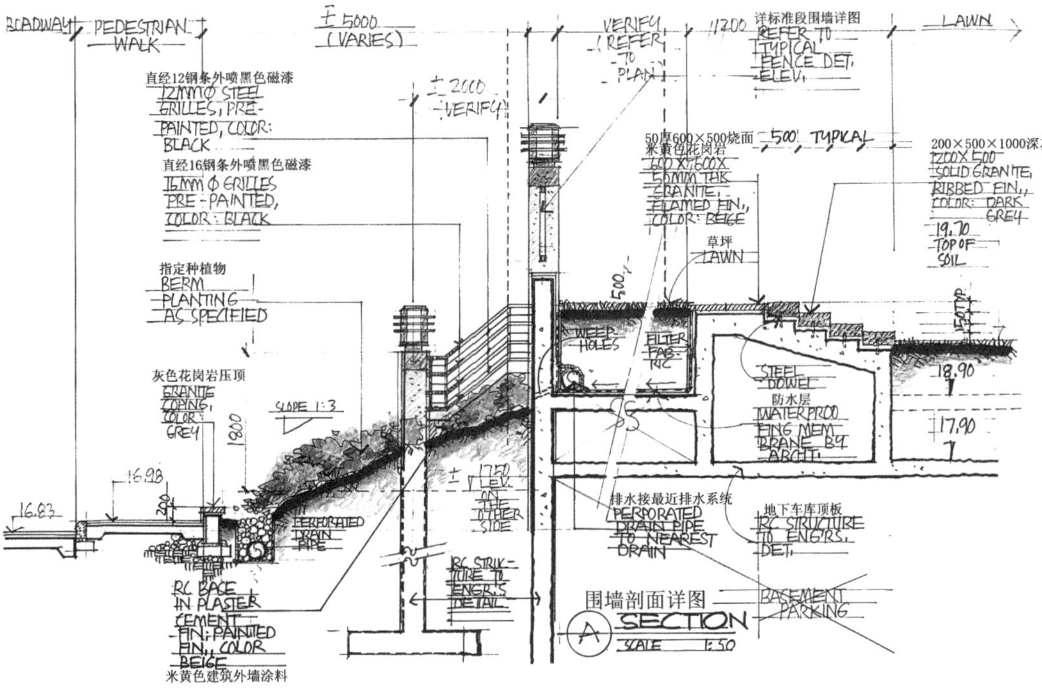
ROADWAY
PEDESTRIAN WALK
±5000 (VARIES)
±2000 VERIFY
VERIFY (REFER TO PLAN)
11200
详标准段围墙详图
REFER TO TYPICAL FENCE DET. ELEV.
LAWN
直经12钢条外喷黑色磁漆
12MM Ø STEEL GRILLES, PRE-PAINTED, COLOR: BLACK
直经16钢条外喷黑色磁漆
16MM Ø GRILLES PRE-PAINTED, COLOR: BLACK
50厚600×500烧面米黄色花岗岩
600×600X 50MM THK GRANITE, FLAMED FIN., COLOR: BEIGE
500 TYPICAL
200×500×1000深灰色花岗岩
1200×500 SOLID GRANITE, RIBBED FIN., COLOR: DARK GREY
19.70 TOP OF SOIL
草坪
LAWN
指定种植物
BERM PLANTING AS SPECIFIED
WEEP HOLES
FILTER FAB. RIC
STEEL DOWEL
防水层
WATERPROOFING MEMBRANE BY ARCHIT.
18.90
17.90
灰色花岗岩压顶
GRANITE COPING, COLOR: GREY
SLOPE 1:3
1800
16.98
16.83
PERFORATED DRAIN PIPE
±17.50 LEV. ON THE OTHER SIDE
排水接最近排水系统
PERFORATED DRAIN PIPE TO NEAREST DRAIN
地下车库顶板
RC STRUCTURE TO ENG'RS. DET.
RC BASE IN PLASTER CEMENT FIN. PAINTED FIN., COLOR BEIGE
米黄色建筑外墙涂料
RC STRUCTURE TO ENG'RS DETAIL
BASEMENT PARKING
围墙剖面详图
A SECTION
SCALE 1:50

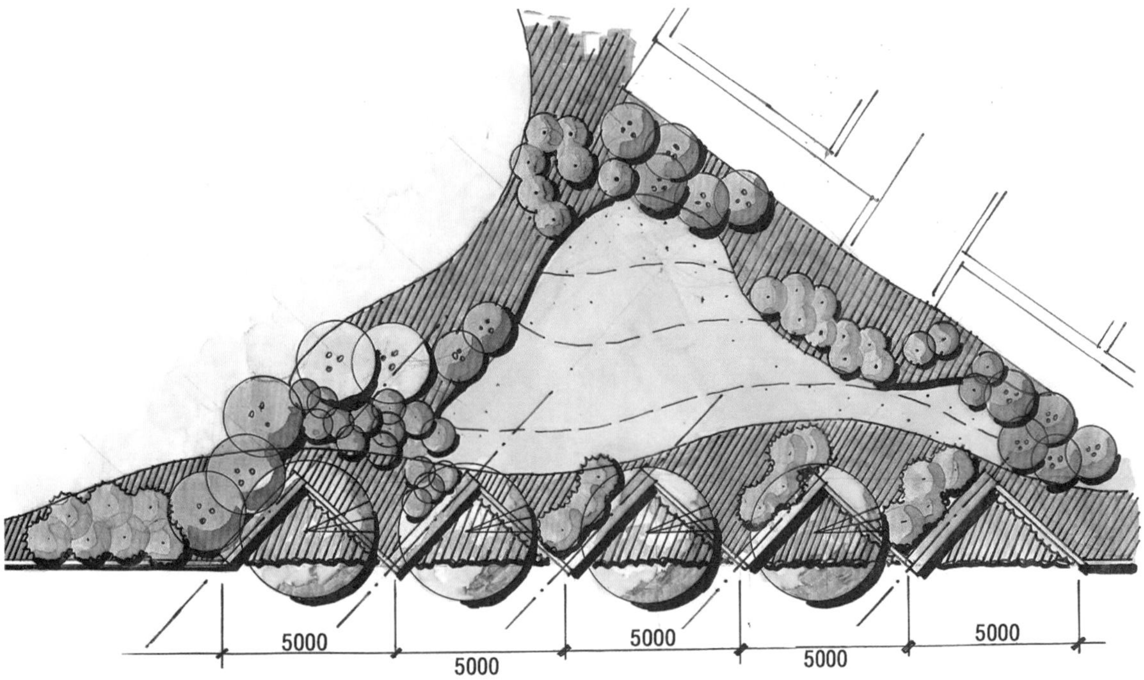
5000
5000
5000
5000
5000

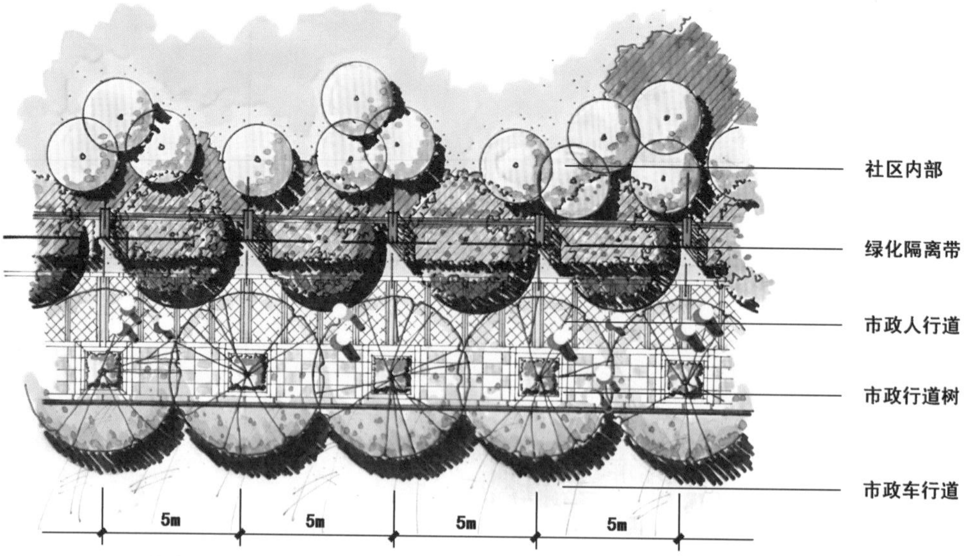
社区内部
绿化隔离带
市政人行道
市政行道树
市政车行道
5m
5m
5m
5m

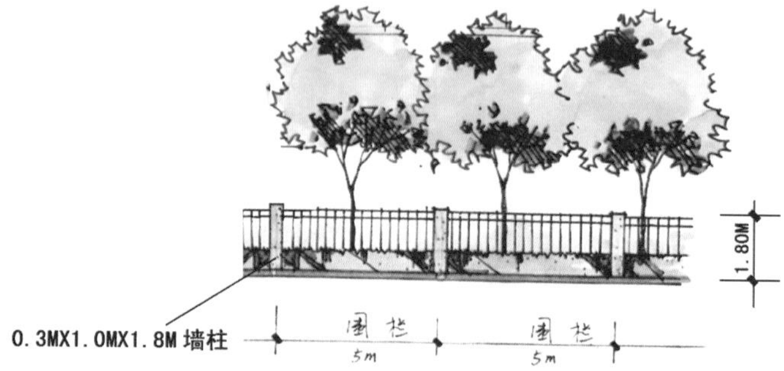
1.80M
0.3MX1.0MX1.8M 墙柱
围栏
5m
围栏
5m

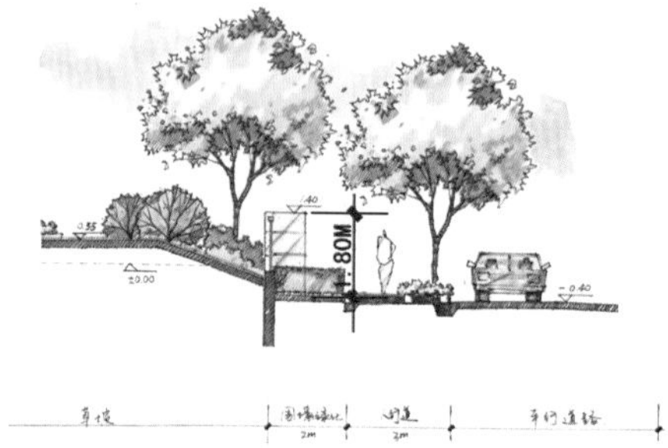

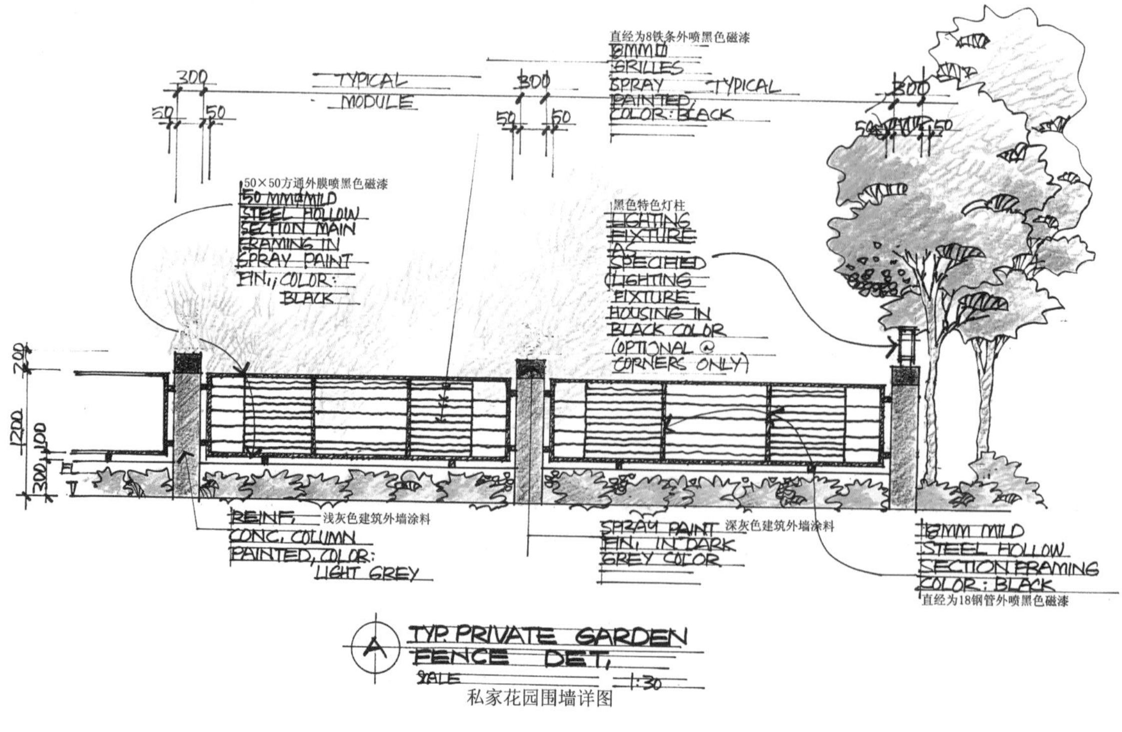
300
TYPICAL
MODULE
300
300
50
50
50
50
50
50
直经为8铁条外喷黑色磁漆
10MMØ
GRILLES
SPRAY
PAINTED,
COLOR: BLACK
TYPICAL
50×50方通外膜喷黑色磁漆
50 MM MILD
STEEL HOLLOW
SECTION MAIN
FRAMING IN
SPRAY PAINT
FIN.; COLOR:
BLACK
黑色特色灯柱
LIGHTING
FIXTURE
AS
SPECIFIED
(LIGHTING
FIXTURE
HOUSING IN
BLACK COLOR
(OPTIONAL @
CORNERS ONLY)
200
1200
100
300
REINF.
浅灰色建筑外墙涂料
CONC. COLUMN
PAINTED, COLOR:
LIGHT GREY
SPRAY PAINT
深灰色建筑外墙涂料
FIN. IN DARK
GREY COLOR
18MM MILD
STEEL HOLLOW
SECTION FRAMING
COLOR: BLACK
直经为18钢管外喷黑色磁漆
A
TYP. PRIVATE GARDEN
FENCE DET.
SCALE 1:30

私家花园围墙详图

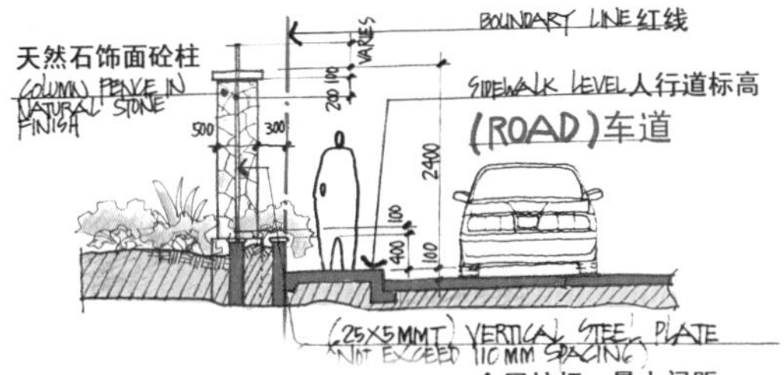
天然石饰面砼柱
COLUMN FENCE IN NATURAL STONE FINISH
BOUNDARY LINE 红线
SIDEWALK LEVEL 人行道标高
(ROAD) 车道
(25X5MMT) VERTICAL STEEL PLATE NOT EXCEED 110MM SPACING
25mm x 5mm 金属柱杆，最大间距不大于110mm

剖面图-B（围墙）

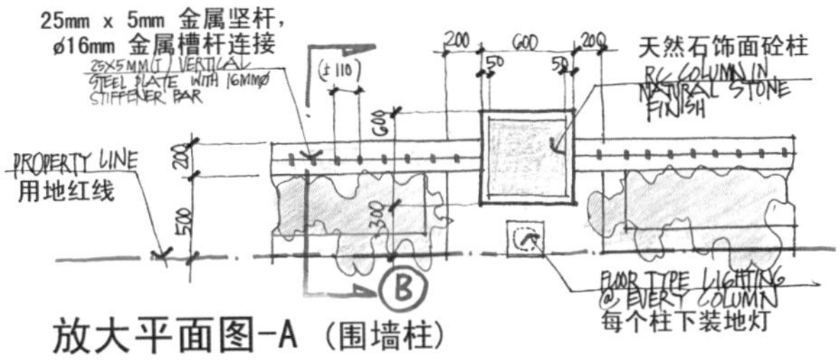
25mm x 5mm 金属竖杆，ø16mm 金属槽杆连接
25X5MMT VERTICAL STEEL PLATE WITH 16MMø STIFFENER BAR
PROPERTY LINE
用地红线
天然石饰面砼柱
RC COLUMN IN NATURAL STONE FINISH
FLOOR TYPE LIGHTING @ EVERY COLUMN
每个柱下装地灯

放大平面图-A（围墙柱）

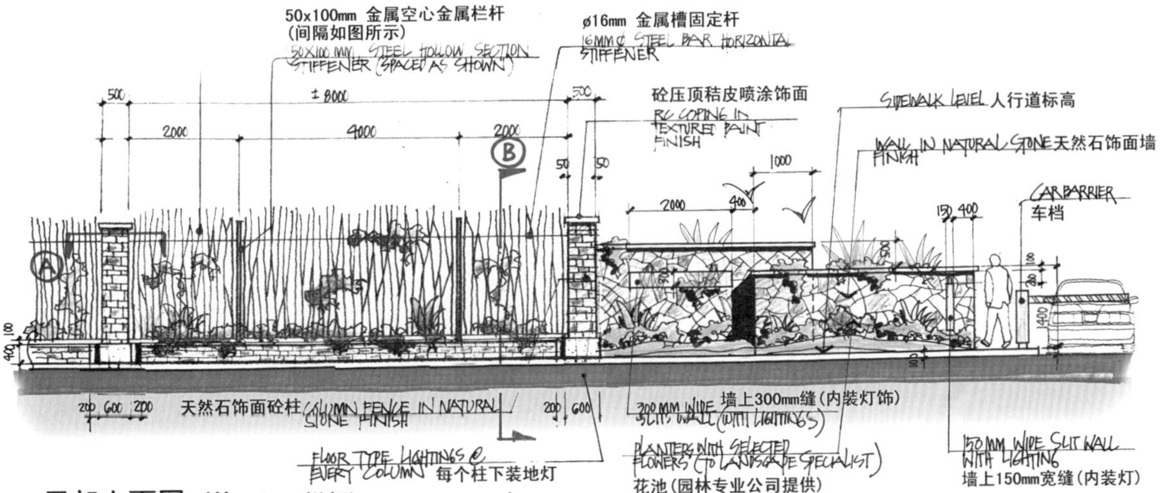
50x100mm 金属空心金属栏杆（间隔如图所示）
50X100 MM STEEL HOLLOW SECTION STIFFENER (SPACED AS SHOWN)
ø16mm 金属槽固定杆
16MMø STEEL BAR HORIZONTAL STIFFENER
砼压顶粘皮喷涂饰面
RC COPING IN TEXTURED PAINT FINISH
SIDEWALK LEVEL 人行道标高
WALL IN NATURAL STONE FINISH 天然石饰面墙
CAR BARRIER
车档
天然石饰面砼柱 COLUMN FENCE IN NATURAL STONE FINISH
300MM WIDE SLIT WALL (WITH LIGHTINGS)
墙上300mm缝（内装灯饰）
FLOOR TYPE LIGHTINGS @ EVERY COLUMN 每个柱下装地灯
PLANTERS WITH SELECTED FLOWERS (TO LANDSCAPE SPECIALIST)
花池（园林专业公司提供）
150MM WIDE SLIT WALL WITH LIGHTING
墙上150mm宽缝（内装灯）

局部立面图-1H（入口栏杆）

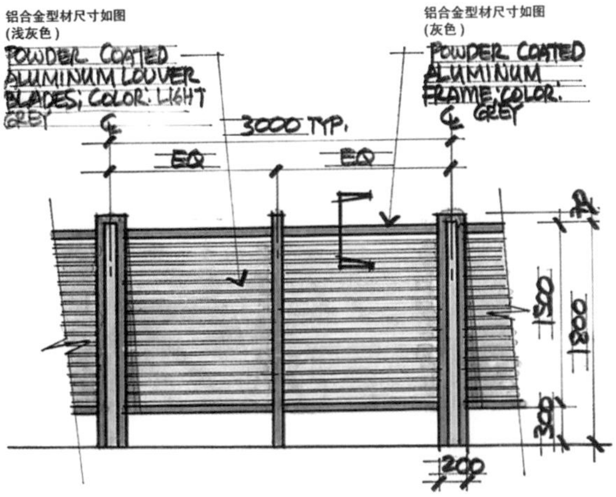
铝合金型材尺寸如图（浅灰色）
POWDER COATED ALUMINUM LOUVER BLADES; COLOR: LIGHT GREY
铝合金型材尺寸如图（灰色）
POWDER COATED ALUMINUM FRAME; COLOR: GREY
3000 TYP.
EQ
EQ

TYPICAL ELEVATION
SCALE
立面图

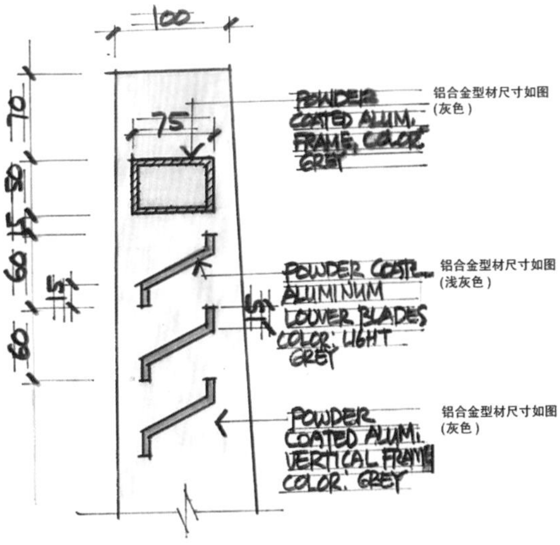
铝合金型材尺寸如图（灰色）
POWDER COATED ALUM. FRAME, COLOR GREY
铝合金型材尺寸如图（浅灰色）
POWDER COATED ALUMINUM LOUVER BLADES COLOR: LIGHT GREY
铝合金型材尺寸如图（灰色）
POWDER COATED ALUMI. VERTICAL FRAME COLOR: GREY

DETAIL SECTION
SCALE
节点详图

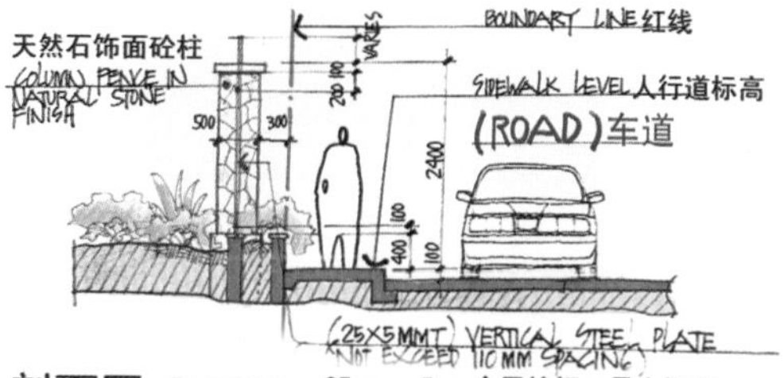

剖面图-B(围墙)

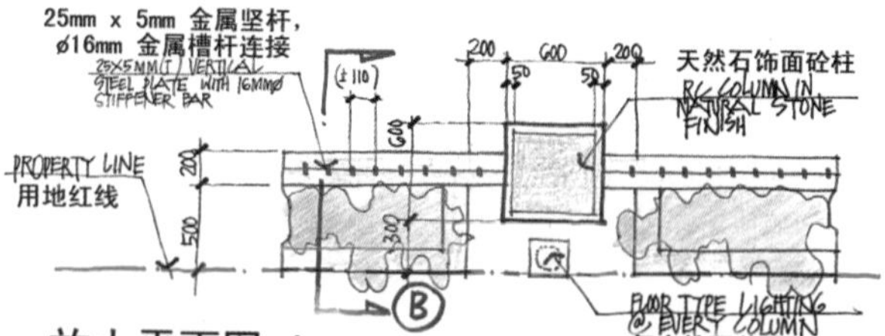

放大平面图-A（围墙柱）

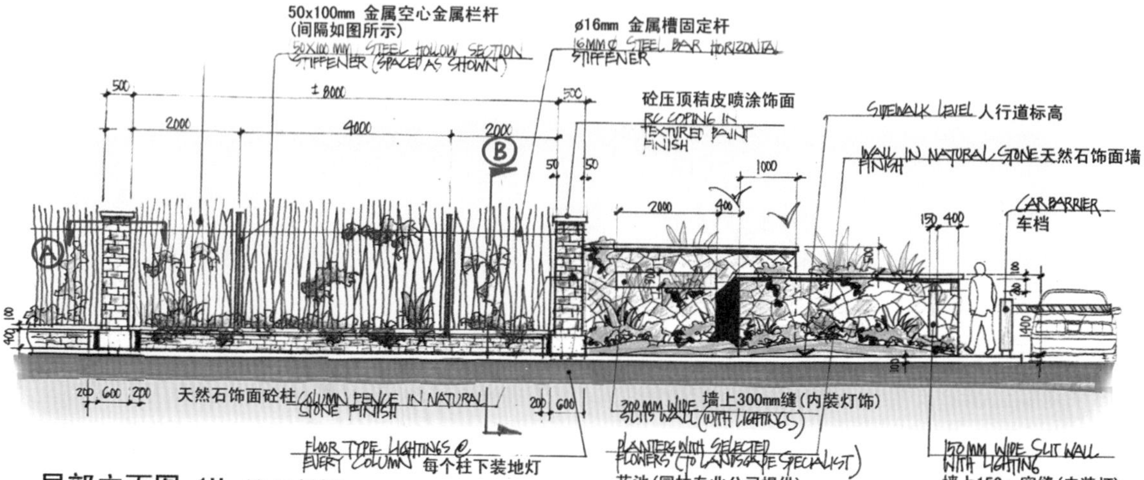

局部立面图-1H（入口栏杆）

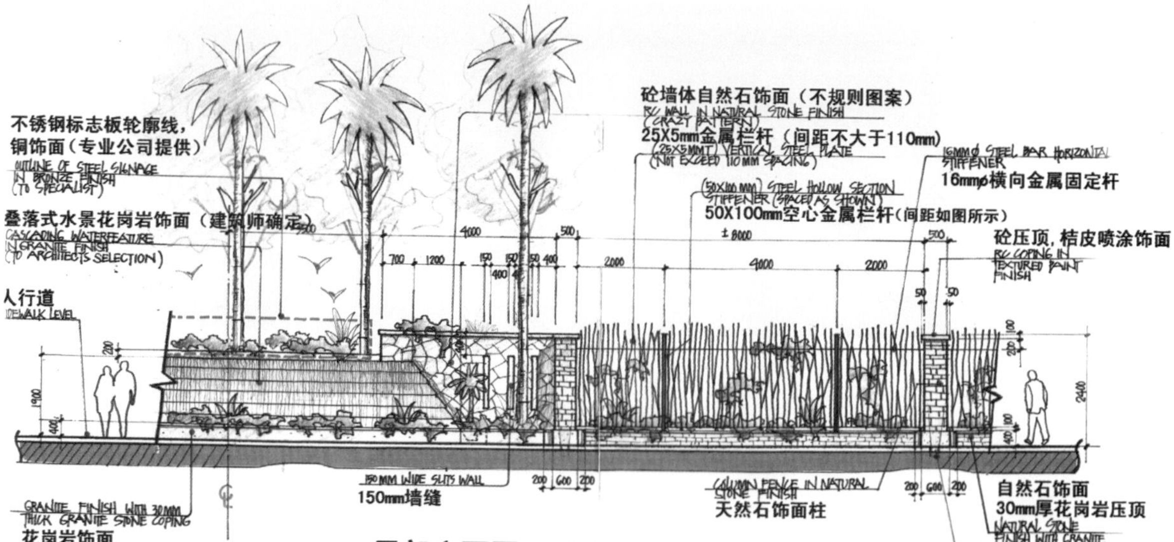

局部立面图 - 2B 水景/栏杆

SELECTED LIGHTINGS (FLOOR TYPE) @ EVERY COLUMN
地灯（每个柱子一个）

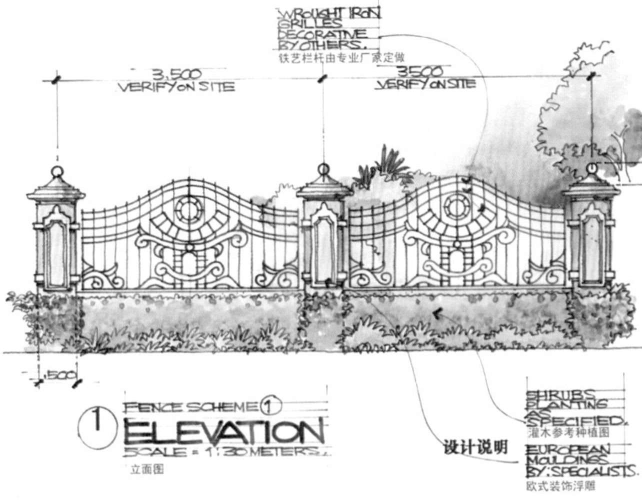
WROUGHT IRON GRILLES DECORATIVE BY OTHERS.
铁艺栏杆由专业厂家定做
3,500 VERIFY ON SITE
3500 VERIFY ON SITE
2100
.500
2100
FENCE SCHEME ①
① ELEVATION
SCALE = 1:30 METERS.
立面图
设计说明
SHRUBS PLANTING AS SPECIFIED.
灌木参考种植图
EUROPEAN MOULDINGS BY: SPECIALISTS.
欧式装饰浮雕

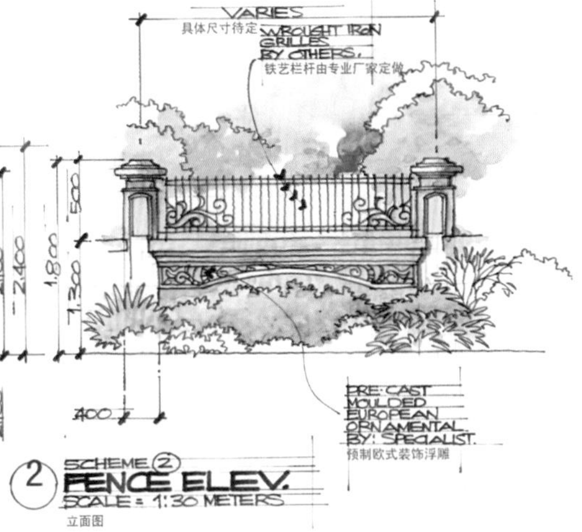
VARIES
具体尺寸待定
WROUGHT IRON GRILLES BY OTHERS.
铁艺栏杆由专业厂家定做
2,400
1,800
500
1,300
400
PRE-CAST MOULDED EUROPEAN ORNAMENTAL. BY: SPECIALIST.
预制欧式装饰浮雕
SCHEME ②
② FENCE ELEV.
SCALE = 1:30 METERS
立面图

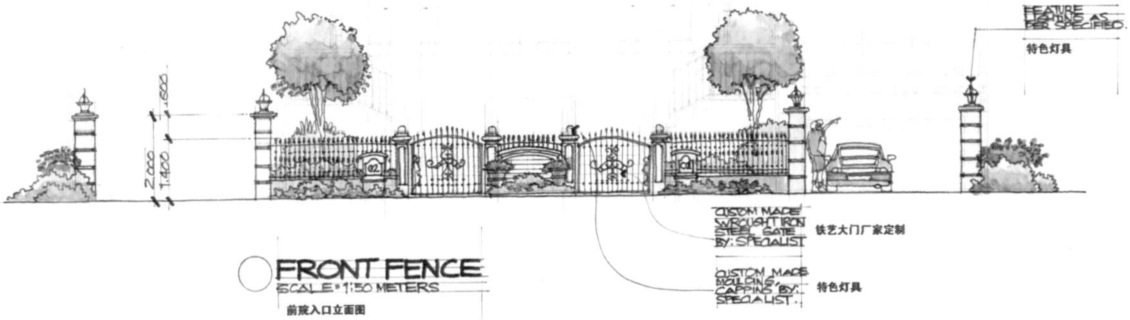
FEATURE LIGHTING AS PER SPECIFIED.
特色灯具
600
2,000
1,400
CUSTOM MADE WROUGHT IRON STEEL GATE BY: SPECIALIST
铁艺大门厂家定制
CUSTOM MADE MOULDING, CAPPING BY: SPECIALIST.
特色灯具
FRONT FENCE
SCALE = 1:50 METERS
前院入口立面图

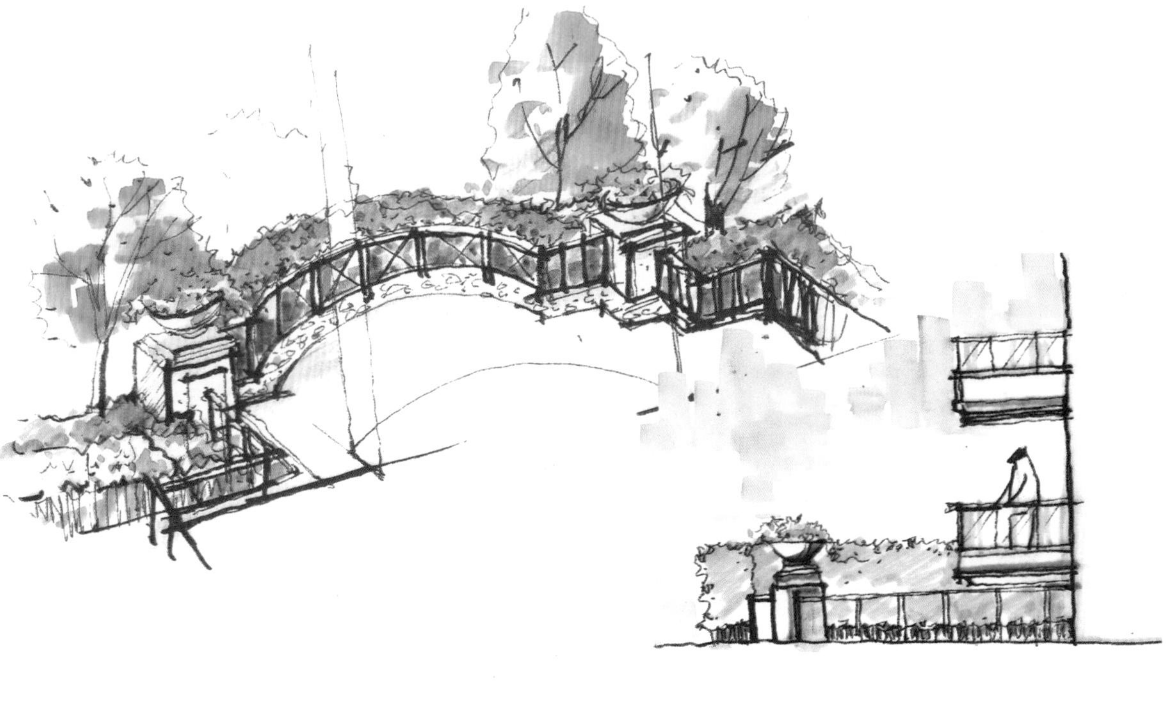

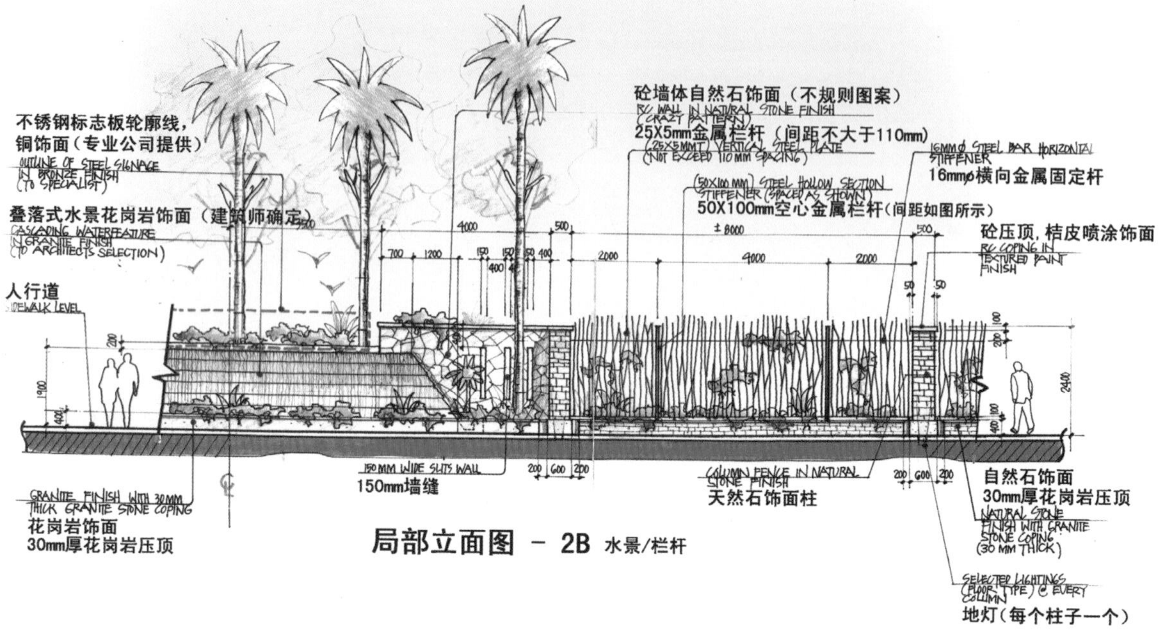

局部立面图 － 2B 水景/栏杆

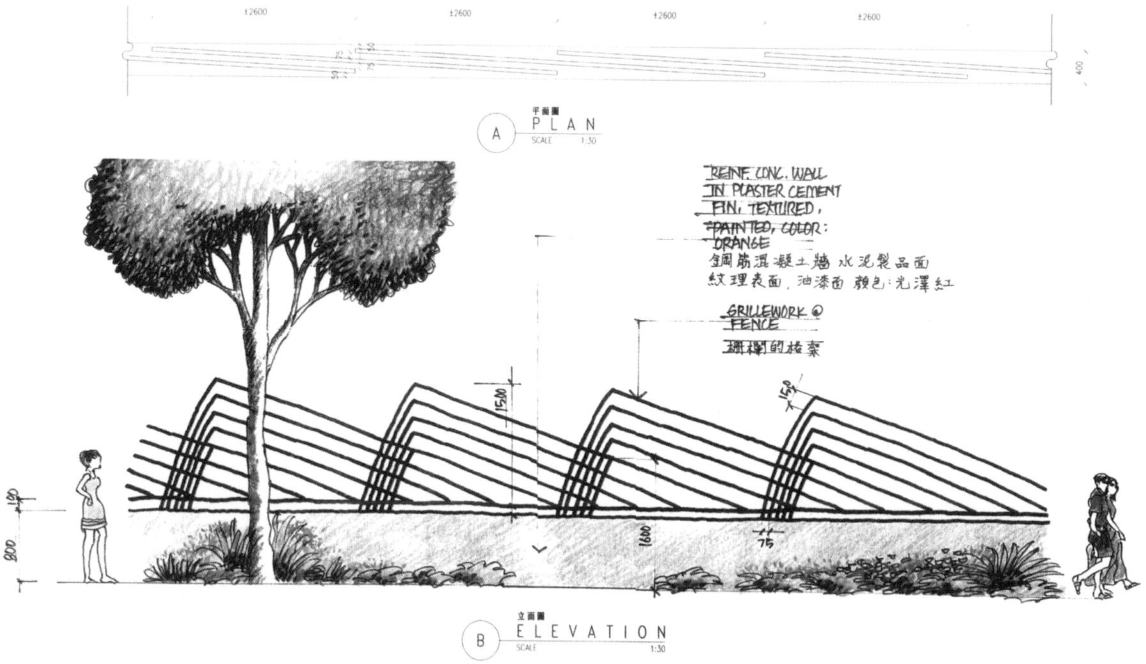

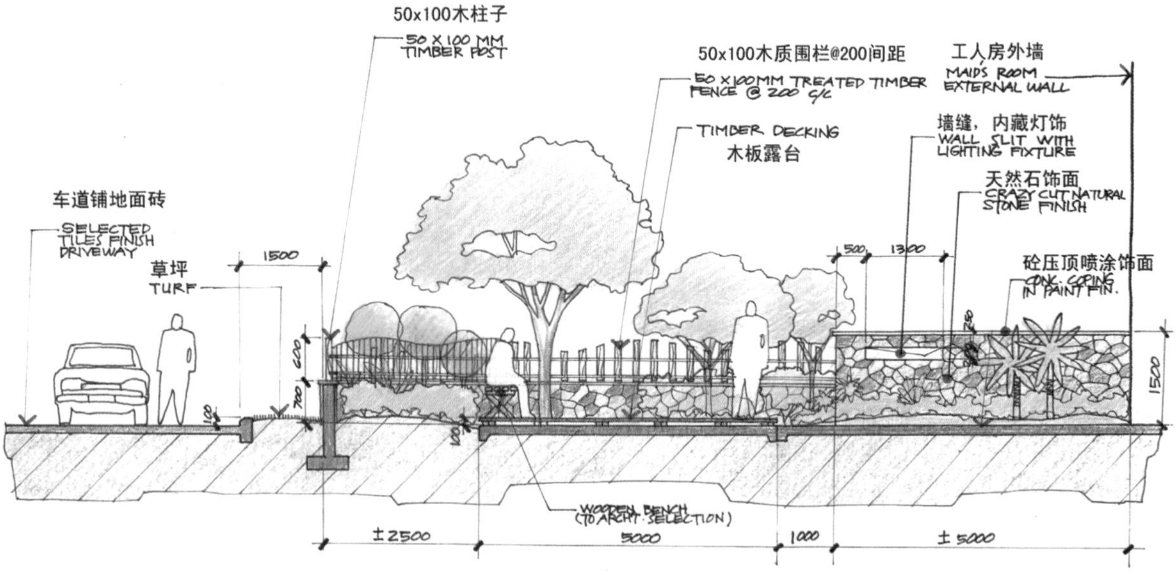
50x100木柱子
50 X 100 MM TIMBER POST
50x100木质围栏@200间距
50 X 100MM TREATED TIMBER FENCE @ 200 C/C
工人房外墙
MAID'S ROOM EXTERNAL WALL
TIMBER DECKING
木板露台
墙缝，内藏灯饰
WALL SLIT WITH LIGHTING FIXTURE
天然石饰面
CRAZY CUT NATURAL STONE FINISH
砼压顶喷涂饰面
CONC. COPING IN PAINT FIN.
车道铺地面砖
SELECTED TILES FINISH DRIVEWAY
草坪
TURF
WOODEN BENCH (TO ARCHT SELECTION)
1500
600
700
100
500
1300
1500
±2500
5000
1000
±5000

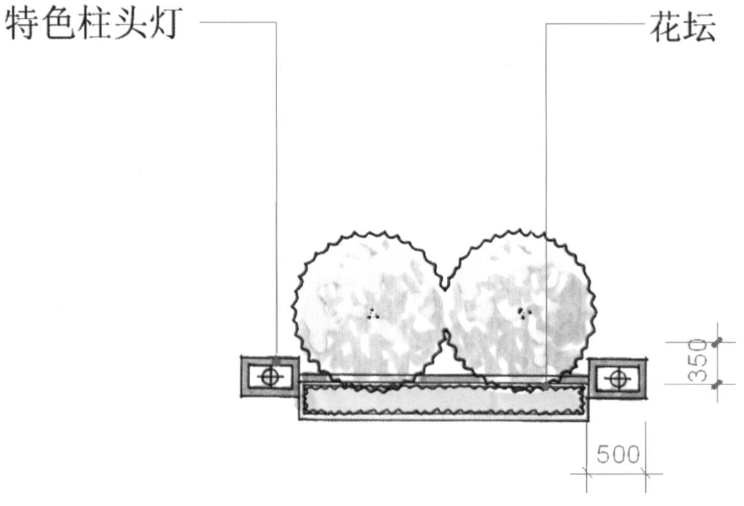
特色柱头灯
花坛
350
500

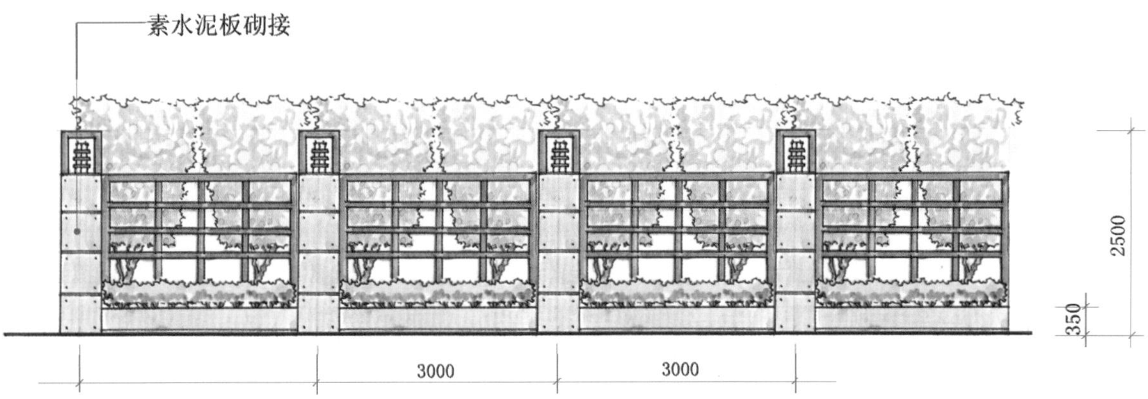
素水泥板砌接
2500
350
3000
3000

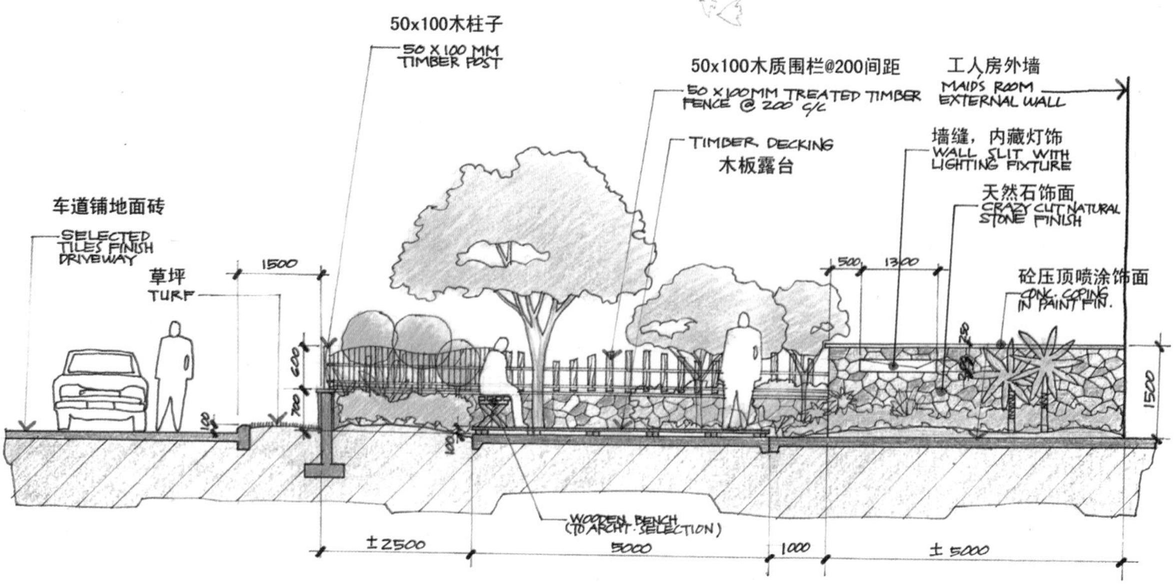
50x100木柱子
50 X 100 MM TIMBER POST
50x100木质围栏@200间距
50 X 100MM TREATED TIMBER FENCE @ 200 C/C
工人房外墙
MAID'S ROOM EXTERNAL WALL
TIMBER DECKING
木板露台
墙缝，内藏灯饰
WALL SLIT WITH LIGHTING FIXTURE
天然石饰面
CRAZY CUT NATURAL STONE FINISH
车道铺地面砖
SELECTED TILES FINISH DRIVEWAY
草坪
TURF
砼压顶喷涂饰面
CONC. COPING IN PAINT FIN.
1500
500
1300
600
700
1500
WOODEN BENCH (TO ARCHT. SELECTION)
±2500
5000
1000
±5000

剖面 - 4A

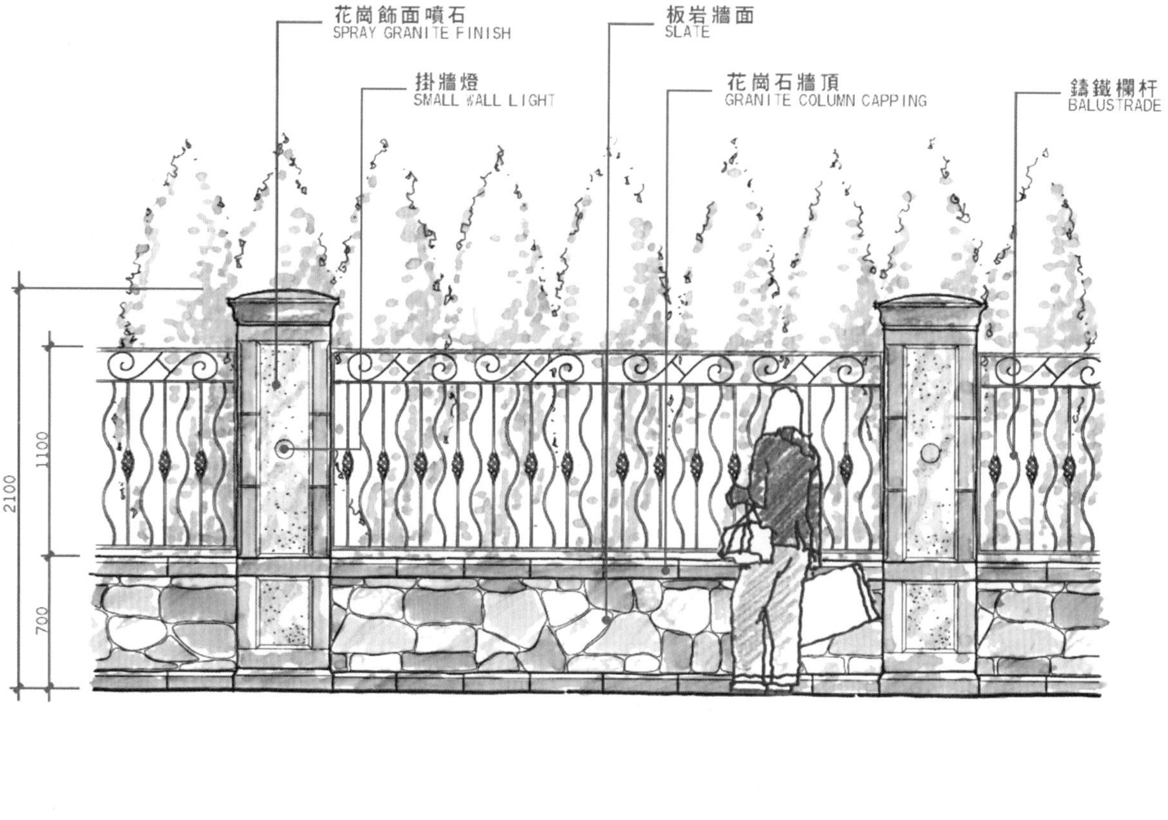
花崗飾面噴石
SPRAY GRANITE FINISH
板岩牆面
SLATE
掛牆燈
SMALL WALL LIGHT
花崗石牆頂
GRANITE COLUMN CAPPING
鑄鐵欄杆
BALUSTRADE
2100
1100
700

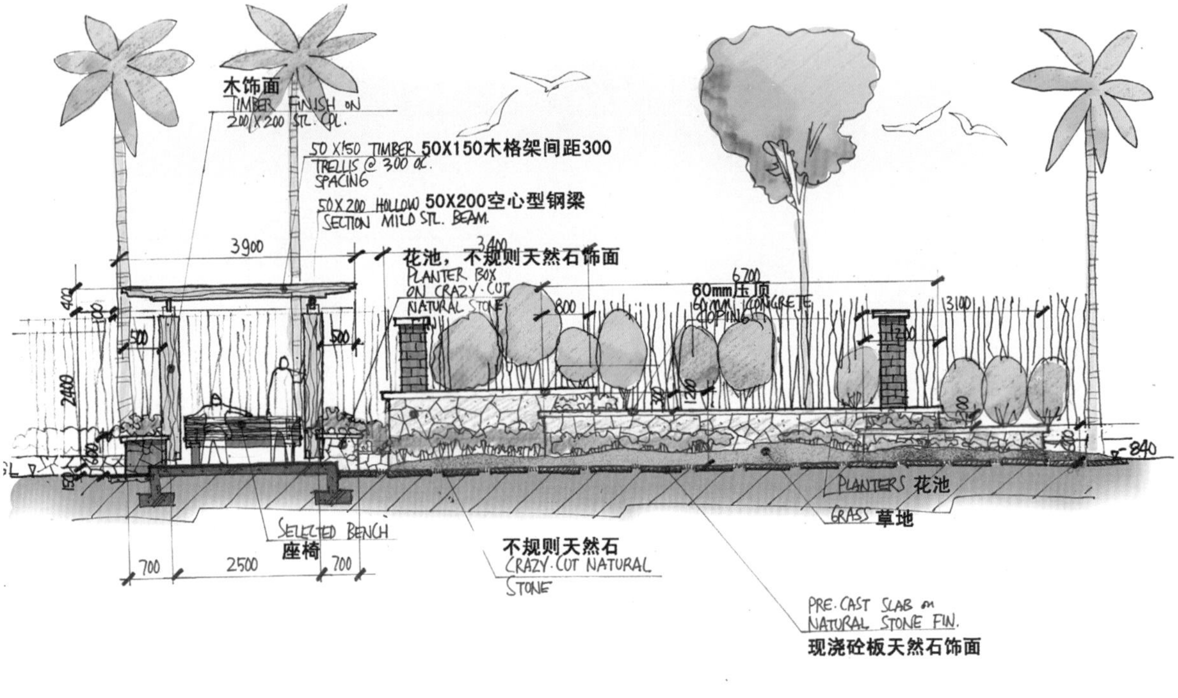

木饰面
TIMBER FINISH ON
200X200 STL. COL.
50X150 TIMBER 50X150木格架间距300
TRELLIS @ 300 OC.
SPACING
50X200 HOLLOW 50X200空心型钢梁
SECTION MILD STL. BEAM.
3900
3400
花池，不规则天然石饰面
PLANTER BOX
ON CRAZY-CUT
NATURAL STONE
60mm压顶
60MM CONCRETE
COPING
6700
3100
800
500
500
2400
840
PLANTERS 花池
GRASS 草地
SELECTED BENCH
座椅
700
2500
700
不规则天然石
CRAZY-CUT NATURAL
STONE
PRE-CAST SLAB on
NATURAL STONE FIN.
现浇砼板天然石饰面

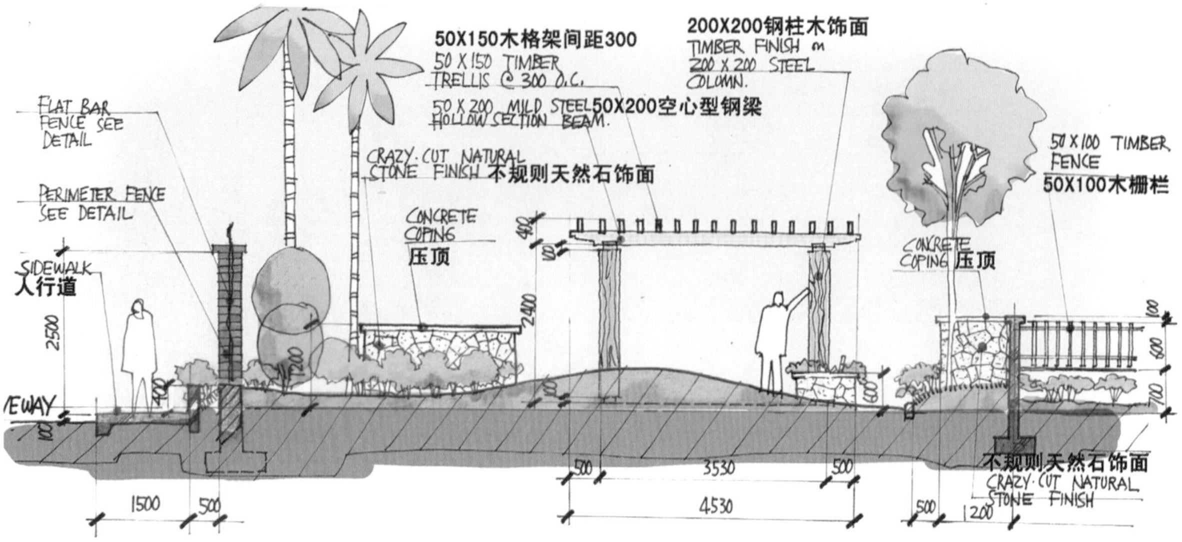
50X150木格架间距300
50 X 150 TIMBER TRELLIS @ 300 O.C.
50 X 200 MILD STEEL HOLLOW SECTION BEAM.
50X200空心型钢梁
200X200钢柱木饰面
TIMBER FINISH on 200 X 200 STEEL COLUMN.
FLAT BAR FENCE SEE DETAIL
PERIMETER FENCE SEE DETAIL
SIDEWALK
人行道
CRAZY-CUT NATURAL STONE FINISH 不规则天然石饰面
CONCRETE COPING
压顶
CONCRETE COPING 压顶
50 X 100 TIMBER FENCE
50X100木栅栏
不规则天然石饰面
CRAZY-CUT NATURAL STONE FINISH
2500
2400
1500
500
3530
4530
1200

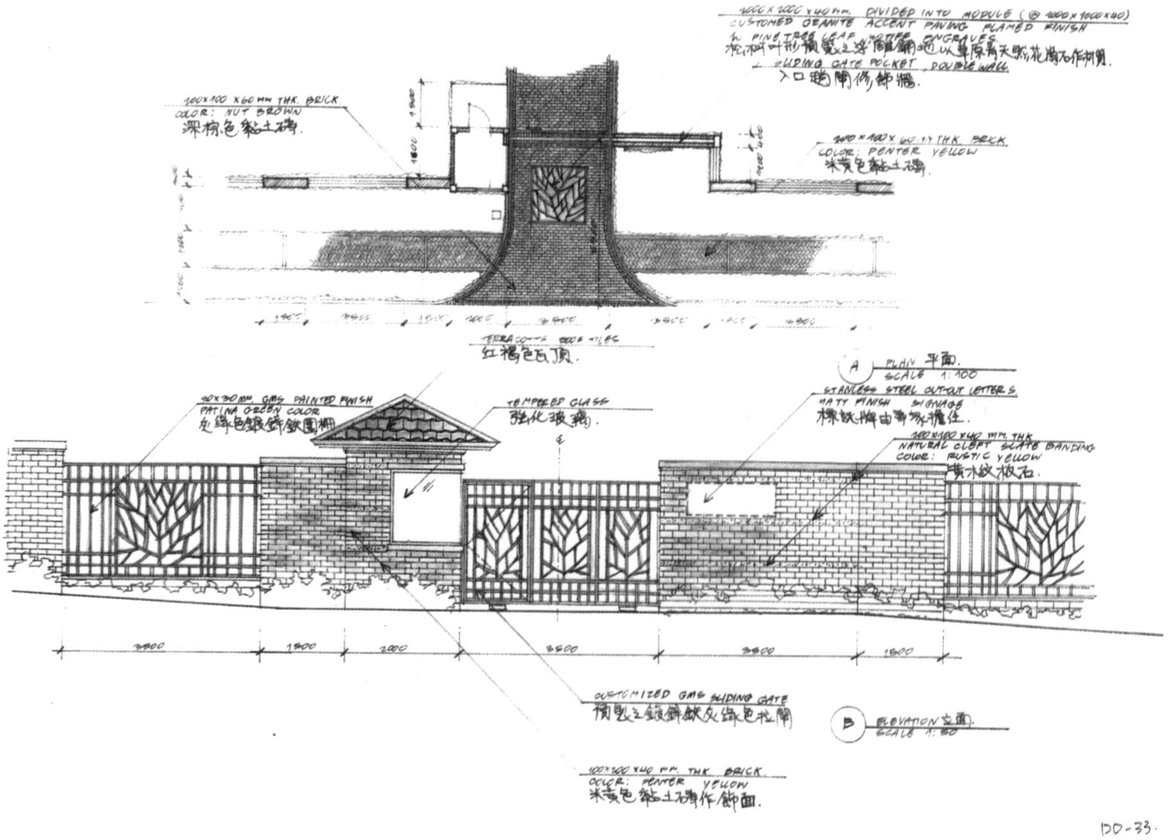
CUSTOMED GRANITE ACCENT PAVING FLAMED FINISH
SLIDING GATE POCKET, DOUBLE WALL
入口趟閘佈飾牆.
COLOR: NUT BROWN
深棕色黏土磚.
COLOR: PENTER YELLOW
米黄色黏土磚.
TERRACOTTA ROOF TILES
红褐色瓦顶.
PLAN 平面.
SCALE 1:100
STAINLESS STEEL CUT-OUT LETTERS
MATT FINISH SIGNAGE
PATINA GREEN COLOR
TEMPERED GLASS
強化玻璃.
NATURAL CLEFT SLATE BANDING
COLOR: RUSTIC YELLOW
黄木紋板石.
ELEVATION 立面.
DD-33.

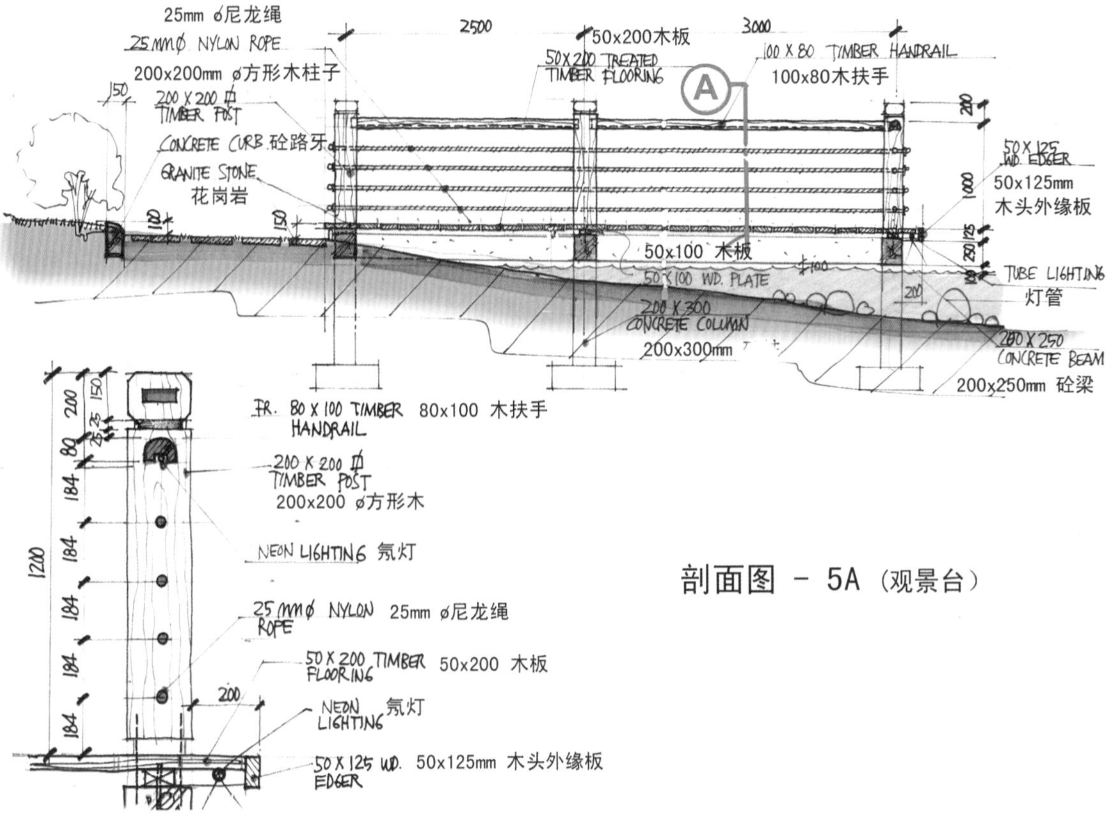
25mm ø尼龙绳
25MMØ NYLON ROPE
200x200mm ø方形木柱子
200 X 200 Ø TIMBER POST
CONCRETE CURB 砼路牙
GRANITE STONE 花岗岩
2500
50x200木板
50X200 TREATED TIMBER FLOORING
3000
100 X 80 TIMBER HANDRAIL
100x80木扶手
A
50X125 WD. EDGER
50x125mm 木头外缘板
1000
50x100 木板
50X100 WD. PLATE
TUBE LIGHTING
灯管
200 X 300 CONCRETE COLUMN
200x300mm
200 X 250 CONCRETE BEAM
200x250mm 砼梁
FR. 80 X 100 TIMBER HANDRAIL
80x100 木扶手
200 X 200 Ø TIMBER POST
200x200 ø方形木
NEON LIGHTING 氖灯
1200
184
25MMØ NYLON ROPE
25mm ø尼龙绳
50 X 200 TIMBER FLOORING
50x200 木板
NEON LIGHTING 氖灯
50 X 125 WD. EDGER
50x125mm 木头外缘板

剖面图 - 5A（观景台）

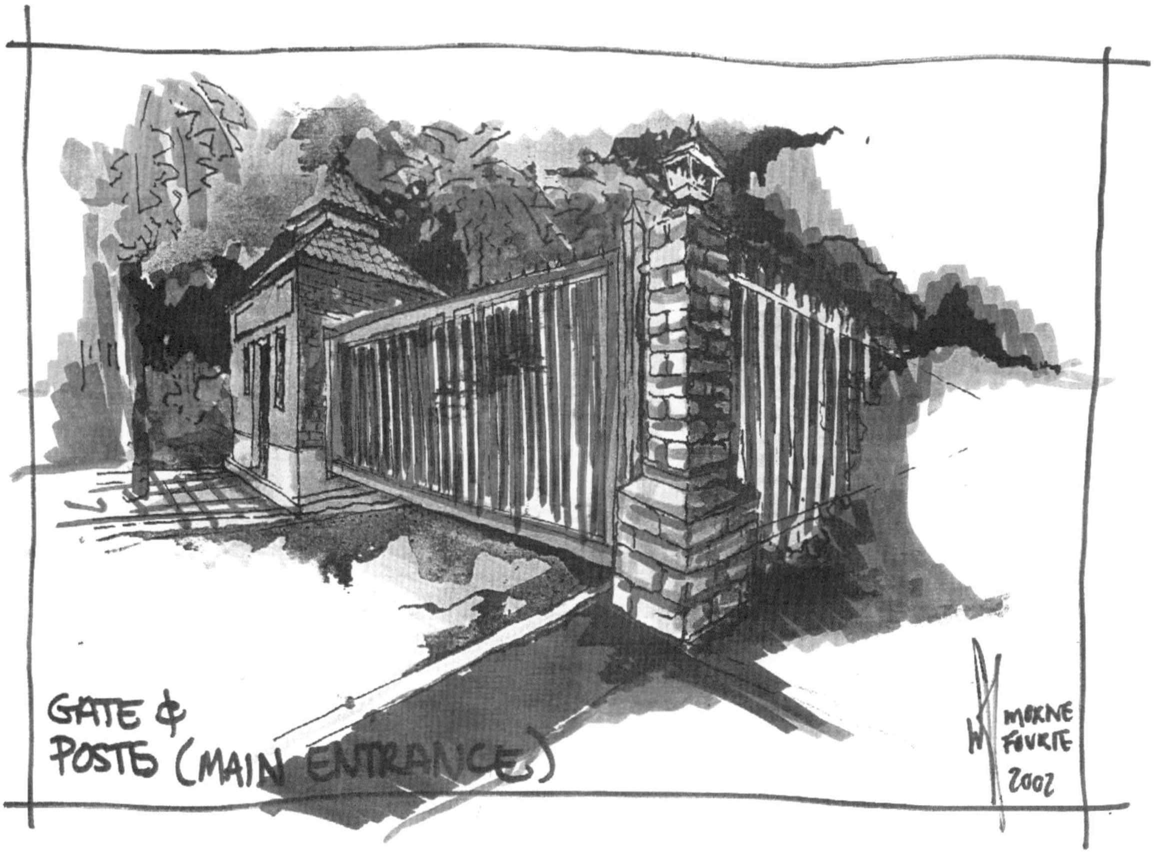
GATE & POSTS (MAIN ENTRANCE)
MORNE FOURIE
2002

1200
1000
200
11.00
2000
450X450浅灰色
微晶石贴面
白色光源小射灯
橙色喷漆
800
150
6800
商铺
D 剖立面
200
70
30
300
细钢丝拉索
8.75
8.60
8.45
8.30
8.15
8.00
7.85
7.70
7.55
7.40
7.25
7.10
6.95
6.80
扶手
2700
立面外围栏杆

黄木纹
细钢丝拉索围栏
8.75
8.60
8.45
8.30
8.15
8.00
7.85
7.70
7.55
7.40
7.25
7.10
6.95
6.80
450
450X450浅灰色
火烧面芝麻灰
950
1800
2000
1200
1400

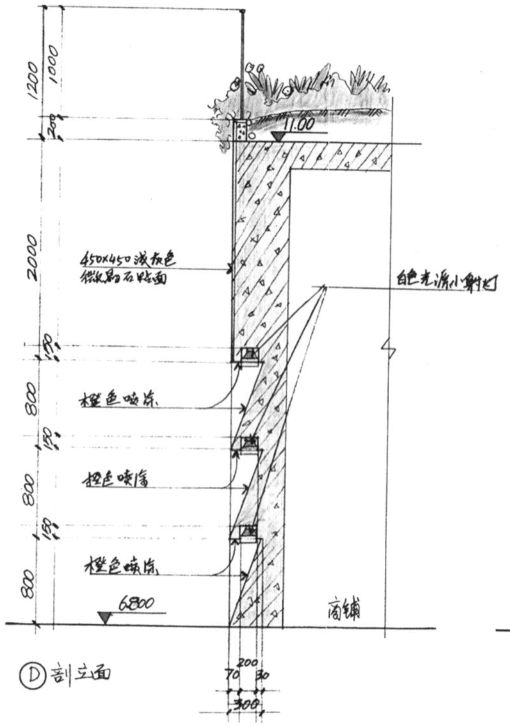

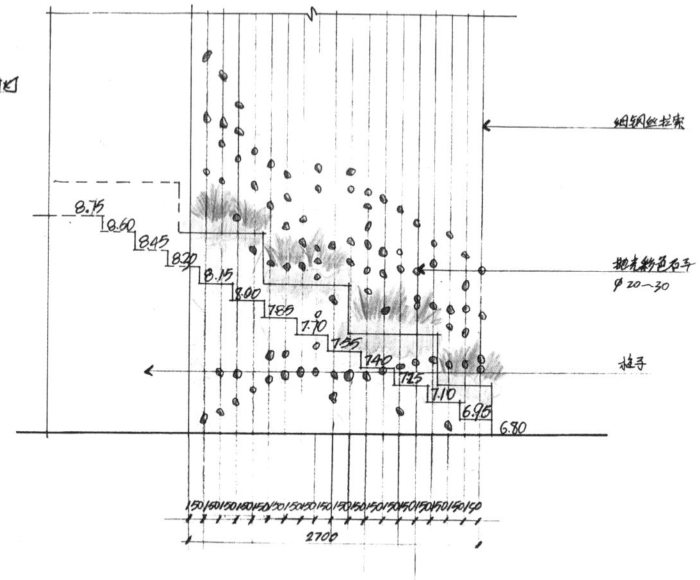

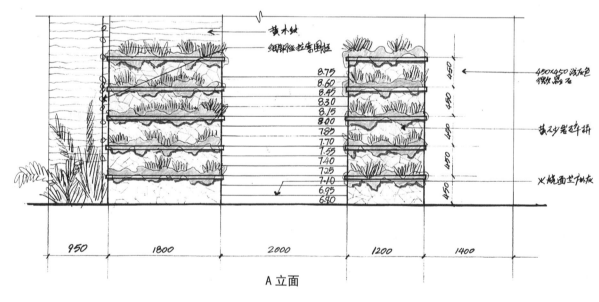

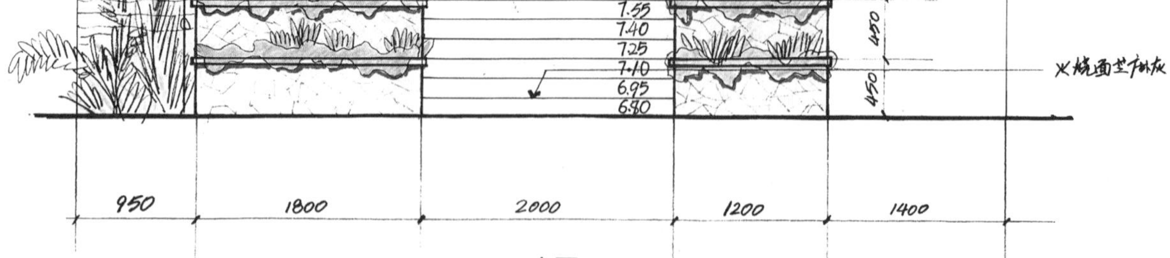

A 立面

私家花园围栏效果

私家花园围栏大样

围墙大样一

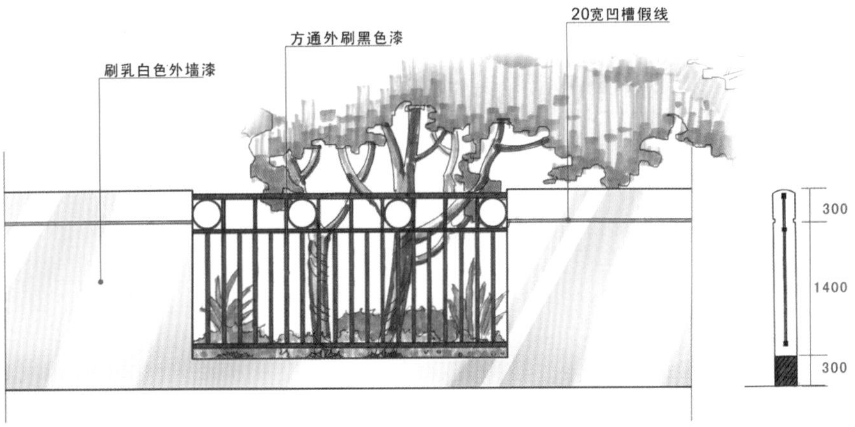

围墙大样二

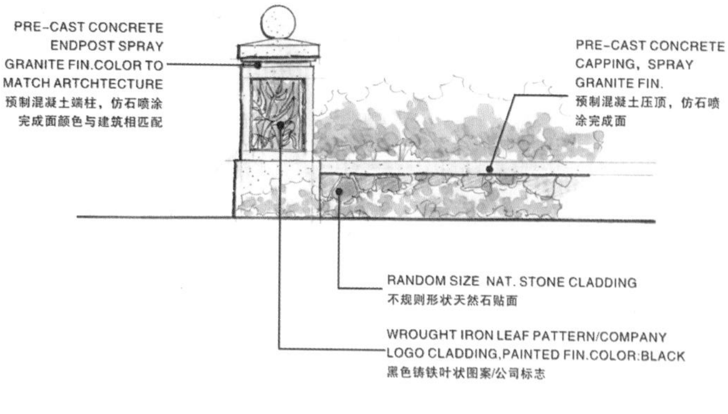
PRE-CAST CONCRETE ENDPOST SPRAY GRANITE FIN.COLOR TO MATCH ARTCHTECTURE
预制混凝土端柱，仿石喷涂完成面颜色与建筑相匹配
PRE-CAST CONCRETE CAPPING, SPRAY GRANITE FIN.
预制混凝土压顶，仿石喷涂完成面
RANDOM SIZE NAT. STONE CLADDING
不规则形状天然石贴面
WROUGHT IRON LEAF PATTERN/COMPANY LOGO CLADDING,PAINTED FIN.COLOR:BLACK
黑色铸铁叶状图案/公司标志

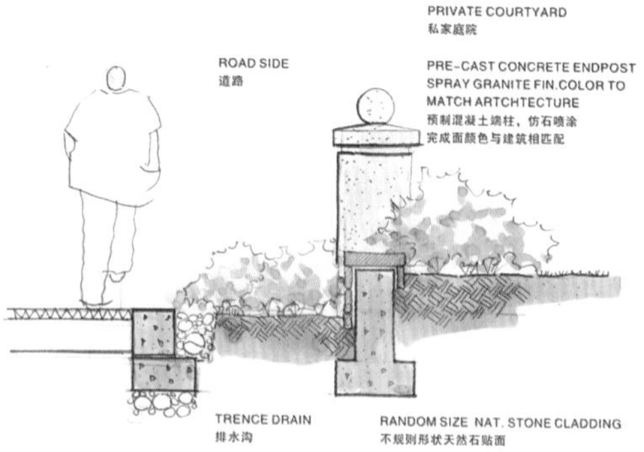
PRIVATE COURTYARD
私家庭院
ROAD SIDE
道路
PRE-CAST CONCRETE ENDPOST SPRAY GRANITE FIN.COLOR TO MATCH ARTCHTECTURE
预制混凝土端柱，仿石喷涂完成面颜色与建筑相匹配
TRENCE DRAIN
排水沟
RANDOM SIZE NAT. STONE CLADDING
不规则形状天然石贴面

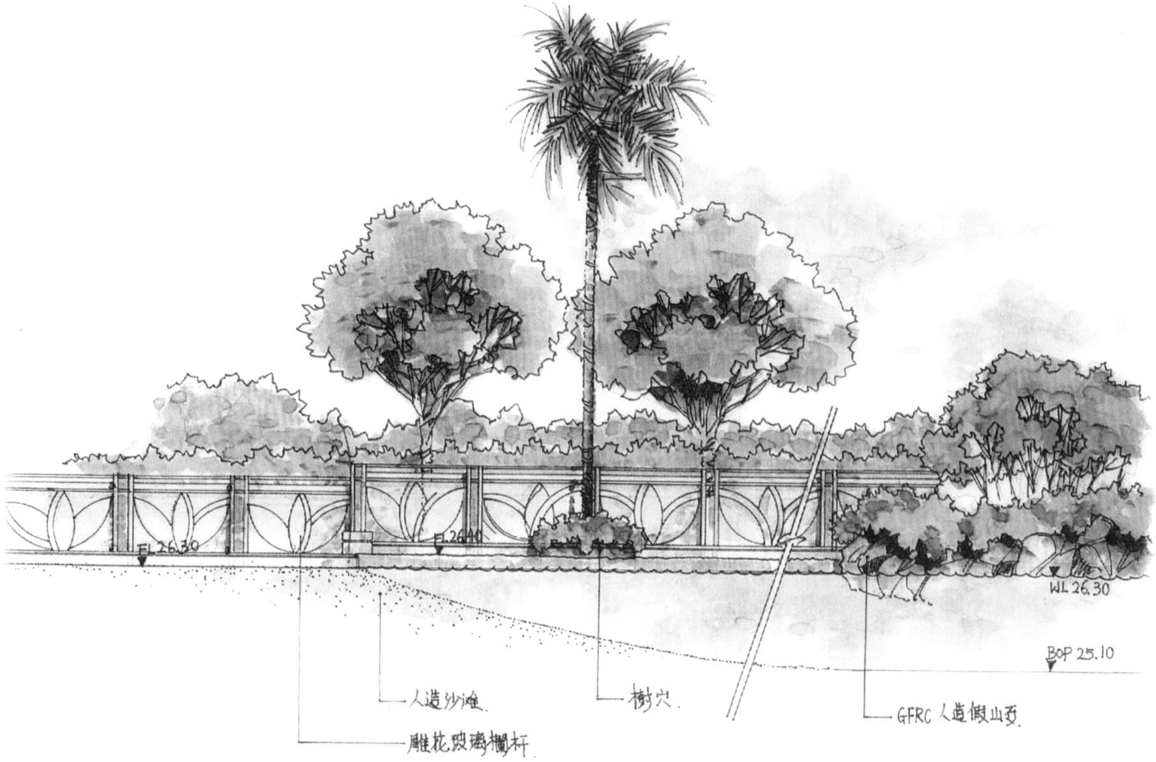
EL 26.30
WL 26.30
BOP 25.10
人造沙滩
樹穴
GFRC 人造假山石
雕花玻璃欄杆

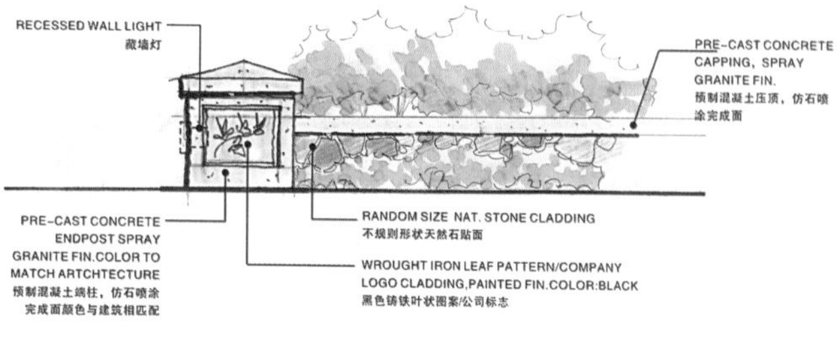
RECESSED WALL LIGHT
藏墙灯
PRE-CAST CONCRETE CAPPING, SPRAY GRANITE FIN.
预制混凝土压顶，仿石喷涂完成面
PRE-CAST CONCRETE ENDPOST SPRAY GRANITE FIN.COLOR TO MATCH ARTCHTECTURE
预制混凝土端柱，仿石喷涂完成面颜色与建筑相匹配
RANDOM SIZE NAT. STONE CLADDING
不规则形状天然石贴面
WROUGHT IRON LEAF PATTERN/COMPANY LOGO CLADDING,PAINTED FIN.COLOR:BLACK
黑色铸铁叶状图案/公司标志

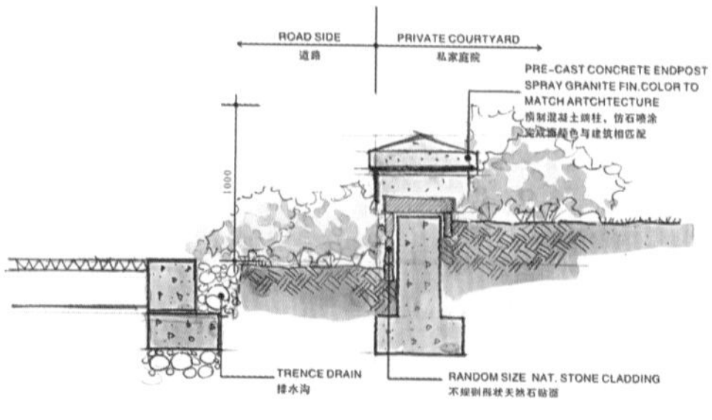
ROAD SIDE
道路
PRIVATE COURTYARD
私家庭院
PRE-CAST CONCRETE ENDPOST SPRAY GRANITE FIN.COLOR TO MATCH ARTCHTECTURE
预制混凝土端柱，仿石喷涂完成面颜色与建筑相匹配
1000
TRENCE DRAIN
排水沟
RANDOM SIZE NAT. STONE CLADDING
不规则形状天然石贴面

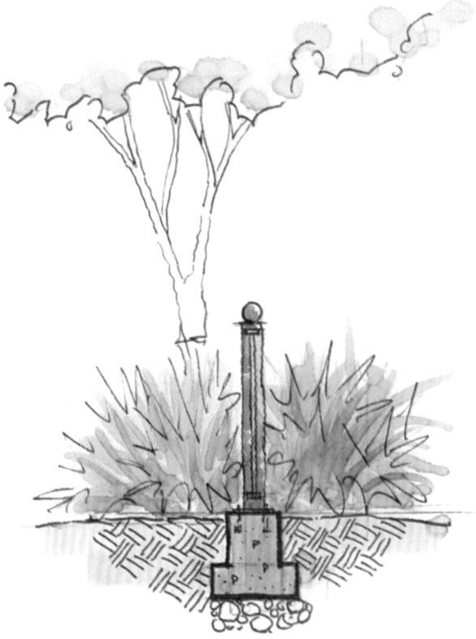

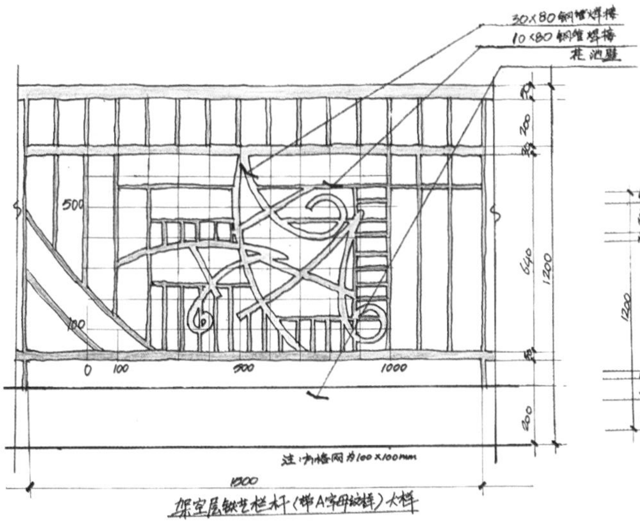

架空层铁艺栏杆（带A字母纹样）大样

比例 1:10

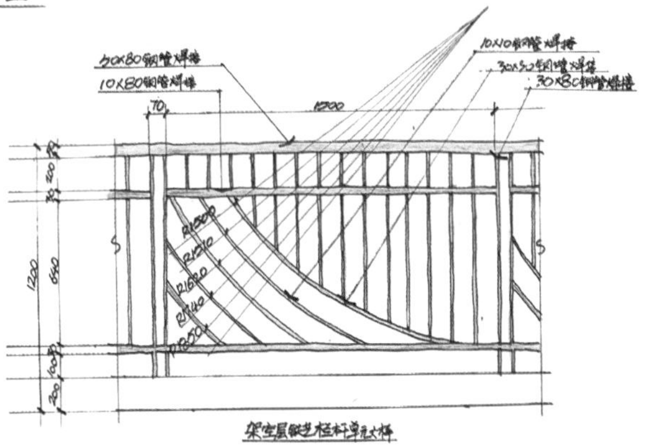

架空层铁艺栏杆单元大样

比例 1:15

架空层铁艺栏杆立面

比例 1:50

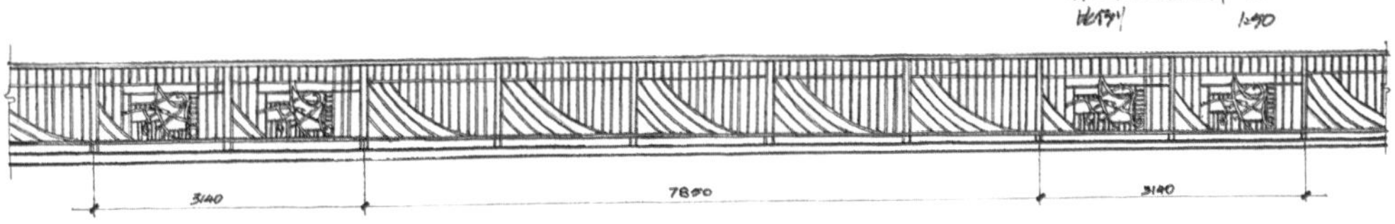

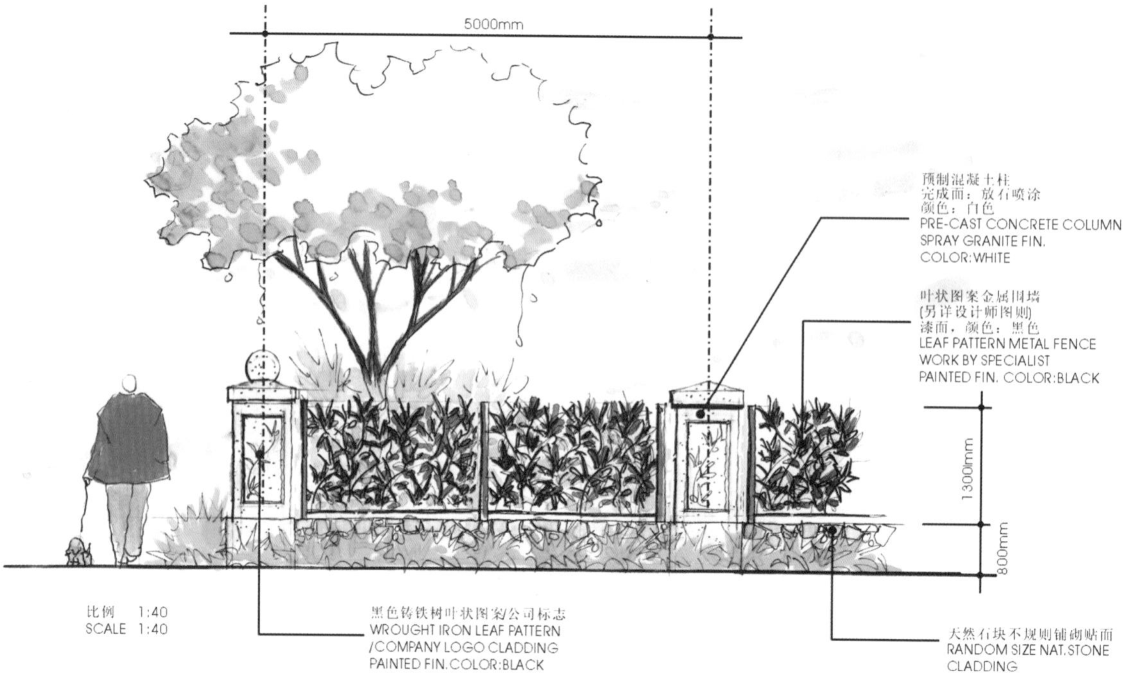
5000mm
预制混凝土柱
完成面：放石喷涂
颜色：白色
PRE-CAST CONCRETE COLUMN
SPRAY GRANITE FIN.
COLOR:WHITE
叶状图案金属围墙
(另详设计师图则)
漆面，颜色：黑色
LEAF PATTERN METAL FENCE
WORK BY SPECIALIST
PAINTED FIN. COLOR:BLACK
1300mm
800mm
比例 1:40
SCALE 1:40
黑色铸铁树叶状图案/公司标志
WROUGHT IRON LEAF PATTERN
/COMPANY LOGO CLADDING
PAINTED FIN. COLOR:BLACK
天然石块不规则铺砌贴面
RANDOM SIZE NAT. STONE
CLADDING

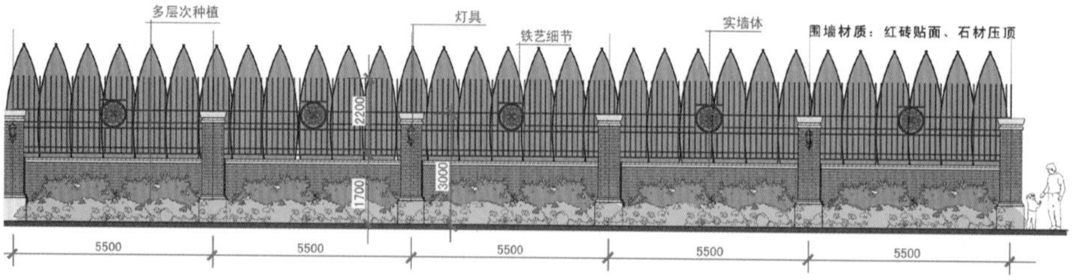
多层次种植
灯具
铁艺细节
实墙体
围墙材质：红砖贴面、石材压顶
2200
1700
3000
5500
5500
5500
5500
5500

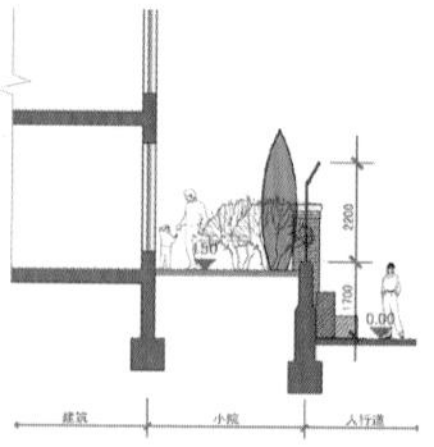
2200
1700
0.00
建筑
小院
人行道

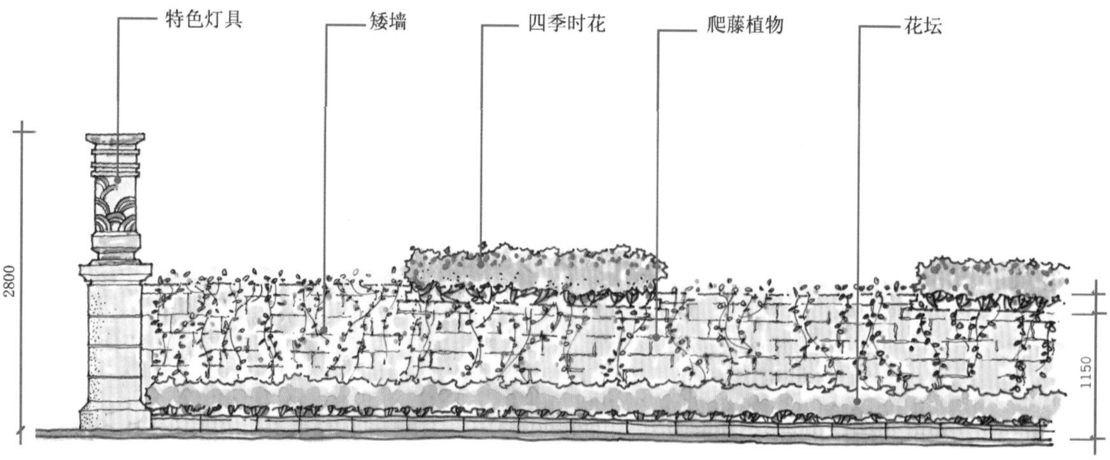
特色灯具
矮墙
四季时花
爬藤植物
花坛
2800
1150

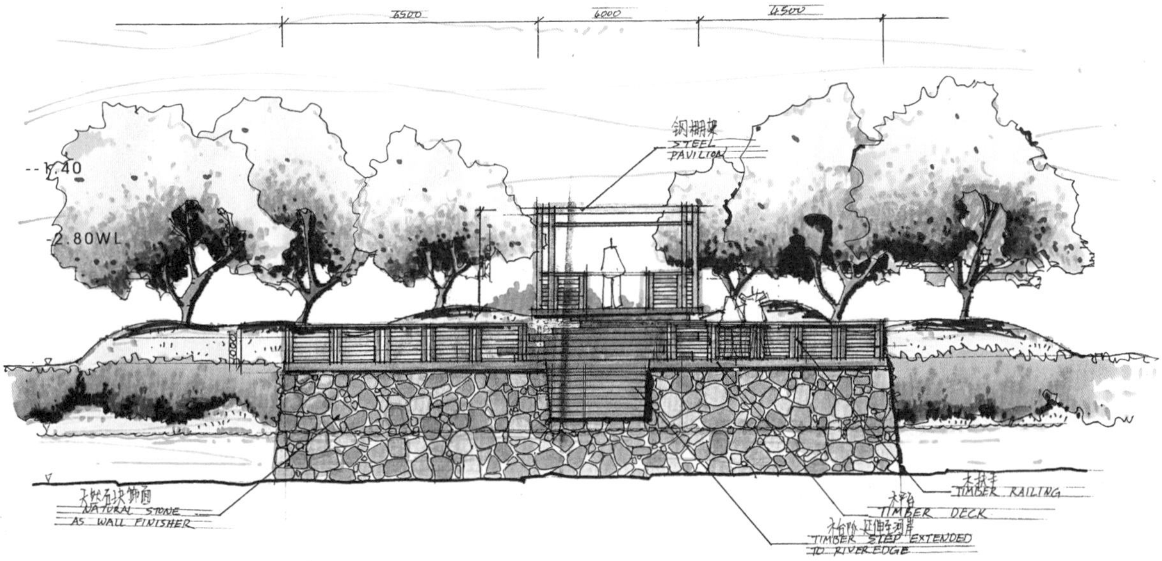

钢棚架
STEEL PAVILION
天然石块饰面
NATURAL STONE AS WALL FINISHER
木扶手
TIMBER RAILING
木平台
TIMBER DECK
TIMBER STEP EXTENDED TO RIVEREDGE

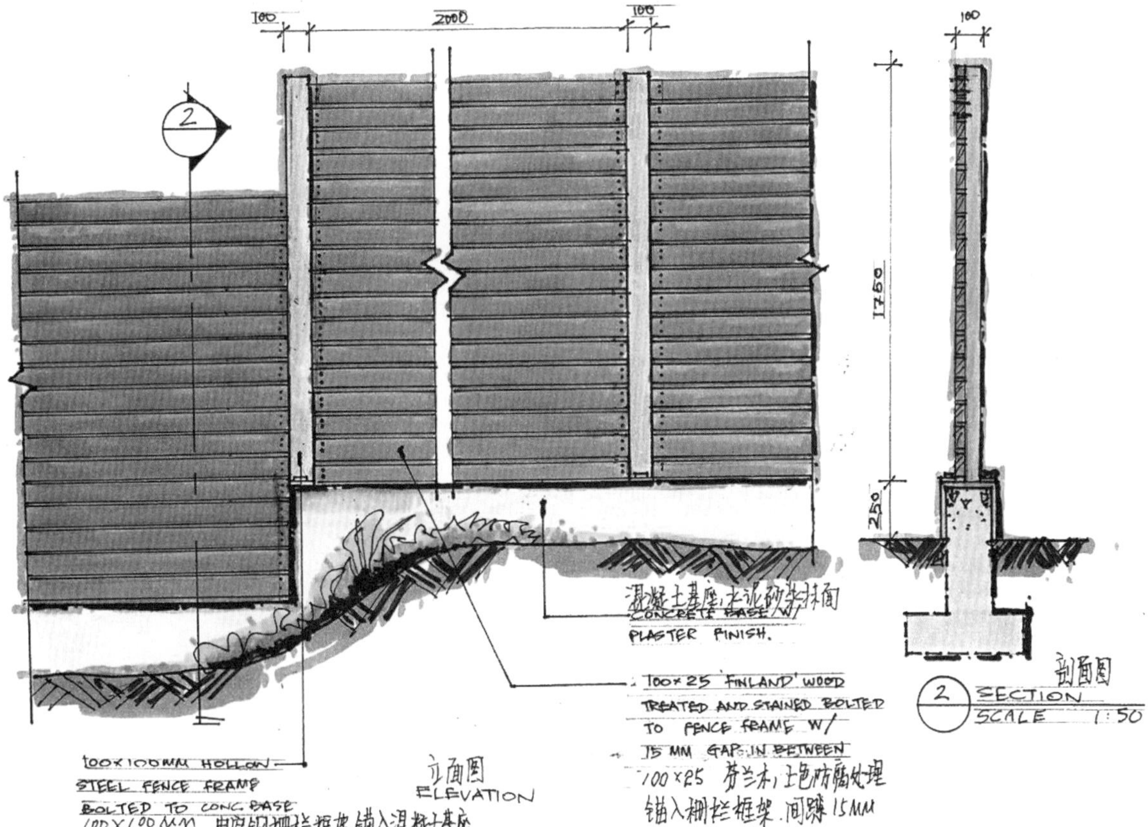

100
2000
100
1750
250
混凝土基座，水泥砂浆抹面
CONCRETE BASE W/ PLASTER FINISH.
100×25 'FINLAND' WOOD TREATED AND STAINED BOLTED TO FENCE FRAME W/ 15 MM GAP IN BETWEEN
100×25 芬兰木，上色防腐处理 锚入栅栏框架，间隙15MM
100×100MM HOLLOW STEEL FENCE FRAME BOLTED TO CONC. BASE
100×100MM 中空钢栅栏框架，锚入混凝土基座
立面图
ELEVATION
剖面图
2 SECTION SCALE 1:50
墙详图
1 WALL DETAILS SCALE 1:50

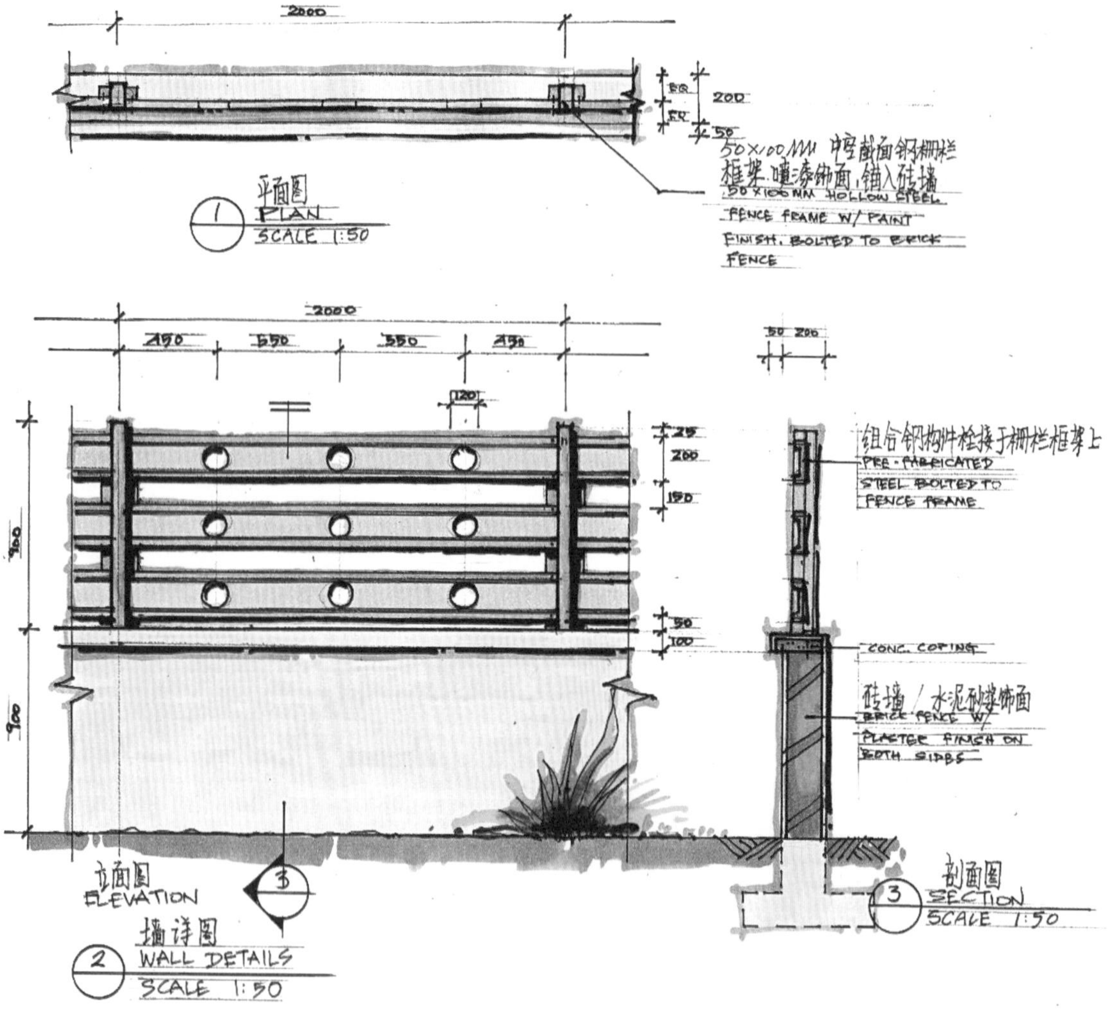
2000
200
50
50×100MM 中空截面钢栅栏框架，喷漆饰面，锚入砖墙
50×100MM HOLLOW STEEL FENCE FRAME W/ PAINT FINISH, BOLTED TO BRICK FENCE
1 平面图 PLAN SCALE 1:50
2000
450
550
550
450
120
25
200
150
50
100
900
900
50 200
组合钢构件栓接于栅栏框架上
PRE-FABRICATED STEEL BOLTED TO FENCE FRAME
CONC. COPING
砖墙/水泥砂浆饰面
BRICK FENCE W/ PLASTER FINISH ON BOTH SIDES
立面图 ELEVATION
3
3 剖面图 SECTION SCALE 1:50
2 墙详图 WALL DETAILS SCALE 1:50

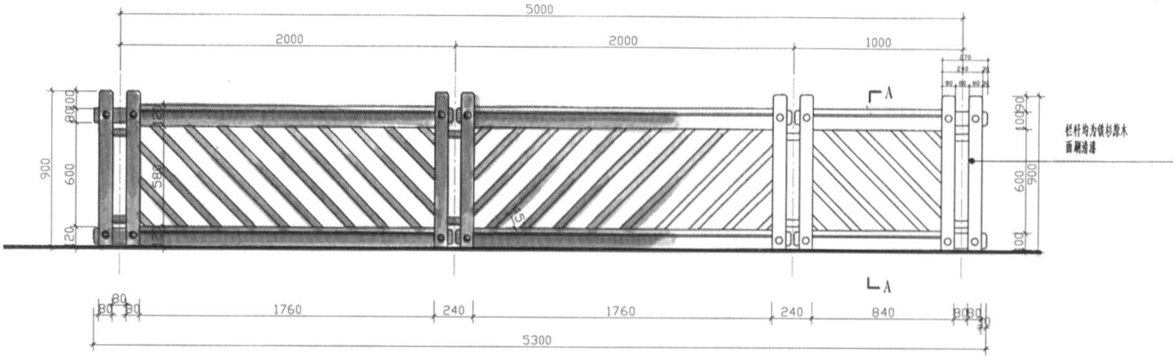
5000
2000
2000
1000
900
600
A
栏杆均为铁杉原木
面刷清漆
80
80
1760
240
1760
240
840
5300

① 栏杆立面图 1:20

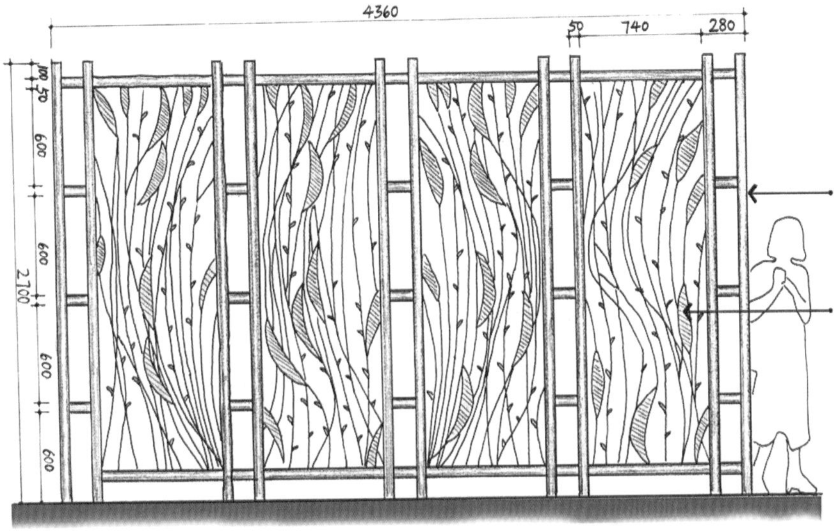

立面 F　1:20

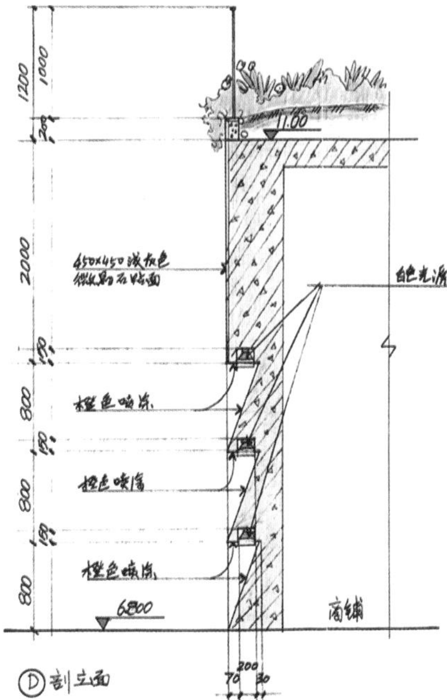

Ⓓ剖立面

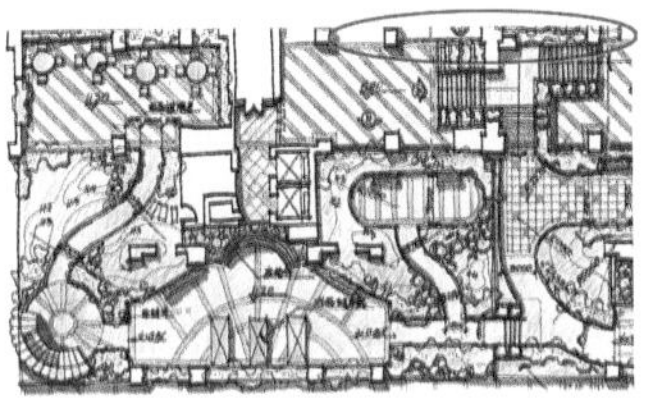

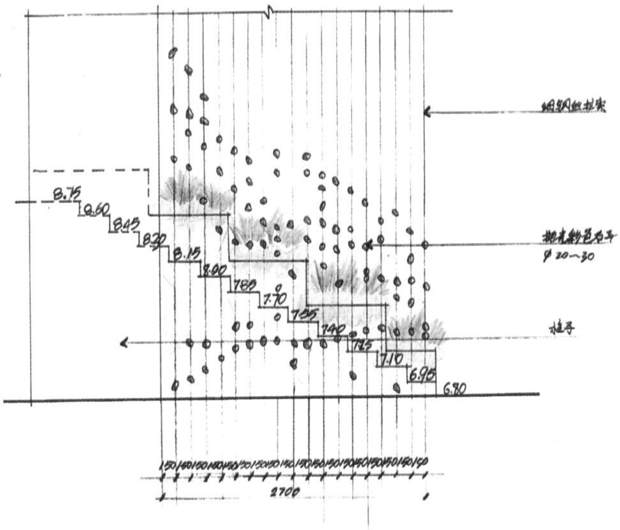

③立面外围栏杆

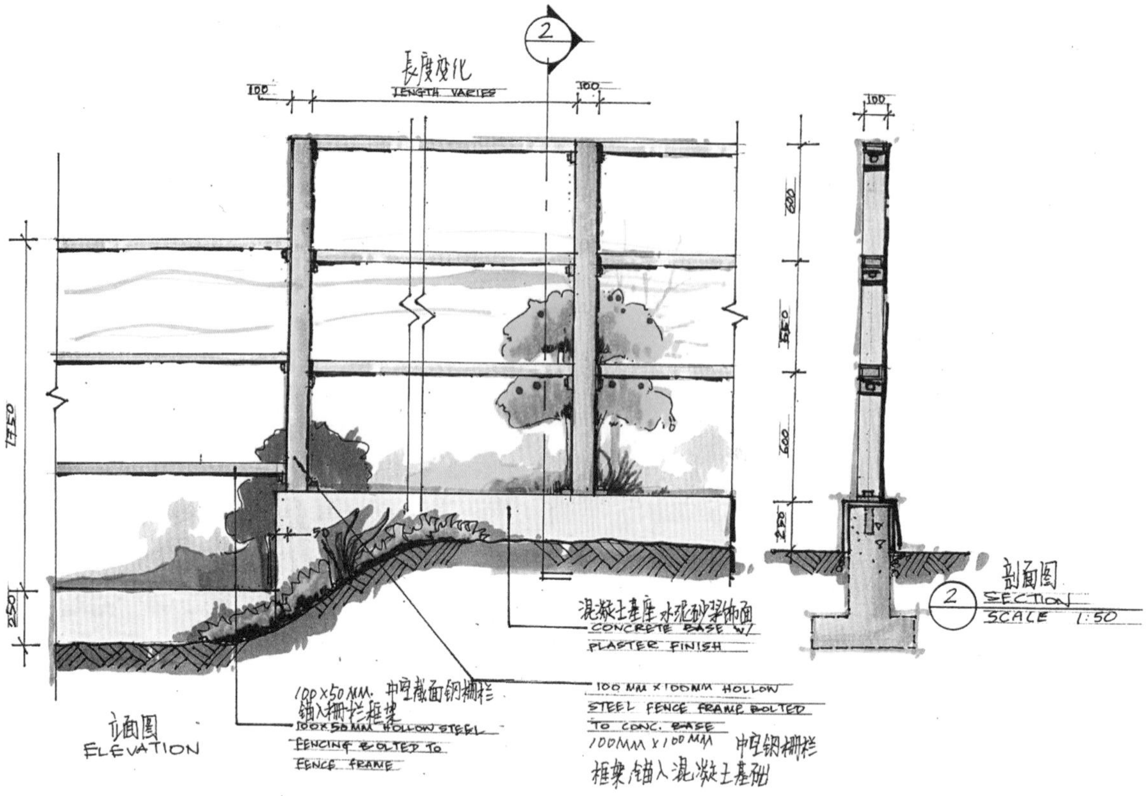

2
长度变化
LENGTH VARIES
100
100
1750
250
50
800
550
600
250
混凝土基座 水泥砂浆饰面
CONCRETE BASE W/
PLASTER FINISH
100×50MM 中空截面钢栅栏
锚入栅栏框架
100×50MM HOLLOW STEEL
FENCING BOLTED TO
FENCE FRAME
100MM×100MM HOLLOW
STEEL FENCE FRAME BOLTED
TO CONC. BASE
100MM×100MM 中空钢栅栏
框架锚入混凝土基础
立面图
ELEVATION
剖面图
2 SECTION
SCALE 1:50

墙详图

1 WALL DETAILS

SCALE 1:50

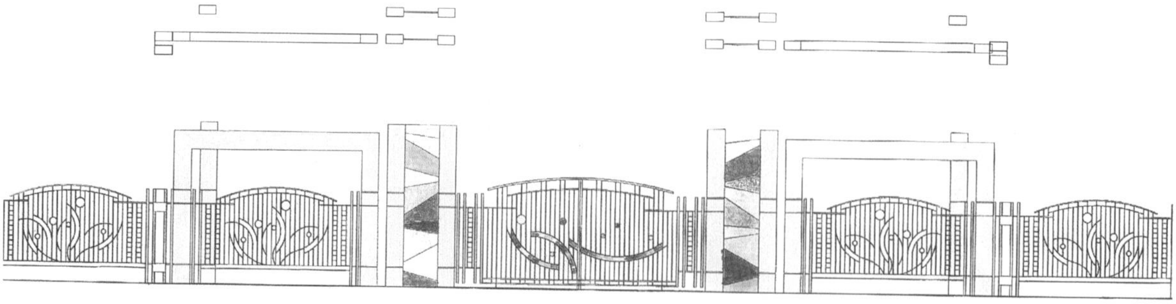

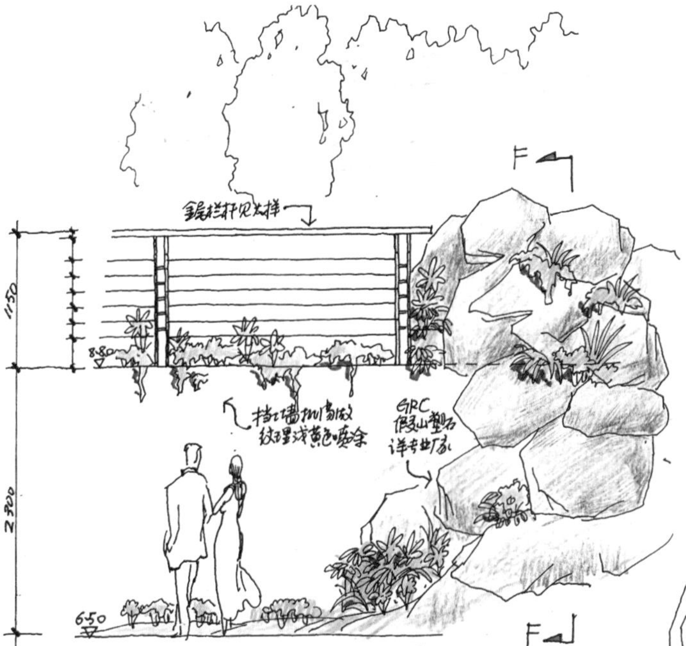

M立面图 1:30
GRC假山塑石详专业厂家
F-F剖面 1:20
GFRC假山塑石见专业厂家

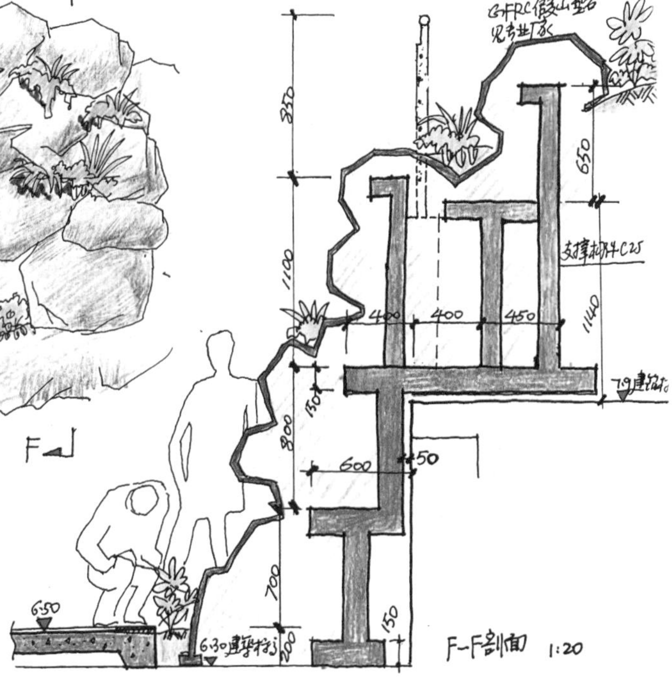

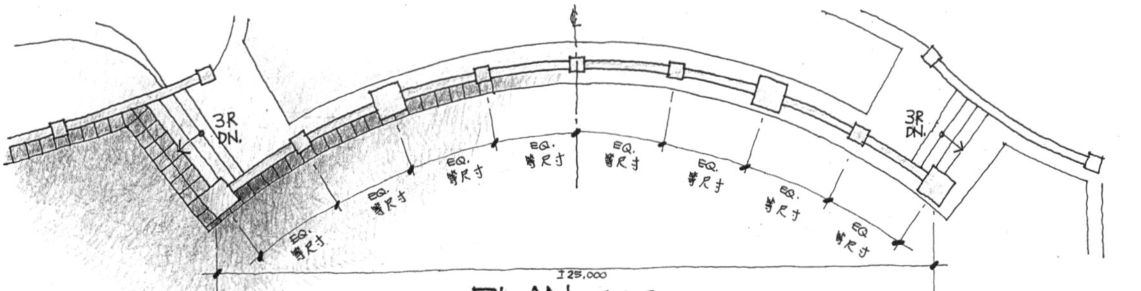

A PLAN 平面图
SCALE 1:100M
BRONZE WINGED FIGURE SCULPTURE BY SPECIALIST DETAIL
METAL RAILING SPRAY PAINTED FINISH
NATURAL GRANITE WALL CLADDING, FLAMED FINISH COLOUR-BROWN
NATURAL GRANITE CLADDING, FLAMED FIN. COLOUR-BIEGE
立面图
B ELEVATION
SCALE 1:100M.

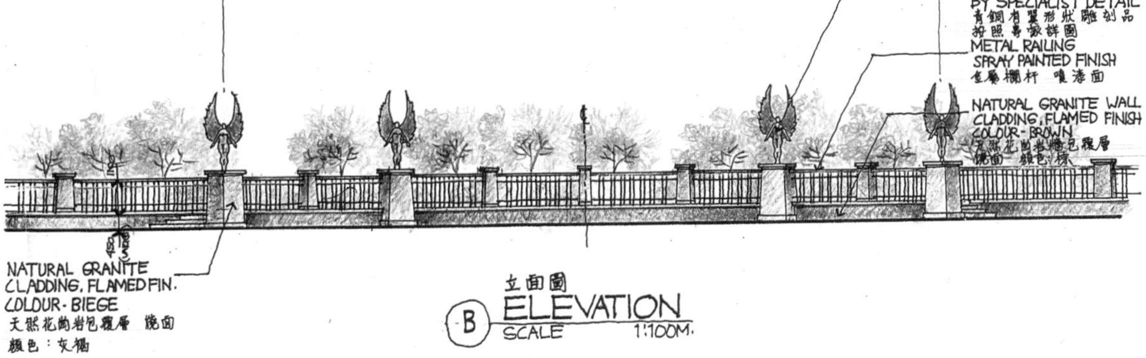

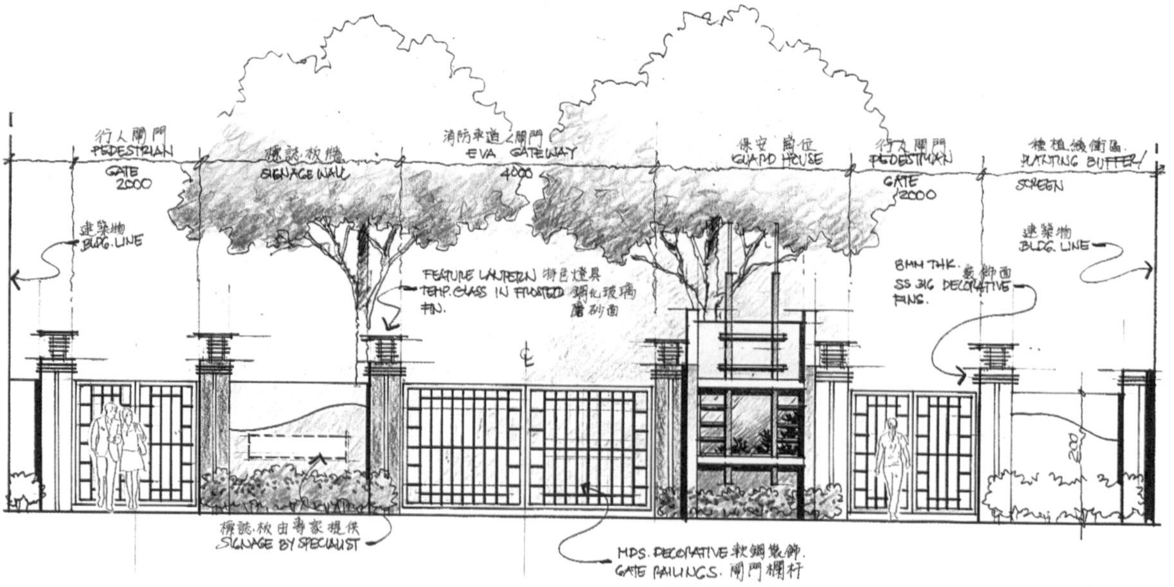

行人閘門 PEDESTRIAN GATE 2000
標誌枚牆 SIGNAGE WALL
消防車道之閘門 EVA GATEWAY 4000
保安崗位 GUARD HOUSE
行人閘門 PEDESTRIAN GATE 2000
種植緩衝區 PLANTING BUFFER/SCREEN
建築物 BLDG. LINE
建築物 BLDG. LINE
FEATURE LANTERN 特色燈具 TEMP. GLASS IN FROSTED FIN. 鋼化玻璃磨砂面
8MM THK. SS 316 DECORATIVE FINS. 裝飾面
標誌枚由專家提供 SIGNAGE BY SPECIALIST
MDS. DECORATIVE 軟鋼裝飾 GATE RAILINGS. 閘門欄杆

5000
500
500
500X500X100 PILASTER 半露柱蓋 WALL CAP, IMPERIAL BLACK 帝皇黑 IN POLISHED FIN. 光面
300X600 X 100 MM THK. RUSTIC YELLOW IN BUSH HAMMERED FIN. WALL CAP. 牆蓋 銹黃色 手打面

PLAN VIEW 平面圖

SCALE 比例 1:25

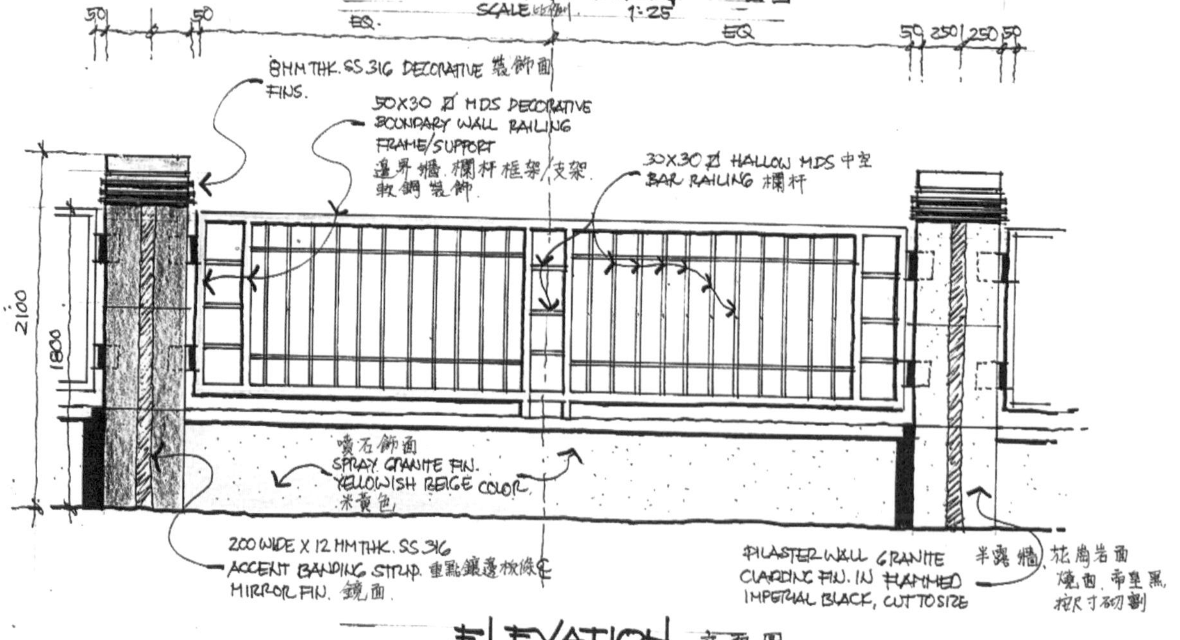

50 50 EQ. EQ 50 250 250 50
8MM THK. SS 316 DECORATIVE 裝飾面 FINS.
50X30 ▱ MDS DECORATIVE BOUNDARY WALL RAILING FRAME/SUPPORT 邊界牆 欄杆框架/支架 軟鋼裝飾
30X30 ▱ HALLOW MDS 中空 BAR RAILING 欄杆
2100
1800
噴石飾面 SPRAY GRANITE FIN. YELLOWISH BEIGE COLOR 米黃色
200 WIDE X 12 MM THK. SS 316 ACCENT BANDING STRIP 重點鑲邊板條 MIRROR FIN. 鏡面
PILASTER WALL GRANITE CLADDING FIN. IN FLAMMED IMPERIAL BLACK, CUT TO SIZE 半露牆 花崗岩面 燒面 帝皇黑 按尺寸切割

ELEVATION 立面圖

SCALE 比例 1:25

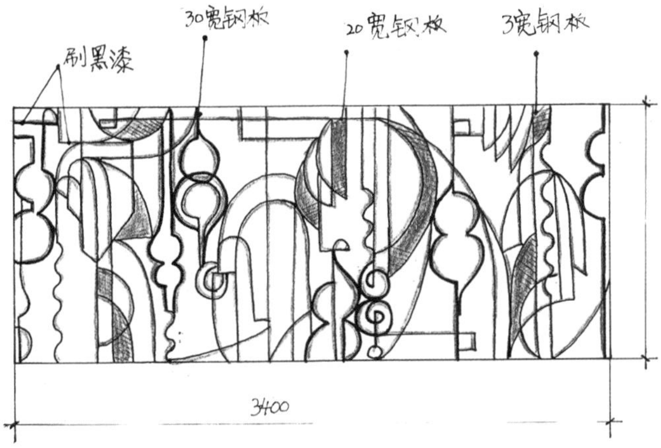

“寿”铁艺大样

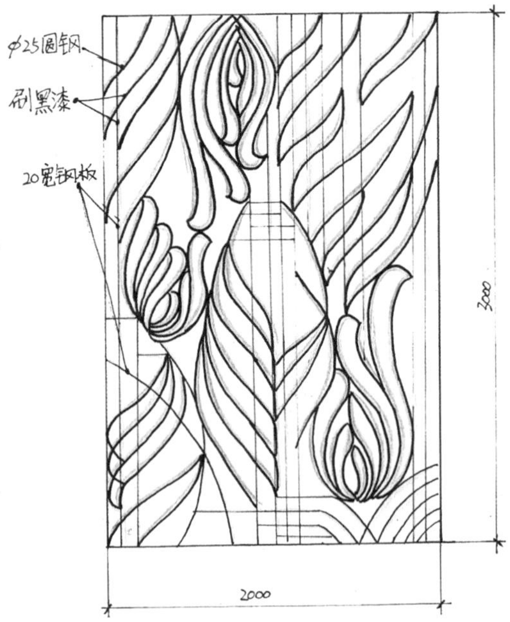

“福”铁艺大样

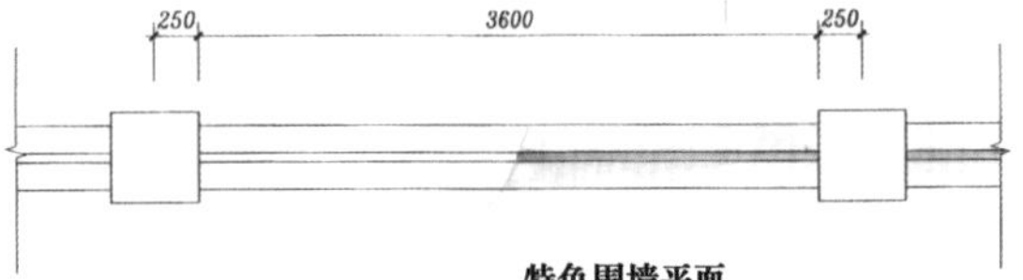

特色围墙平面

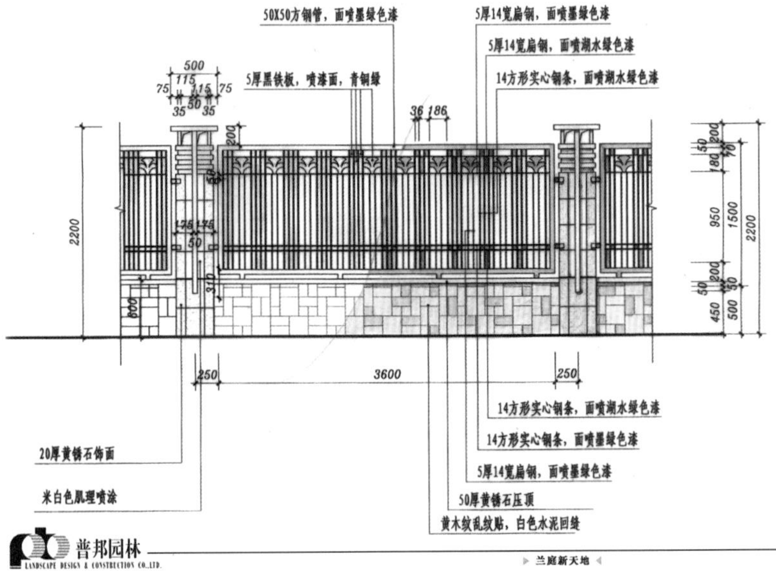

特色围墙立面

特色围墙平立面图

SCALE 1:30

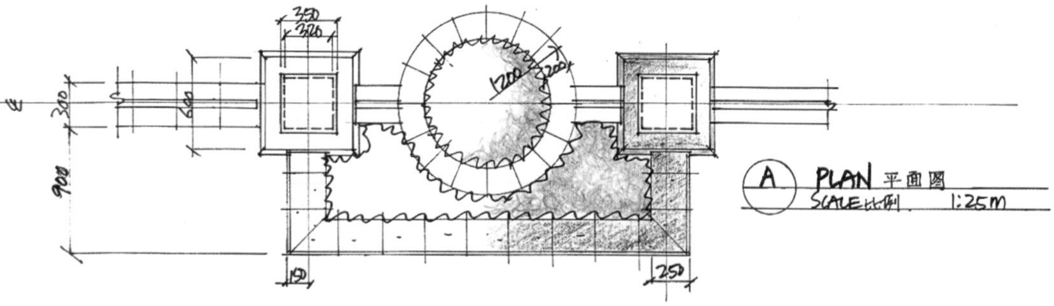

PLAN 平面图
SCALE 比例 1:25M
ELEVATION 立面图
SCALE 比例 1:25m.
CUSTOMED STAINLESS STEEL MATTED FINISH WITH WHITE GLASS DIFFUSED FEATURE LANTERN
30X10 MM THK STAINLESS STEEL FLAT BAR FRAMES HAIRLINE FINISH
6MM THK TEMPERED GLASS BLUE LIGHT BACKPAINTED, LIGHTED COLUMN, DIFFUSSED
12 MM THK STAINLESS STEEL BASEPLATE ANCHORED TO CONCRETE PLINTH
20 MM THK HONED FINISH NAT. CHINESE GRANITE CLADDING COLOR: ROYAL RED
TEXTURED PAINTED FIN. PLASTERED WALL. COLOR: BEIGE
纹理油漆面 灰泥墙，米色
TYPICAL FENCE DETAIL
标准围墙详图
20MM THK CRAZY CUT CLEFT FINISH SLATE. COLOR: MARSH GREEN
40X300X300 MM THK FLAMED FINISH GRANITE BORDER COLOR: ROYAL RED
FFL

φ25圆钢
3宽钢板
刷黑漆
30宽钢板
孔距10钢网
1400
3400

“禄”铁艺大样

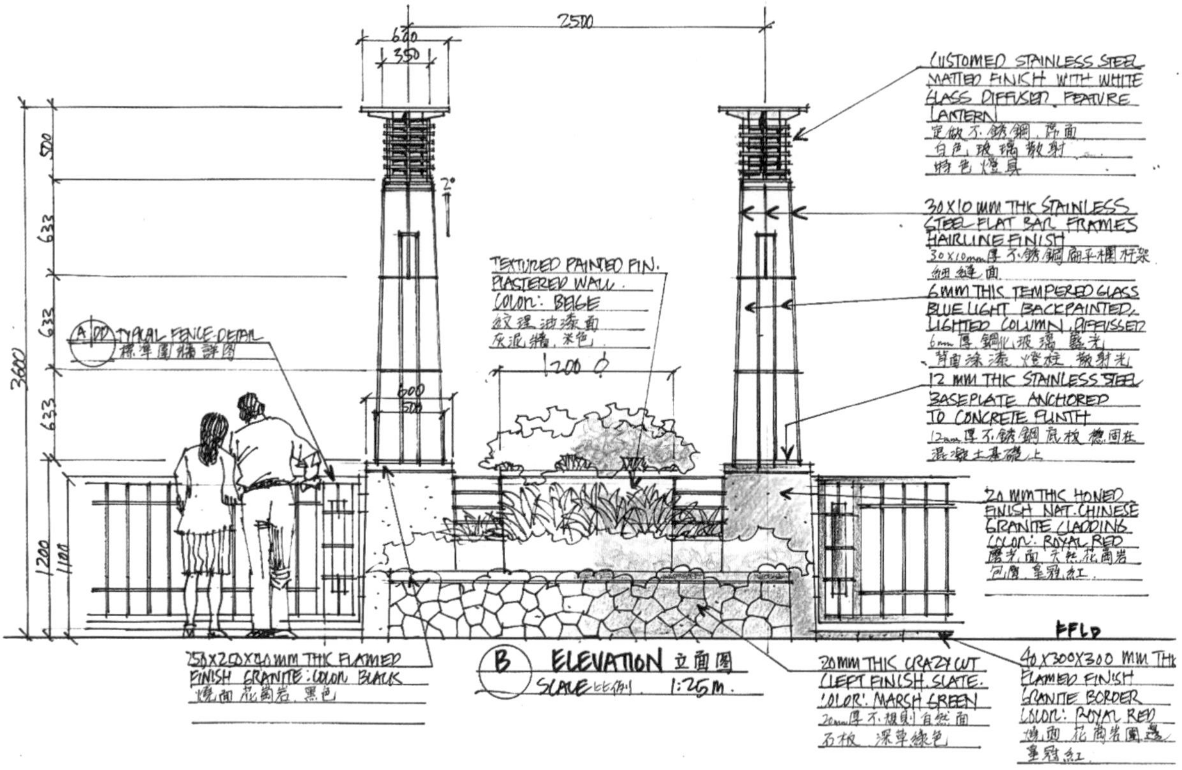

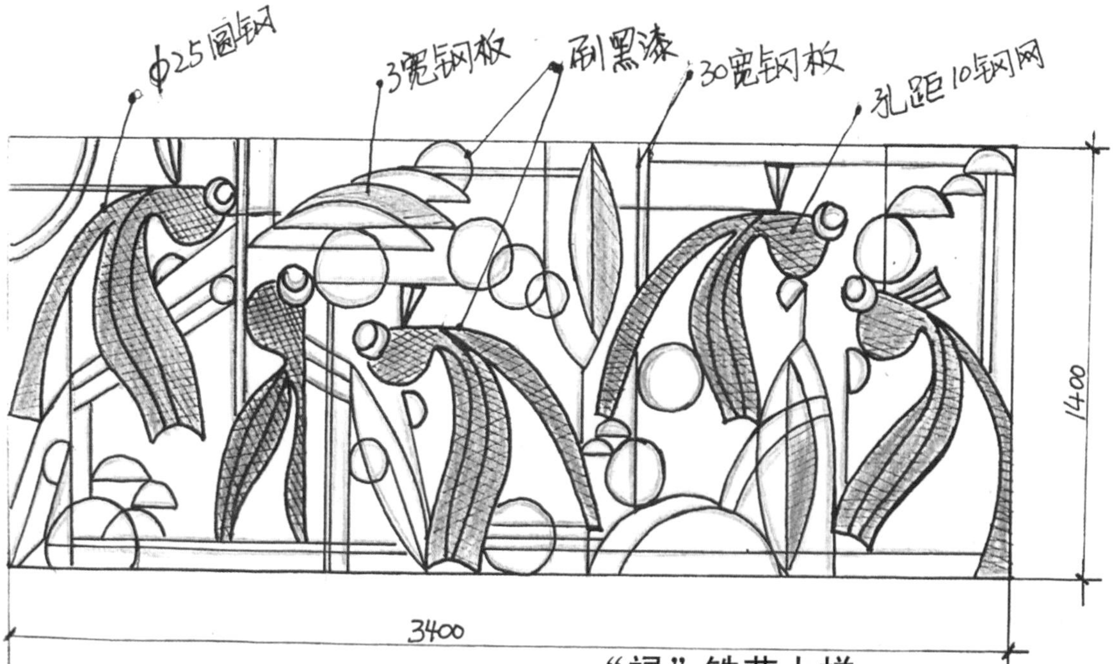

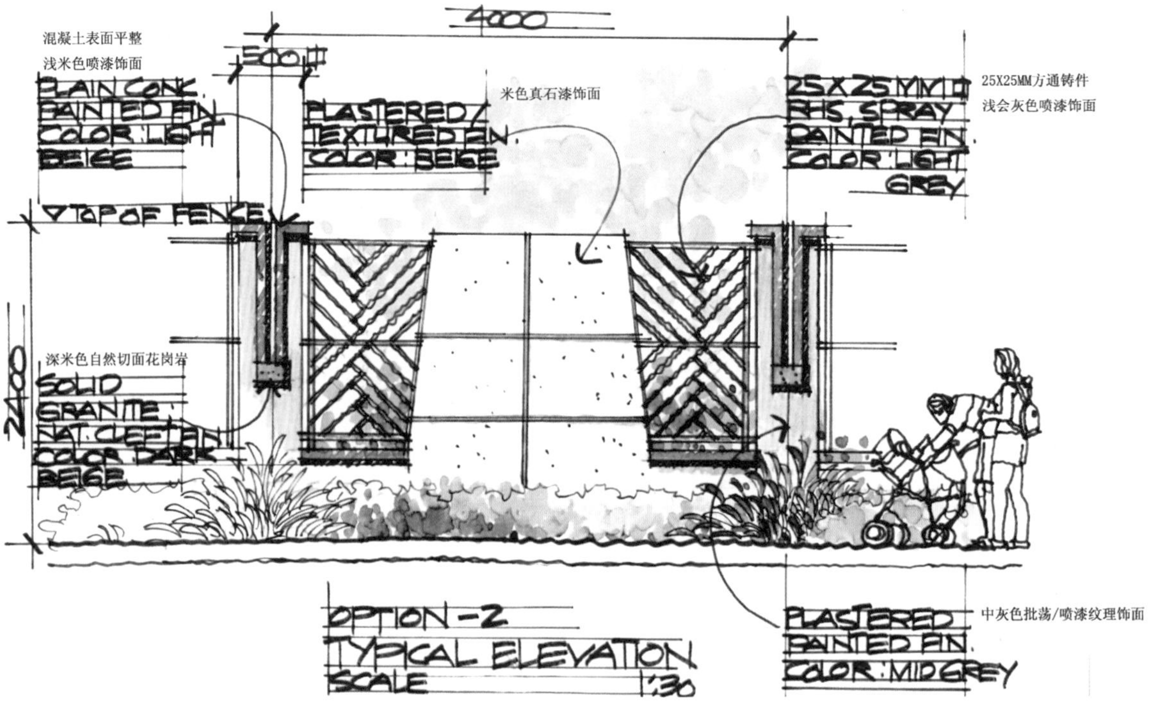
4000
500
混凝土表面平整
浅米色喷漆饰面
PLAIN CONC. PAINTED FIN. COLOR: LIGHT BEIGE
PLASTERED/ TEXTURED FIN. COLOR: BEIGE
米色真石漆饰面
25X25MM RHS, SPRAY PAINTED FIN. COLOR: LIGHT GREY
25X25MM方通铸件
浅会灰色喷漆饰面
TOP OF FENCE
2400
深米色自然切面花岗岩
SOLID GRANITE NAT. CLEFT FIN. COLOR: DARK BEIGE
PLASTERED PAINTED FIN. COLOR: MID GREY
中灰色批荡/喷漆纹理饰面
OPTION -2
TYPICAL ELEVATION
SCALE 1:30
典型围墙立面图

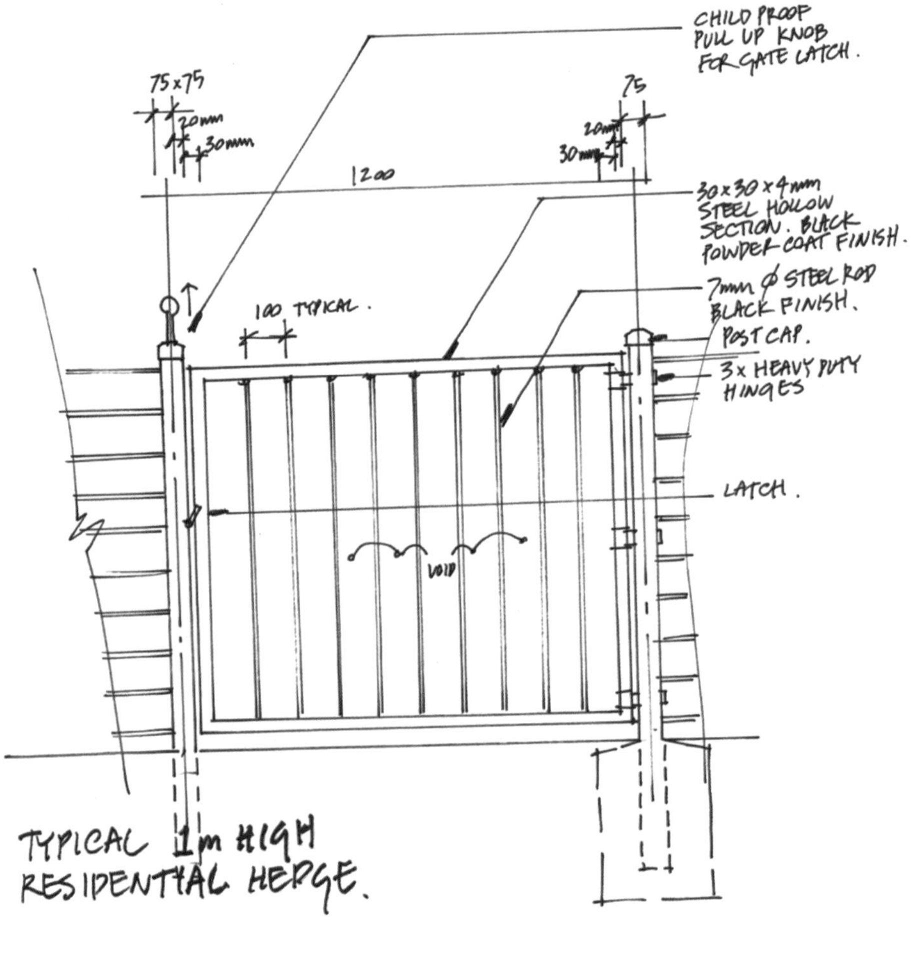
CHILD PROOF PULL UP KNOB FOR GATE LATCH.
75x75
20mm
30mm
75
20mm
30mm
1200
30x30x4mm STEEL HOLLOW SECTION. BLACK POWDER COAT FINISH.
7mm ϕ STEEL ROD BLACK FINISH.
100 TYPICAL.
POST CAP.
3x HEAVY DUTY HINGES
LATCH.
VOID
TYPICAL 1m HIGH RESIDENTIAL HEDGE.

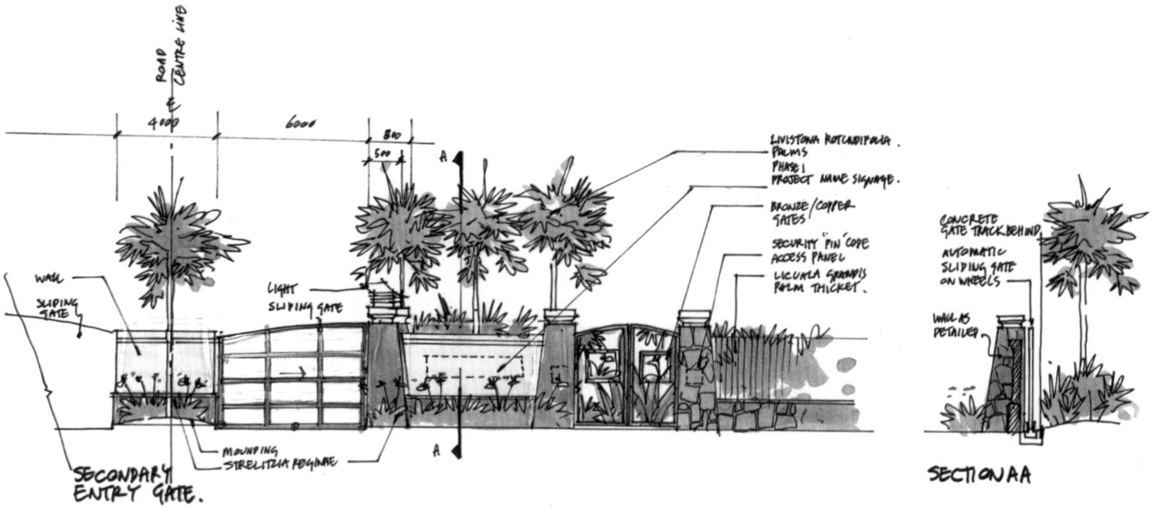

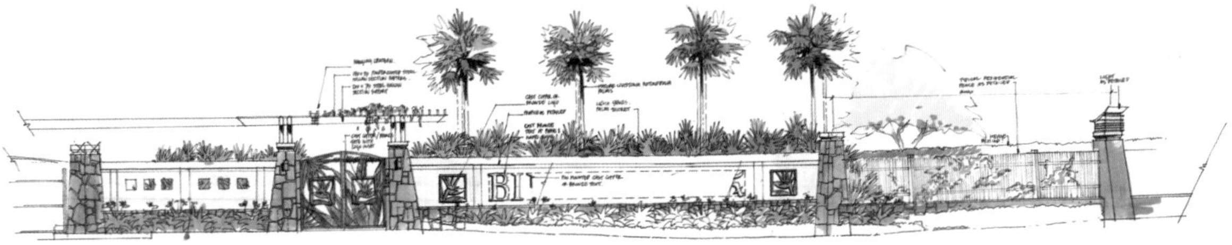

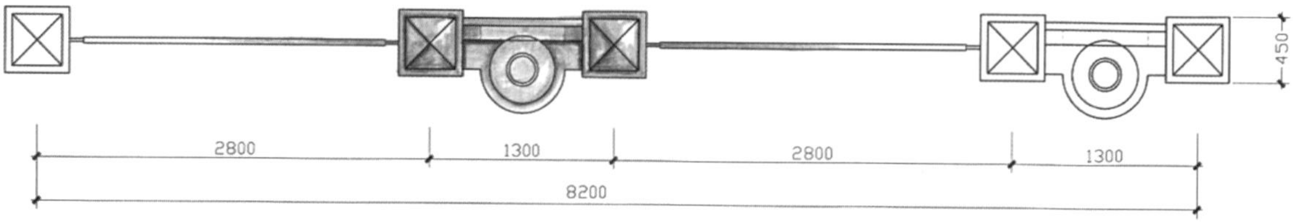

① 入口围墙平面图1:30

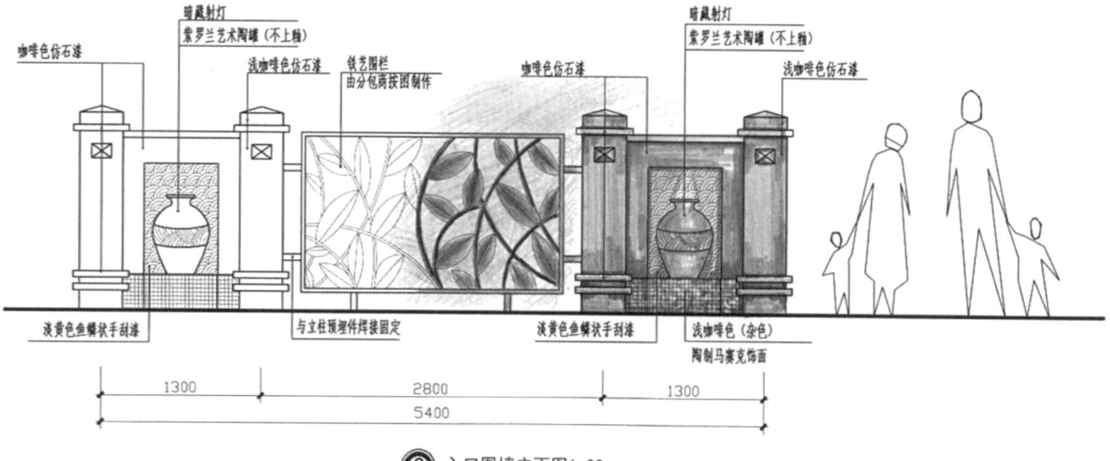

② 入口围墙立面图1:30

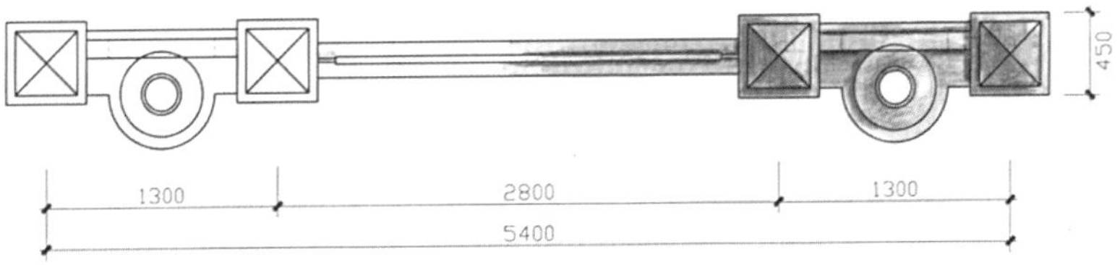

① 入口围墙平面图1:30

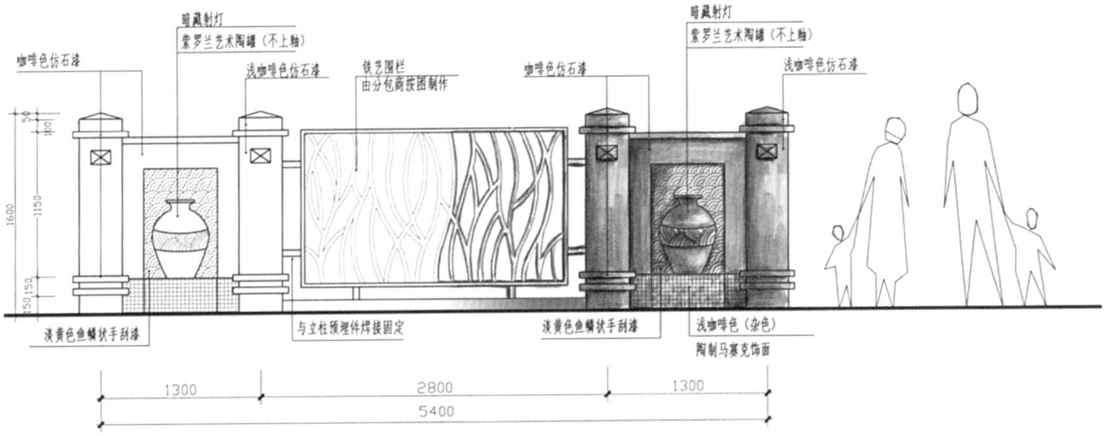

② 入口围墙立面图1:30

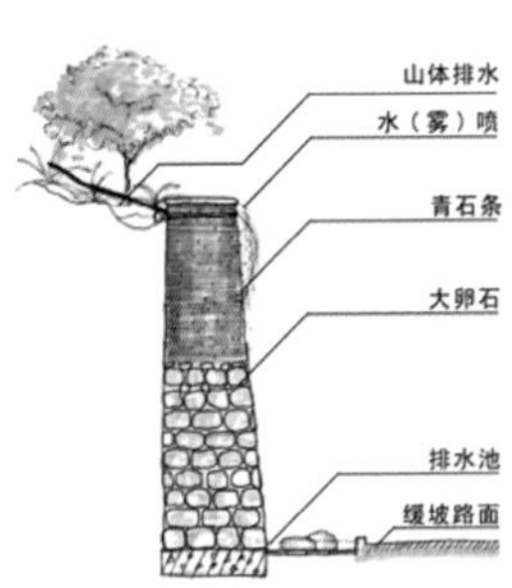

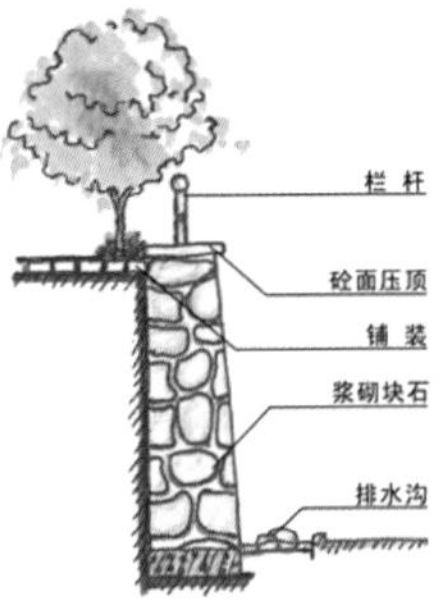

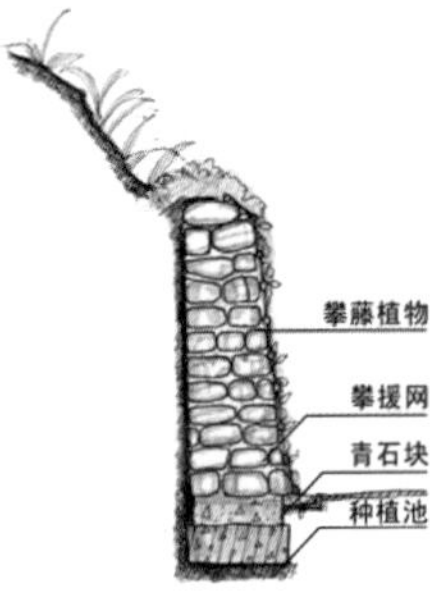

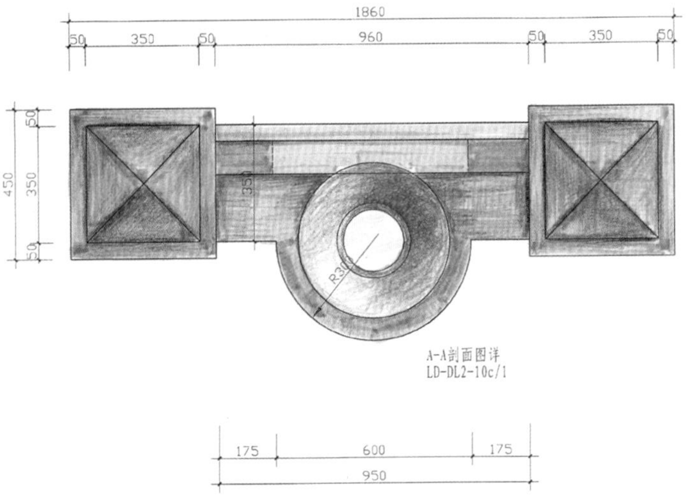

① 围墙立柱墙立面图1:10

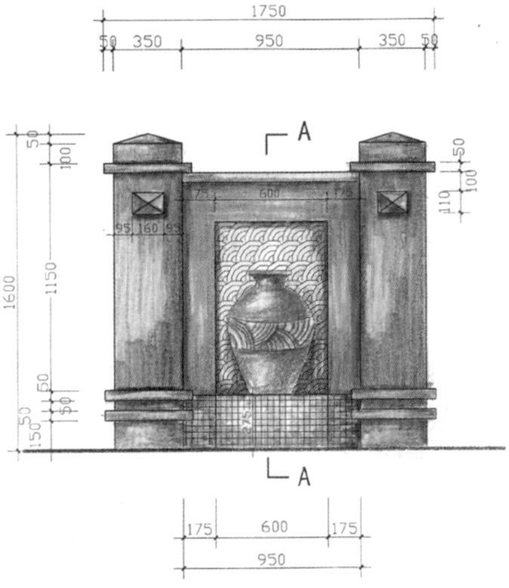

② 围墙立柱墙立面图1:20

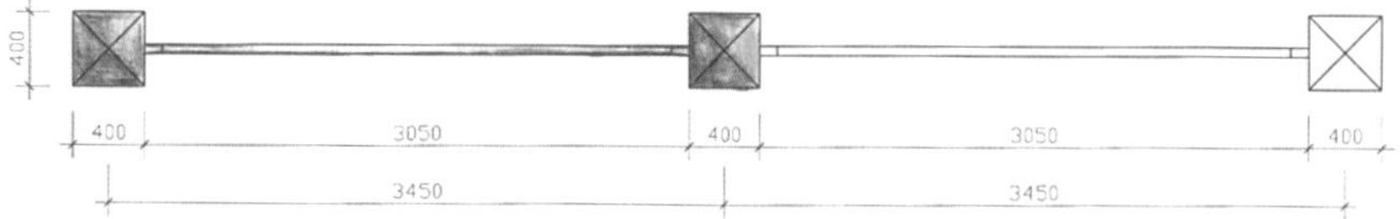

① 标准围墙平面图 1:30

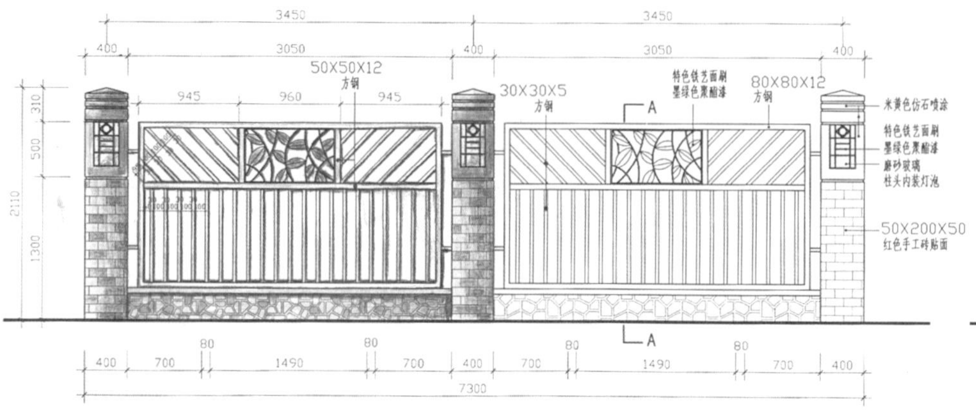

② 标准围墙立面图 1:30

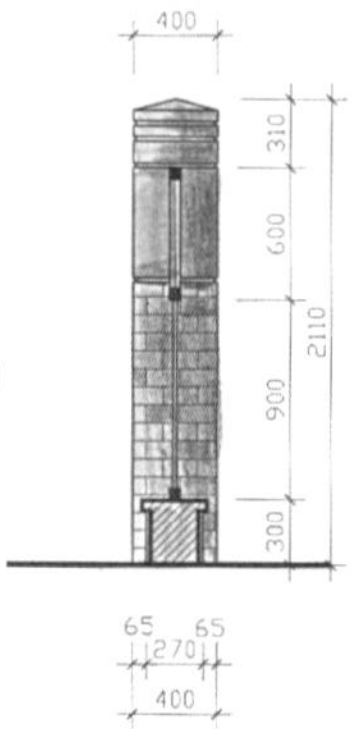

③ A-A剖面图 1:30

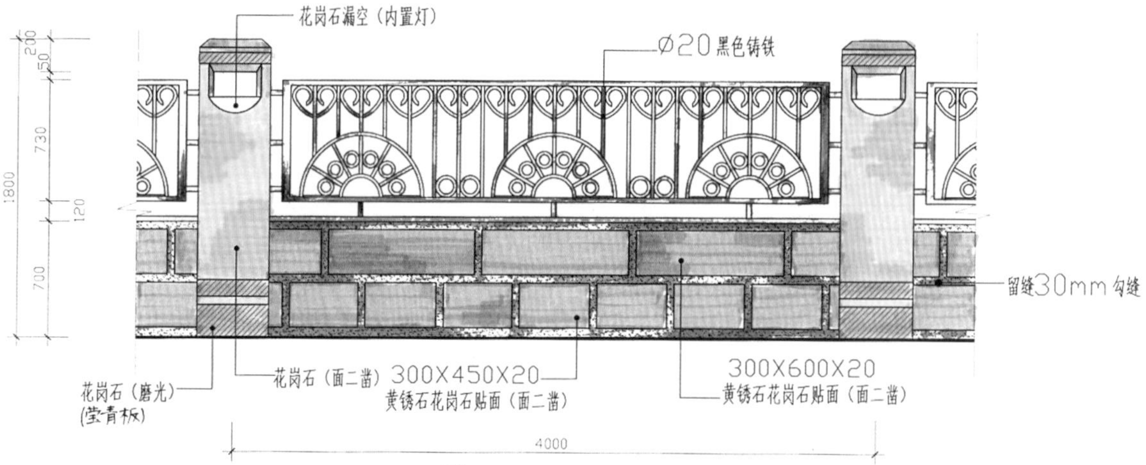

① 围墙立面图 1:20

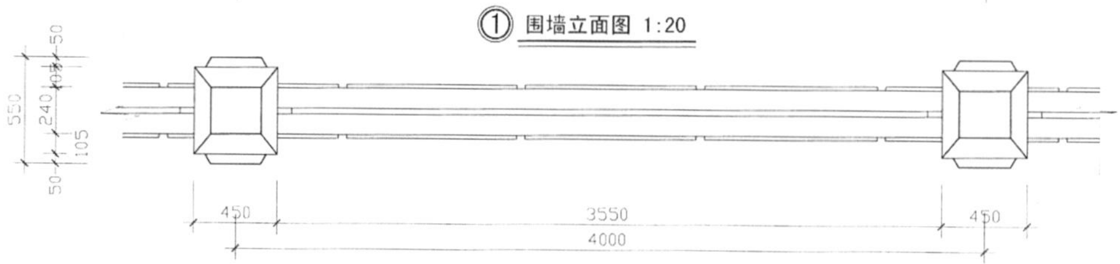

② 围墙平面图 1:20

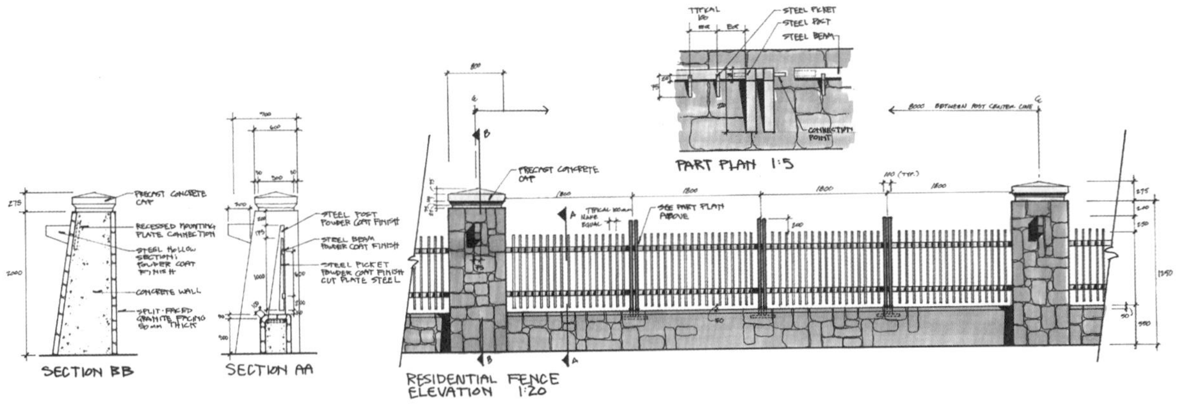
PRECAST CONCRETE CAP
RECESSED MOUNTING PLATE CONNECTION
STEEL HOLLOW SECTION; POWDER COAT FINISH
CONCRETE WALL
SPLIT-FACED GRANITE FACING 50mm THICK
SECTION BB
STEEL POST POWDER COAT FINISH
STEEL BEAM POWDER COAT FINISH
STEEL PICKET POWDER COAT FINISH CUT PLATE STEEL
SECTION AA
STEEL PICKET
STEEL POST
STEEL BEAM
CONNECTION POINT
PART PLAN 1:5
8000 BETWEEN POST CENTER LINE
PRECAST CONCRETE CAP
SEE PART PLAN ABOVE
RESIDENTIAL FENCE ELEVATION 1:20

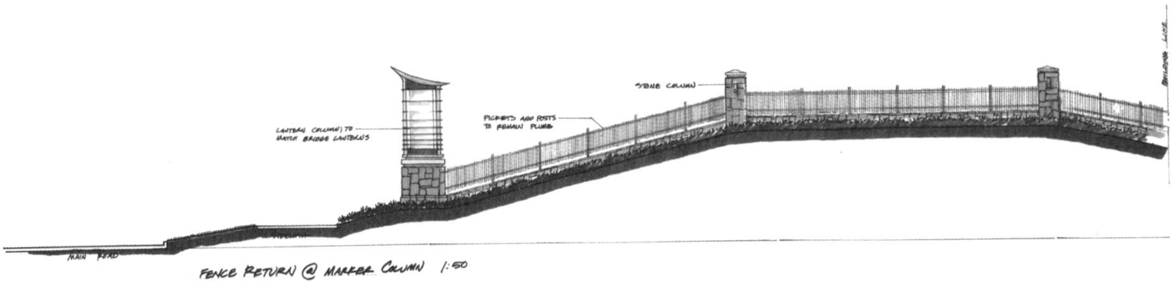
STONE COLUMN
PICKETS AND POSTS TO REMAIN PLUMB
MAIN ROAD
FENCE RETURN @ MARKER COLUMN 1:50

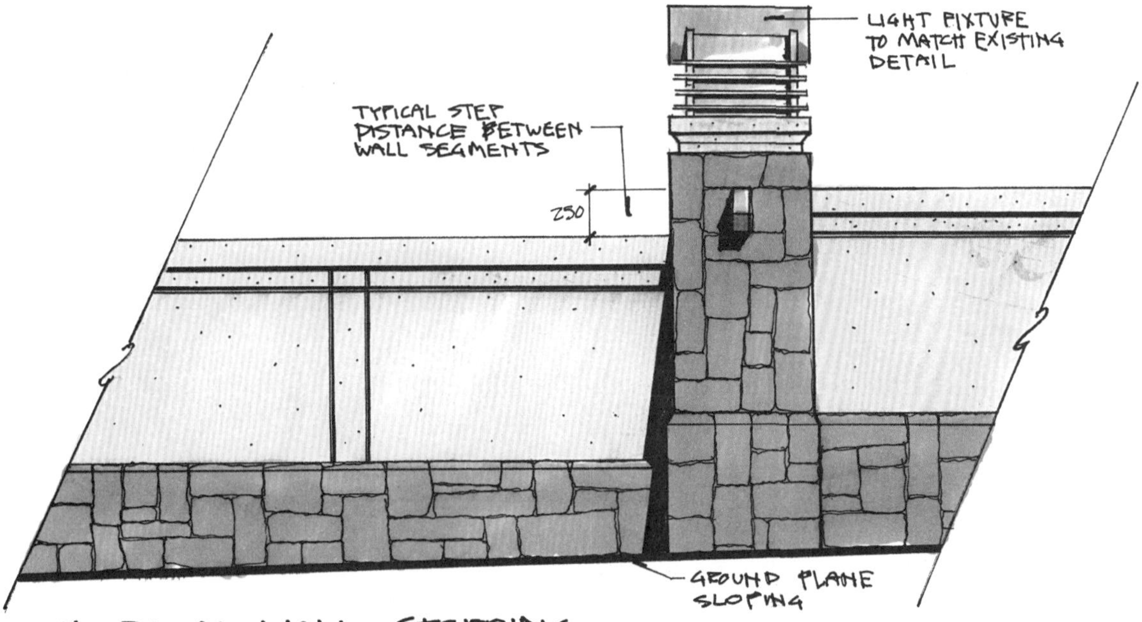
LIGHT FIXTURE TO MATCH EXISTING DETAIL
TYPICAL STEP DISTANCE BETWEEN WALL SEGMENTS
250
GROUND PLANE SLOPING
TYPICAL WALL STEPPING DETAIL 1:20

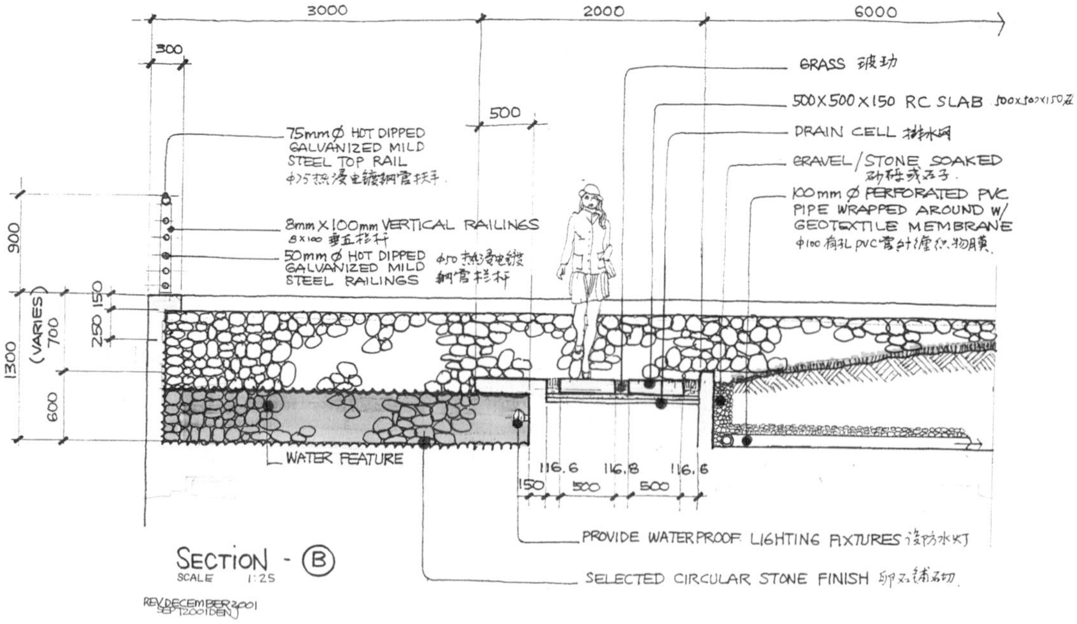
3000
2000
6000
300
500
GRASS 玻功
500X500X150 RC SLAB
DRAIN CELL 排水网
GRAVEL/STONE SOAKED
100mm Ø PERFORATED PVC PIPE WRAPPED AROUND W/ GEOTEXTILE MEMBRANE
75mm Ø HOT DIPPED GALVANIZED MILD STEEL TOP RAIL
8mm X 100mm VERTICAL RAILINGS
50mm Ø HOT DIPPED GALVANIZED MILD STEEL RAILINGS
900
1300
(VARIES) 700
600
250
150
WATER FEATURE
116.6
116.8
116.6
150
500
500
PROVIDE WATERPROOF LIGHTING FIXTURES
SELECTED CIRCULAR STONE FINISH
SECTION - B
SCALE 1:25

泳池栏杆立面示意图

泳池栏杆剖面示意图

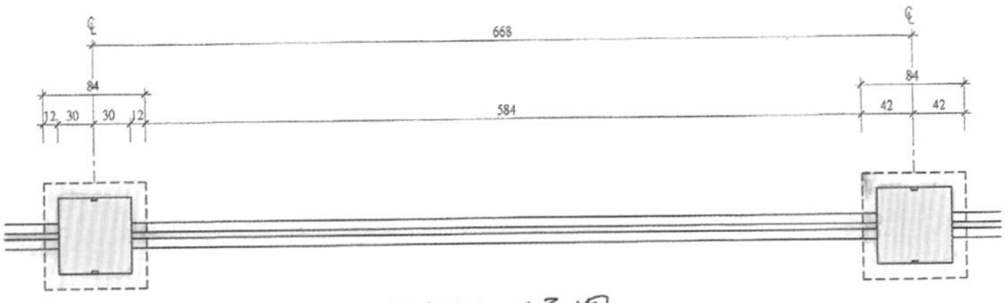

社區欄杆平面圖
S:1/50

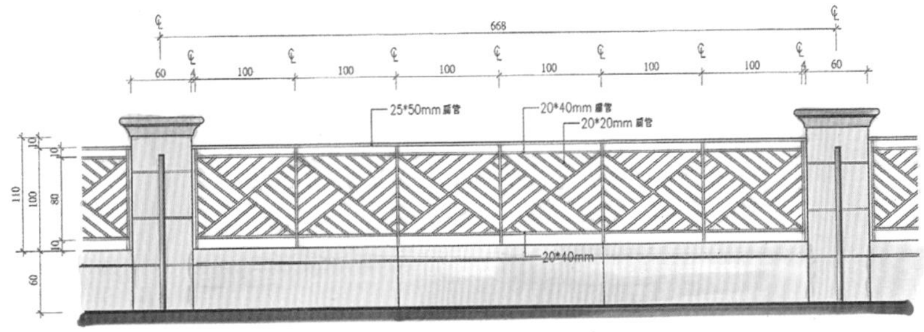

社區欄杆平面圖
S:1/60

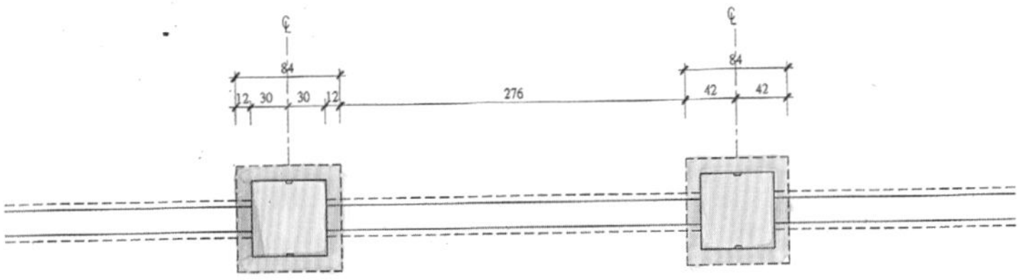

社區圍牆平面圖
S:1/30

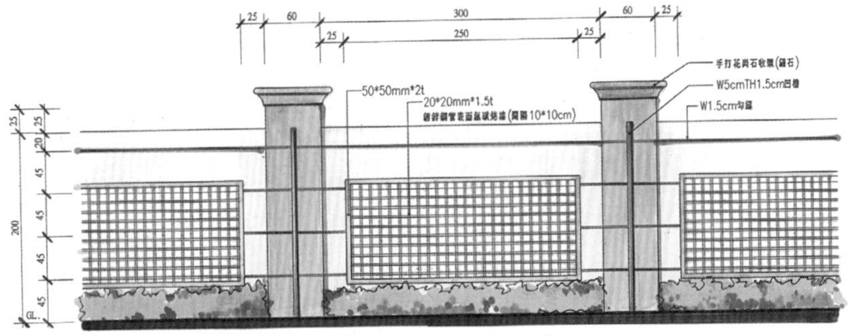

社區圍牆正立面圖
S:1/30

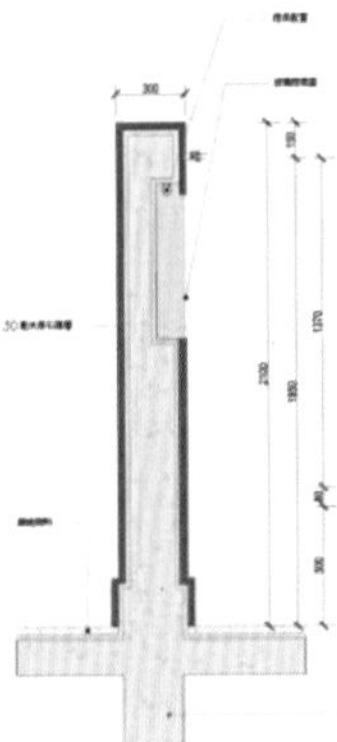

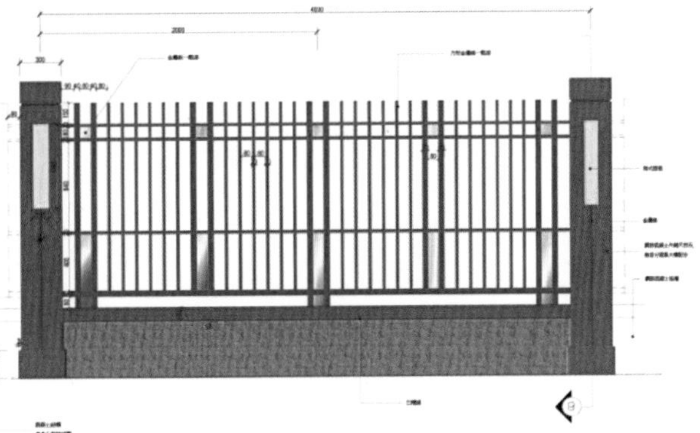

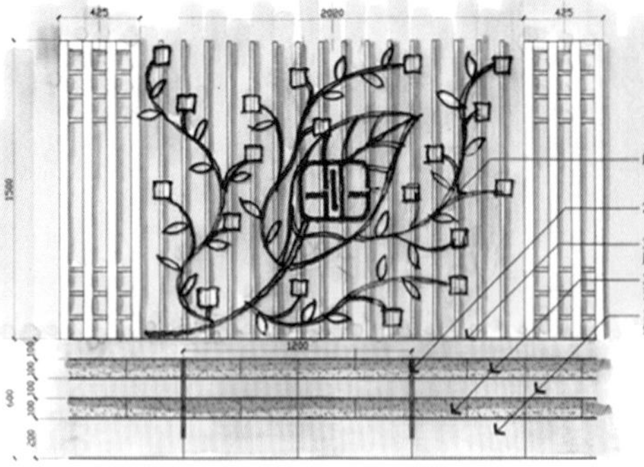
425
2600
425
1200
鍍鋅鋼鐵條
20毫米闊不銹鋼條板, 1200毫米中/中
300 x300x100毫MM厚，天然花崗石庄頂
磨光面, 顏色 :珍珠黑
300 x100x20MM厚，天然花崗石 燒面, 顏色 :珍珠黑
天然切割面, 顏色 :珍珠黑
100/200 X600X20MM厚，天然花崗石
燒面, 顏色 :珍珠黑

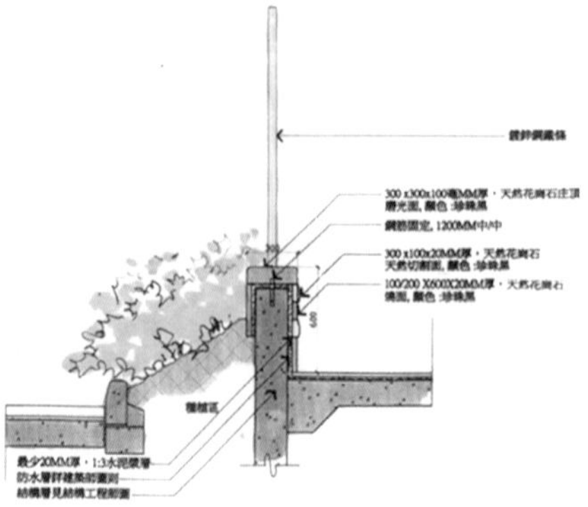
鍍鋅鋼鐵條
300 x300x100毫MM厚，天然花崗石庄頂
磨光面, 顏色 :珍珠黑
鋼筋固定, 1200MM中/中
300 x100x20MM厚，天然花崗石
天然切割面, 顏色 :珍珠黑
100/200 X600X20MM厚，天然花崗石
燒面, 顏色 :珍珠黑

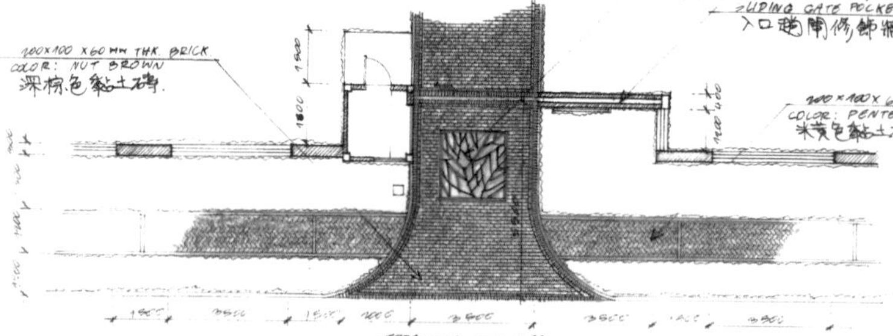
DIVIDED INTO MODULE (@ 1000 x 1000 x 40)
CUSTOMED GRANITE ACCENT PAVING FLAMED FINISH
MOTIFE ENGRAVES
SLIDING GATE POCKET DOUBLE WALL
200x100 x 60 MM THK BRICK
COLOR: NUT BROWN
200 x 100 x 60 MM THK BRICK
COLOR: PENTER YELLOW
TERRACOTTA ROOF TILES
紅褐色瓦頂
A PLAN 平面
SCALE 1:100

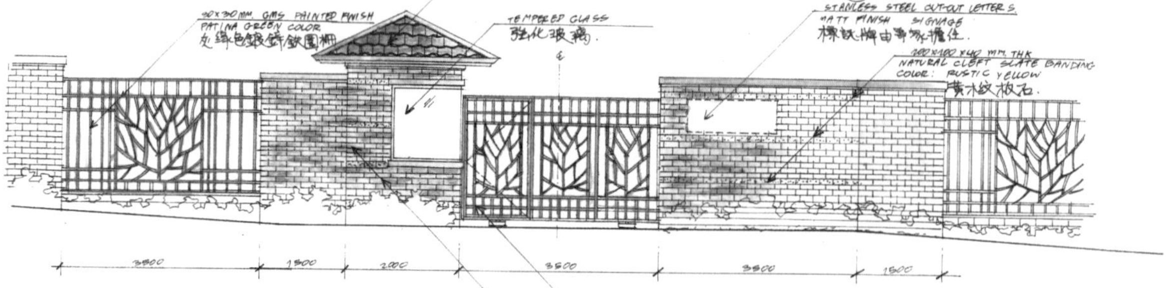
TEMPERED GLASS
強化玻璃
STAINLESS STEEL CUT-OUT LETTERS
NATURAL CLEFT SLATE BANDING
COLOR: RUSTIC YELLOW
3500
1500
2000
3500
3500
1500
CUSTOMIZED GMS SLIDING GATE
B ELEVATION 立面
SCALE 1:50
100x100 x40 MM THK BRICK
COLOR: PENTER YELLOW

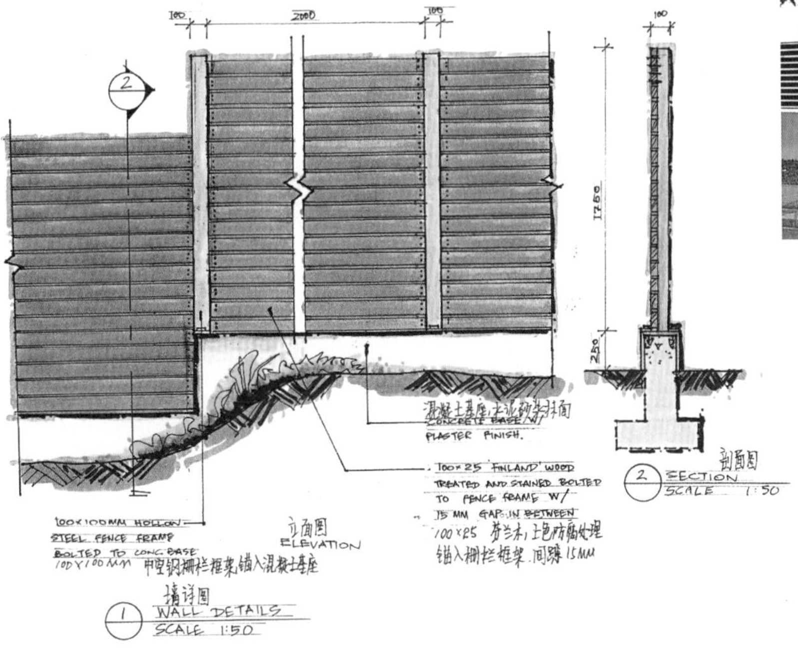
100
2000
100
1750
250
混凝土基座，水泥砂浆抹面
CONCRETE BASE W/
PLASTER FINISH.
100×25 FINLAND WOOD
TREATED AND STAINED BOLTED
TO FENCE FRAME W/
15 MM GAP IN BETWEEN
100×25 芬兰木，上色防腐处理
锚入栅栏框架，间隙15MM
100×100MM HOLLOW
STEEL FENCE FRAME
BOLTED TO CONC BASE
100×100MM 中空钢栅栏框架，锚入混凝土基座
立面图
ELEVATION
剖面图
2 SECTION
SCALE 1:50
墙详图
1 WALL DETAILS
SCALE 1:50

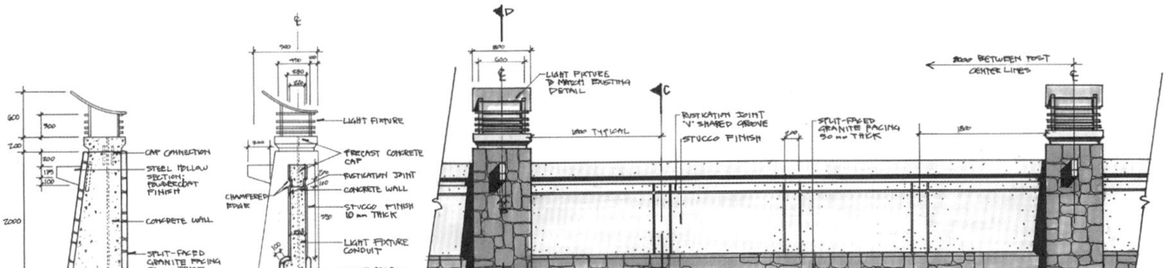
CAP CONNECTION
STEEL HOLLOW SECTION; POWDERCOAT FINISH
CONCRETE WALL
SPLIT-FACED GRANITE FACING 50 mm THICK
SECTION DD
LIGHT FIXTURE
PRECAST CONCRETE CAP
RUSTICATION JOINT
CONCRETE WALL
CHAMFERED EDGE
STUCCO FINISH 10 mm THICK
LIGHT FIXTURE CONDUIT
SPLIT-FACED GRANITE FACING 50 mm THICK
SECTION CC
LIGHT FIXTURE TO MATCH EXISTING DETAIL
TYPICAL
RUSTICATION JOINT 'V' SHAPED GROOVE
STUCCO FINISH
SPLIT-FACED GRANITE FACING 50 mm THICK
BETWEEN POST CENTER LINES
BOUNDARY FENCE
ELEVATION 1:20

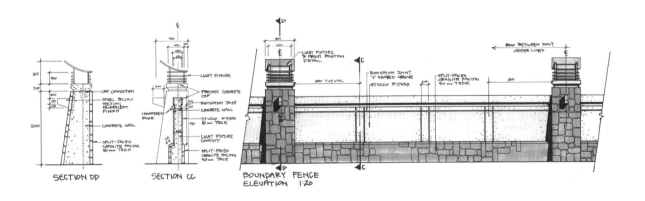

地下室
350x350結構柱
砌磚
樹池
(由管理處管理)
花槽(由管理處管理)
四號路圍牆平面
特色花盆
FEATURE POT
雕浮雕

長度變化
LENGTH VARIES
立面圖
ELEVATION
剖面圖
SECTION
SCALE 1:50
混凝土基座 水泥砂漿飾面
CONCRETE BASE W/
PLASTER FINISH
100MM X100MM HOLLOW
STEEL FENCE FRAME BOLTED
TO CONC. BASE
100MM X 100MM 中空鋼欄杆
框架 錨入混凝土基礎
100X50MM 中空截面鋼柵欄
錨入柵欄框架
100X50MM HOLLOW STEEL
FENCING BOLTED TO
FENCE FRAME
墙详图
WALL DETAILS
SCALE 1:50

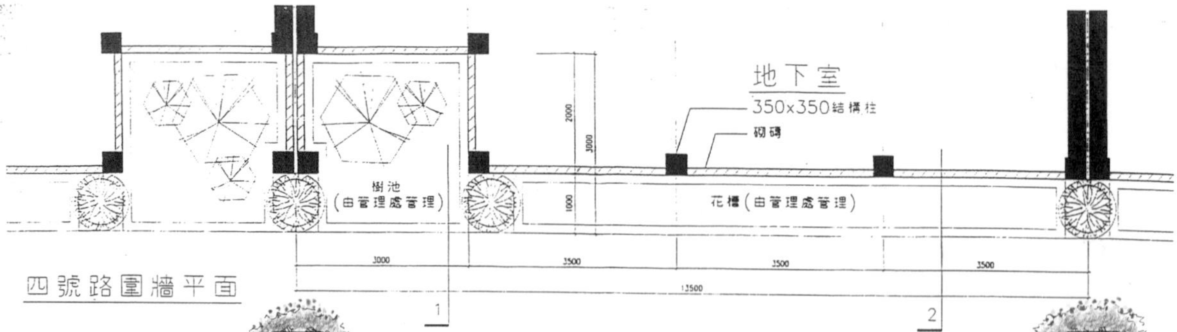

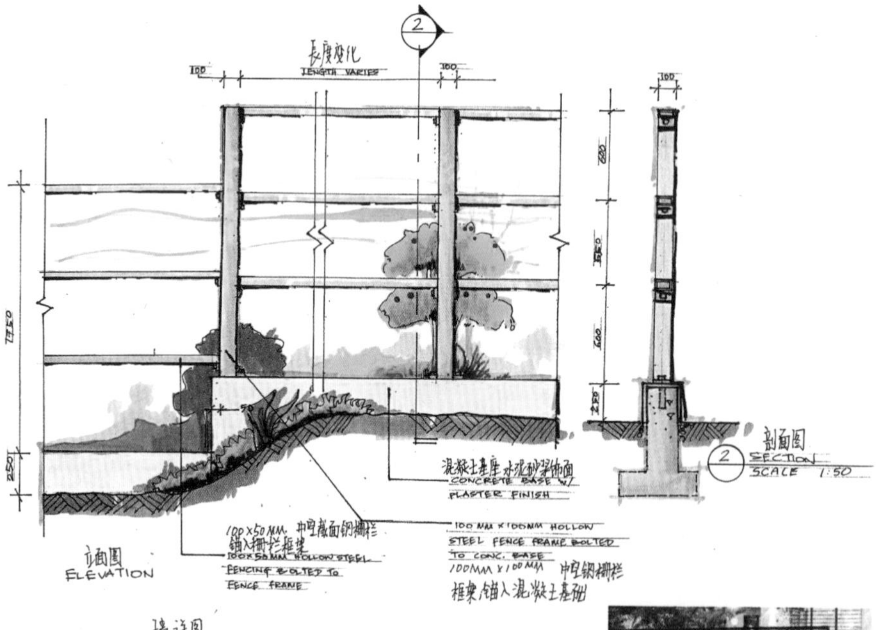

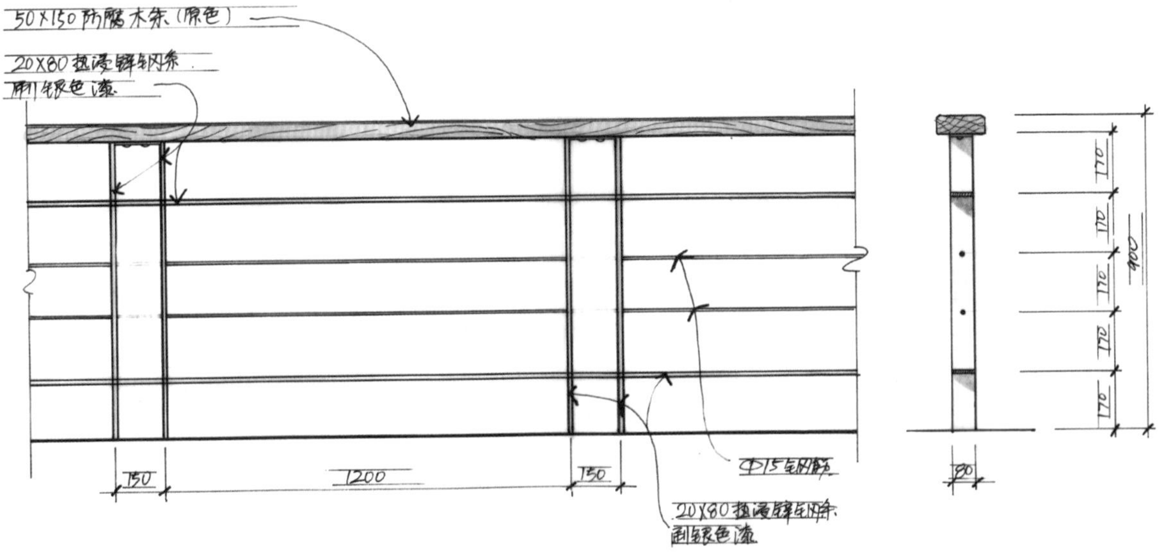
50×150防腐木条(原色)
20×80热浸锌钢条
刷银色漆
150
1200
150
Φ15钢筋
20×80热浸锌钢条
刷银色漆
170
170
170
170
170
80

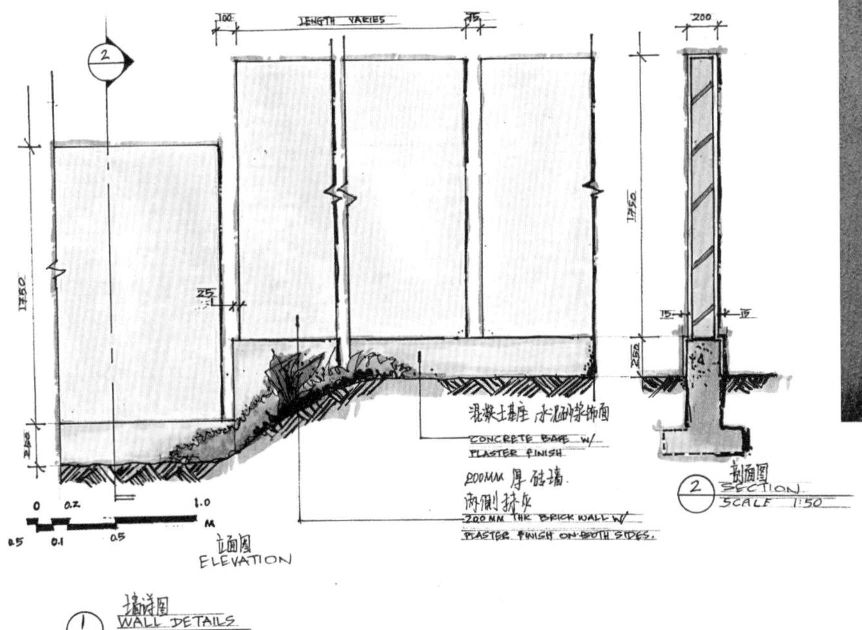
LENGTH VARIES
200
混凝土基座 水泥砂浆饰面
CONCRETE BASE W/ PLASTER FINISH
200MM厚砖墙
两侧抹灰
200 MM THK BRICK WALL W/ PLASTER FINISH ON BOTH SIDES.
立面图
ELEVATION
剖面图
SECTION
SCALE 1:50
墙详图
WALL DETAILS
SCALE 1:50

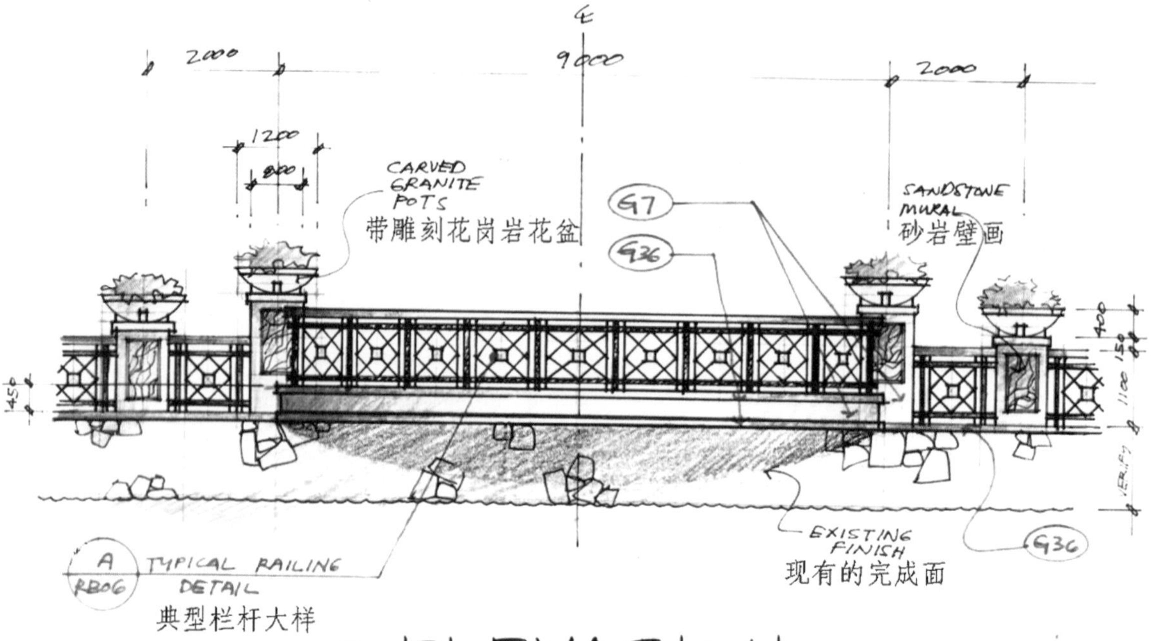

B ELEVATION
SCALE: 立面图 1:75

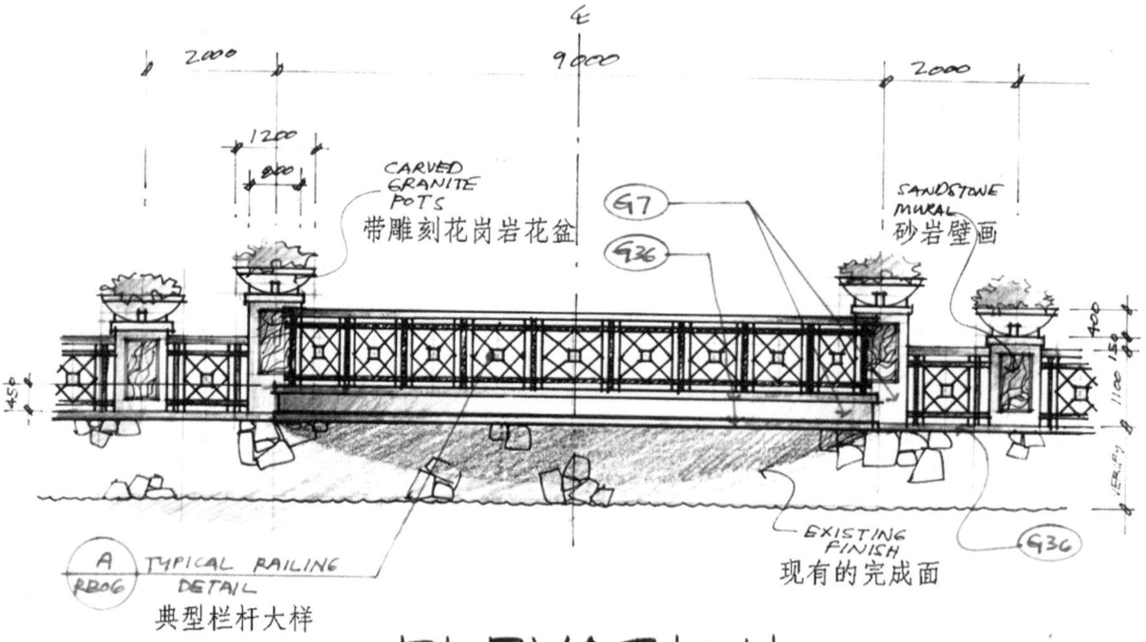

B ELEVATION
SCALE: 立面图 1:75

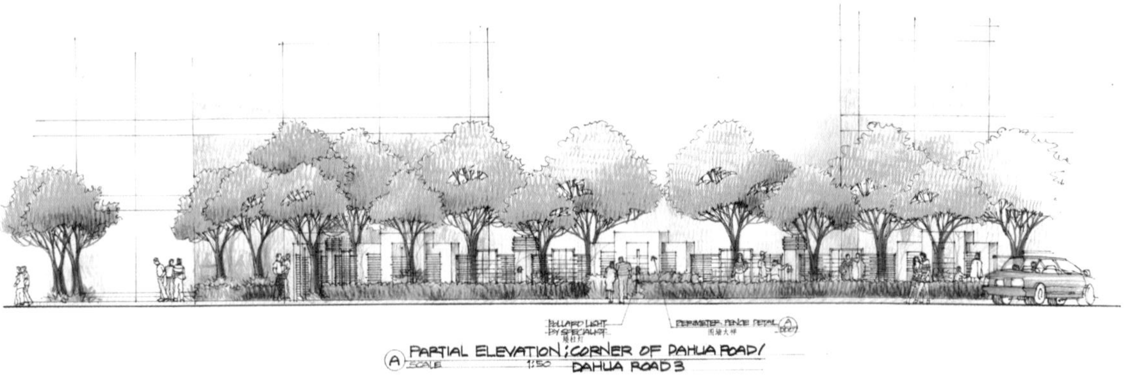

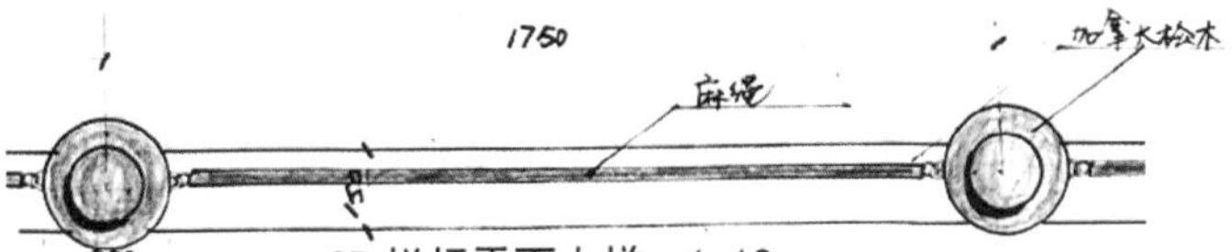

2B 栏杆平面大样 1:10

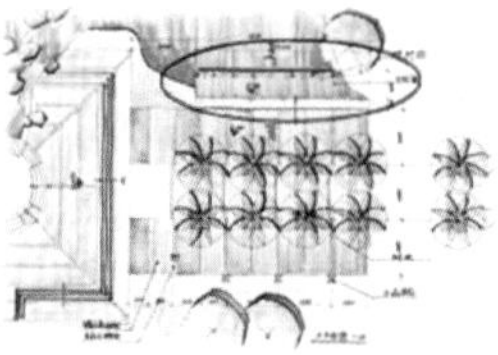

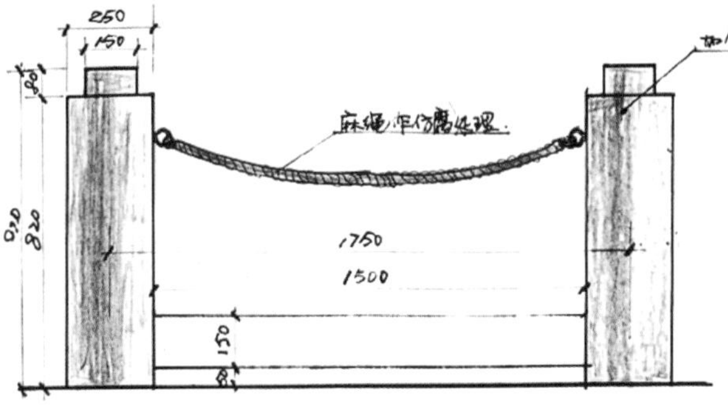

2B 栏杆大样 1:10

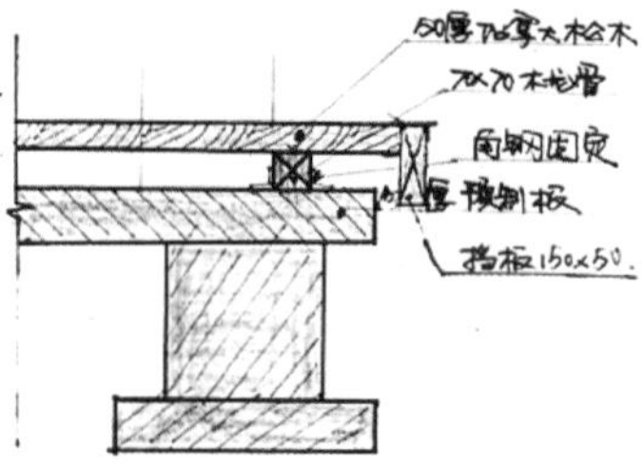

木龙骨安装大样 1:10

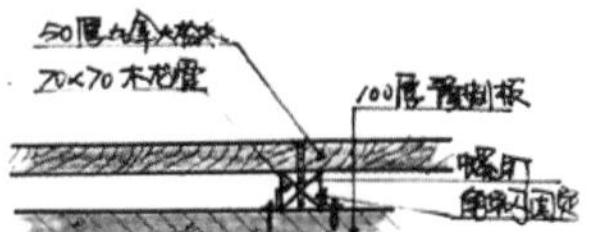

木龙骨安装大样 1:10

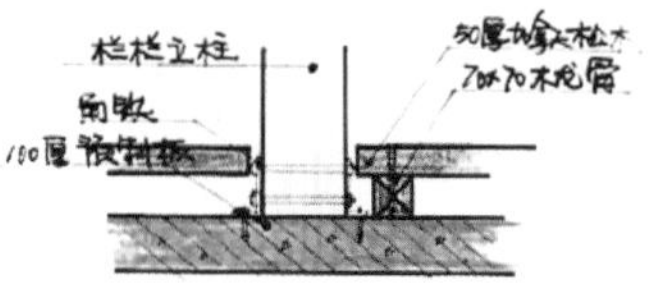

栏杆立柱安装大样 1:10

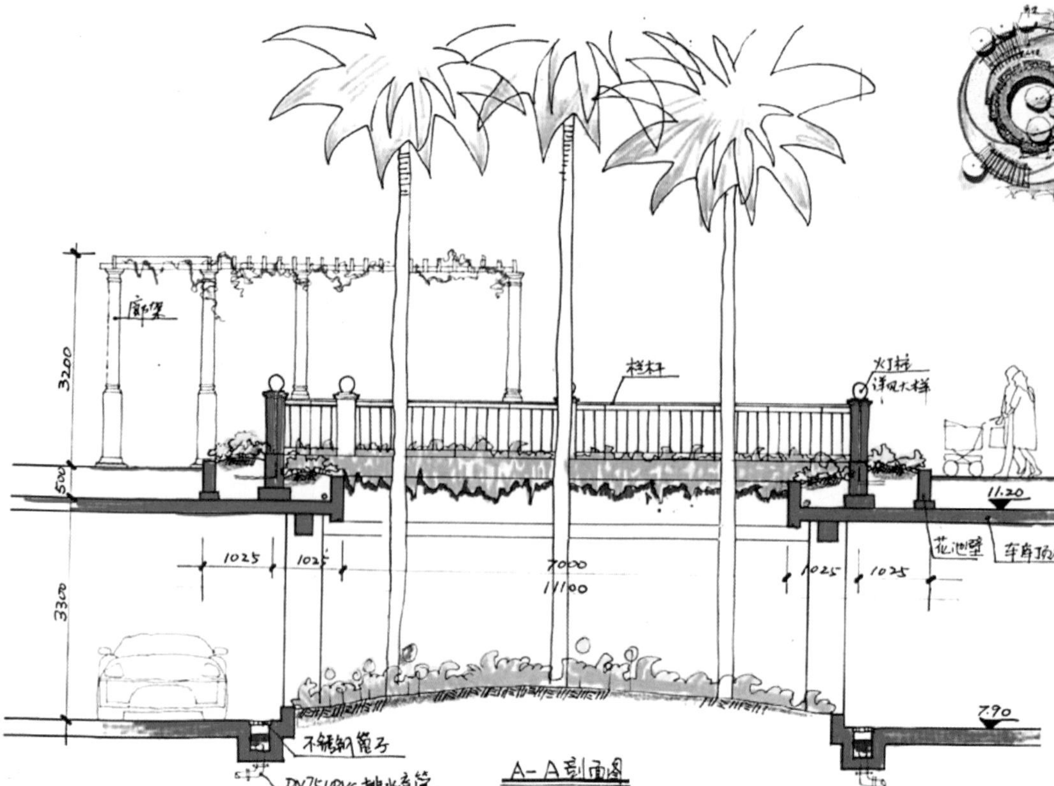

A-A剖面图 1:50

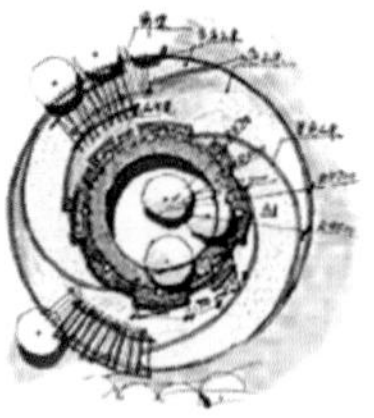

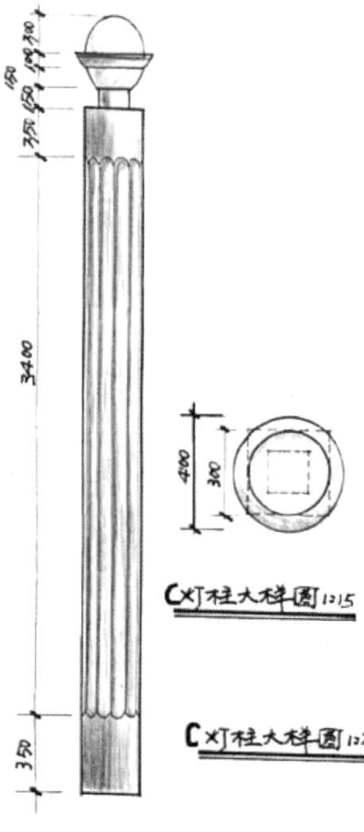

C灯柱大样图 1:15

C灯柱大样图 1:30

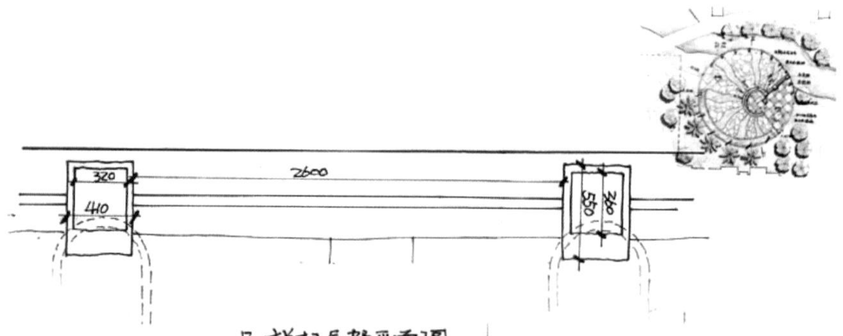

B 栏杆局部平面图 1:20

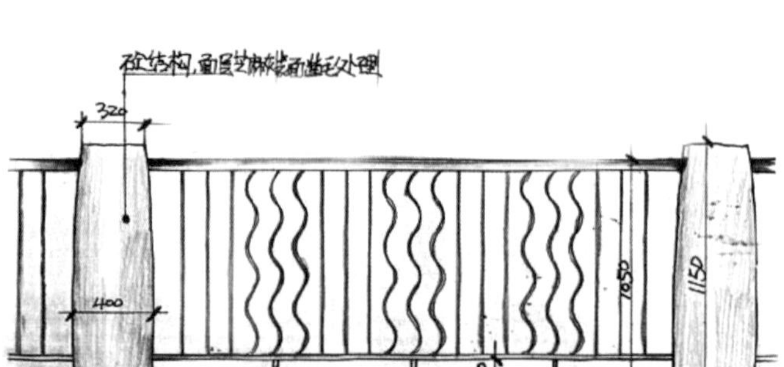

B栏杆局部立面图 1:20

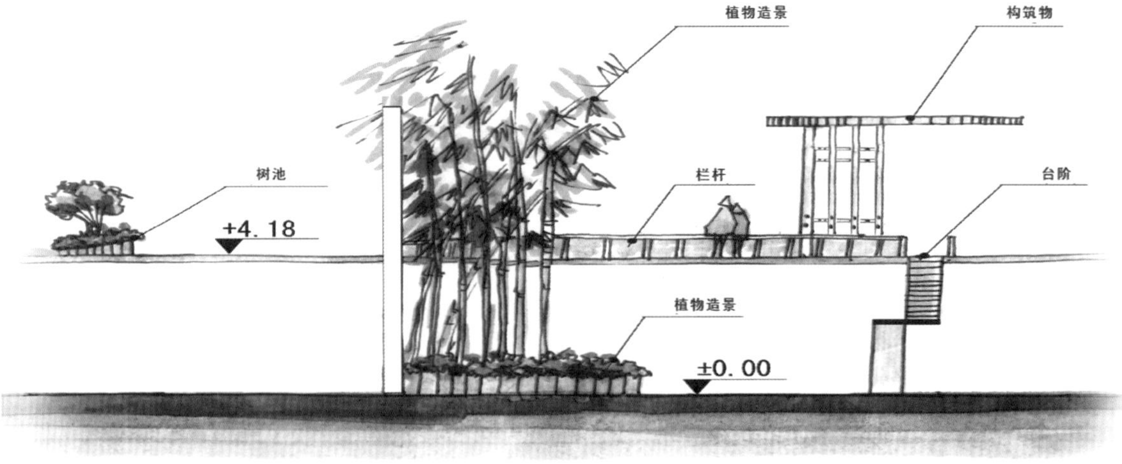
植物造景
构筑物
树池
+4.18
栏杆
台阶
植物造景
±0.00

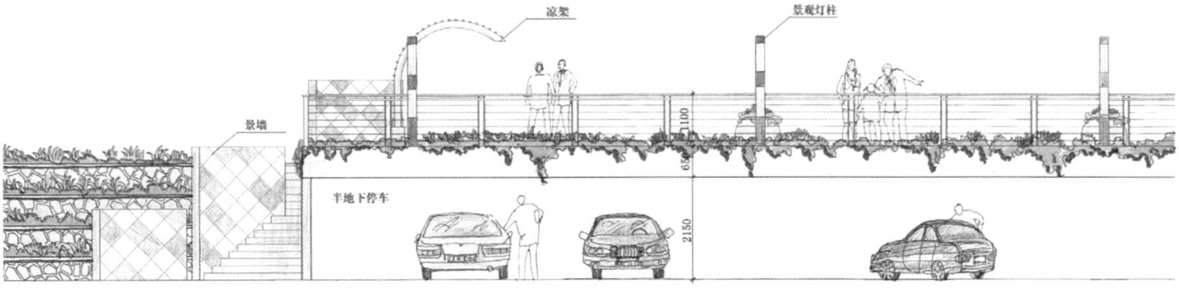
凉架
景观灯柱
景墙
1100
650
丰地下停车
2150

4 桥

一、园桥定义及功能

园林中的桥，可以联系风景点的水陆交通，组织游览线路，变换观赏视线，点缀水景，增加水面层次，兼有交通和艺术欣赏的双重作用。园桥在造园艺术上的价值，往往超过交通功能。园桥一般有如下功能：联系两岸或水面交通；引导游览路线；点缀水面景色；划分和组织水景空间；增加风景层次等作用。

二、桥位的选择

1. 桥位应与园林道路系统配合，联系游览路线与观景点，方便交通；

2. 注意景观要求，水面的分隔或聚合与水体面积大小密切相关；考虑景观要求，水面大的应选窄处架桥，水面小的要注意水面分割，使水体分而不断，使环境空间增加层次有扩大空间的效果。

3. 注意与水路通航与桥上的通行；

4. 根据交通情况的要求，如桥上是否通车，桥下是否通航，载重能力和与净空高度，并与环境造景统一效果等，选择合适的形式；

5. 考虑结构的经济合理性；

6. 考虑园桥结构的经济、合理，根据水体的宽窄、水位的深浅、水流大小与地质基础条件考虑。

三、园桥的类型及应用

1. 汀步

又称河步、跳墩子，虽然这是最原始的过水形式，早被新技术所代替，但在园林中尚可应用发挥有情趣的跨水小景，使人走在汀步上，有脚下清流游鱼可数的近水亲切感。汀步最适合浅滩小溪、跨度不大的水面。也可结合滚水坝体设置过坝得汀步，但要注意安全。汀步形式有以下几种：

（1）自然式

用天然石材自然式布置。设在自然石矶或假山石驳岸，最容易取得协调效果，又是最简便的步行过水形式。

（2）整形式

有圆形、方形，或塑造荷叶等水生植物造型，可用石材雕凿或耐水材料砌塑而成。

（3）汀步应用注意事项

1）过水用汀步为了安全，间距不可过大，高度能出水面即可，不宜过高；

2）表面应平整、防滑；

3）基础也要求稳固；

4）要注意到北方冬季冻结时的景观效果，限于行人量不大的通路使用。

2. 梁桥

梁桥以受弯为主的主梁作为承重构件的桥梁。主梁可以是实腹梁或桁架梁。实腹梁构造简单，制造、架设和维修均较方便，广泛用于中、小跨度桥梁，但在材料利用上不够经济。桁架梁的杆件承受轴向力，材料能充分利用，自重较轻，跨越能力大，多用于建造大跨度桥梁。

梁桥的分类：

（1）按平面划分：单跨平桥；多跨平桥；

（2）按材料分：木架桥，由简单独木桥到木板桥有结构简单、施工方便、就地取材等优点，只是易腐蛀，不耐久。为了耐久，有用钢筋水泥塑制仿木桥；

3. 拱桥

拱桥是人用小块石材建造大跨度工程的创造，在我国很早就有拱券的利用。河北省赵县著名的赵州桥是建于 1300 多年前的隋朝，主圆拱跨度 32m 多，拱上叠砌四个小拱，不但省料、减轻自重，同时在洪水期增加流量减少阻力，造型十分优美，为我国桥梁史上的伟大杰作。拱桥的形式多样，有单拱、三拱到连续多拱，方便园林不同环境的要求而选用。

4. 浮桥

浮桥是在较宽水面通行的简单和临时性办法。它可以免去做桥墩基础等工程措施，它只用船或浮筒替代桥墩上架梁板用绳索拉固就成通行的浮桥。

5. 吊桥

在急流深涧，高山峡谷，桥下不便建墩的条件，如我国西南地区最宜建吊桥。近代科学技术发展新的高强耐拉材料的生产，使吊桥有可能创造以前无法建造的大跨度，轻巧的悬索吊桥随着我国科技发展，今后必将出现更多的具有优美曲线、轻巧的吊桥。

6. 亭桥与廊桥

这类具有交通作用又有游憩功能与造景效果的桥，很适合园林要求。如北京颐和园西堤上建有的风桥、镜桥、练桥、柳桥等事桥。这些桥在长堤游览线上起着点景休息作用，在远观上打破长堤水平线构图，有对比造景、分割水面层次作用。扬州瘦西湖上的五亭桥是瘦西湖长轴上主景建筑。

四、桥的组成

桥由两大部分组成：横跨水上的梁或拱和承担它的荷载桥台基础两部分。水面宽时用梁、拱跨度有限制，水中可设桥墩支撑，使梁每个分段跨度减短。

1. 上部结构

桥的上部结构是桥的主体。要考虑当地水文地质和技术条件选择适合要求跨度载重的材料与结构，评比坚固、经济、美观的设计方案。桥梁也是过水道路的延续，所以桥梁上部也有路面，在梁拱承重结构上设路面层、基层、防水层。

2. 桥台、桥墩支撑部分

要使得桥坚固耐久，耐水流冲刷，就要有坚固的桥台、桥墩基础。设计时根据水文地质条件与适当的材料构造来完成。桥台、桥墩要有深入地基的基础，上面要用耐水流冲刷材料，又尽量减少对水流的阻力，常做成 45 度分水金刚墙。

五、设计要点

1. 桥的选型、体量应与园林环境、水体大小协调

大型水面空间开阔为突出水景效果，常取多孔拱桥，以桥的体量与水体相称。如北京颐和园的十七孔桥；小水面，常以单跨平桥或折桥，使人能接近水面。如南京瞻园小曲桥；而平静小水面及小溪流，常设贴近水面的小桥，或汀步过水，使人接近水面，远观也不使空间割断。

2. 桥的栏杆是丰富桥体造型的重要因素

栏杆的高度要充分考虑安全需要，也要与桥体大小宽度相协调。如苏州园林小桥一般只设低的坐凳栏杆，其造型很简洁。甚至有些小桥只设单面栏杆或不设栏杆以突出桥的轻快造型。

3. 桥与岸相接处，要处理得当

桥与岸常以灯具、雕塑、山石、花木丰富桥体与岸壁的衔接，桥头装饰有显示桥位、增加安全的作用，因此这些装饰物兼有引导交通的作用，绝不可阻碍交通。

4. 充分考虑桥上与桥下的交通要求

桥体尺度除应考虑水体大小、道路宽度及造景效果外，还要满足功能上通车、行船的高度、坡度要求；为满足人流集散与停留观景等要求；常设置桥廊及桥头小广场。

5. 桥的照明

桥上灯具，具有良好的桥体装饰效果，在夜间游园更有指示桥的位置及照明的作用。灯具可结合桥的体形、栏杆及其他装饰物，统一设置，使其更好突出桥的景观效果，尤其夜间的景观。

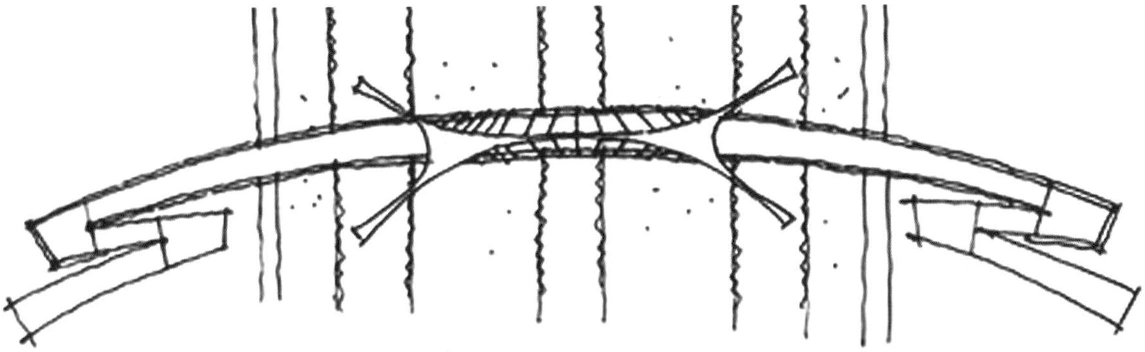

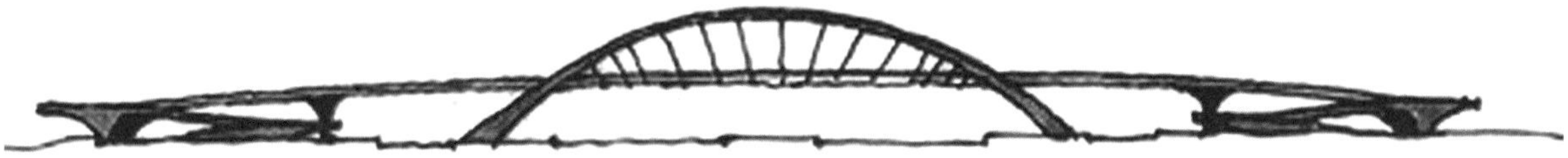

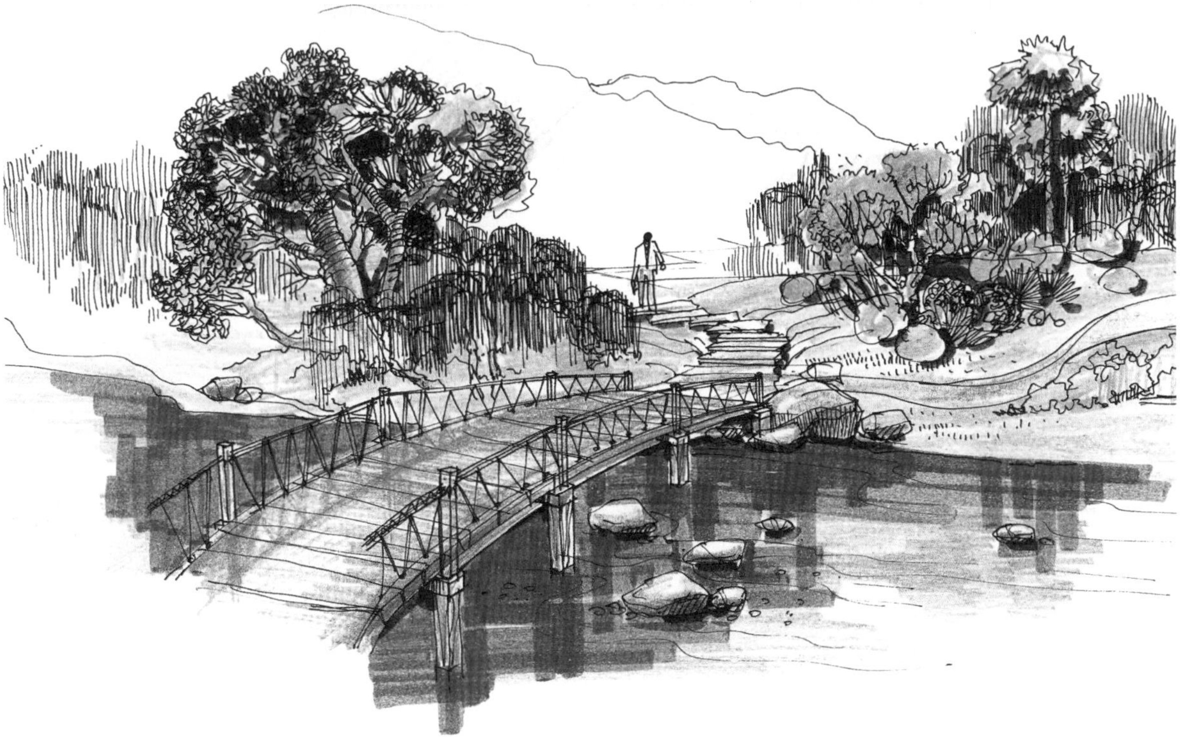

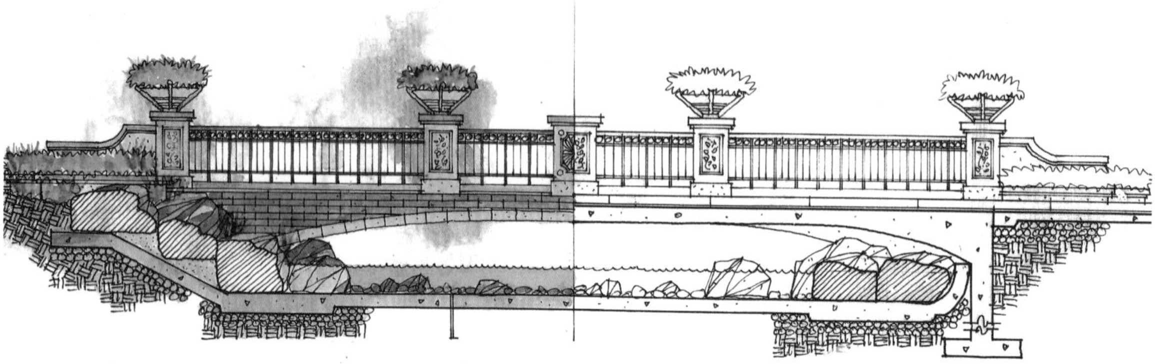

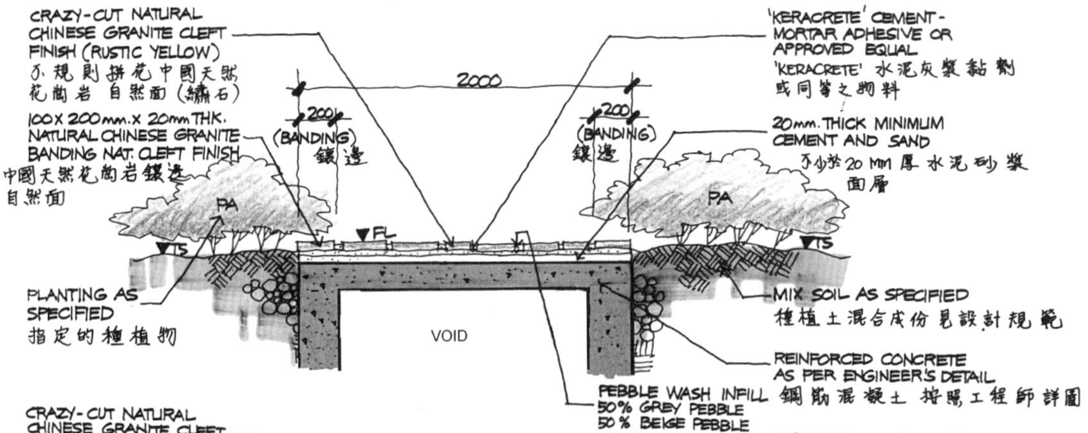
CRAZY-CUT NATURAL CHINESE GRANITE CLEFT FINISH (RUSTIC YELLOW)
不規則拼花中國天然花崗岩 自然面(鏽石)
2000
200 (BANDING) 鑲邊
200 (BANDING) 鑲邊
'KERACRETE' CEMENT-MORTAR ADHESIVE OR APPROVED EQUAL
'KERACRETE' 水泥灰漿黏劑或同等之物料
100 X 200 mm. X 20 mm THK. NATURAL CHINESE GRANITE BANDING NAT. CLEFT FINISH
中國天然花崗岩鑲邊 自然面
20 mm. THICK MINIMUM CEMENT AND SAND
不少於20 MM 厚 水泥砂漿 面層
PA
PA
FL
TS
TS
PLANTING AS SPECIFIED
指定的種植物
VOID
MIX SOIL AS SPECIFIED
種植土混合成份見設計規範
REINFORCED CONCRETE AS PER ENGINEER'S DETAIL
鋼筋混凝土 按照工程師詳圖
PEBBLE WASH INFILL 50% GREY PEBBLE 50% BEIGE PEBBLE
洗水石填充
50% 灰卵石
50% 灰楊卵石

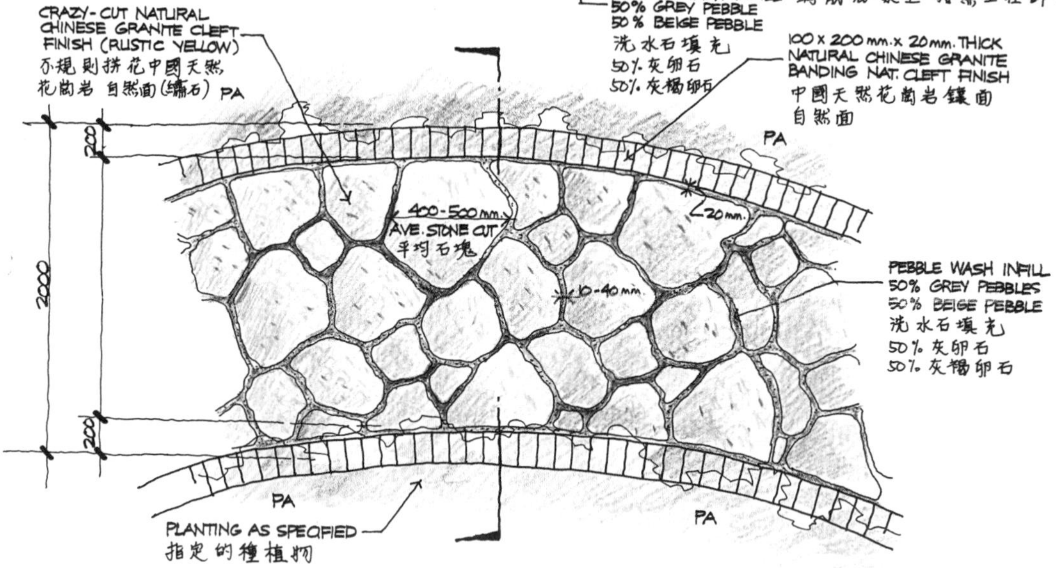
CRAZY-CUT NATURAL CHINESE GRANITE CLEFT FINISH (RUSTIC YELLOW)
不規則拼花中國天然花崗岩 自然面(鏽石)
PA
100 X 200 mm. X 20 mm. THICK NATURAL CHINESE GRANITE BANDING NAT. CLEFT FINISH
中國天然花崗岩鑲面 自然面
200
2000
200
400-500 mm. AVE. STONE CUT
平均石塊
20 mm.
10-40 mm.
PEBBLE WASH INFILL 50% GREY PEBBLES 50% BEIGE PEBBLE
洗水石填充
50% 灰卵石
50% 灰楊卵石
PA
PA
PA
PLANTING AS SPECIFIED
指定的種植物

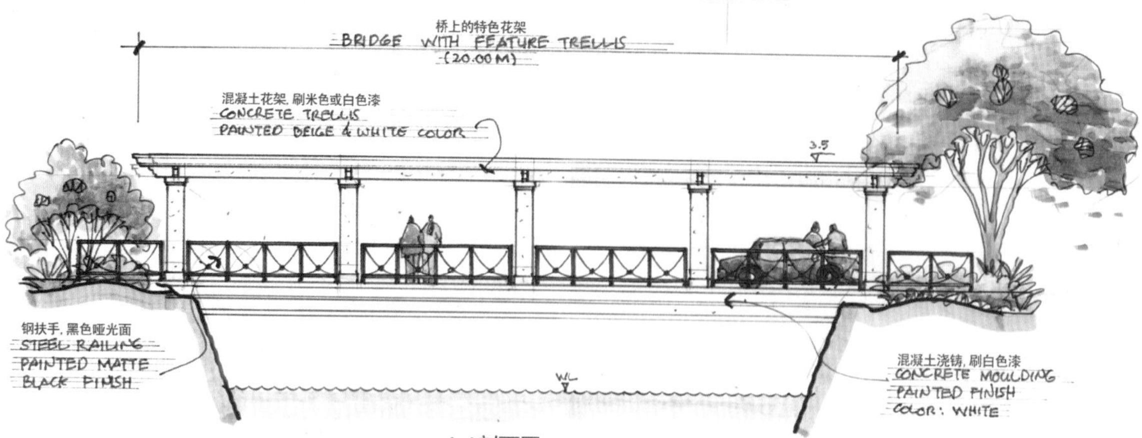
桥上的特色花架
BRIDGE WITH FEATURE TRELLIS
(20.00 M)
混凝土花架，刷米色或白色漆
CONCRETE TRELLIS
PAINTED BEIGE & WHITE COLOR
3.5
钢扶手，黑色哑光面
STEEL RAILING
PAINTED MATTE
BLACK FINISH
混凝土浇铸，刷白色漆
CONCRETE MOULDING
PAINTED FINISH
COLOR: WHITE
WL

A-A剖面图

SECTION @ A-A

SCALE 1:100

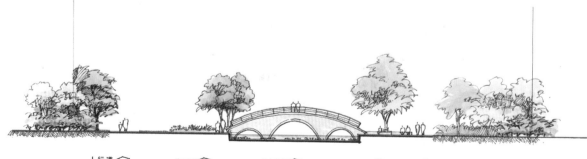

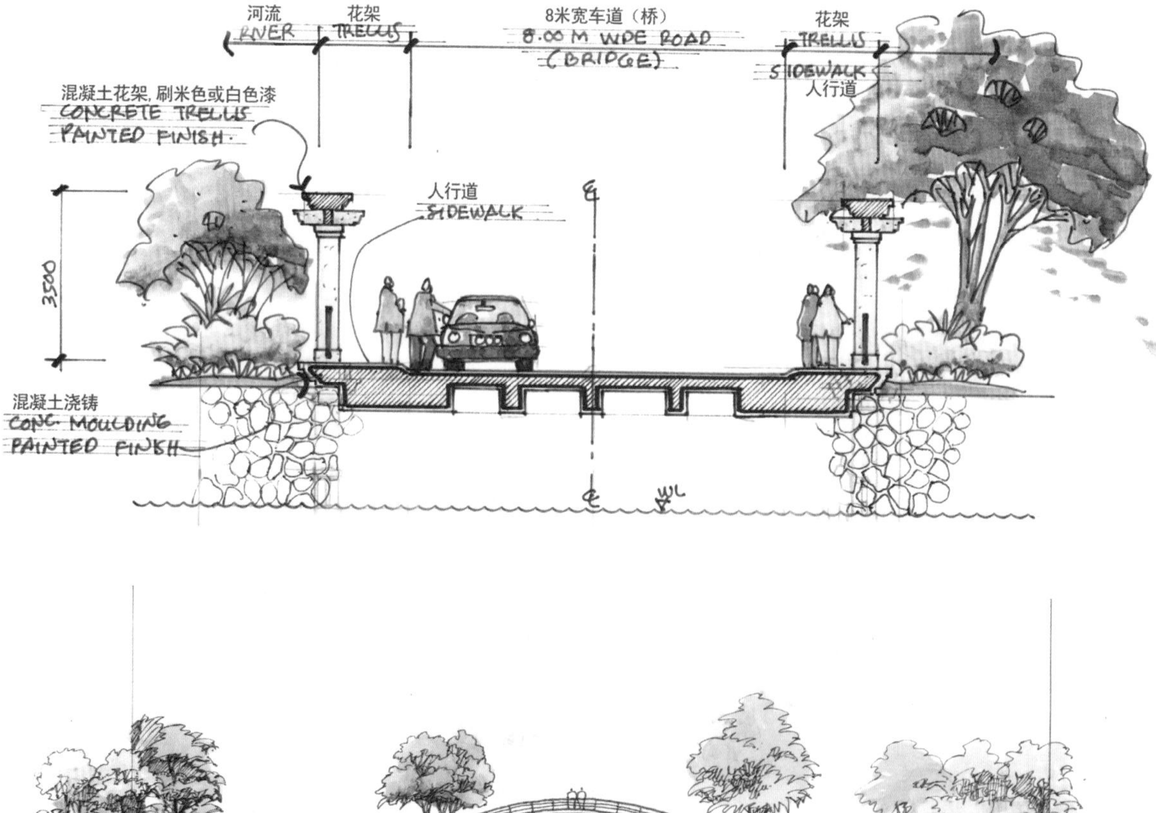
河流
RIVER
花架
TRELLIS
8米宽车道（桥）
8.00 M WIDE ROAD
(BRIDGE)
花架
TRELLIS
SIDEWALK
人行道
混凝土花架，刷米色或白色漆
CONCRETE TRELLIS
PAINTED FINISH
人行道
SIDEWALK
3500
混凝土浇铸
CONC. MOULDING
PAINTED FINISH
WL

人行道
桥头广场
特色拱桥
桥头广场
人行道

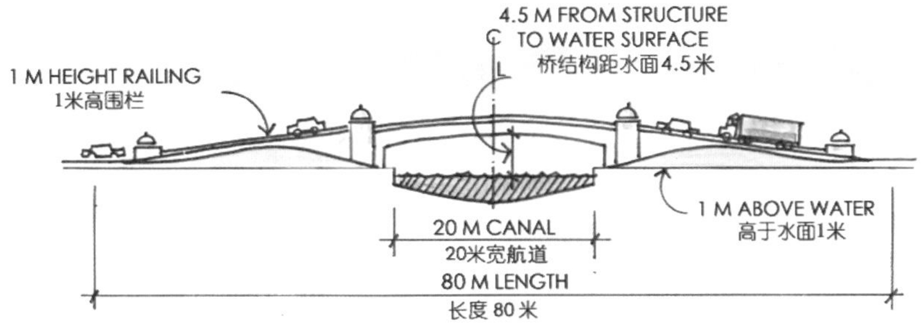
4.5 M FROM STRUCTURE
TO WATER SURFACE
桥结构距水面4.5米
1 M HEIGHT RAILING
1米高围栏
1 M ABOVE WATER
高于水面1米
20 M CANAL
20米宽航道
80 M LENGTH
长度 80 米

花岗石压顶
花岗岩线条
天然花岗岩
花坛柱
天然文化石饰面

桥

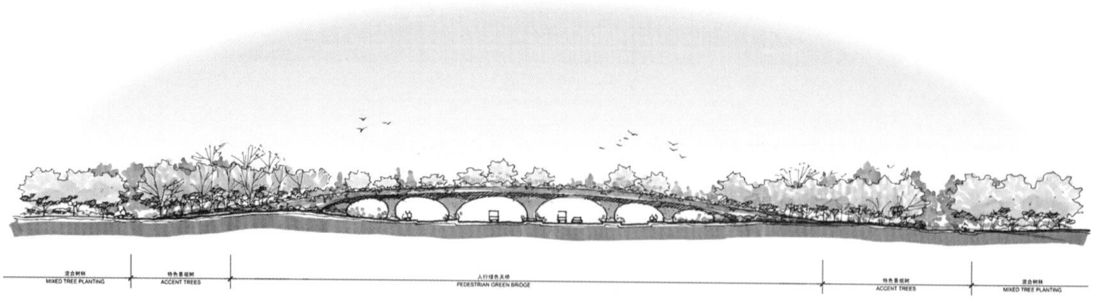

混合树林
MIXED TREE PLANTING
特色景观树
ACCENT TREES
人行绿色天桥
PEDESTRIAN GREEN BRIDGE
特色景观树
ACCENT TREES
混合树林
MIXED TREE PLANTING

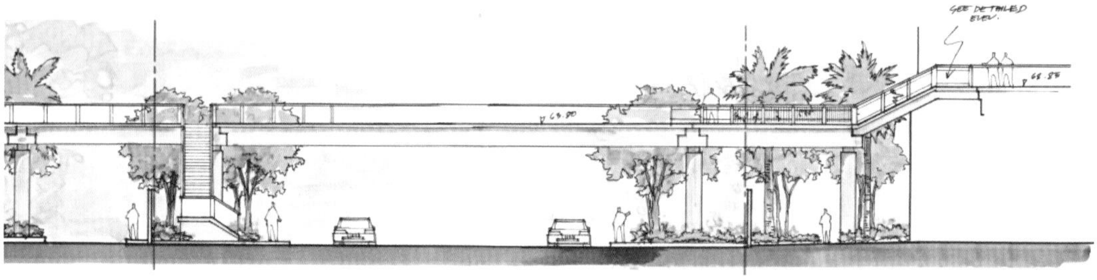

SEE DETAILED ELEV.

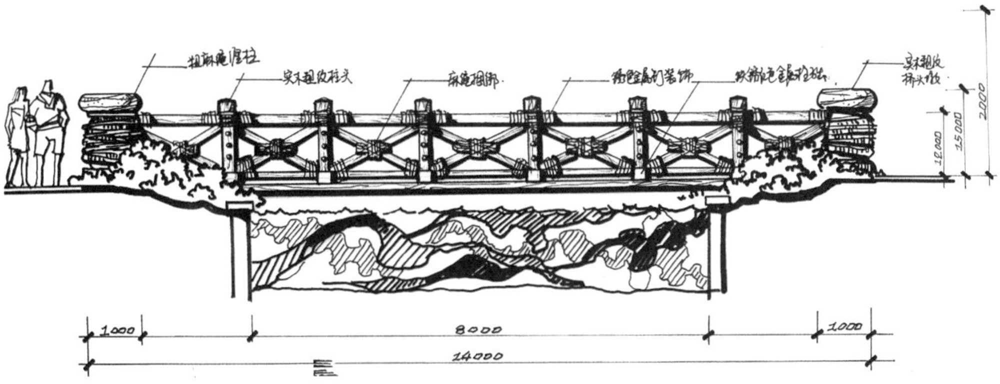

热带区入口桥立面图

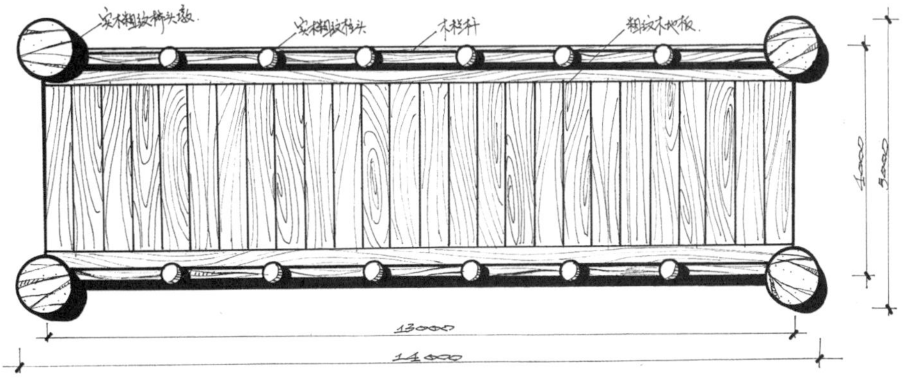

热带区入口桥平面图

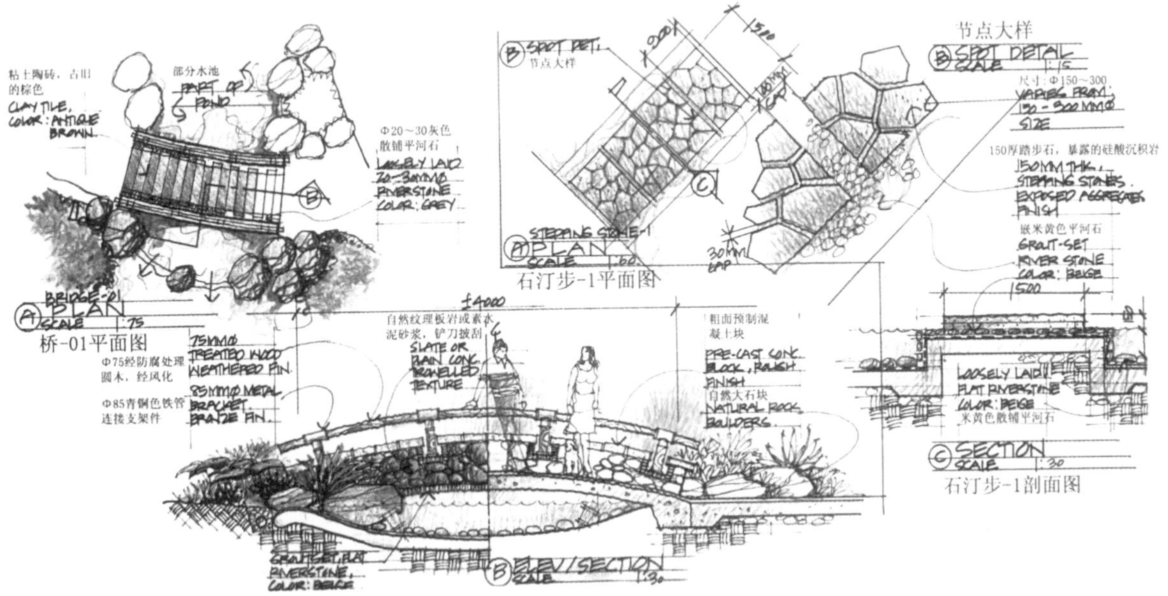

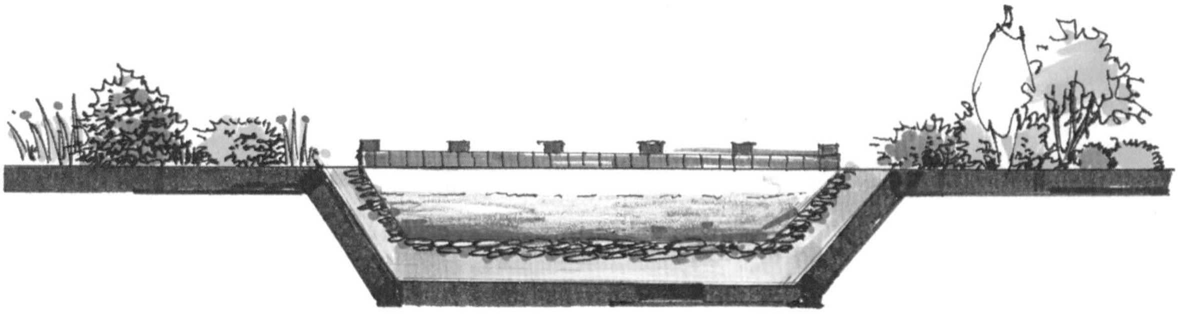

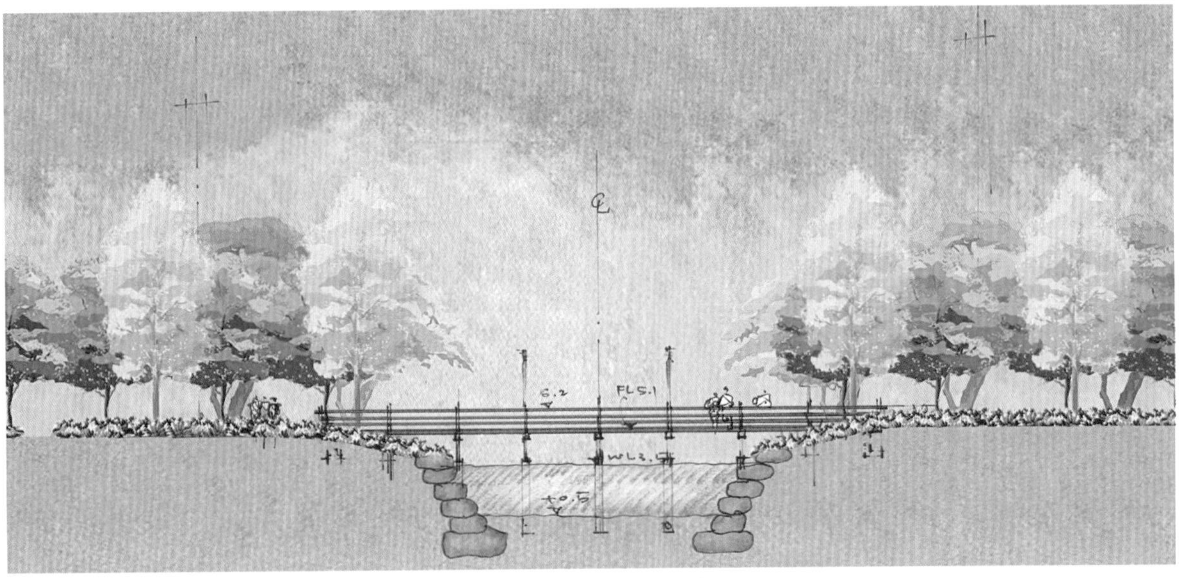
6.2
FL5.1
WL3.5
+0.5

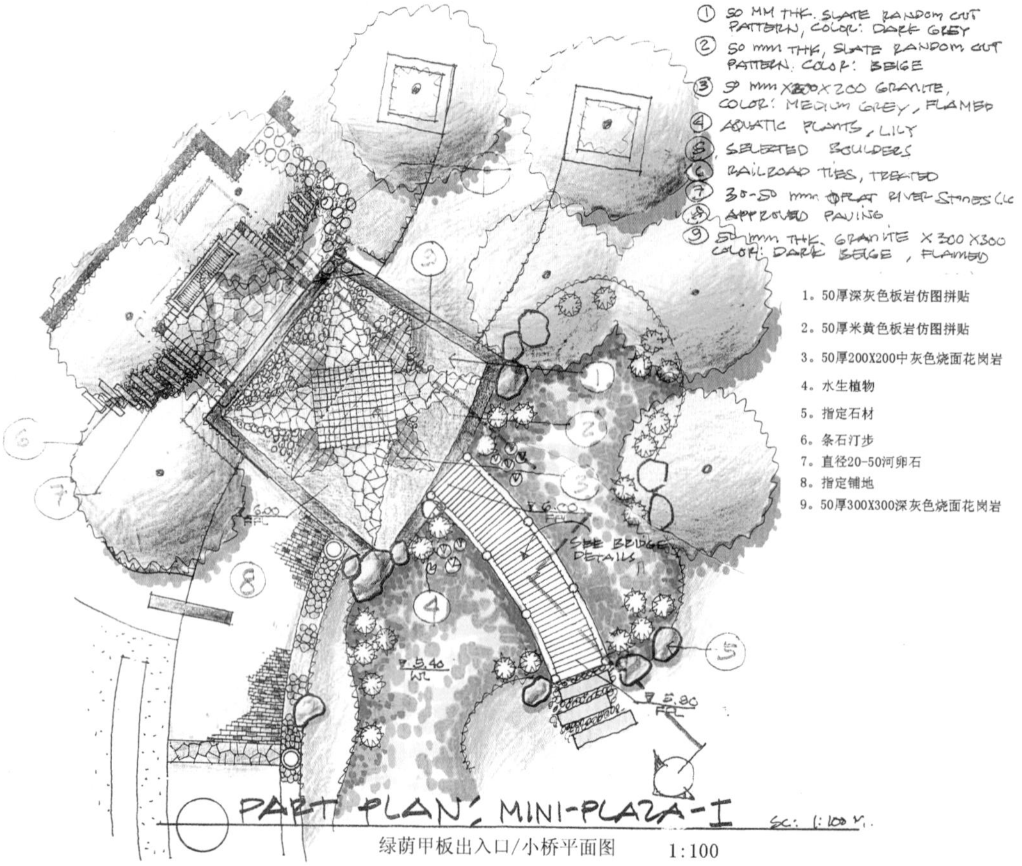

1。50厚深灰色板岩仿图拼贴
2。50厚米黄色板岩仿图拼贴
3。50厚200X200中灰色烧面花岗岩
4。水生植物
5。指定石材
6。条石汀步
7。直径20-50河卵石
8。指定铺地
9。50厚300X300深灰色烧面花岗岩

绿荫甲板出入口/小桥平面图　1:100

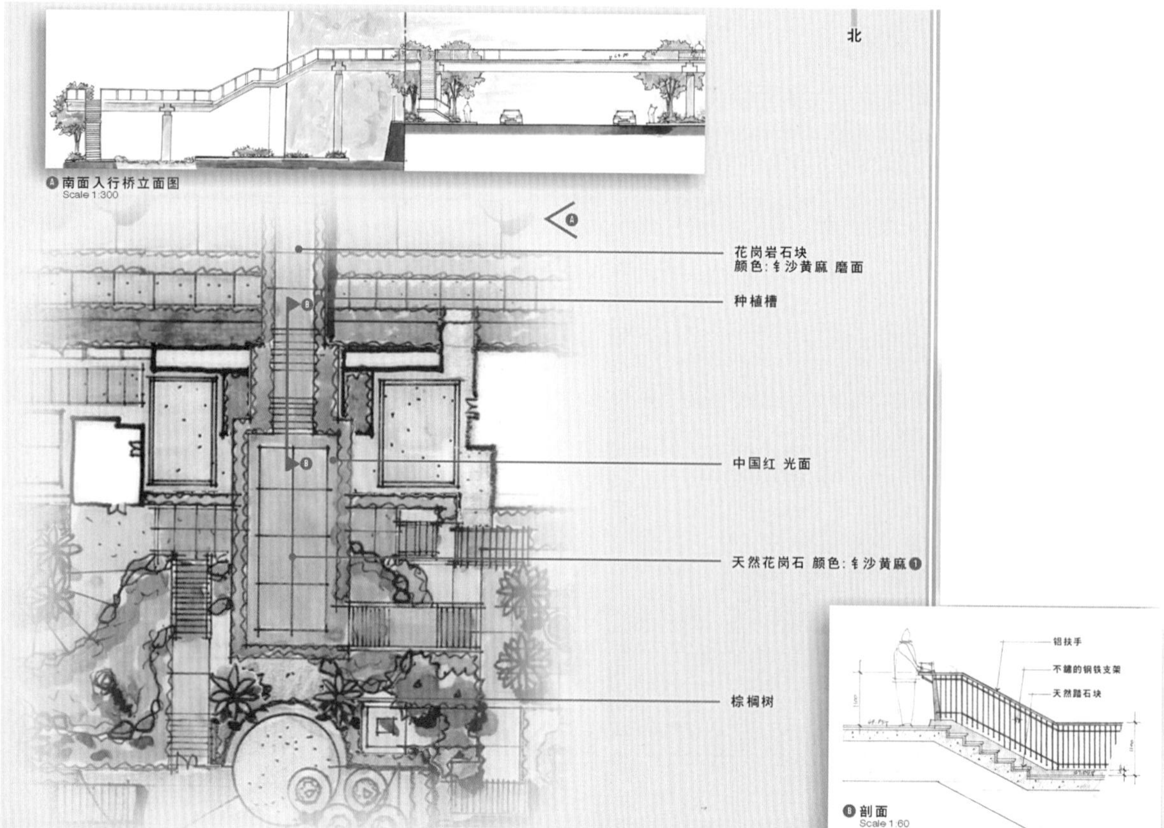

Ⓐ 南面入行桥立面图
Scale 1:300

Ⓑ 剖面
Scale 1:60

SECTION A—A
1:100

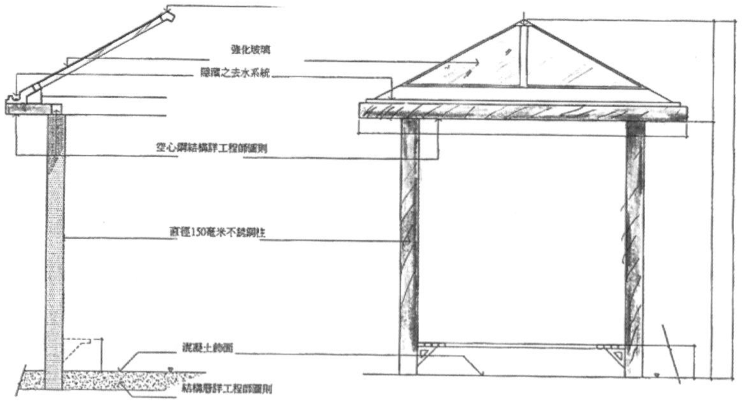

A 南面入行桥立面图
Scale 1:300

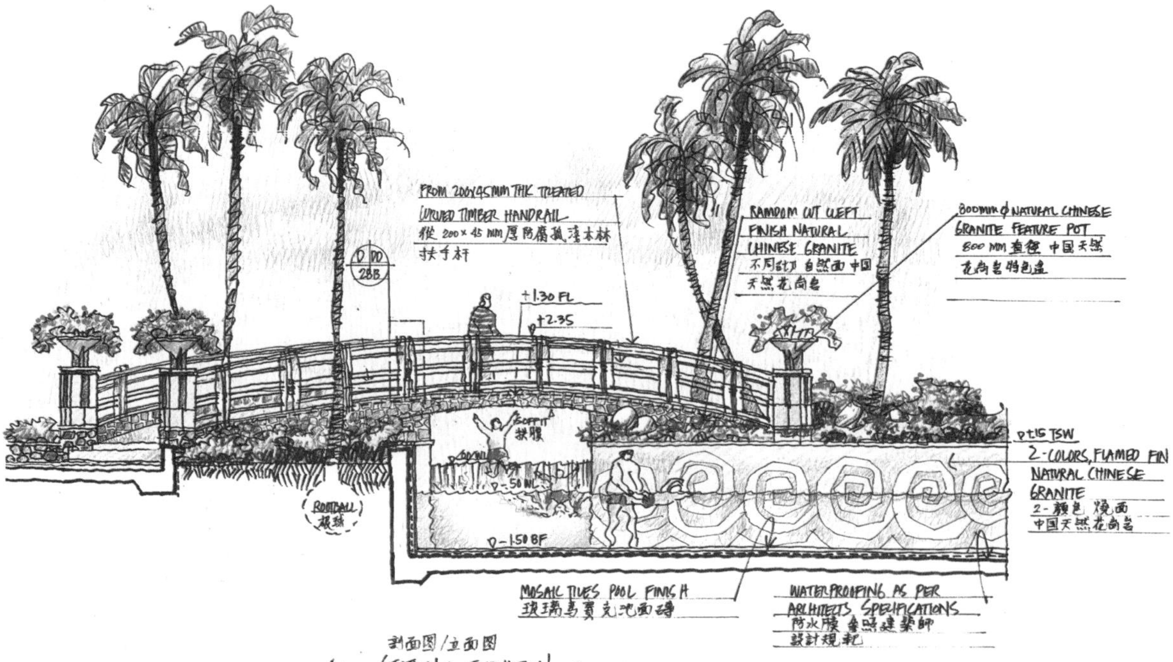

剖面图/立面图
A1 SECTION \ ELEVATION SCALE 1:50

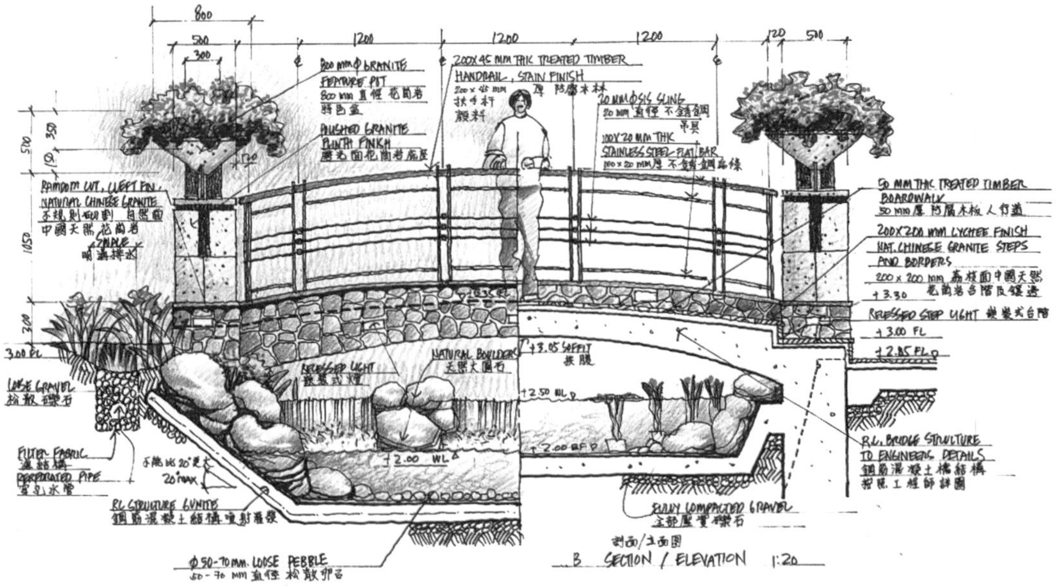
800 MM Φ GRANITE FEATURE POT
800 MM 直径 花岗岩 特色盆
200X45 MM THK TREATED TIMBER HANDRAIL, STAIN FINISH
POLISHED GRANITE PLINTH FINISH
20 MM Φ S/S SLING
100X20 MM THK STAINLESS STEEL FLAT BAR
50 MM THK TREATED TIMBER BOARDWALK
200X200 MM LYCHEE FINISH NAT. CHINESE GRANITE STEPS AND BORDERS
RECESSED STEP LIGHT
RECESSED LIGHT
NATURAL BOULDERS
R.C. BRIDGE STRUCTURE TO ENGINEERS DETAILS
FULLY COMPACTED GRAVEL
LOOSE GRAVEL
FILTER FABRIC
PERFORATED PIPE
RC STRUCTURE GUNITE
Φ50-70MM LOOSE PEBBLE
50-70 MM 直径 松散卵石
剖面/立面图
B SECTION / ELEVATION 1:20

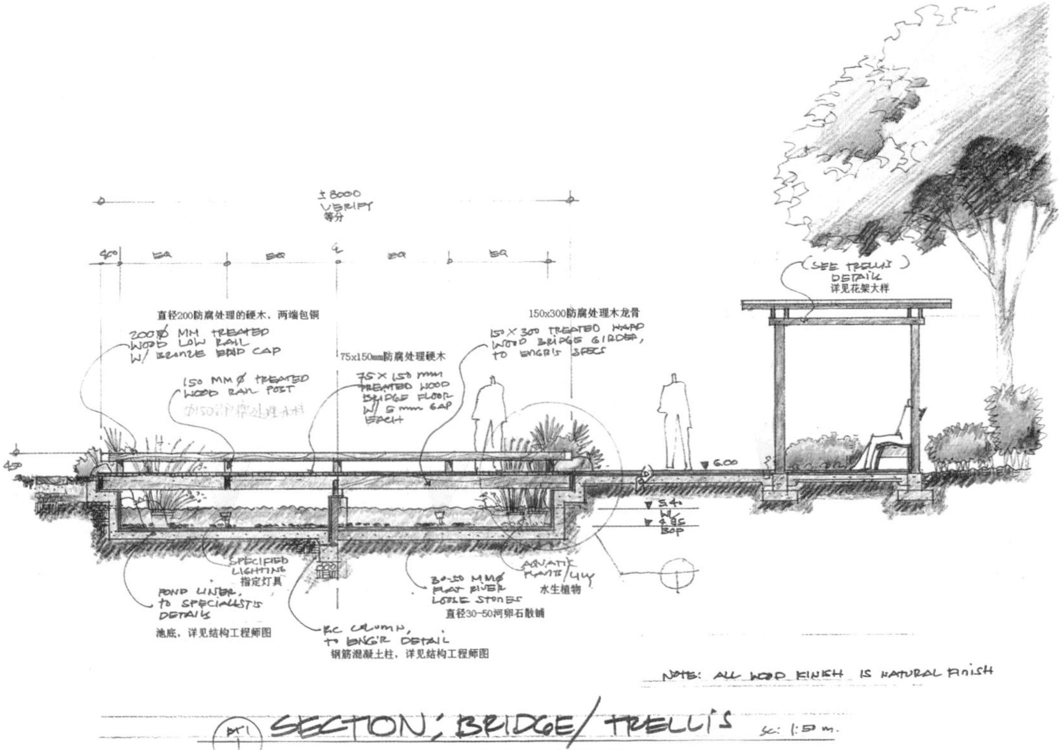
直径200防腐处理的硬木，两端包铜
150x300防腐处理木龙骨
75x150mm防腐处理硬木
详见花架大样
指定灯具
水生植物
直径30-50河卵石散铺
池底，详见结构工程师图
钢筋混凝土柱，详见结构工程师图
NOTE: ALL WOOD FINISH IS NATURAL FINISH
SECTION: BRIDGE/TRELLIS sc: 1:50 m.

绿荫甲板出入口/小桥剖面图 1:50

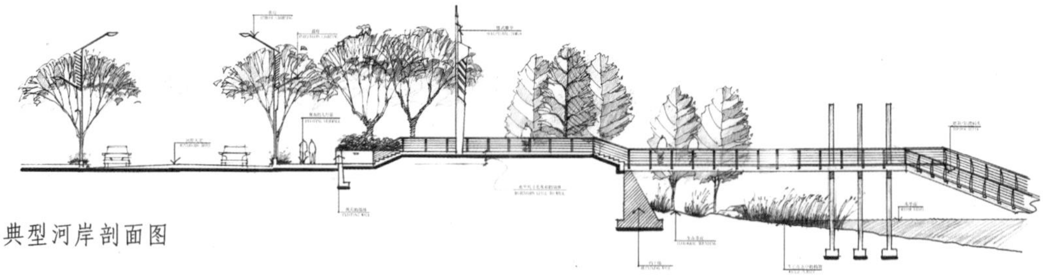

典型河岸剖面图

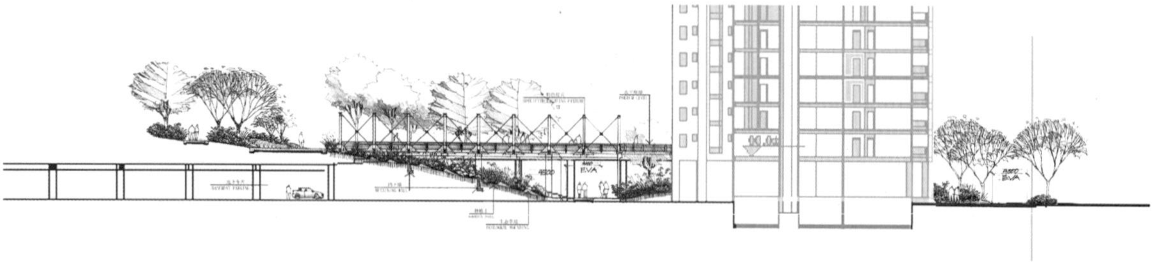

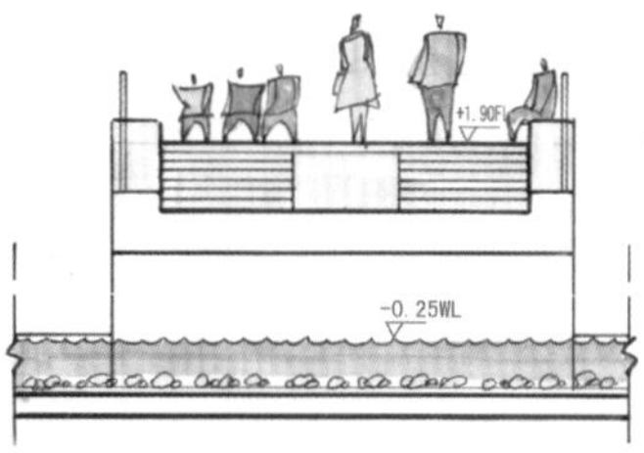

剖面图
比例：1/75

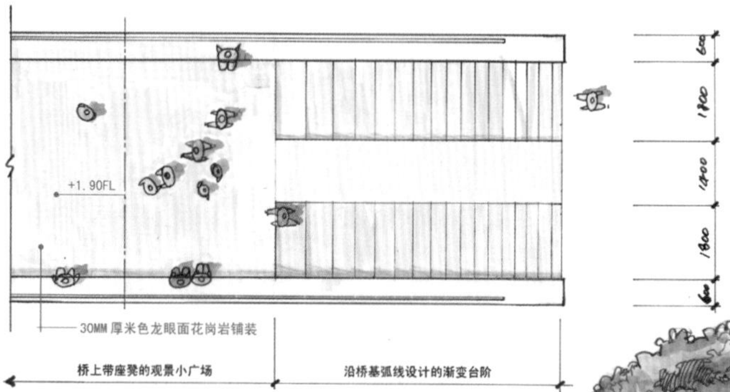

平面图
比例：1/75

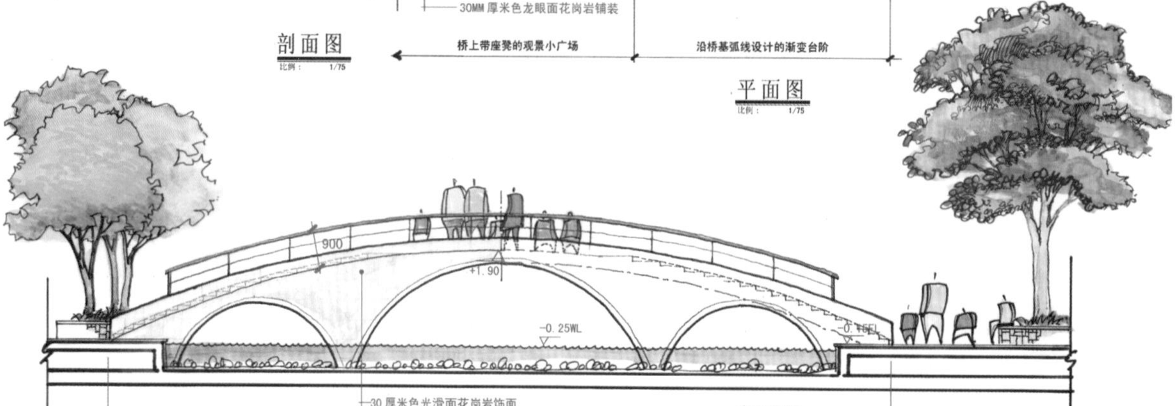

立面图
比例：1/75

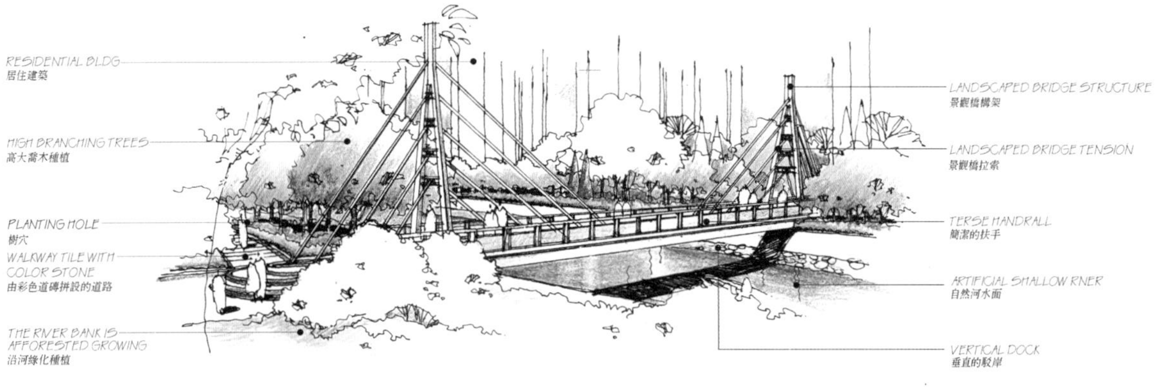

RESIDENTIAL BLDG
居住建築
HIGH BRANCHING TREES
高大喬木種植
PLANTING HOLE
樹穴
WALKWAY TILE WITH COLOR STONE
由彩色道磚拼設的道路
THE RIVER BANK IS AFFORESTED GROWING
沿河綠化種植
LANDSCAPED BRIDGE STRUCTURE
景觀橋構架
LANDSCAPED BRIDGE TENSION
景觀橋拉索
TERSE HANDRAIL
簡潔的扶手
ARTIFICIAL SHALLOW RIVER
自然河水面
VERTICAL DOCK
垂直的駁岸

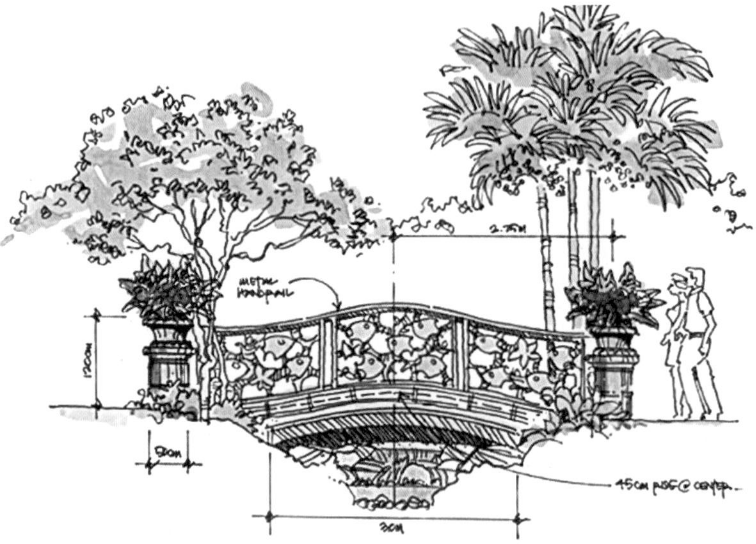

METAL HANDRAIL
2.75M
120CM
50CM
45CM RISE @ CENTER

BRIDGE NEAR GARDEN TERRACE

ELEVATION SCALE 1:25M

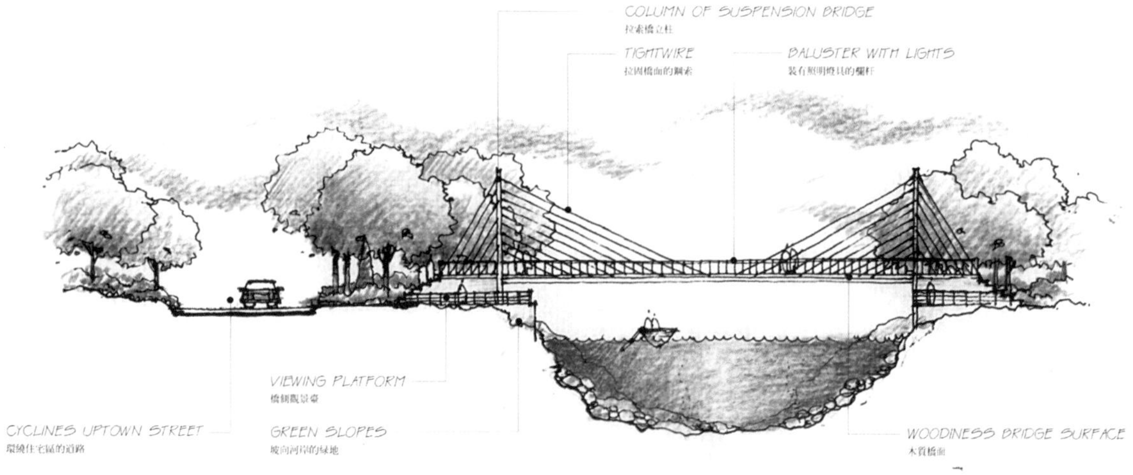

COLUMN OF SUSPENSION BRIDGE
拉索橋立柱
TIGHTWIRE
拉固橋面的鋼索
BALUSTER WITH LIGHTS
裝有照明燈具的欄杆
VIEWING PLATFORM
橋側觀景臺
CYCLINES UPTOWN STREET
環繞住宅區的道路
GREEN SLOPES
坡向河岸的綠地
WOODINESS BRIDGE SURFACE
木質橋面

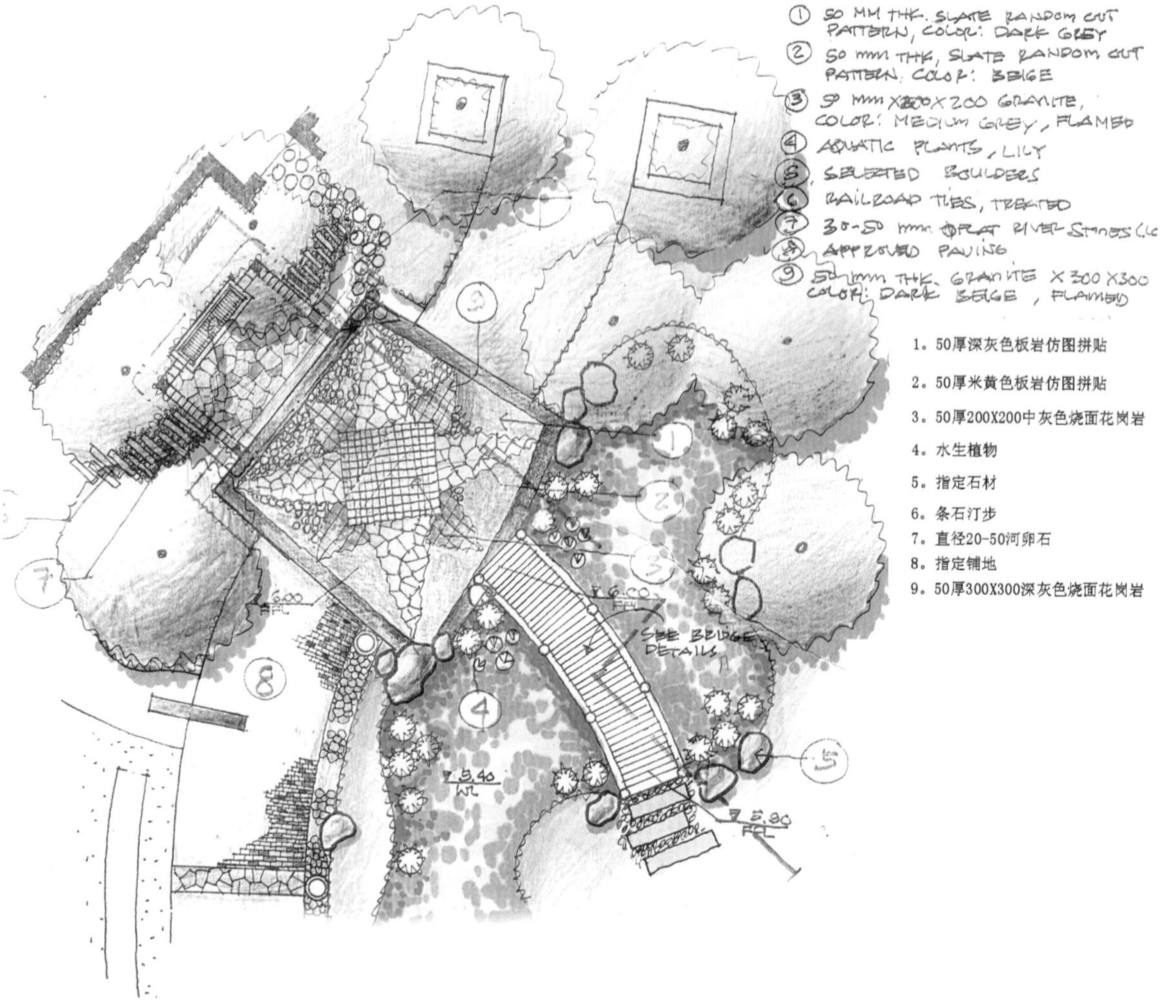

① 50 MM THK. SLATE RANDOM CUT PATTERN, COLOR: DARK GREY
② 50 mm THK, SLATE RANDOM CUT PATTERN. COLOR: BEIGE
③ 50 mm X200X200 GRANITE, COLOR: MEDIUM GREY, FLAMED
④ AQUATIC PLANTS, LILY
⑤ SELECTED BOULDERS
⑥ RAILROAD TIES, TREATED
⑦ 30-50 mm FLAT RIVER STONES
⑧ APPROVED PAVING
⑨ 50 mm THK. GRANITE X300X300 COLOR: DARK BEIGE, FLAMED
1. 50厚深灰色板岩仿图拼贴
2. 50厚米黄色板岩仿图拼贴
3. 50厚200X200中灰色烧面花岗岩
4. 水生植物
5. 指定石材
6. 条石汀步
7. 直径20-50河卵石
8. 指定铺地
9. 50厚300X300深灰色烧面花岗岩
SEE BRIDGE DETAILS

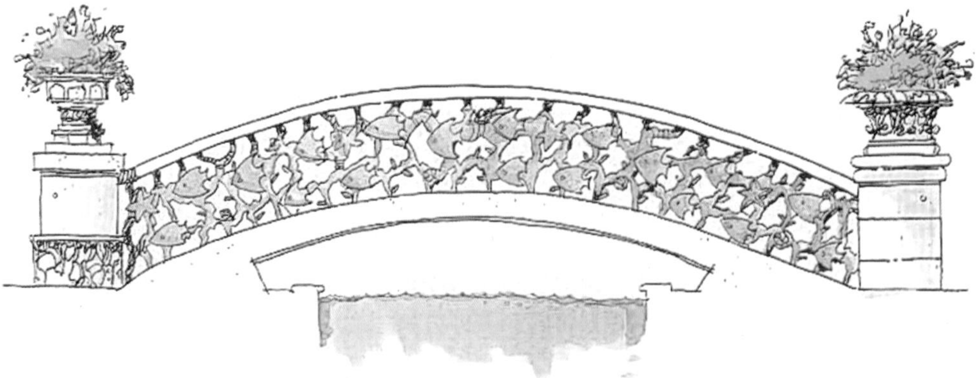

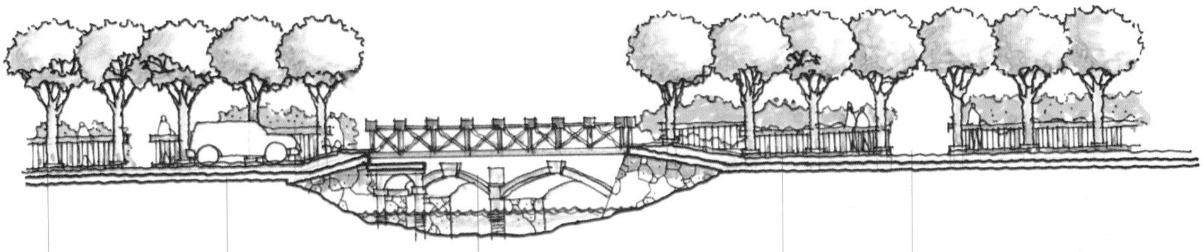
FEATURE TREE
行道景观树
VEHICLE
车行道
STONE BRIDGE
石桥
FENCE
景观围墙
PATHWAY
步行休闲小径

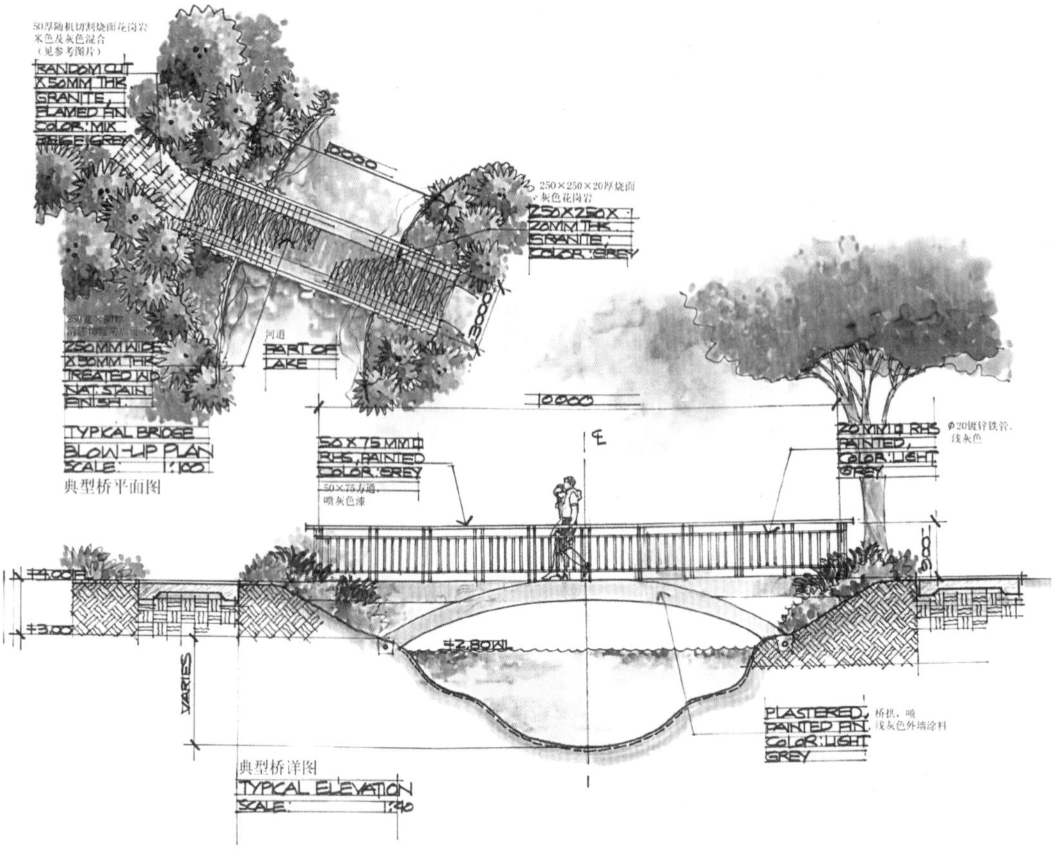

50厚随机切割烧面花岗岩
米色及灰色混合
（见参考图片）
RANDOM CUT X 50MM THK GRANITE, FLAMED FIN. COLOR: MIX BEIGE/GREY
10000
250×250×20厚烧面灰色花岗岩
250X250X 20MM THK. GRANITE, COLOR: GREY
3000
250MM WIDE X 30MM THK. TREATED WD. NAT. STAIN FINISH.
河道
PART OF LAKE
TYPICAL BRIDGE BLOW-UP PLAN SCALE: 1:100
典型桥平面图
10000
50 X 75 MM RHS, PAINTED COLOR: GREY
50×75方通，喷灰色漆
20 MM RHS PAINTED, COLOR: LIGHT GREY
φ20镀锌铁管，浅灰色
1000
+4.00
+3.00
+2.80WL
VARIES
PLASTERED, PAINTED FIN. COLOR: LIGHT GREY
桥拱，喷浅灰色外墙涂料
典型桥详图
TYPICAL ELEVATION SCALE: 1:40

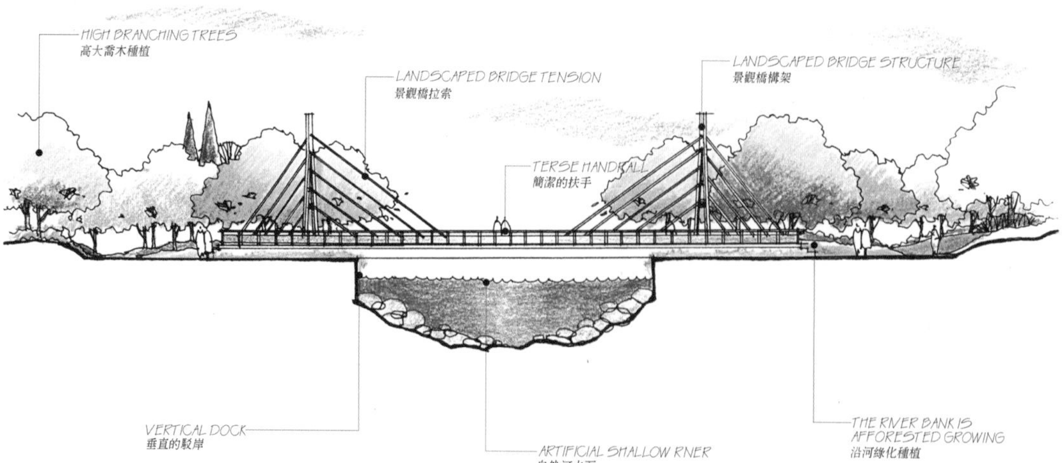

HIGH BRANCHING TREES
高大喬木種植
LANDSCAPED BRIDGE TENSION
景觀橋拉索
TERSE HANDRAIL
簡潔的扶手
LANDSCAPED BRIDGE STRUCTURE
景觀橋構架
VERTICAL DOCK
垂直的駁岸
ARTIFICIAL SHALLOW RIVER
自然河水面
THE RIVER BANK IS AFFORESTED GROWING
沿河綠化種植

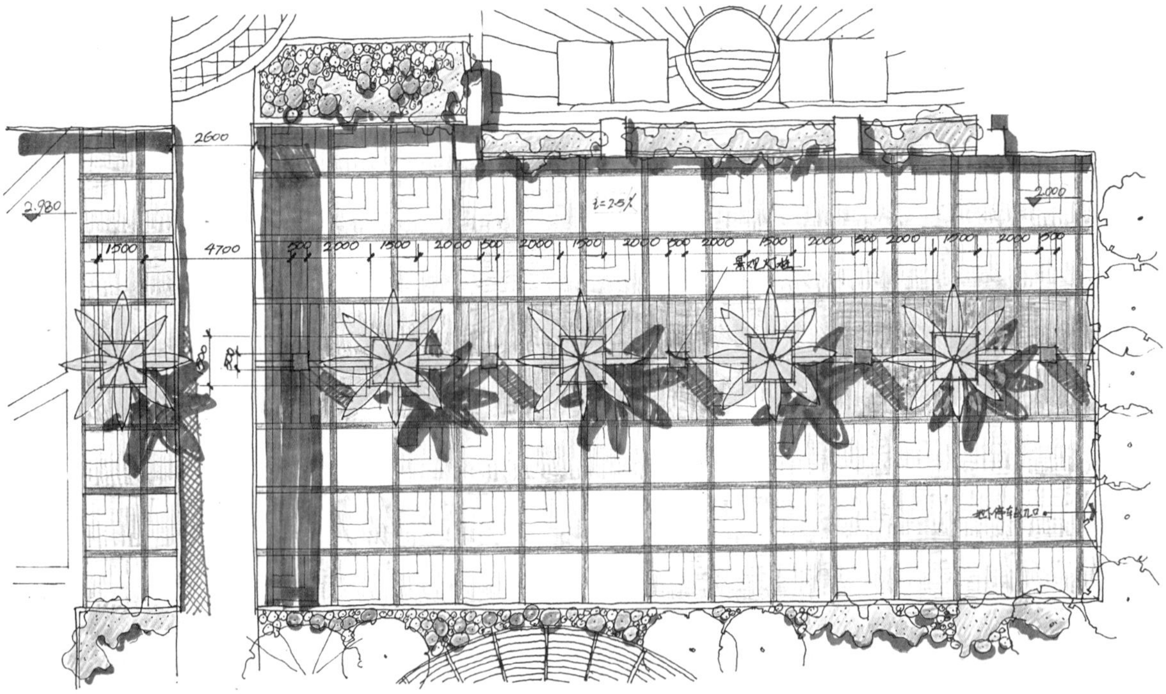

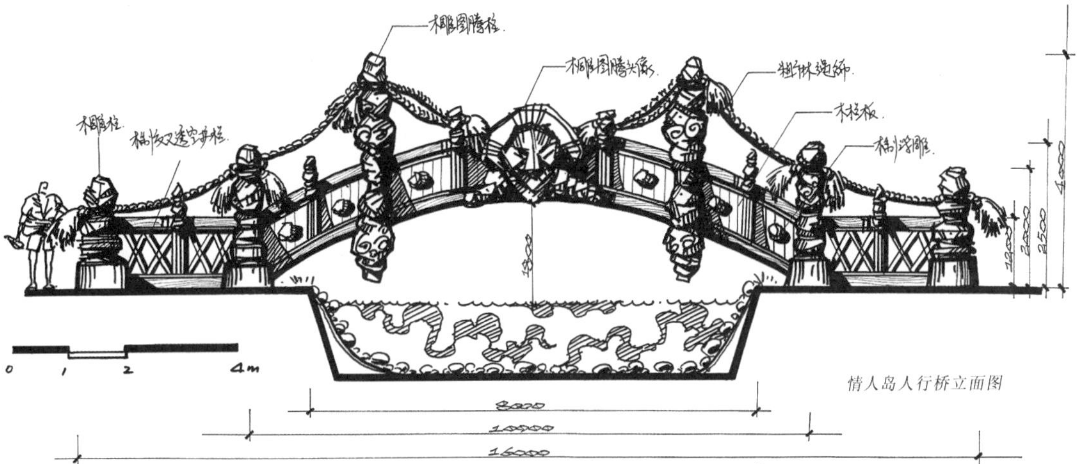

情人岛人行桥立面图

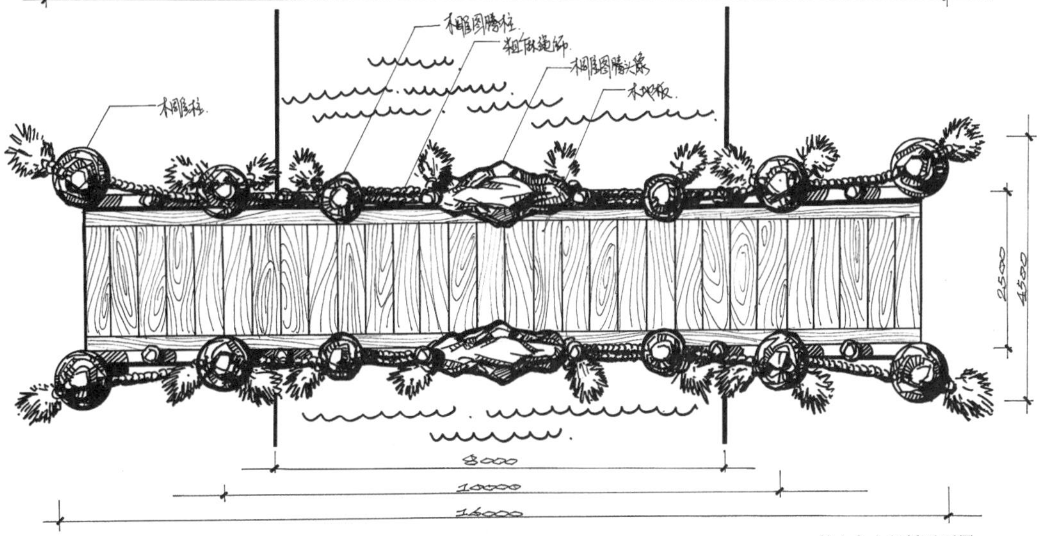

情人岛人行桥平面图

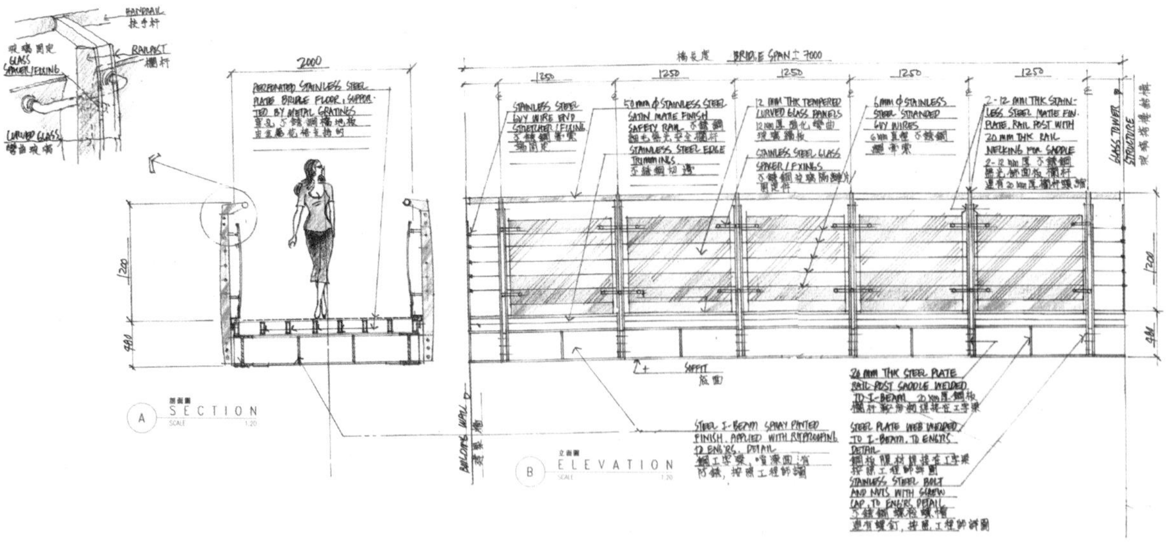
2000
SECTION
1:20
ELEVATION
1:20
BRIDGE SPAN ± 7000
1250
1250
1250
1250
1250
1200
SOFFIT

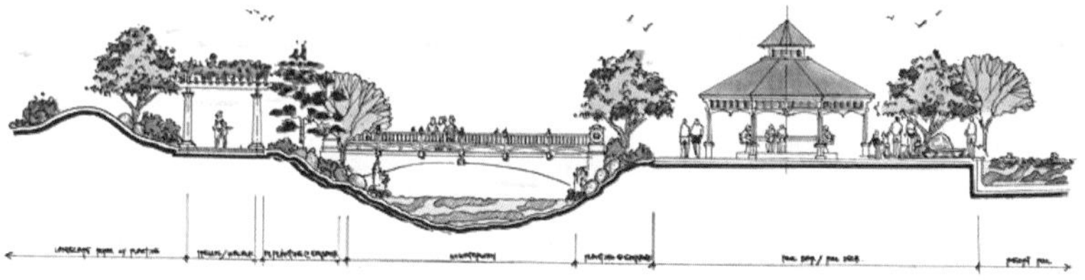

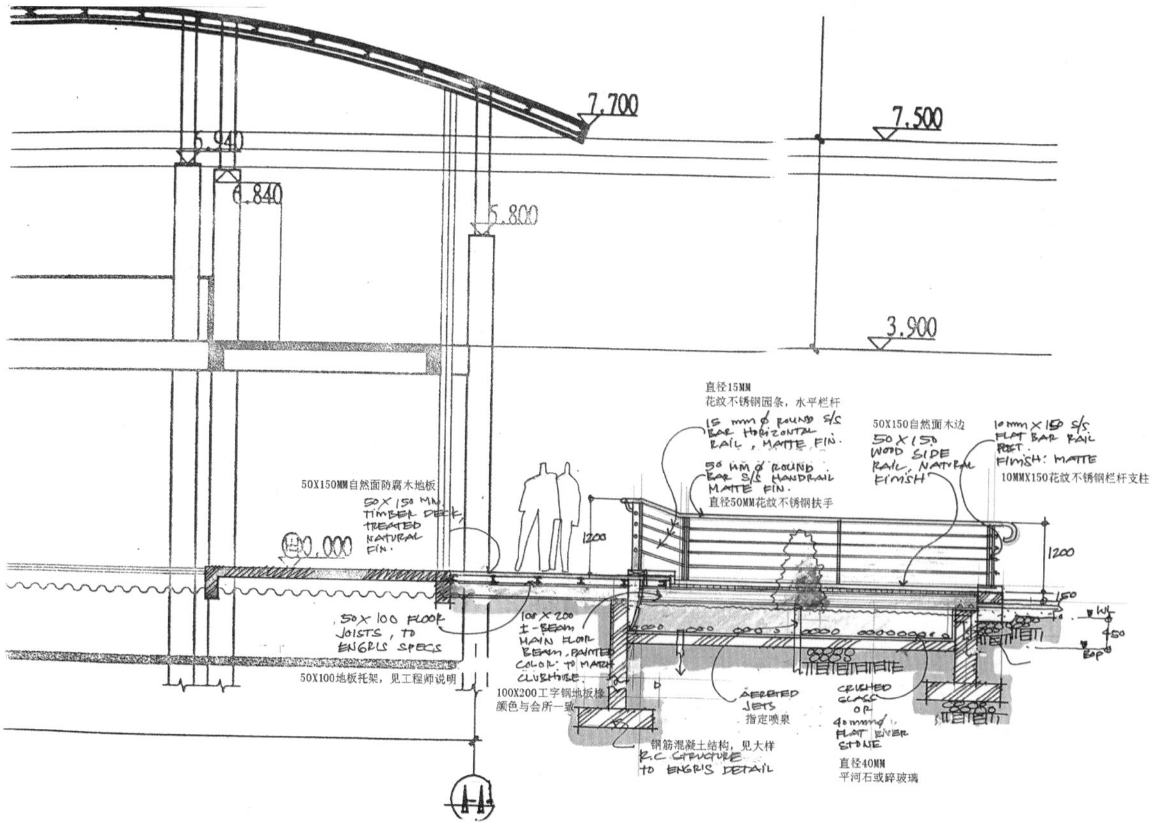
7.700
7.500
6.940
6.840
5.800
3.900
±0.000
直径15MM
花纹不锈钢园条，水平栏杆
15 mm ø ROUND S/S BAR HORIZONTAL RAIL, MATTE FIN.
50 MM ø ROUND BAR S/S HANDRAIL MATTE FIN.
直径50MM花纹不锈钢扶手
50X150自然面木边
50 X 150 WOOD SIDE RAIL, NATURAL FINISH
10 mm X 150 S/S FLAT BAR RAIL POST. FINISH: MATTE
10MMX150花纹不锈钢栏杆支柱
50X150MM自然面防腐木地板
50 X 150 MM TIMBER DECK, TREATED NATURAL FIN.
1200
1200
150
450
50 X 100 FLOOR JOISTS, TO ENGR'S SPECS
50X100地板托架，见工程师说明
100 X 200 I-BEAM MAIN FLOOR BEAM, PAINTED COLOR TO MATCH CLUBHOUSE.
100X200工字钢地板橡
颜色与会所一致
AERATED JETS
指定喷泉
CRUSHED GLASS OR 40mm ø FLAT RIVER STONE
直径40MM
平河石或碎玻璃
钢筋混凝土结构，见大样
R.C STRUCTURE TO ENGR'S DETAIL

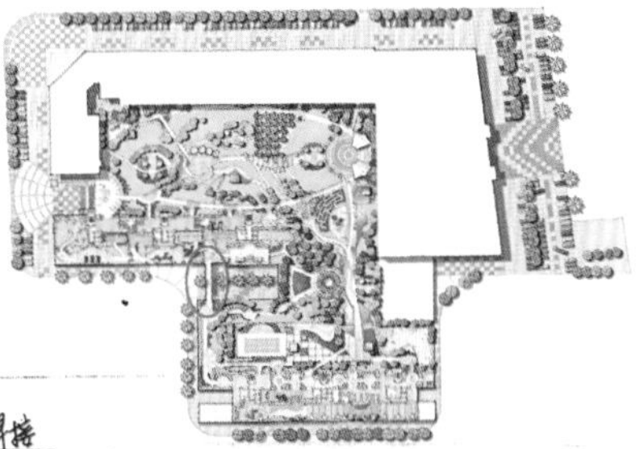

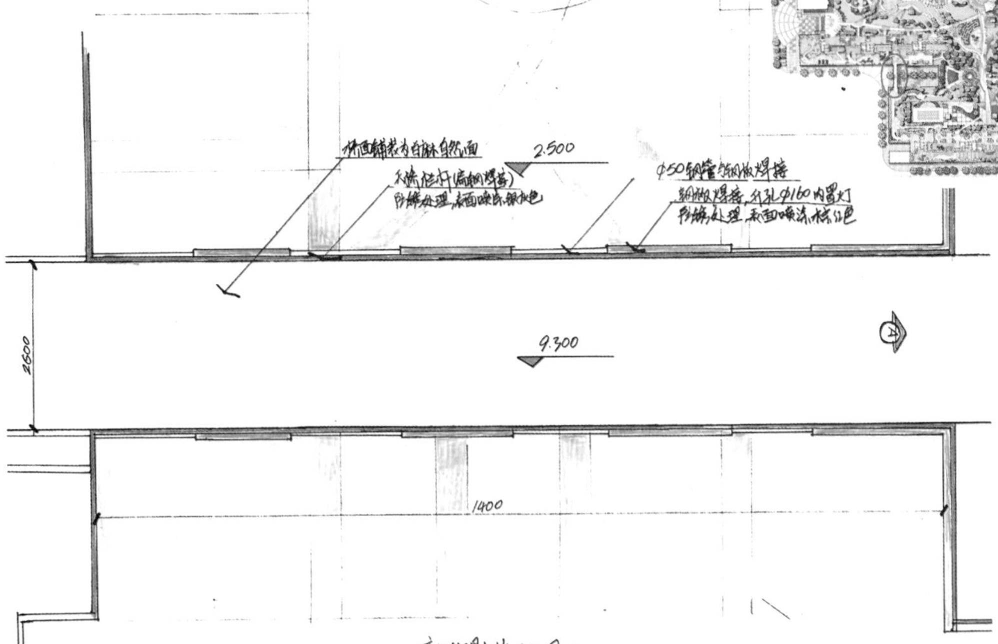

商业街湿地桥平面图
比例 1:50

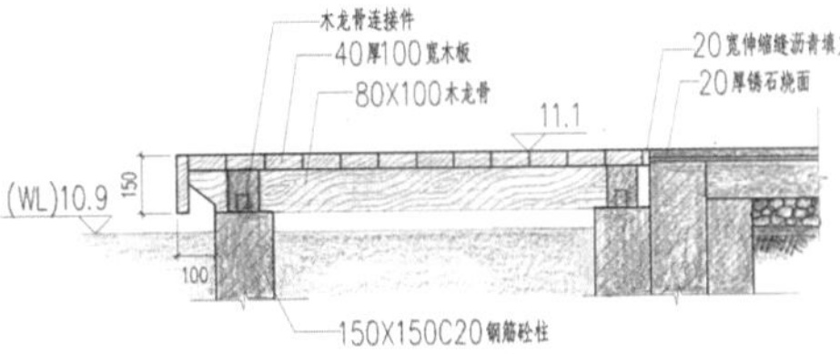

浣溪戏石 B-B 剖面（1：15）

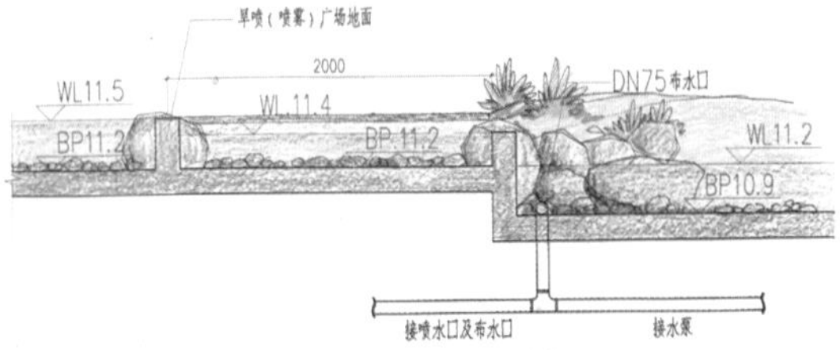

中心广场旱喷 A-A 剖面（1：30）

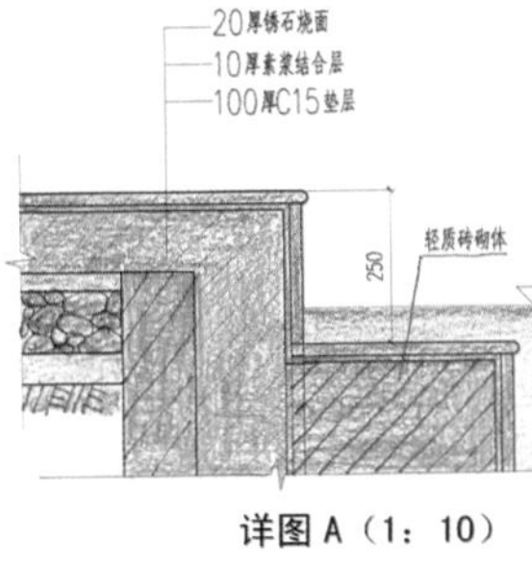

详图 A（1：10）

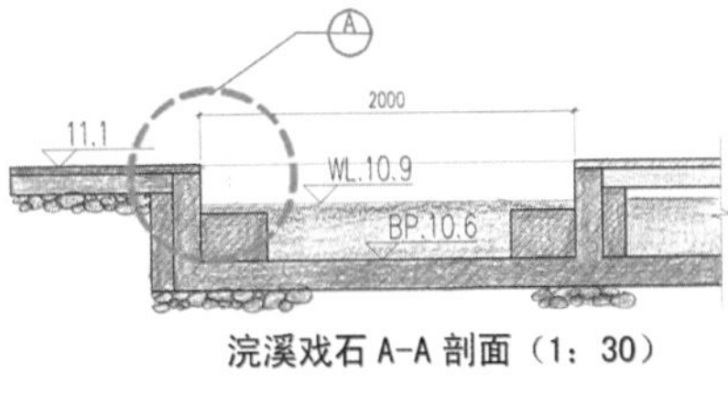

浣溪戏石 A-A 剖面（1：30）

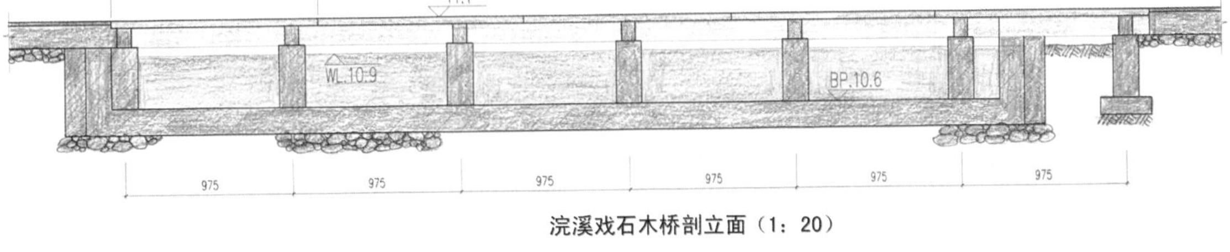

浣溪戏石木桥剖立面（1：20）

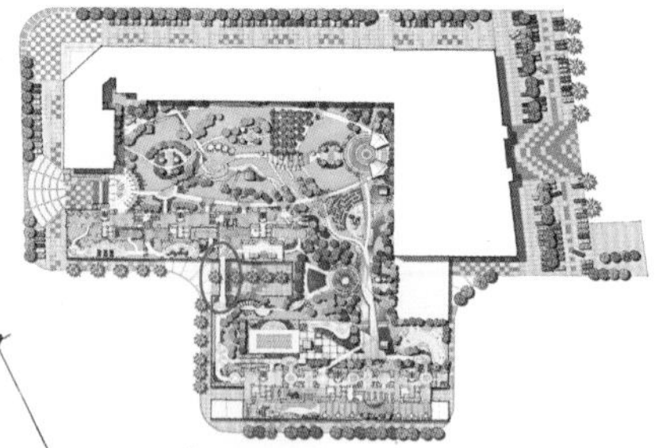

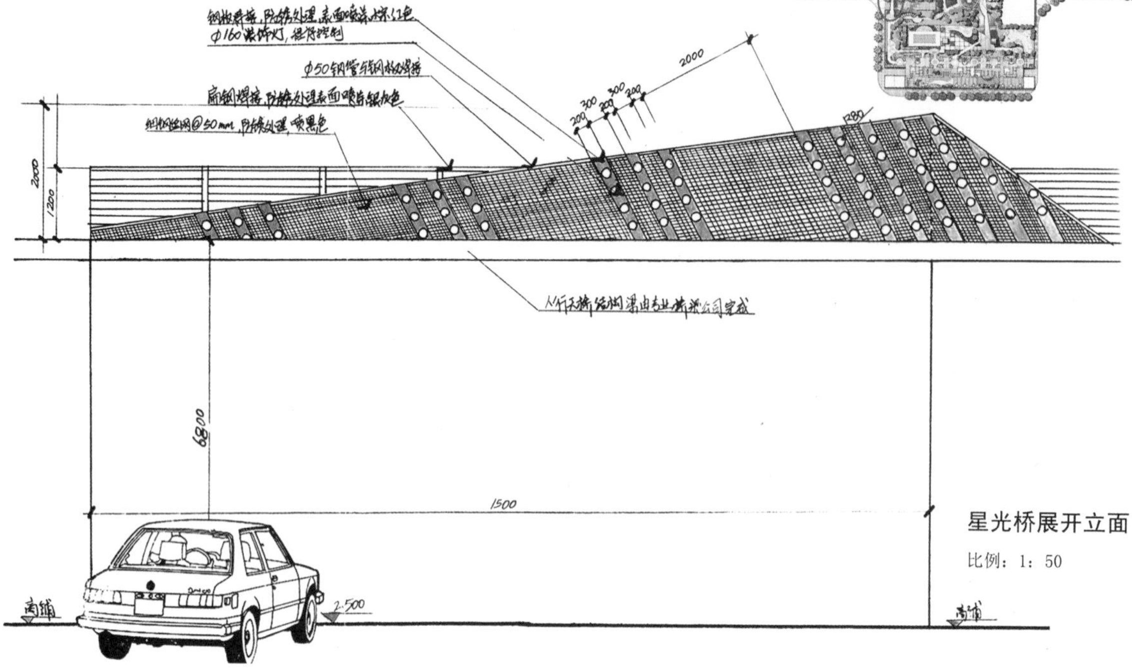

星光桥展开立面

比例：1：50

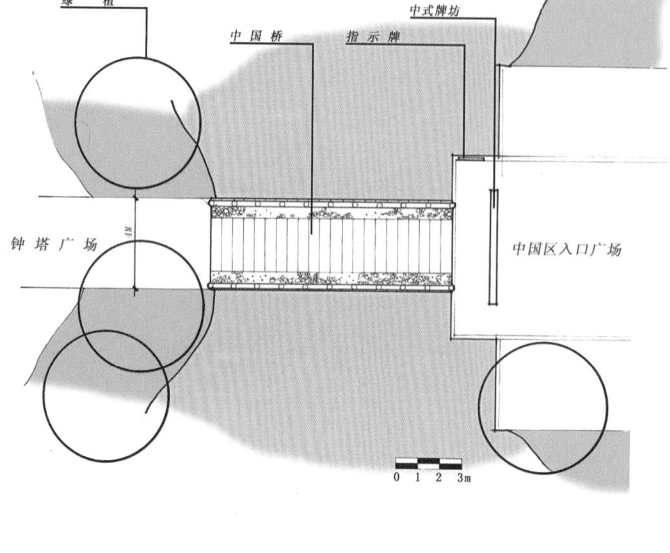

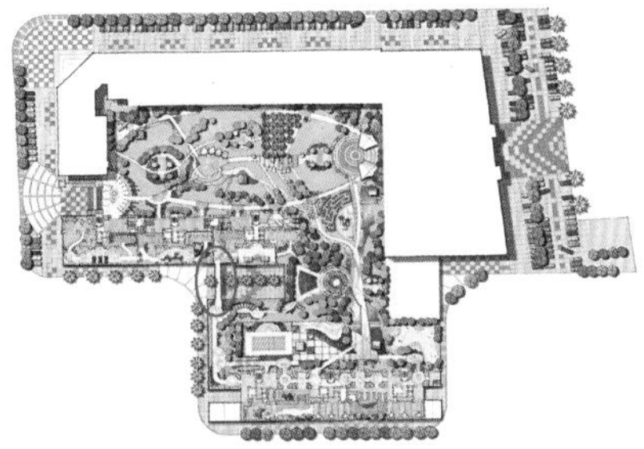

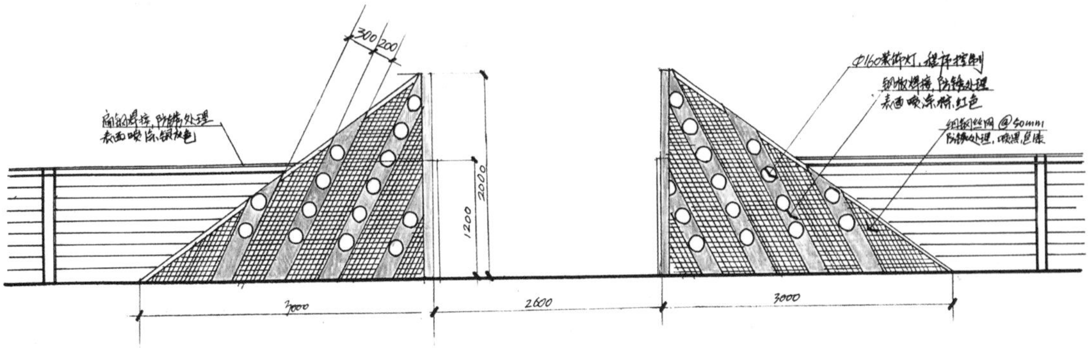
300
200
2000
1200
3000
2600
3000

天桥Ⓐ立面 1:20

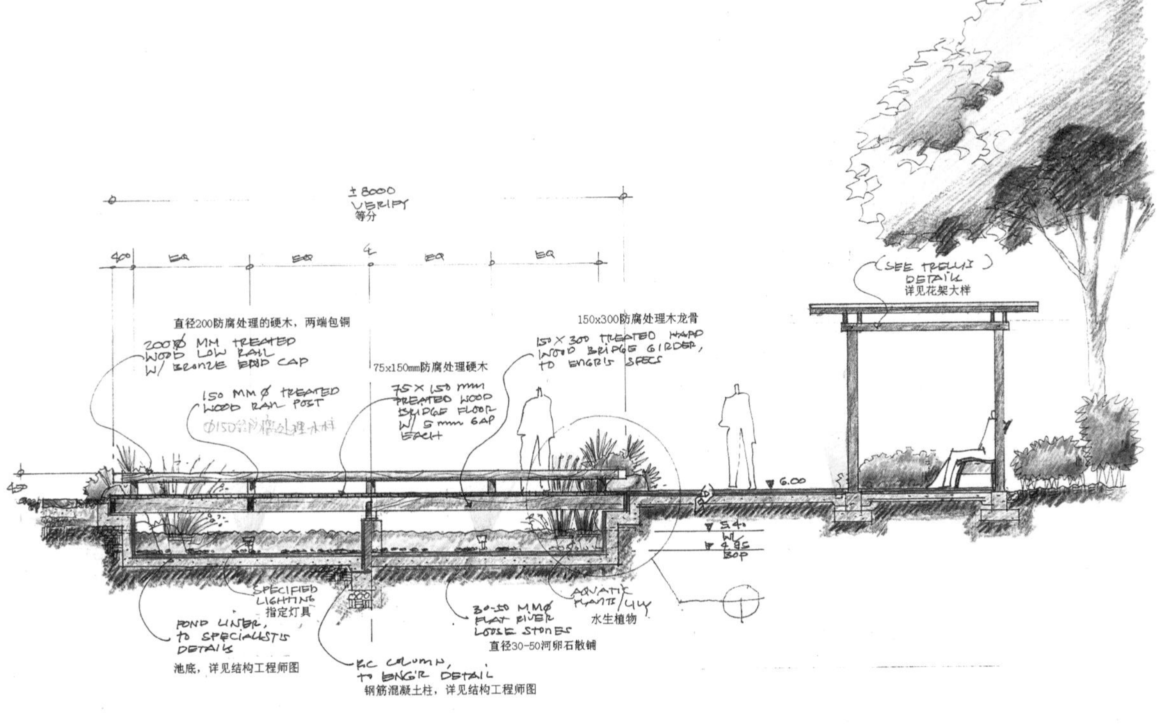
±8000
VERIFY
等分
400
EQ
EQ
EQ
EQ
直径200防腐处理的硬木，两端包铜
200Ø MM TREATED WOOD LOW RAIL W/ BRONZE END CAP
150 MM Ø TREATED WOOD RAIL POST
75x150mm防腐处理硬木
75 X 150 mm TREATED WOOD BRIDGE FLOOR W/ 5 mm GAP EACH
150x300防腐处理木龙骨
150 X 300 TREATED HARD WOOD BRIDGE GIRDER, TO ENGR'S SPECS
SEE TRELLIS DETAIL
详见花架大样
450
6.00
SPECIFIED LIGHTING
指定灯具
POND LINER, TO SPECIALIST'S DETAILS
池底，详见结构工程师图
RC COLUMN, TO ENGR DETAIL
钢筋混凝土柱，详见结构工程师图
30-50 MMØ FLAT RIVER LOOSE STONES
直径30-50河卵石散铺
AQUATIC PLANTS
水生植物

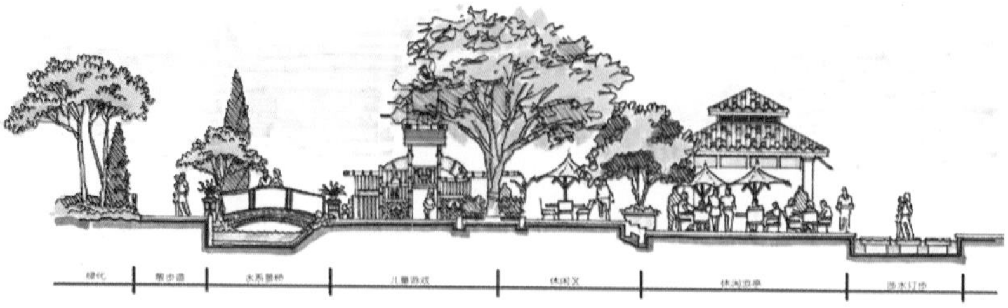
绿化
散步道
水系景桥
儿童游戏
休闲区

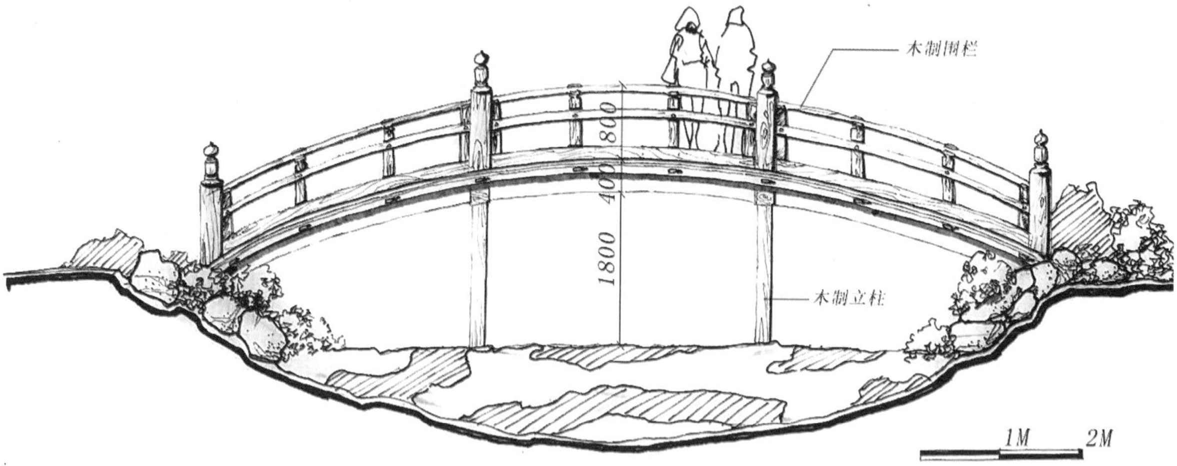
木制围栏
800
400
1800
木制立柱
1M
2M

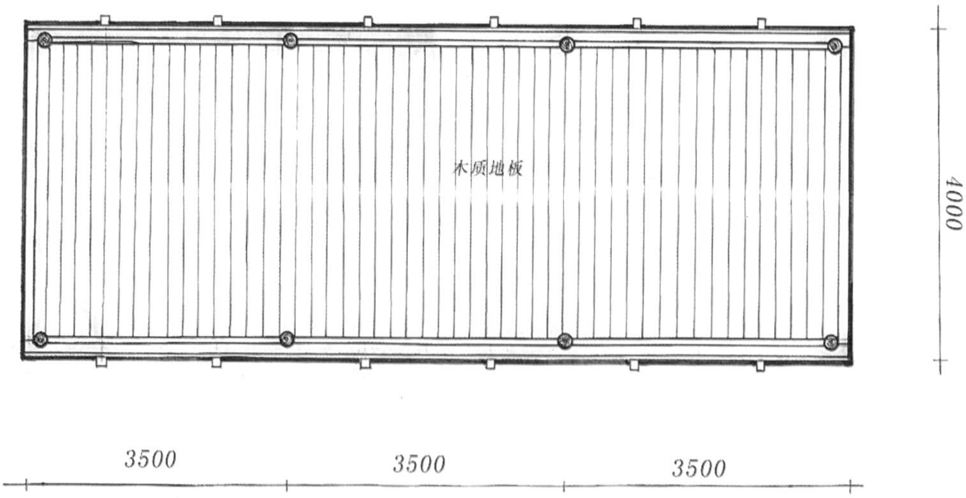
木质地板
4000
3500
3500
3500

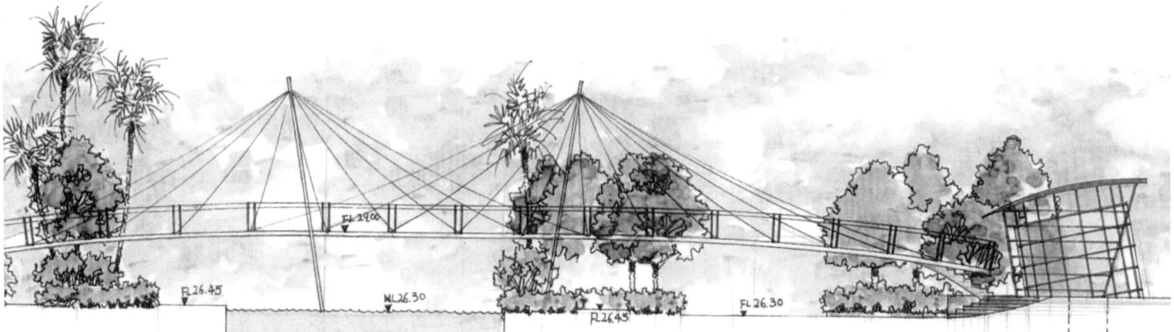
FL29.00
FL26.45
WL26.30
FL26.45
FL26.30

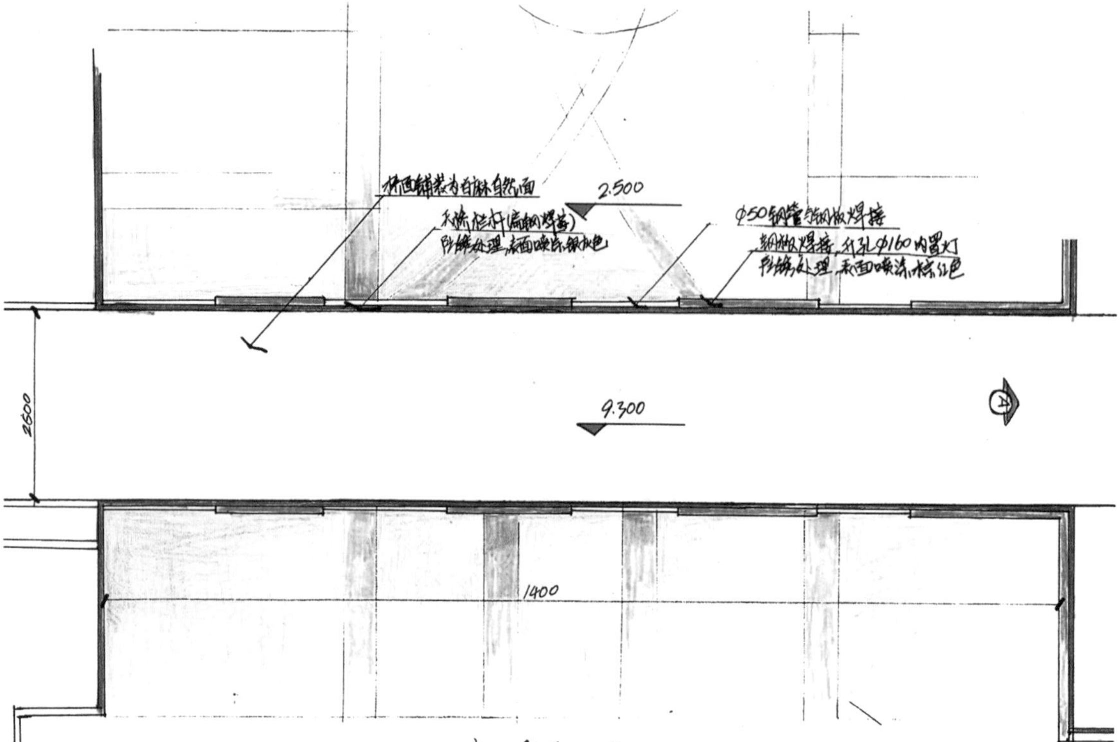

商业街景观桥平面图
比例 1:50

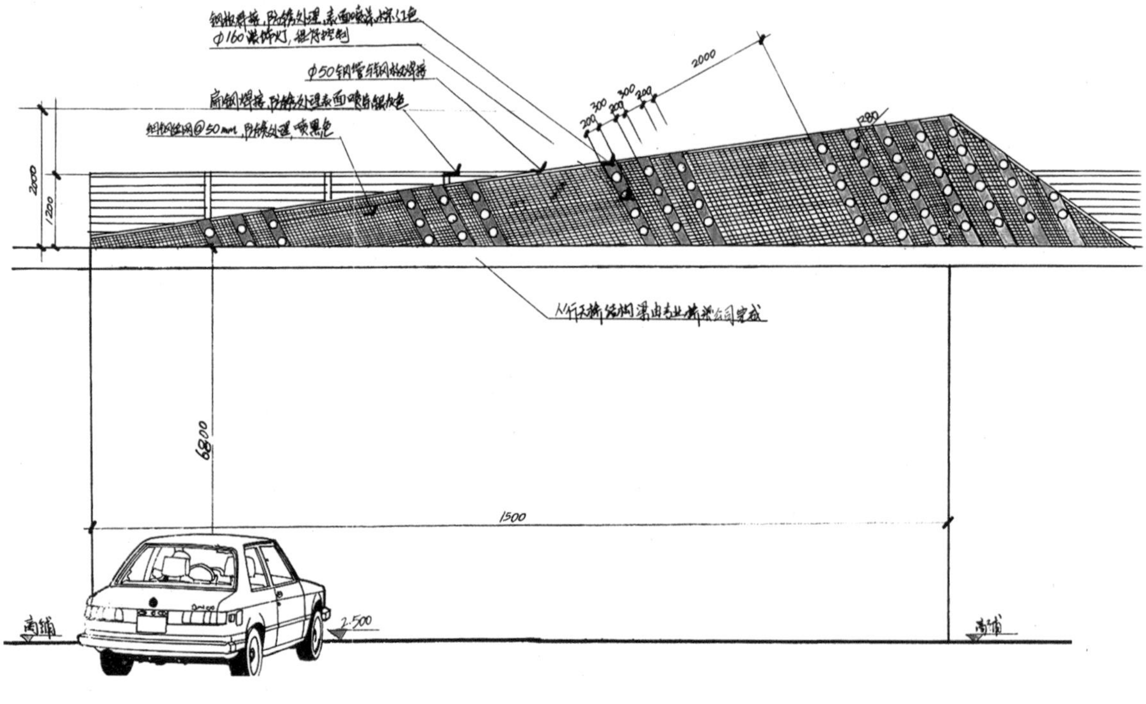

桥面铺装为白麻自然面
2.500
不锈栏杆(扁钢焊接)
防锈处理,表面喷涂银灰色
Φ50钢管与钢板焊接
钢板焊接,开孔Φ160内置灯
防锈处理,表面喷涂棕红色
2600
9.300
1400
商业街星光桥平面图
钢板焊接,防锈处理,表面喷涂棕红色
Φ160装饰灯,程序控制
Φ50钢管与钢板焊接
扁钢焊接,防锈处理表面喷涂银灰色
2000
200
300
200
300
200
1200
6800
1500
2.500

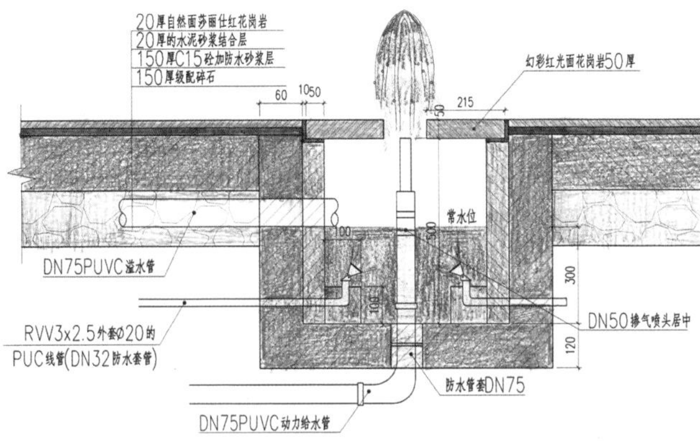

"水语馨声"旱喷泉剖面（1:10）

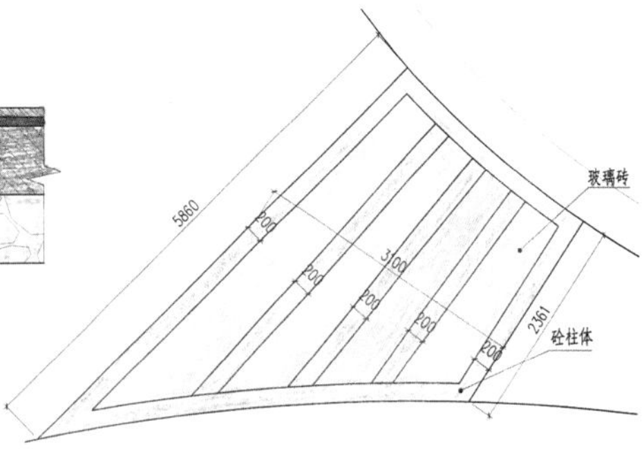

"水语馨声"玻璃桥平面（1:50）

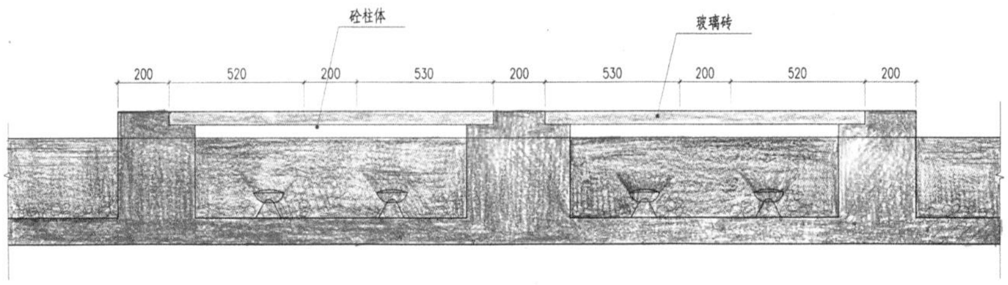

"水语馨声"玻璃桥剖面（1:15）

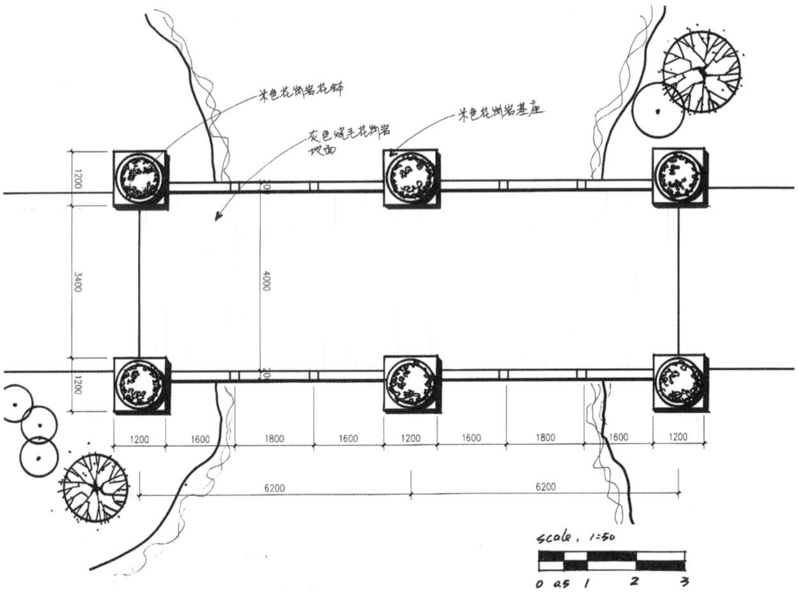

欧洲景区入口桥平面

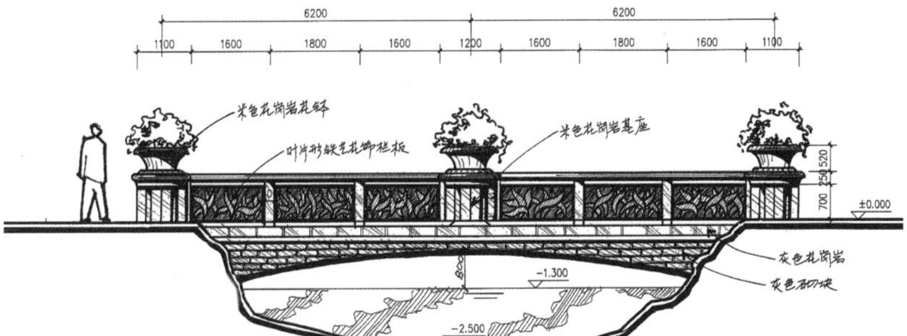

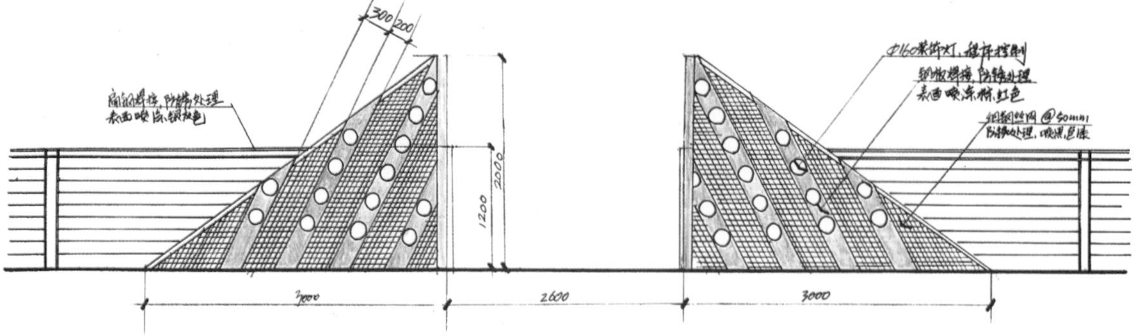
300
200
扁钢焊接，防锈处理
表面喷涂银灰色
2000
1200
2600
3000
Φ160装饰灯，程序控制
钢板焊接，防锈处理
表面喷涂橘红色
细钢丝网 @50mm

天桥Ⓐ立面 1:20

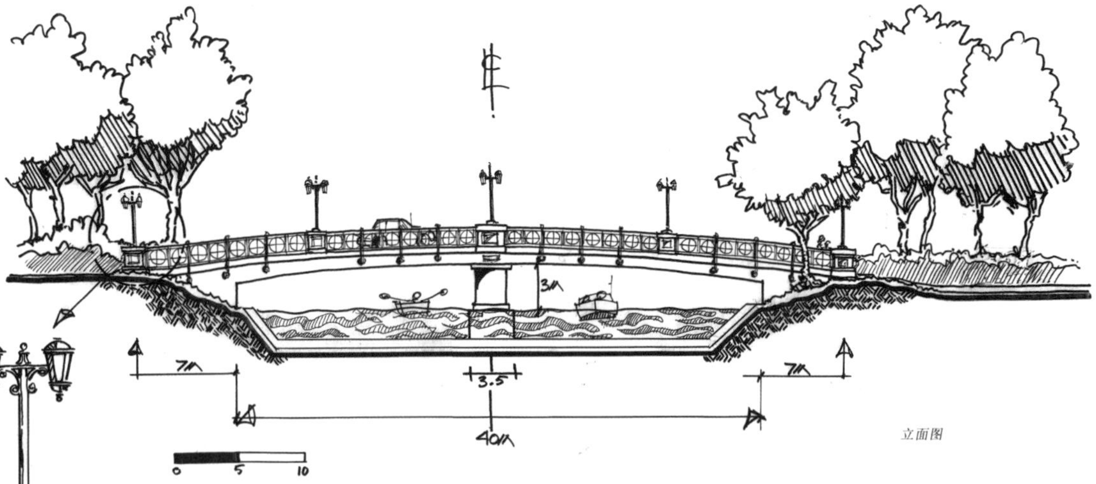

立面图

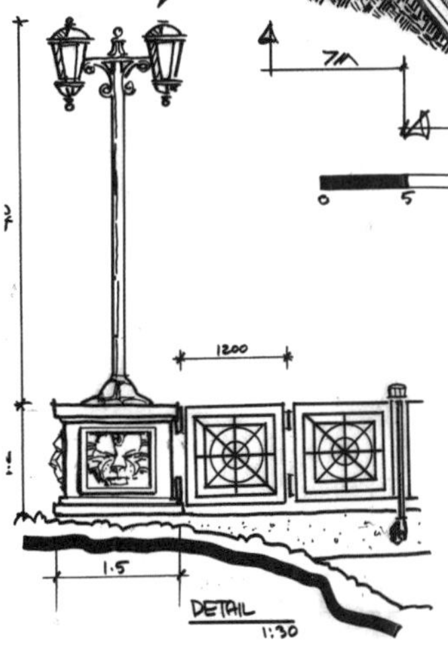

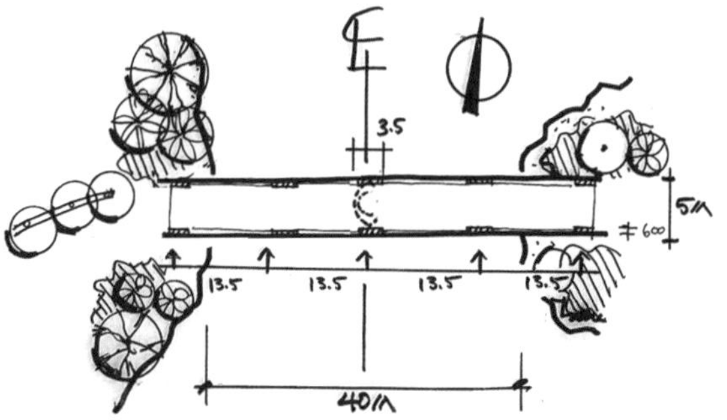

主题铁艺图案大样 1:30　　方格网 500X500

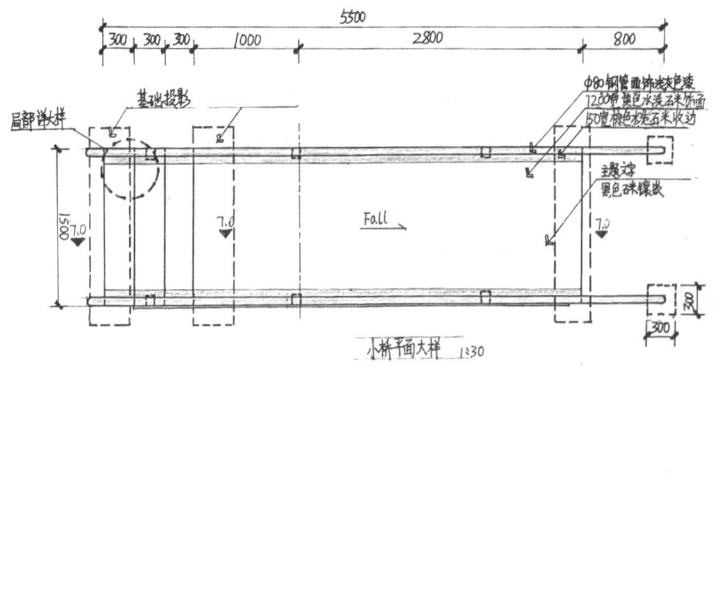

小桥平面大样 1:30

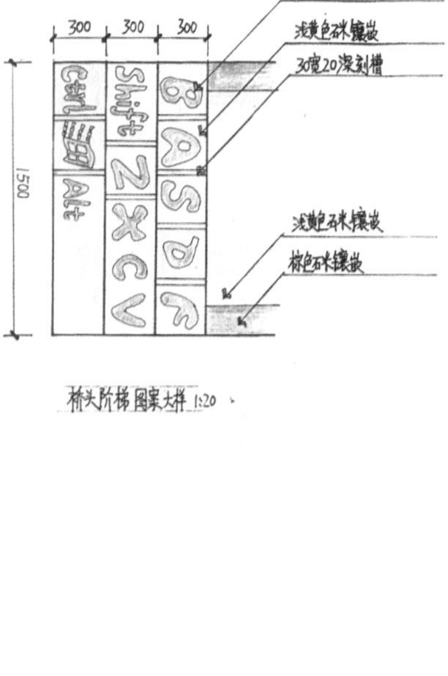

桥头阶梯图案大样 1:20

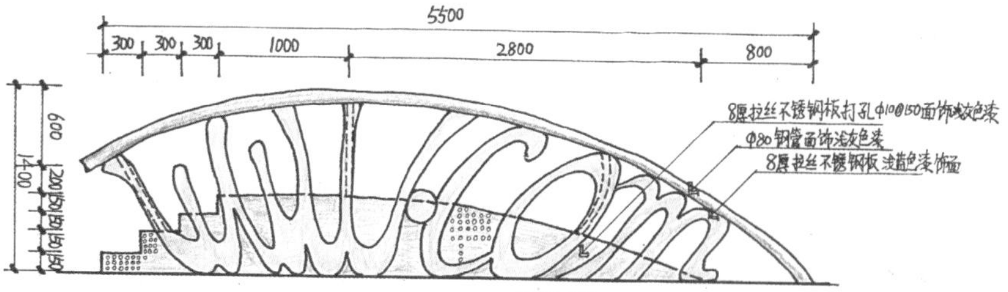

5500
300
1000
2800
800
8厚拉丝不锈钢板打孔Φ10@150面饰浅灰色漆
Φ80钢管面饰浅灰色漆
8厚拉丝不锈钢板浅蓝色漆饰面
小桥立面图 1:30
Φ80钢管面饰浅灰色漆
□60X60热浸镀型面饰浅蓝漆
R5000
钢筋砼桥体水洗石米饰面
扶手与地面连接
轻质材料填充
小桥剖面图A 1:30
1500
1400
180宽轻质砖墙
小桥剖面B 1:30

4500
7000
5500
灯具
LIGHTING
混凝土墙,水泥砂浆抹面
CONCRETE WALL W/ PLASTER FINISH
天然石块,如墙饰面
NATURAL STONE AS WALL FINISHER
木扶手
TIMBER RAILING
广场区
PLAZA AREA
景观墙
WALL FEATURE
木座椅
TIMBER SEAT
木平台
TIMBER DECK
2100
立面图
ELEVATION
SCALE 1:100

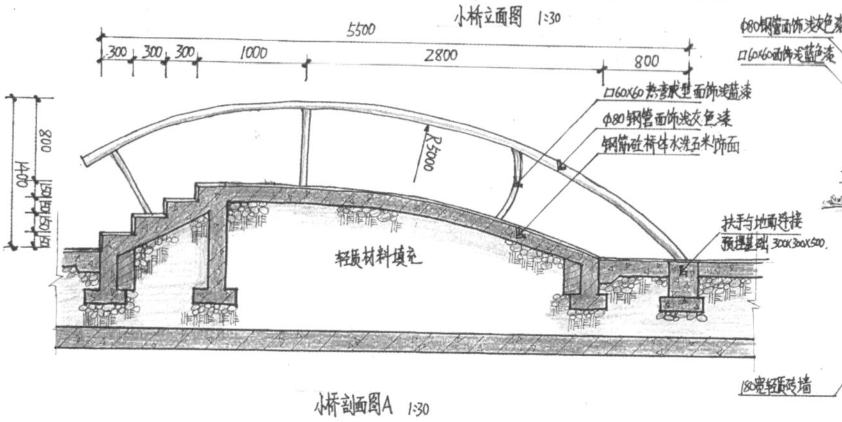

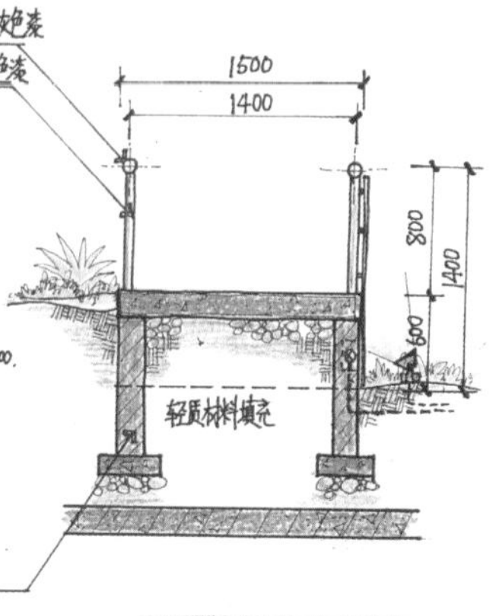

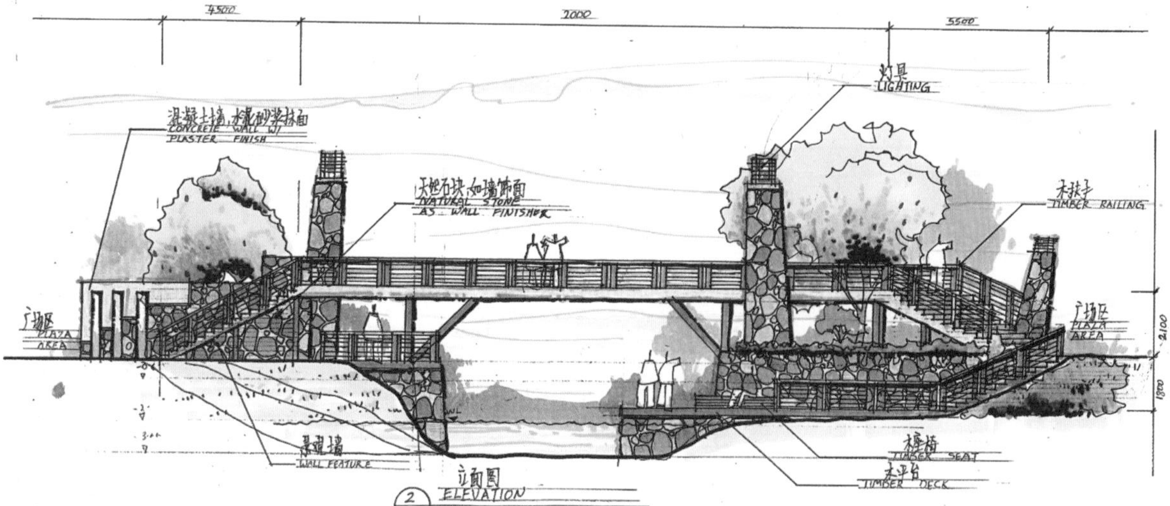

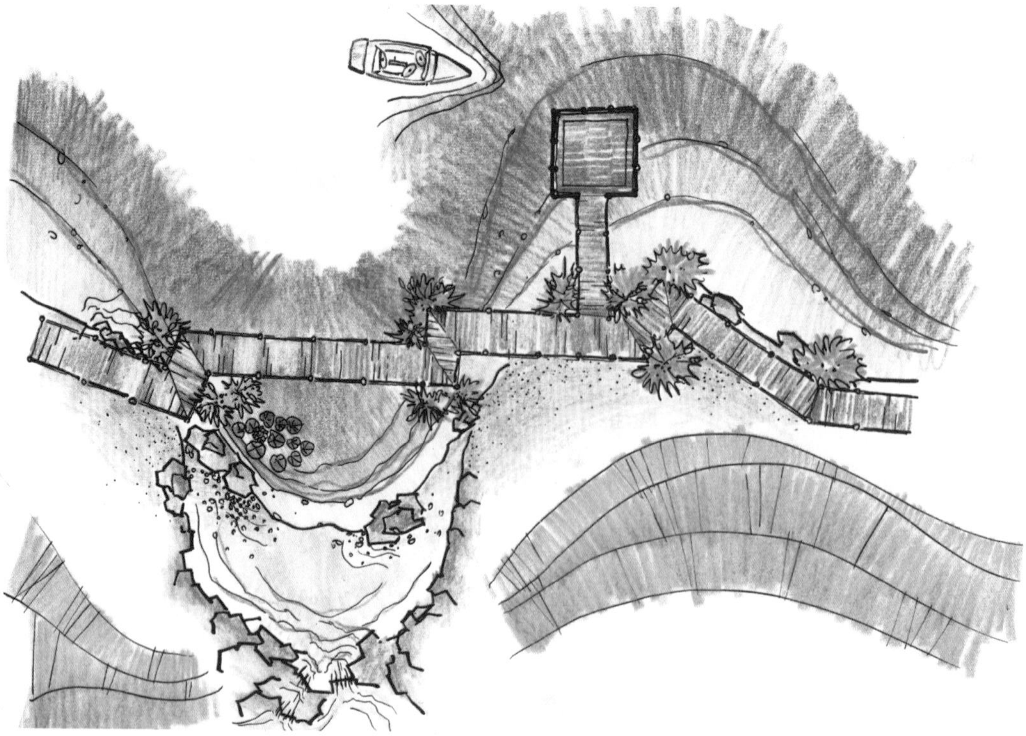

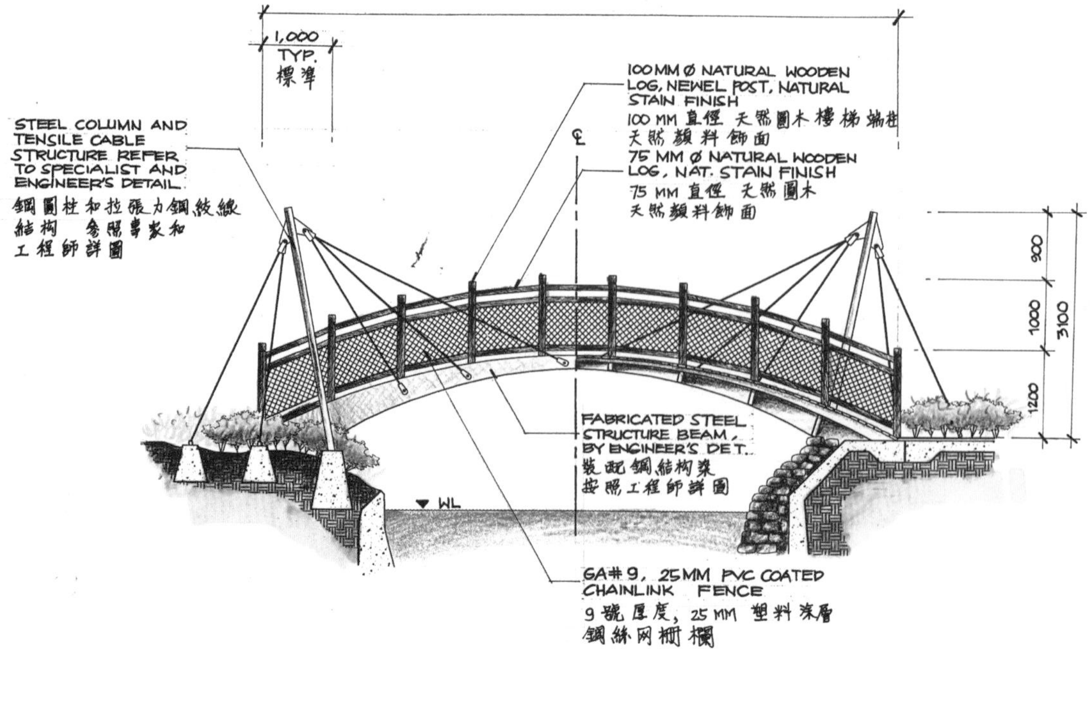
1,000
TYP.
標準
100MM Ø NATURAL WOODEN LOG, NEWEL POST, NATURAL STAIN FINISH
100 MM 直徑 天然圓木樓梯端柱
天然顏料飾面
STEEL COLUMN AND TENSILE CABLE STRUCTURE REFER TO SPECIALIST AND ENGINEER'S DETAIL.
鋼圓柱和拉張力鋼紋纜結构 參照專家和工程師詳圖
℄
75 MM Ø NATURAL WOODEN LOG, NAT. STAIN FINISH
75 MM 直徑 天然圓木
天然顏料飾面
900
1000
3100
1200
FABRICATED STEEL STRUCTURE BEAM, BY ENGINEER'S DET.
裝配鋼結构梁
按照工程師詳圖
WL
GA#9, 25MM PVC COATED CHAINLINK FENCE
9號厚度, 25 MM 塑料涂層
鋼絲网柵欄

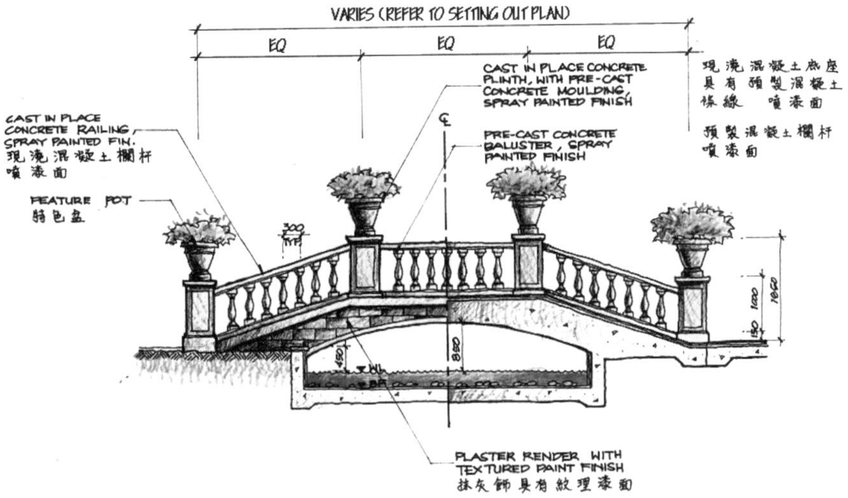
VARIES (REFER TO SETTING OUT PLAN)
EQ
EQ
EQ
CAST IN PLACE CONCRETE PLINTH, WITH PRE-CAST CONCRETE MOULDING, SPRAY PAINTED FINISH
現澆混凝土底座具有預製混凝土條線 噴漆面
預製混凝土欄杆噴漆面
CAST IN PLACE CONCRETE RAILING, SPRAY PAINTED FIN.
現澆混凝土欄杆噴漆面
PRE-CAST CONCRETE BALUSTER, SPRAY PAINTED FINISH
FEATURE POT
特色盆
300 TYP.
1650
1000
150
880
450
WL
BF
PLASTER RENDER WITH TEXTURED PAINT FINISH
抹灰飾具有紋理漆面

立面圖/剖面圖

A ELEVATION / SECTION

SCALE 1:50

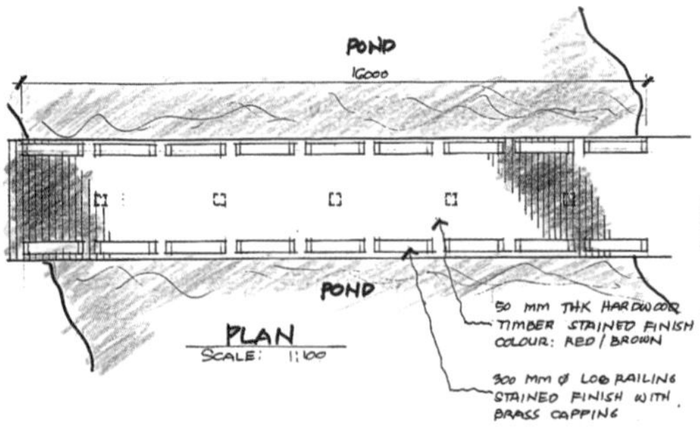
POND
16000
POND
PLAN
SCALE: 1:100
50 MM THK HARDWOOD TIMBER STAINED FINISH COLOUR: RED / BROWN
300 MM Ø LOG RAILING STAINED FINISH WITH BRASS CAPPING

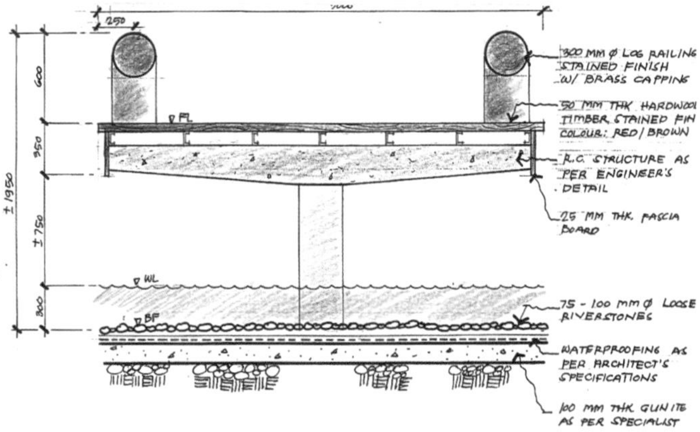
250
600
950
±1950
±750
300
FL
WL
BF
300 MM Ø LOG RAILING STAINED FINISH W/ BRASS CAPPING
50 MM THK HARDWOOD TIMBER STAINED FIN COLOUR: RED / BROWN
R.C. STRUCTURE AS PER ENGINEER'S DETAIL
25 MM THK. FASCIA BOARD
75 - 100 MM Ø LOOSE RIVERSTONES
WATERPROOFING AS PER ARCHITECT'S SPECIFICATIONS
100 MM THK. GUNITE AS PER SPECIALIST

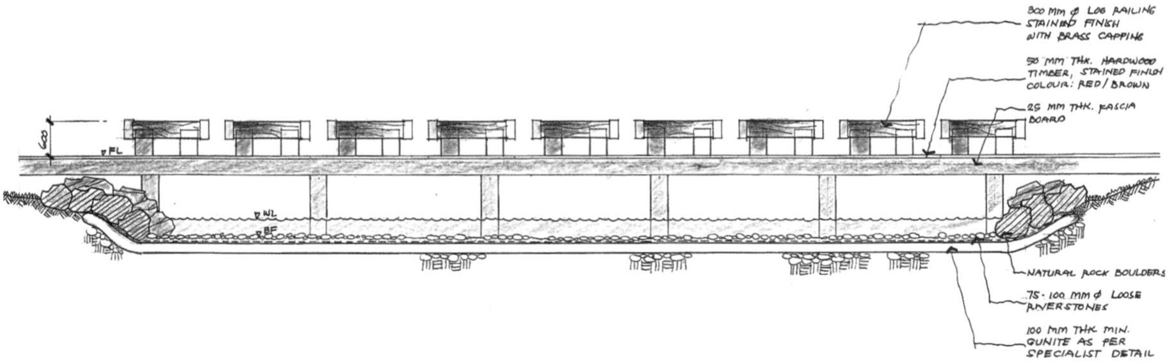
300 MM Ø LOG RAILING STAINED FINISH WITH BRASS CAPPING
50 MM THK. HARDWOOD TIMBER, STAINED FINISH COLOUR: RED / BROWN
25 MM THK. FASCIA BOARD
600
FL
WL
BF
NATURAL ROCK BOULDERS
75 - 100 MM Ø LOOSE RIVERSTONES
100 MM THK MIN. GUNITE AS PER SPECIALIST DETAIL

桥

方木条，经风化
圆木条，经风化
木桥
水位
200毫米
米黄色平间石沙
水灯
天然大石块
天然花岗石砂砖块铺地
庭园灯
预制混合料铺地
颜色：米黄色

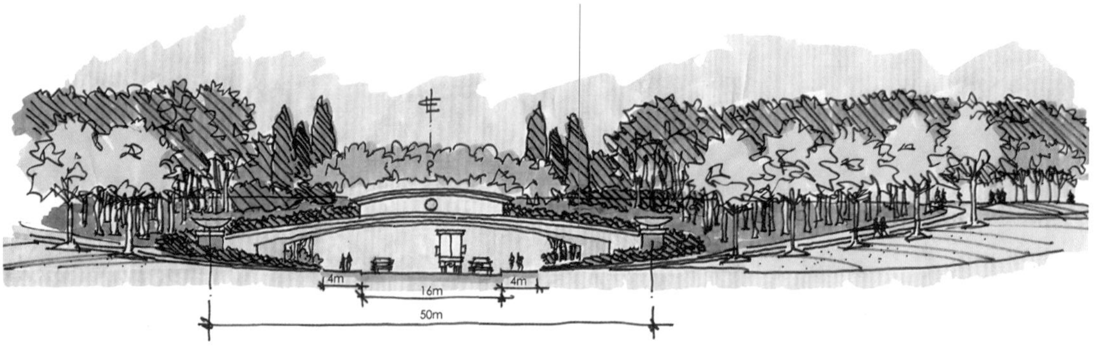
4m
16m
4m
50m

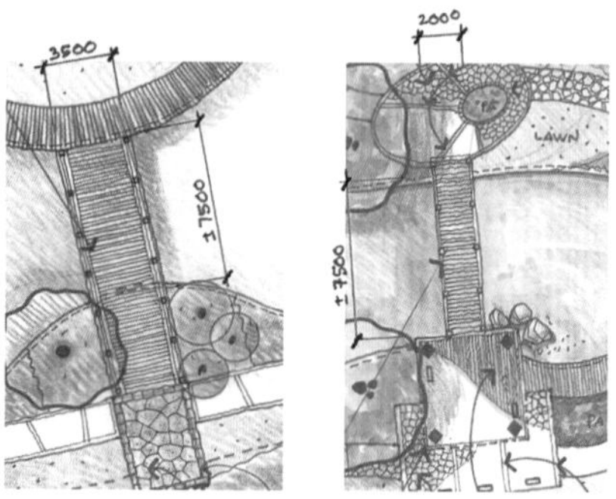
3500
±7500
2000
±7500
LAWN

FEATURE BRIDGE PLAN
SCALE 1:200

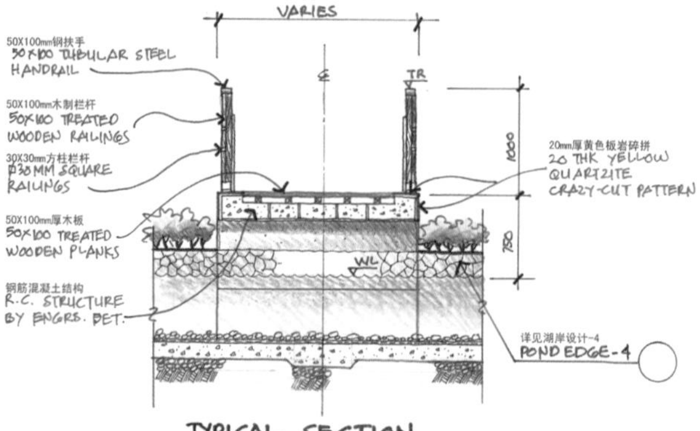
VARIES
50X100mm钢扶手
50X100 TUBULAR STEEL HANDRAIL
50X100mm木制栏杆
50X100 TREATED WOODEN RAILINGS
30X30mm方柱栏杆
Ø30MM SQUARE RAILINGS
50X100mm厚木板
50X100 TREATED WOODEN PLANKS
钢筋混凝土结构
R.C. STRUCTURE BY ENGRS. DET.
TR
1000
750
WL
20mm厚黄色板岩碎拼
20 THK YELLOW QUARTZITE CRAZY-CUT PATTERN
详见湖岸设计-4
POND EDGE-4

TYPICAL SECTION
SCALE 1:30

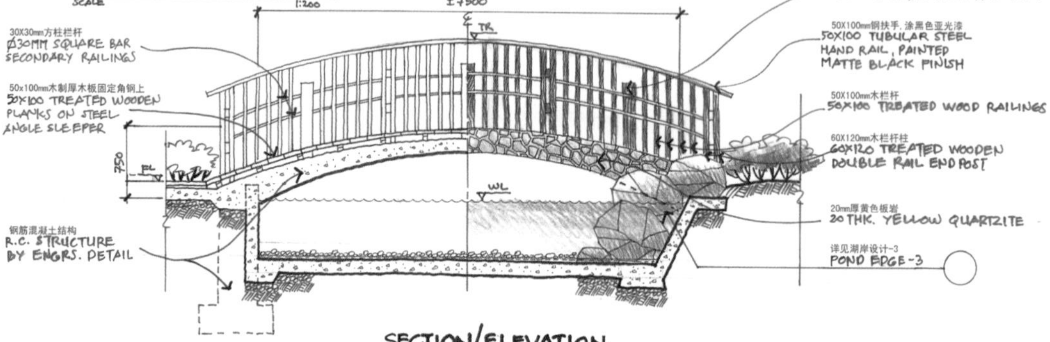
±7500
TR
120X150mm木栏杆柱
120X150 TREATED WOOD RAIL POST
50X100mm钢扶手，涂黑色亚光漆
50X100 TUBULAR STEEL HAND RAIL, PAINTED MATTE BLACK FINISH
50X100mm木栏杆
50X100 TREATED WOOD RAILINGS
60X120mm木栏杆柱
60X120 TREATED WOODEN DOUBLE RAIL END POST
20mm厚黄色板岩
20 THK. YELLOW QUARTZITE
详见湖岸设计-3
POND EDGE-3
30X30mm方柱栏杆
Ø30MM SQUARE BAR SECONDARY RAILINGS
50x100mm木制厚木板固定角钢上
50X100 TREATED WOODEN PLANKS ON STEEL ANGLE SLEEPER
750
FL
WL
钢筋混凝土结构
R.C. STRUCTURE BY ENGRS. DETAIL

SECTION/ELEVATION
SCALE 1:30

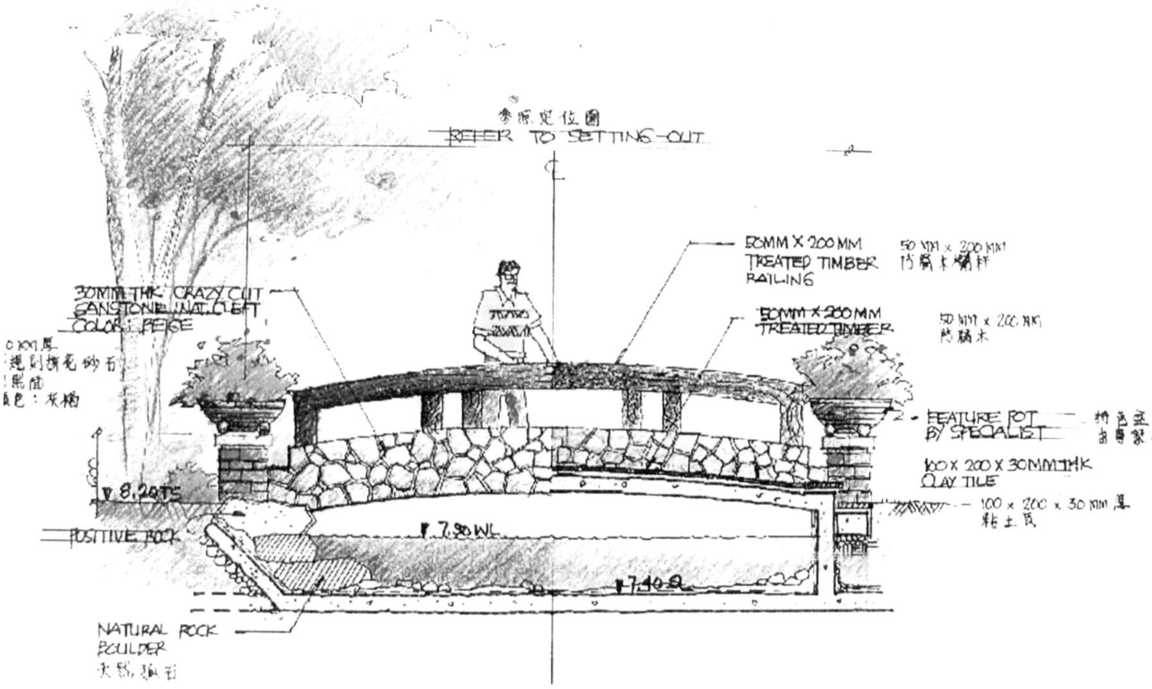
参照定位圖
REFER TO SETTING-OUT
50MM X 200 MM TREATED TIMBER RAILING
30MM THK CRAZY CUT SANDSTONE NAT. CLEFT COLOR: BEIGE
50MM X 200MM TREATED TIMBER
FEATURE POT BY SPECIALIST
100 X 200 X 30MM THK CLAY TILE
8.20TS
POSITIVE ROCK
7.90 WL
NATURAL ROCK BOULDER

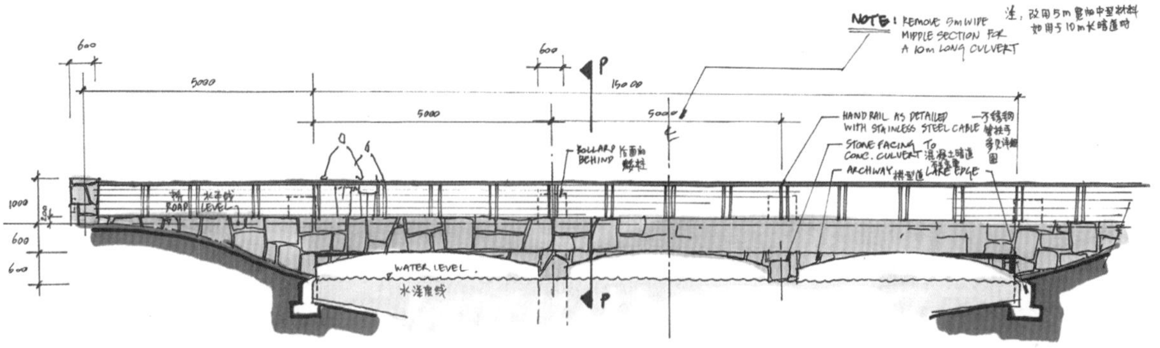
NOTE: REMOVE 5M WIDE MIDDLE SECTION FOR A 10M LONG CULVERT
600
5000
5000
600
15000
HANDRAIL AS DETAILED WITH STAINLESS STEEL CABLE
STONE FACING TO CONC. CULVERT
ARCHWAY
LAKE EDGE
BOLLARD BEHIND
ROAD LEVEL
WATER LEVEL
水位线
1000
600

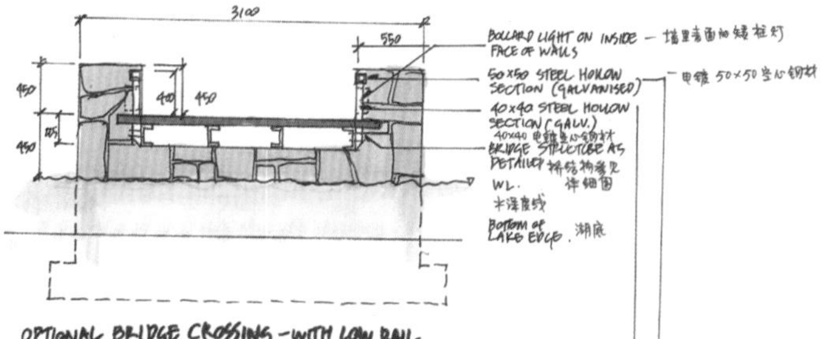
3100
550
450
BOLLARD LIGHT ON INSIDE FACE OF WALLS
50x50 STEEL HOLLOW SECTION (GALVANISED)
40x40 STEEL HOLLOW SECTION (GALV.)
BRIDGE STRUCTURE AS DETAILED
WL.
水位线
BOTTOM OF LAKE EDGE
湖底

OPTIONAL BRIDGE CROSSING - WITH LOW RAIL
SECTION DD. 1:25.

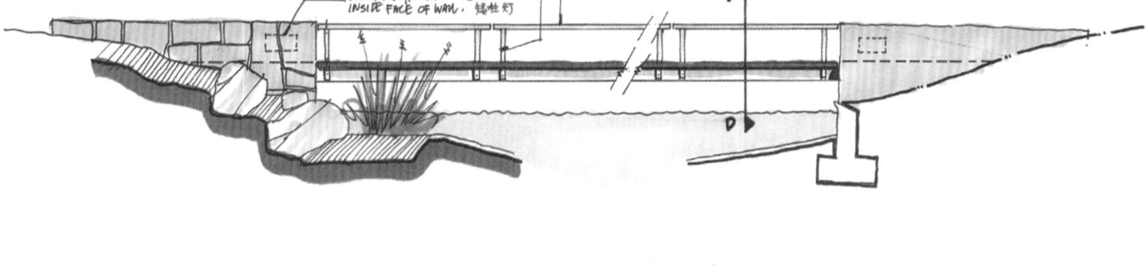
BOLLARD LIGHT ON INSIDE FACE OF WALL
D
D

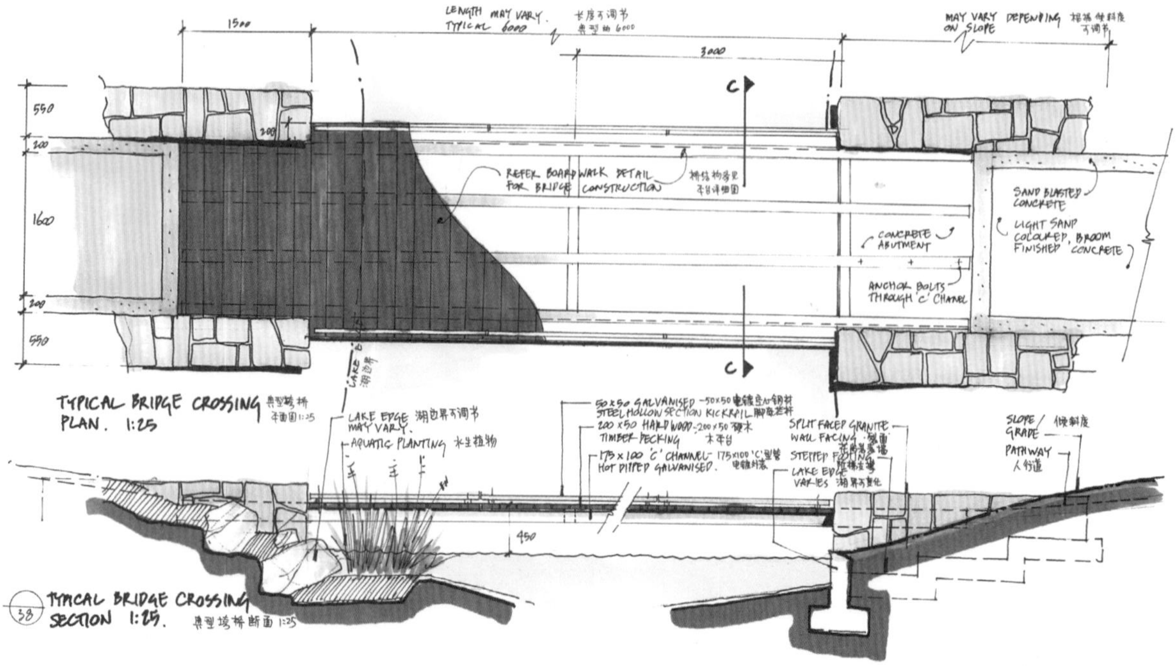
LENGTH MAY VARY. TYPICAL 6000
长度可调节 典型为6000
MAY VARY DEPENDING ON SLOPE
1500
3000
550
200
1600
200
550
C
REFER BOARDWALK DETAIL FOR BRIDGE CONSTRUCTION
SAND BLASTED CONCRETE
LIGHT SAND COLOURED, BROOM FINISHED CONCRETE
CONCRETE ABUTMENT
ANCHOR BOLTS THROUGH 'C' CHANNEL
TYPICAL BRIDGE CROSSING PLAN. 1:25
典型过桥 平面图1:25
LAKE EDGE MAY VARY.
湖边界可调节
AQUATIC PLANTING 水生植物
50×50 GALVANISED STEEL HOLLOW SECTION KICKRAIL
200×50 HARDWOOD TIMBER DECKING
木平台
175×100 'C' CHANNEL HOT DIPPED GALVANISED.
SPLIT FACED GRANITE WALL FACING
STEPPED FOOTING
LAKE EDGE VARIES
SLOPE/GRADE
PATHWAY
人行道
450
38
TYPICAL BRIDGE CROSSING SECTION 1:25.
典型过桥断面 1:25

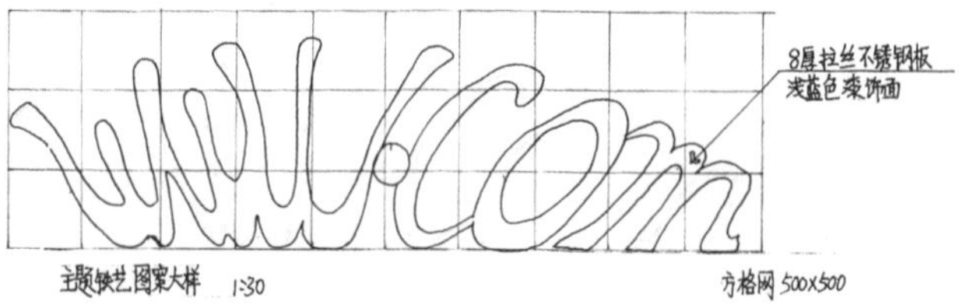
WWW.com
8厚拉丝不锈钢板
浅蓝色漆饰面
主题铁艺图案大样 1:30
方格网 500×500

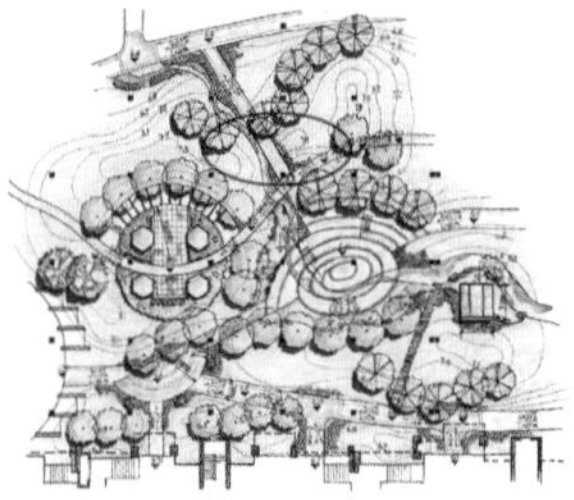

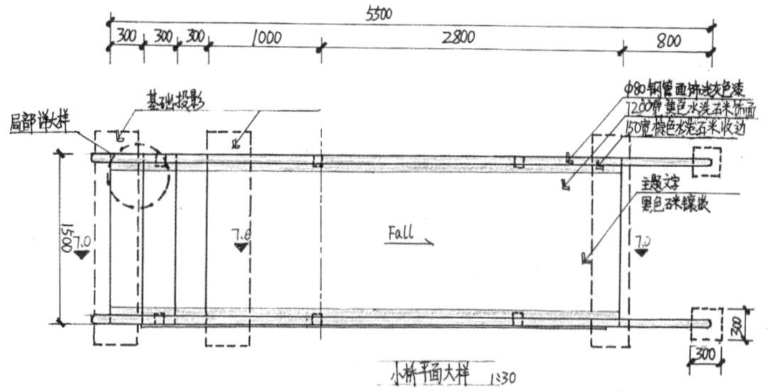
5300
300
300
300
1000
2800
800
局部详大样
基础投影
Φ80钢管面饰浅灰色漆
1200宽黄色水洗石米饰面
50宽棕色水洗石米收边
主题文字
黑色石米镶嵌
Fall
1500
7.0
7.0
7.0
300
300
小桥平面大样 1:30

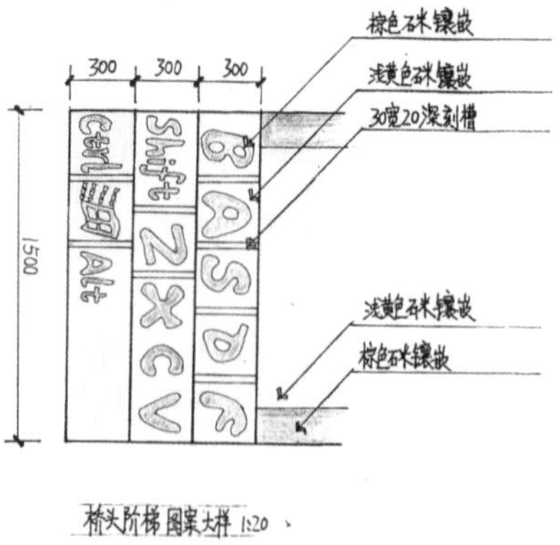
300
300
300
1500
Ctrl
Shift
Alt
Z
X
C
V
B
A
S
D
F
棕色石米镶嵌
浅黄色石米镶嵌
30宽20深刻槽
浅黄色石米镶嵌
棕色石米镶嵌
桥头阶梯图案大样 1:20

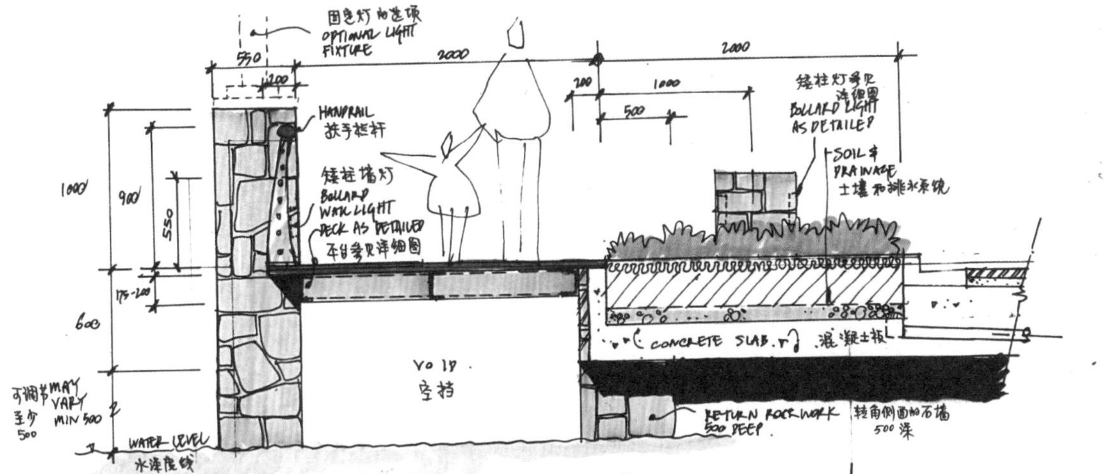

TYPICAL BRIDGE SECTION QQ
典型小桥QQ断面

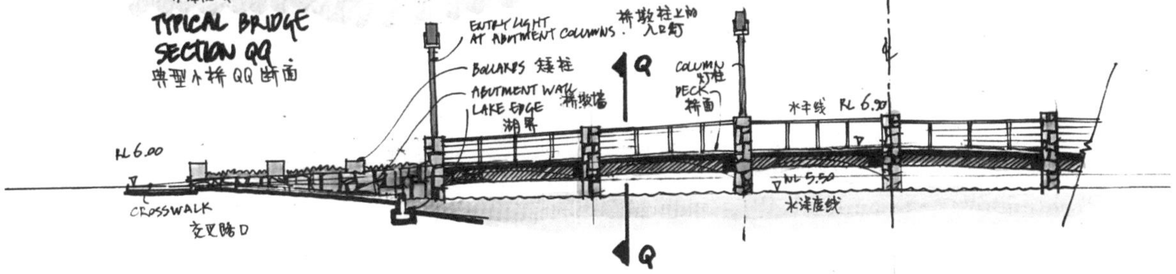

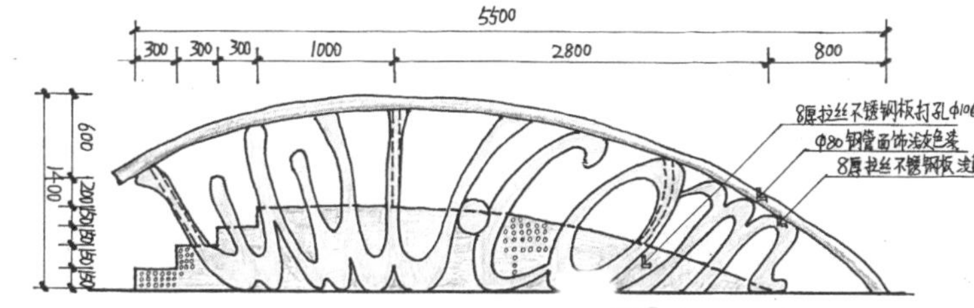

立面图 1:30

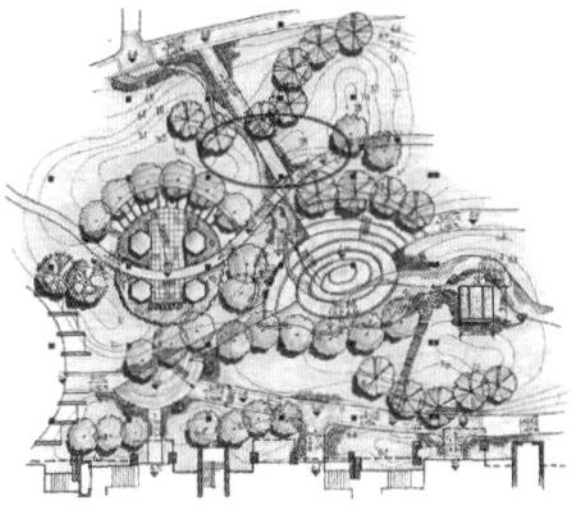

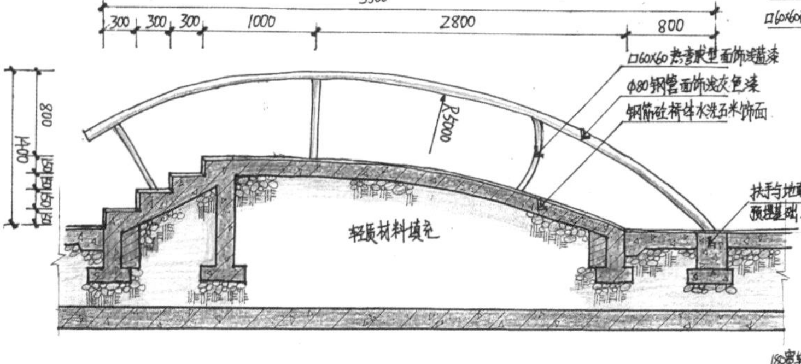

小桥剖面图A 1:30

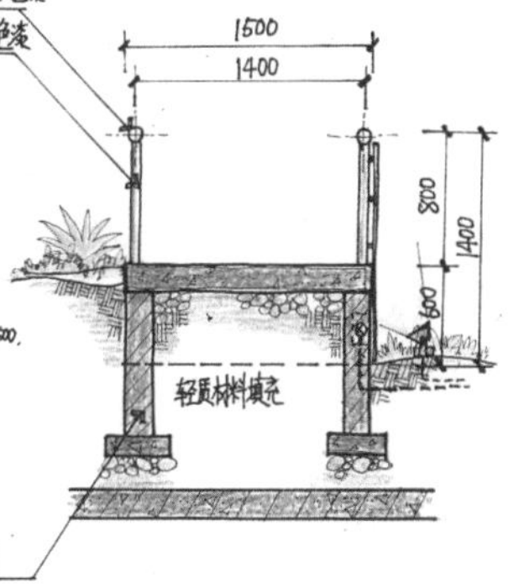

小桥剖面B 1:30

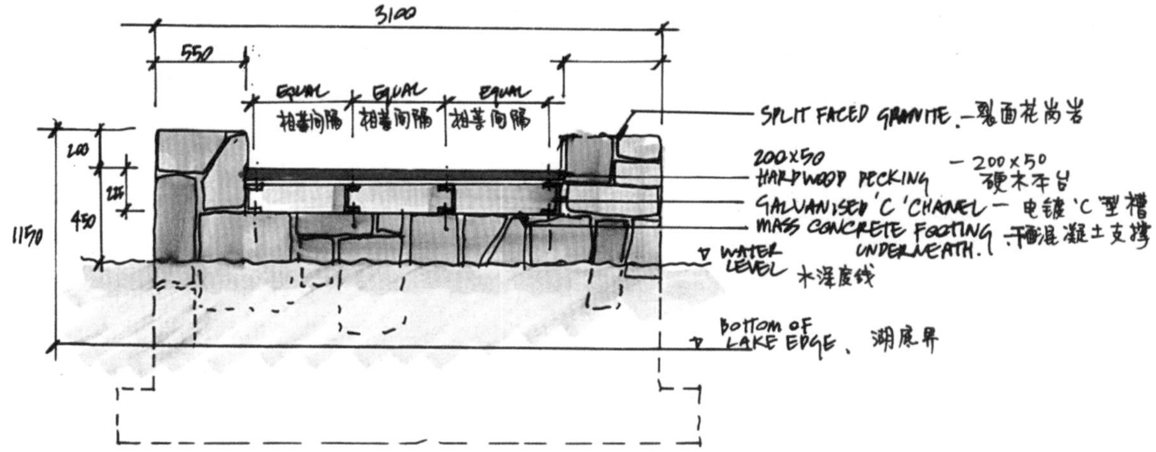
3100
550
EQUAL
EQUAL
EQUAL
相等间隔
相等间隔
相等间隔
200
225
450
1150
SPLIT FACED GRANITE —裂面花岗岩
200X50
HARDWOOD DECKING
—200×50
硬木平台
GALVANISED 'C' CHANEL —电镀'C'型槽
MASS CONCRETE FOOTING
UNDERNEATH.
下面混凝土支撑
WATER
LEVEL 水平度线
BOTTOM OF
LAKE EDGE、湖底界

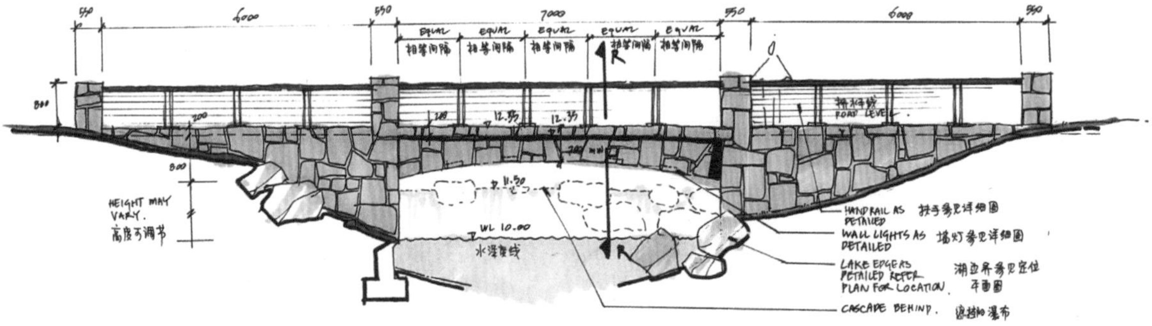
550
6000
550
7000
550
6000
550
EQUAL
EQUAL
EQUAL
EQUAL
EQUAL
相等间隔
相等间隔
相等间隔
相等间隔
相等间隔
HEIGHT MAY
VARY.
高度可调节
WL 10.00
水平度线
HANDRAIL AS
DETAILED
扶手参见详细图
WALL LIGHTS AS
DETAILED
墙灯参见详细图
LAKE EDGE AS
DETAILED REFER
PLAN FOR LOCATION
湖边界参见定位
平面图
CASCADE BEHIND.

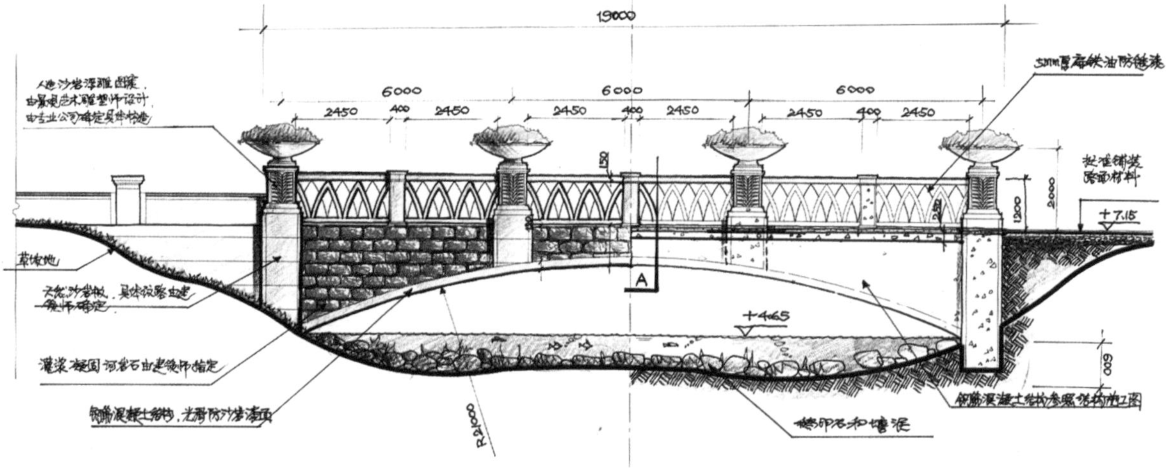
19000
6000
6000
6000
2450
400
2450
2450
400
2450
2450
400
2450
5mm厚瓷铁油防锈漆
拟准铺装
路面材料
1200
2000
+7.15
草坡地
灌浆凝固河岩石由建筑师指定
+4.65
600
R21000
钢筋混凝土结构参照结构施工图

碧浮桥立面图

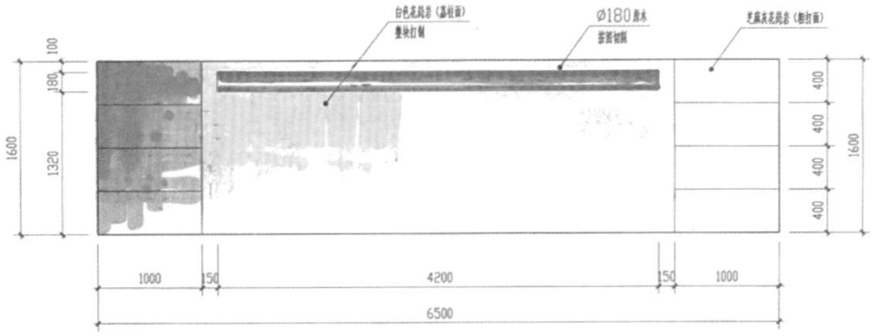

① 石桥平面图 1:30

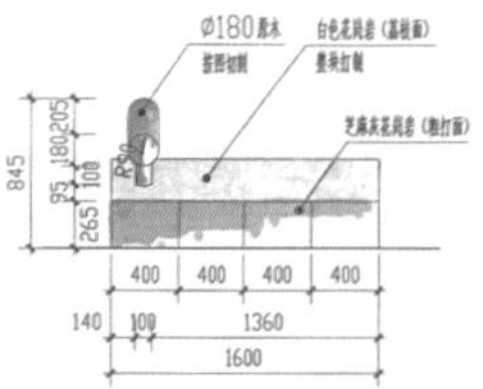

② 石桥侧立面图 1:30

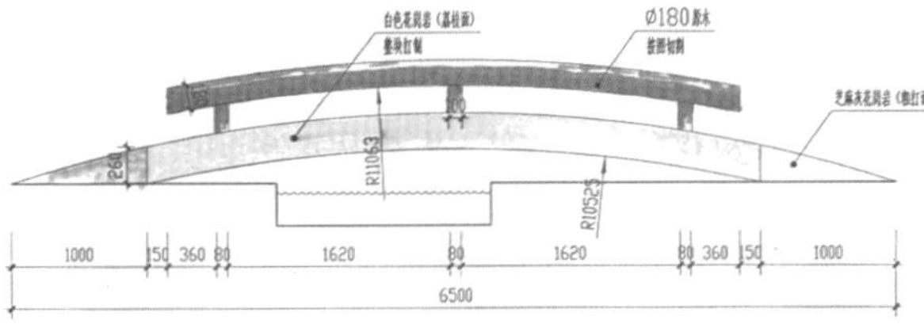

③ 石桥立面图 1:30

设计说明：
1.图中所注尺寸单位为毫米
2.图中所示石构均为整块石材打制

TIMBER BRIDGE

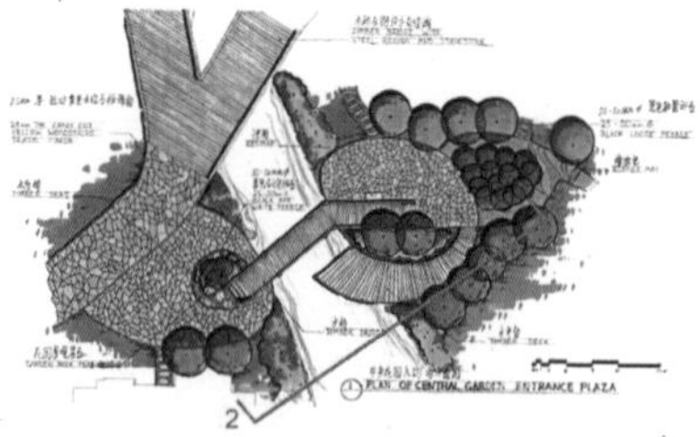

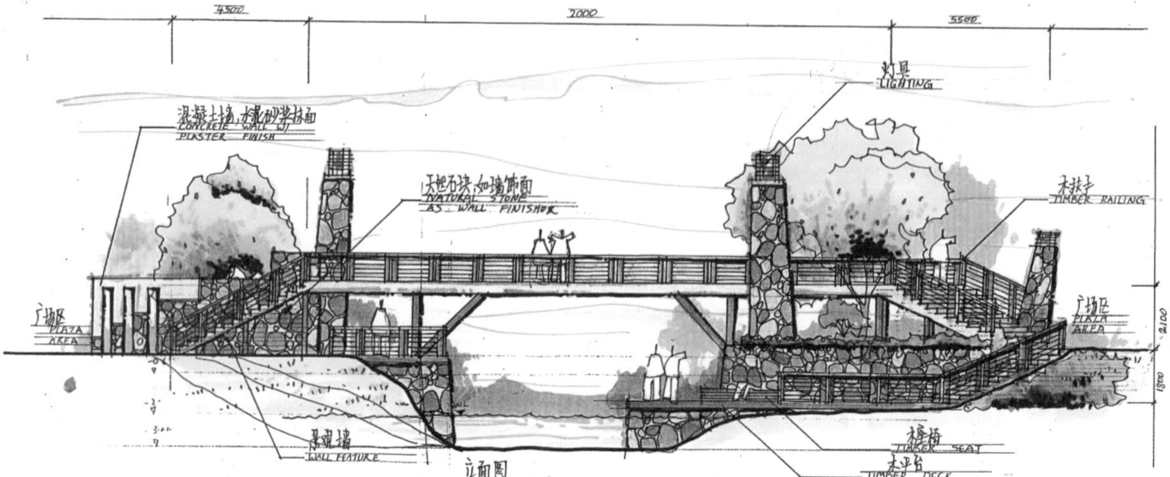

② 立面图 ELEVATION
SCALE 1:100

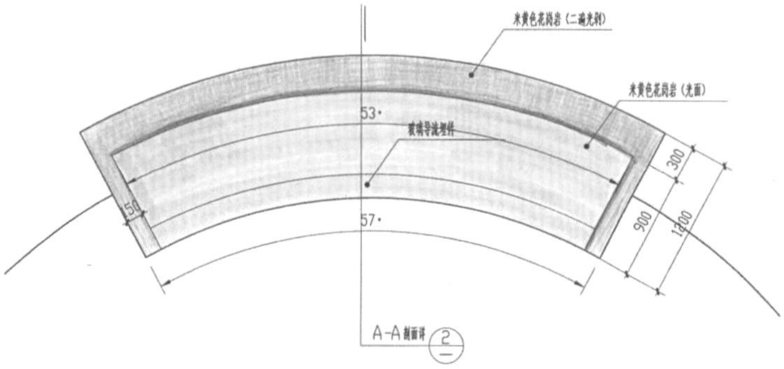

① 水幔平面图 1:30

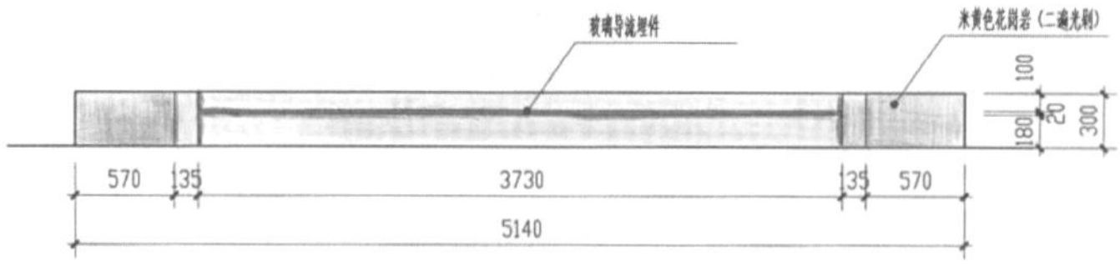

④ 水幔立面图 1:30

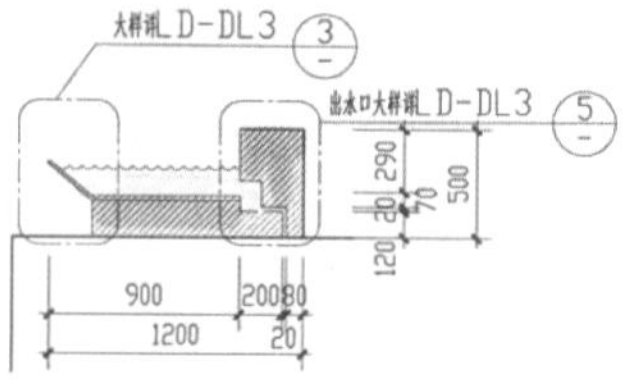

② A-A剖面图 1:30

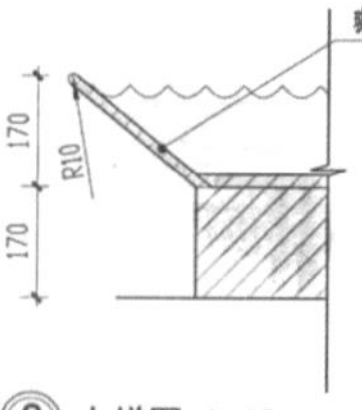

③ 大样图 1:10

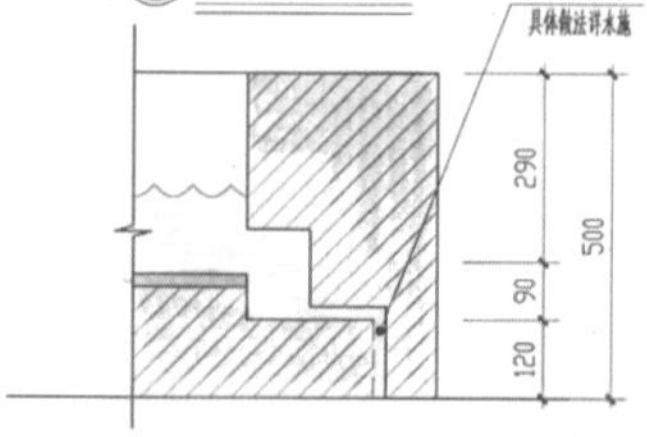

⑤ 出水口大样图 1:10

设计说明:
1.图中所注尺寸单位为毫米

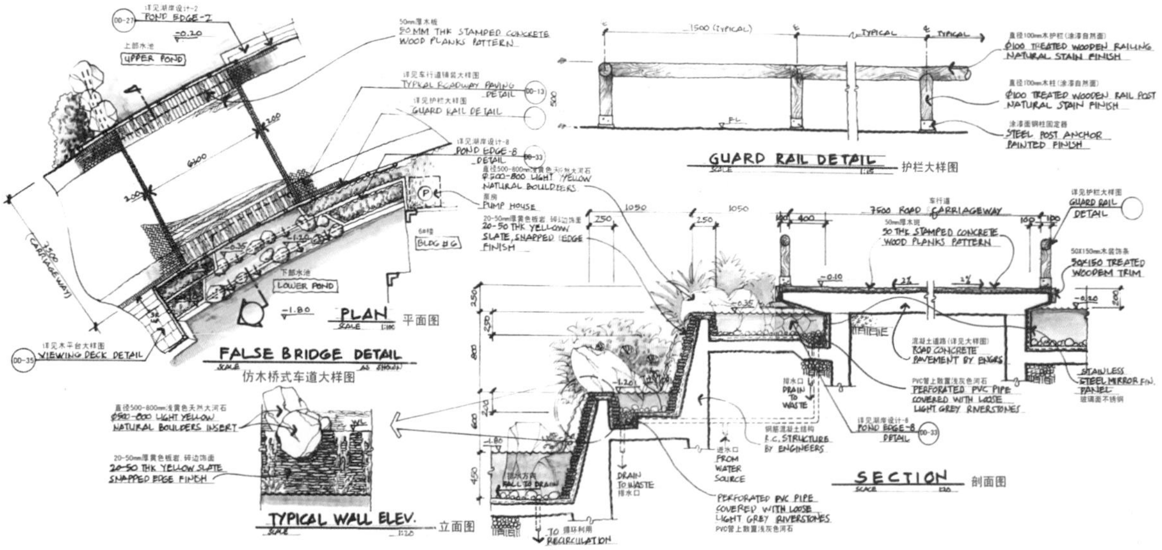

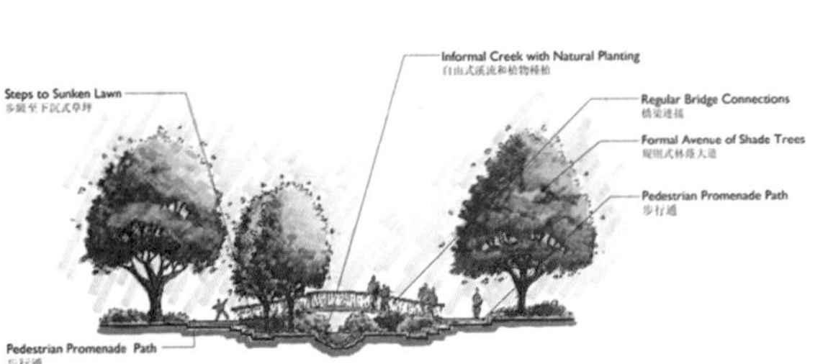

SECTION 1 - TYPICAL PEDESTRIAN PROMENADE AND CENTRAL OPEN SPACE

剖面 1 - 代表性步行道和中心開放空間

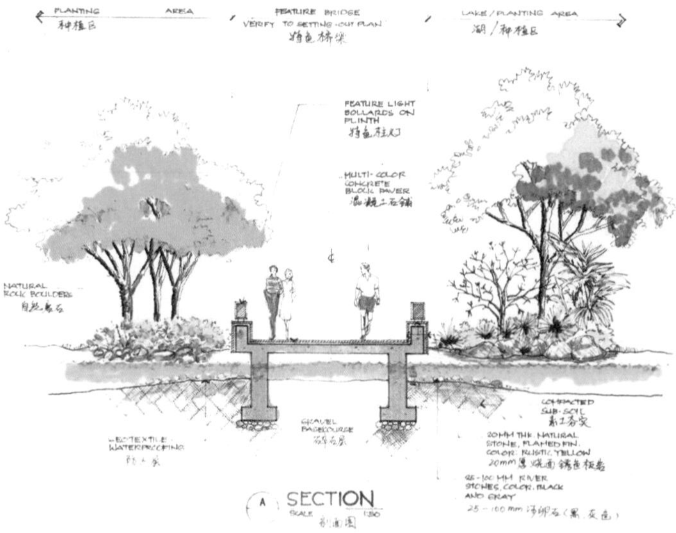

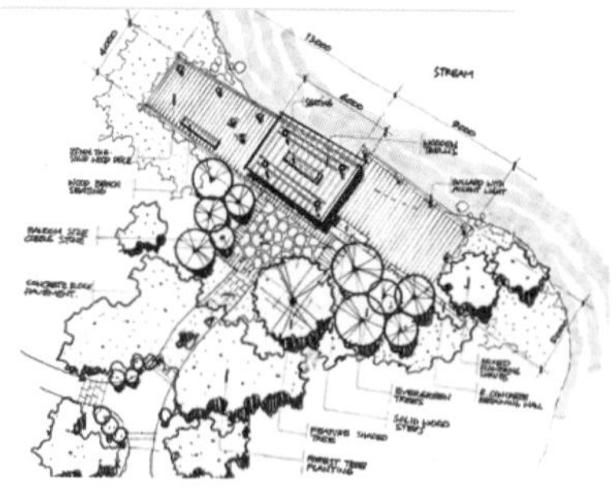

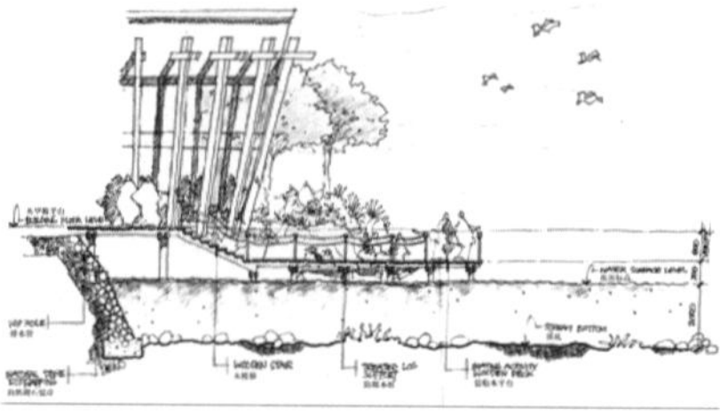

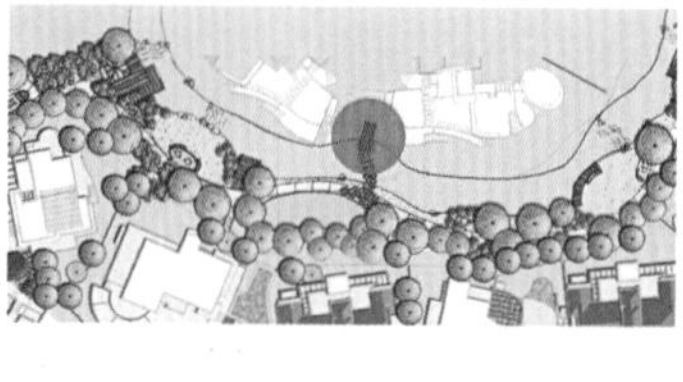

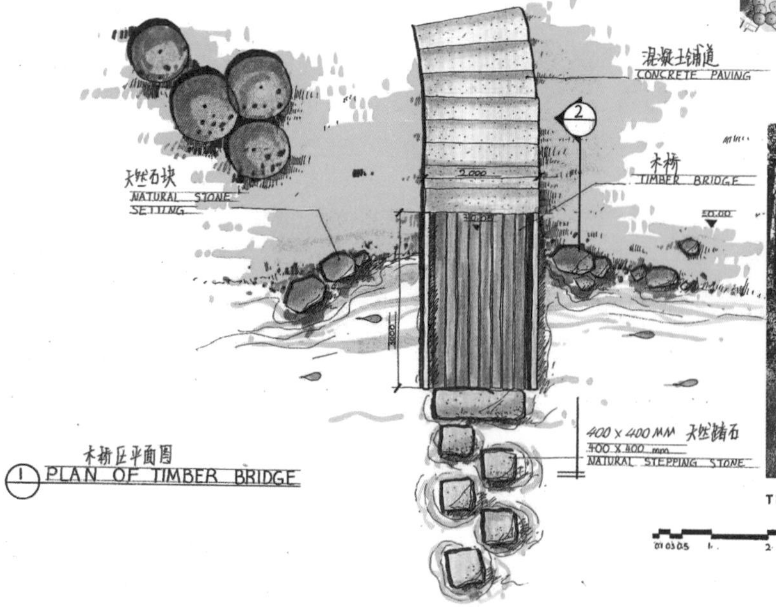

木桥区平面图

1 PLAN OF TIMBER BRIDGE

TIMBER BRIDGE

NATURAL STEPPING STONE

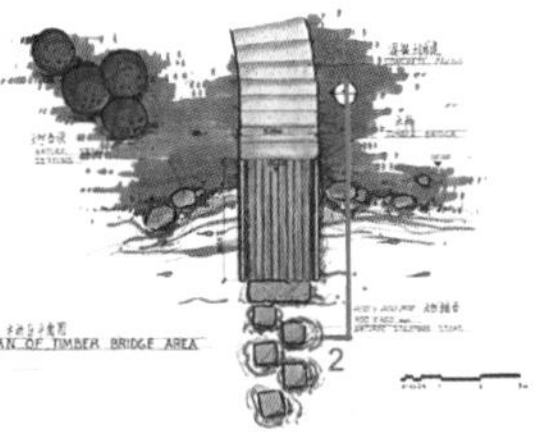
PLAN OF TIMBER BRIDGE AREA
2

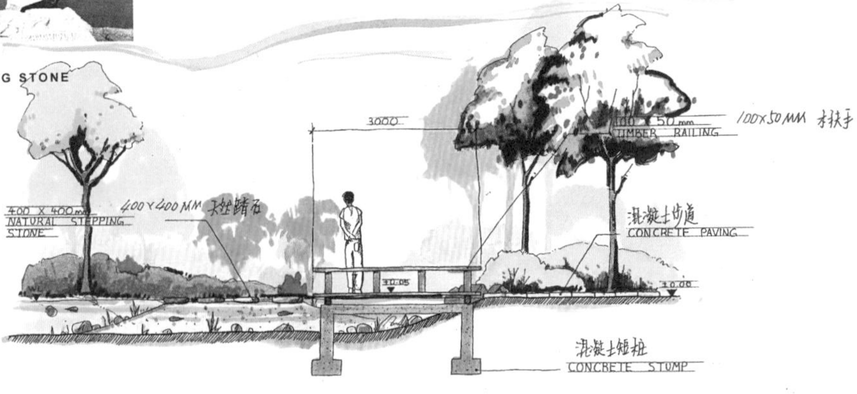
3000
100 X 50 mm
TIMBER RAILING
100X50MM 木扶手
400 X 400 mm
NATURAL STEPPING STONE
400X400MM 天然踏石
混凝土步道
CONCRETE PAVING
混凝土短柱
CONCRETE STUMP

剖立面图
2 SECTION THRU TIMBER BRIDGE

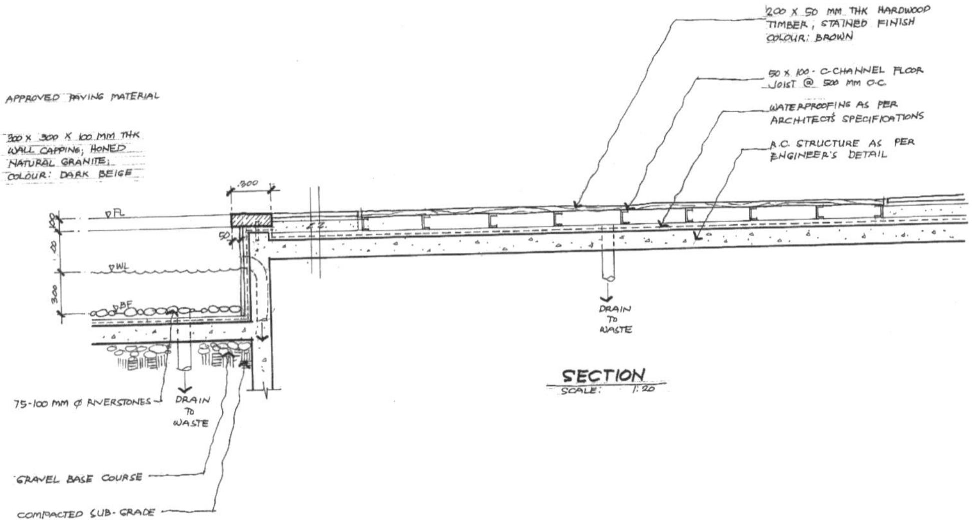
200 X 50 MM THK HARDWOOD TIMBER, STAINED FINISH COLOUR: BROWN
50 X 100 - C-CHANNEL FLOOR JOIST @ 500 MM O.C.
WATERPROOFING AS PER ARCHITECTS SPECIFICATIONS
R.C. STRUCTURE AS PER ENGINEER'S DETAIL
APPROVED PAVING MATERIAL
300 X 300 X 100 MM THK WALL CAPPING; HONED NATURAL GRANITE; COLOUR: DARK BEIGE
300
DRAIN TO WASTE
75-100 MM Ø RIVERSTONES
DRAIN TO WASTE
GRAVEL BASE COURSE
COMPACTED SUB-GRADE
SECTION
SCALE: 1:20

NATURAL STEPPING STONE

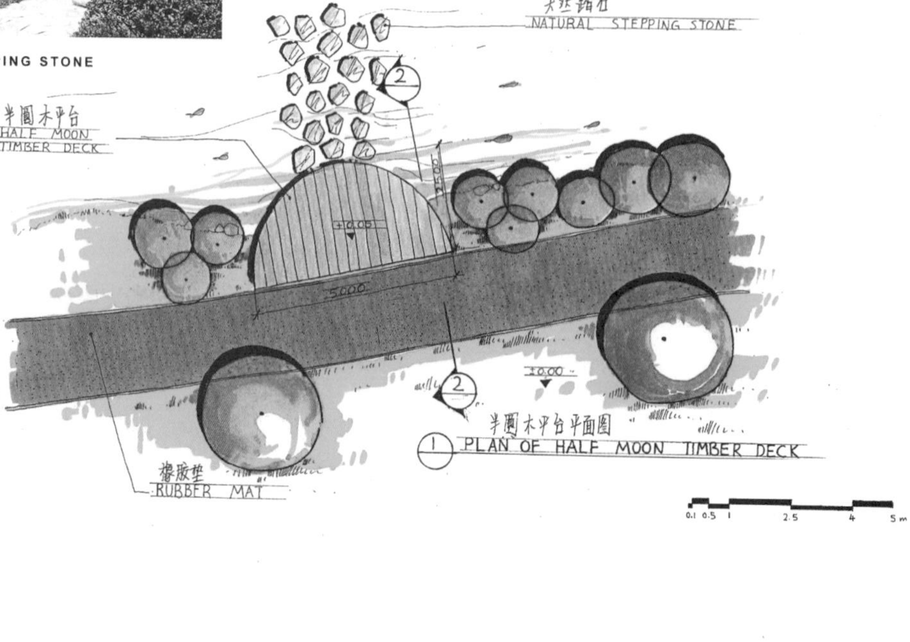

NATURAL STEPPING STONE

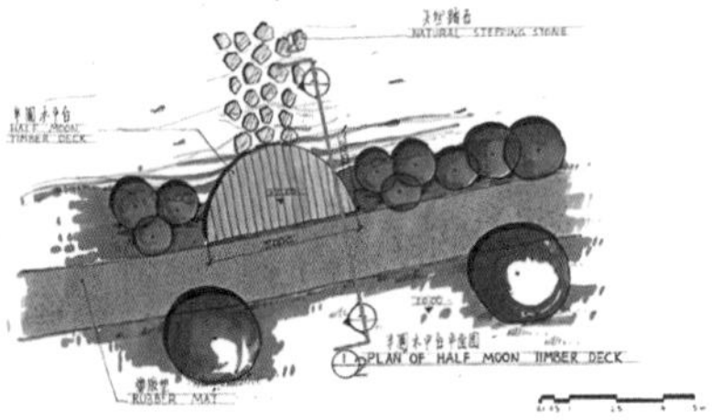

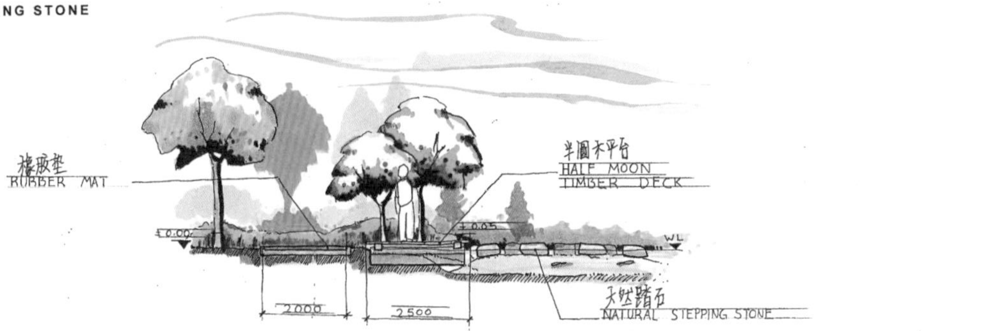

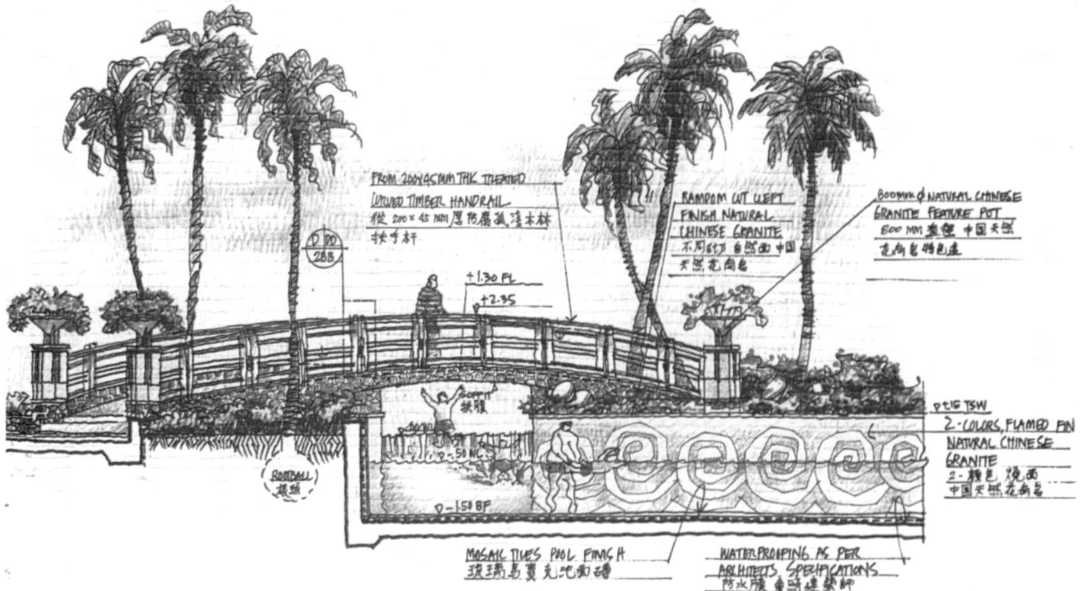

剖面图/立面图

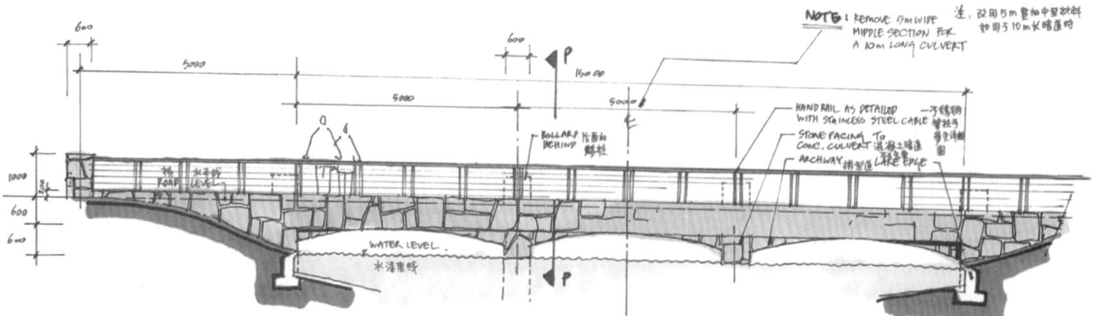

41 TYPICAL 15 m LONG CULVERT
ELEVATION 1:50
OPTIONAL 10m CULVERT.

典型的15 m长暗道（或10 m长）
立面图 1:50

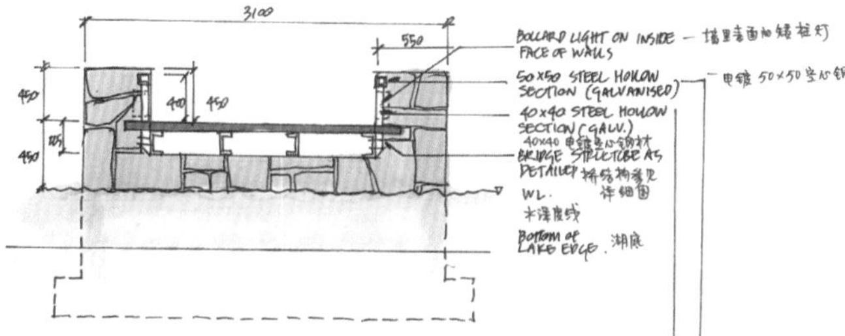

OPTIONAL BRIDGE CROSSING - WITH LOW RAIL
SECTION DD. 1:25.

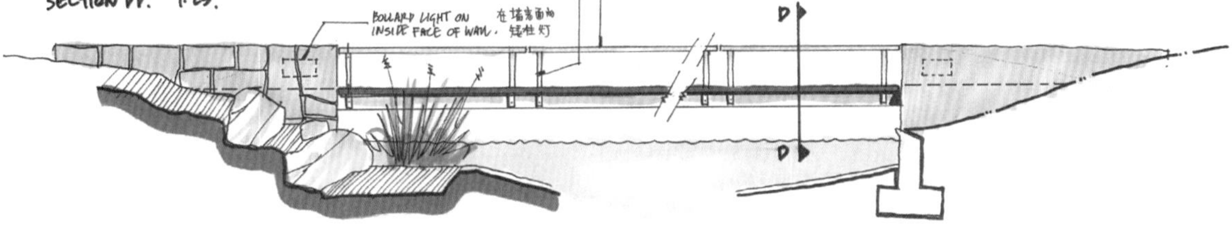

39 OPTION BRIDGE CROSSING - WITH LOW RAIL. 跨桥选项低栏杆
ELEVATION 1:25. 立面图 1:25

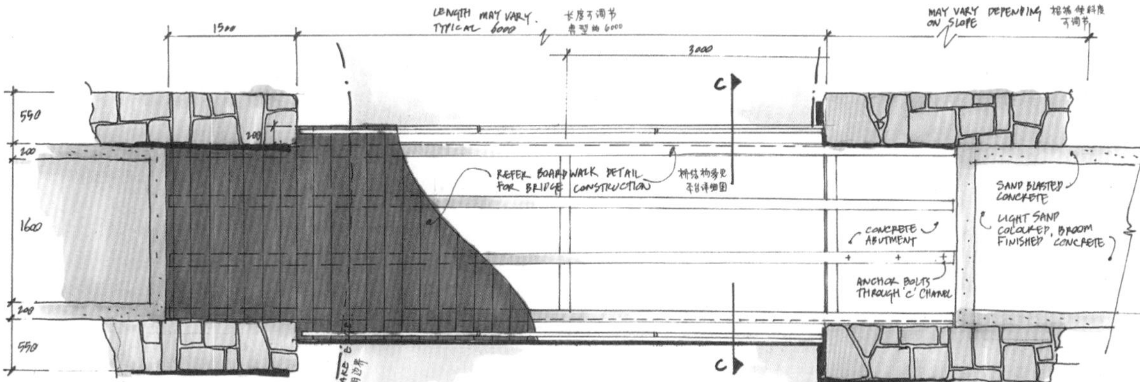

TYPICAL BRIDGE CROSSING 典型跨桥
PLAN. 1:25 平面图 1:25

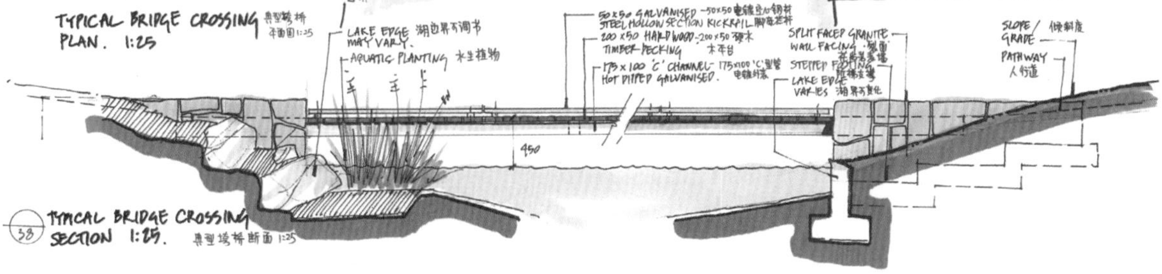

38 TYPICAL BRIDGE CROSSING
SECTION 1:25. 典型跨桥断面 1:25

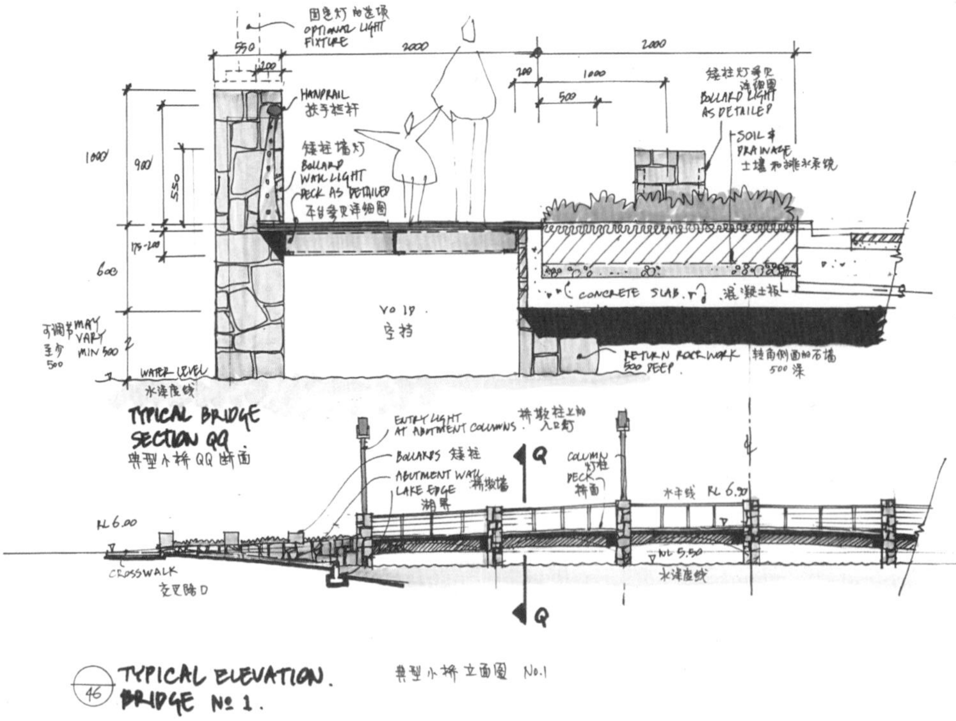
可选灯加选项
OPTIONAL LIGHT FIXTURE
HANDRAIL
扶手栏杆
矮柱墙灯
BOLLARD WALL LIGHT
DECK AS DETAILED
桥面参见详细图
矮柱灯参见详细图
BOLLARD LIGHT AS DETAILED
SOIL & DRAINAGE
土壤和排水系统
VOID
空档
CONCRETE SLAB
混凝土板
WATER LEVEL
水源底线
RETURN ROCKWORK 500 DEEP
转角侧面的石墙 500深
TYPICAL BRIDGE SECTION QQ
典型小桥QQ断面
ENTRY LIGHT AT ABUTMENT COLUMNS
BOLLARDS 矮柱
ABUTMENT WALL
LAKE EDGE 湖界
COLUMN 灯柱
DECK 桥面
CROSSWALK
交叉路口
水源底线
TYPICAL ELEVATION. BRIDGE № 1.
典型小桥立面图 No.1

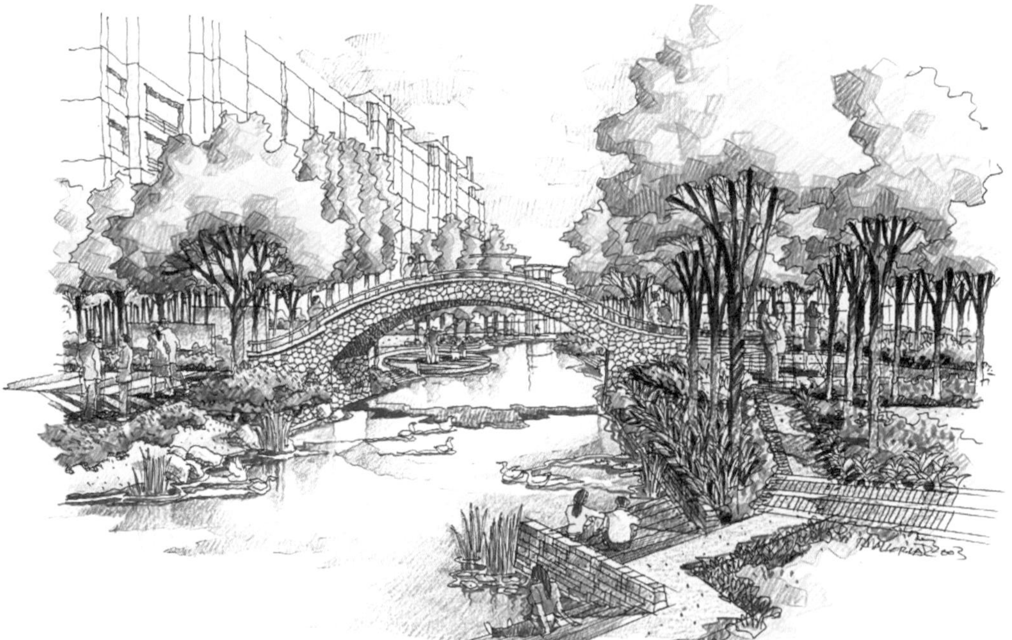

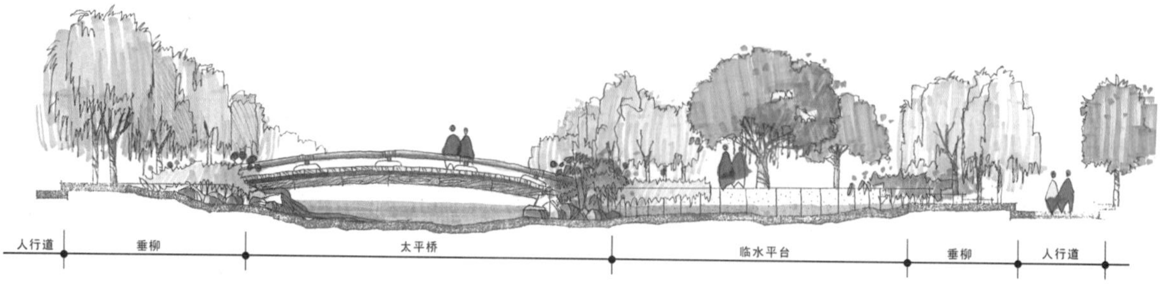
人行道
垂柳
太平桥
临水平台
垂柳
人行道

A-A 剖面

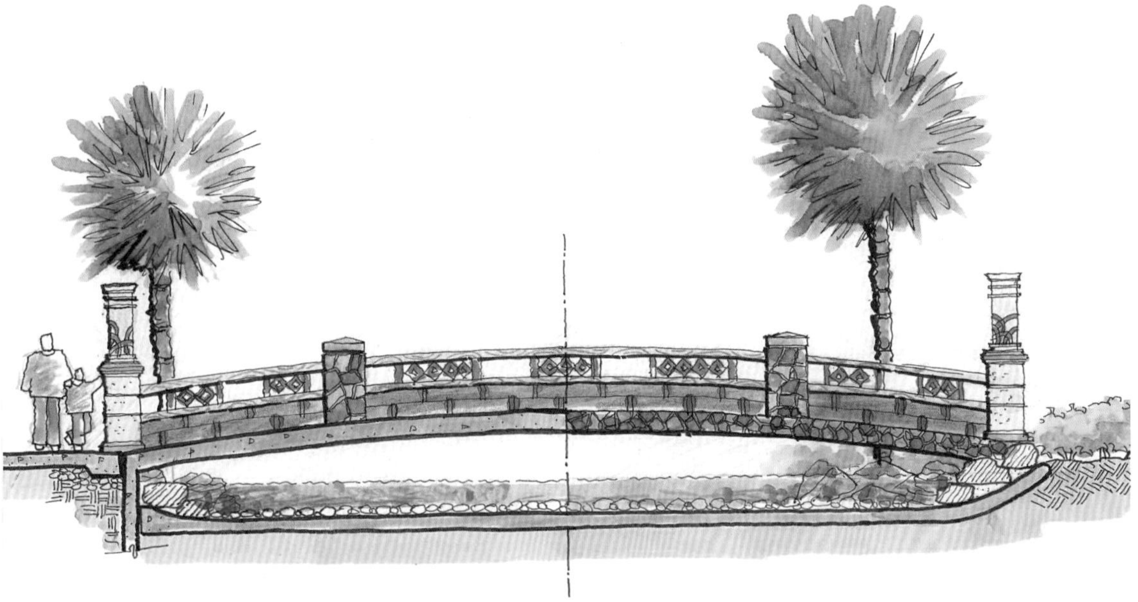

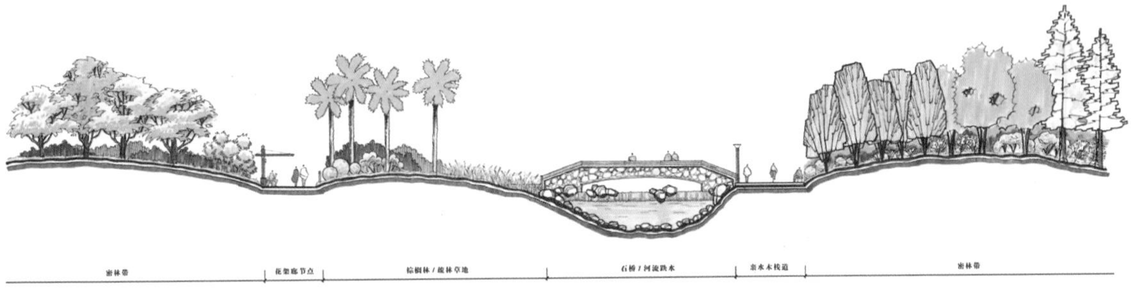

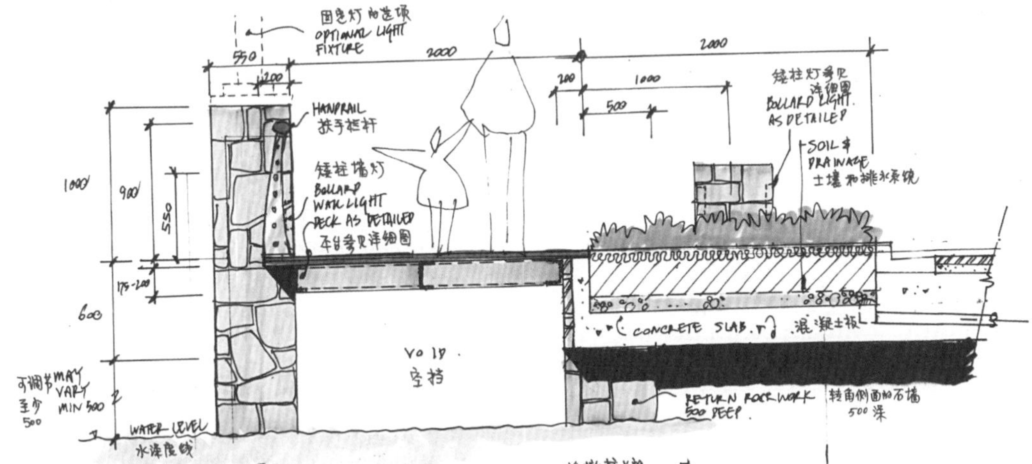

TYPICAL BRIDGE SECTION QQ
典型小桥 QQ 断面

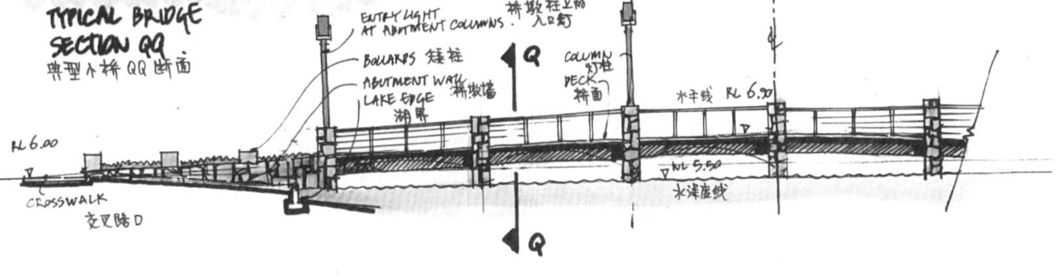

(46) **TYPICAL ELEVATION. BRIDGE No 1.**
典型小桥立面图 No.1

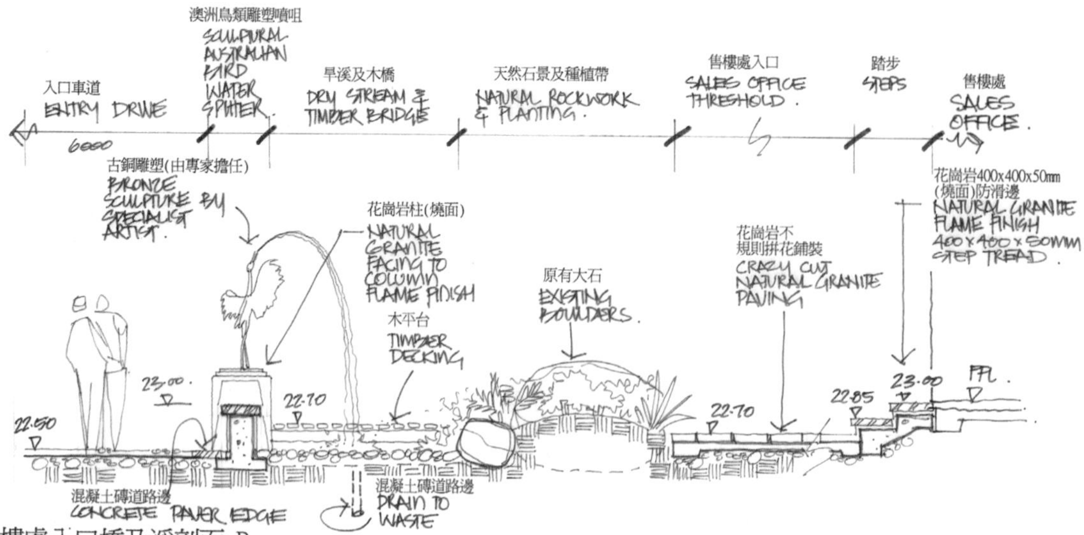

入口車道
ENTRY DRIVE
6000
澳洲鳥類雕塑噴咀
SCULPTURAL AUSTRALIAN BIRD WATER SPITTER
旱溪及木橋
DRY STREAM & TIMBER BRIDGE
天然石景及種植帶
NATURAL ROCKWORK & PLANTING.
售樓處入口
SALES OFFICE THRESHOLD.
踏步
STEPS
售樓處
SALES OFFICE.
古銅雕塑(由專家擔任)
BRONZE SCULPTURE BY SPECIALIST ARTIST.
花崗岩柱(燒面)
NATURAL GRANITE FACING TO COLUMN FLAME FINISH
木平台
TIMBER DECKING
原有大石
EXISTING BOULDERS.
花崗岩不規則拼花鋪裝
CRAZY CUT NATURAL GRANITE PAVING
花崗岩400x400x50mm (燒面)防滑邊
NATURAL GRANITE FLAME FINISH 400 x 400 x 50MM STEP TREAD.
22.50
23.00
22.70
22.70
22.85
23.00
FFL.
混凝土磚道路邊
CONCRETE PAVER EDGE
混凝土磚道路邊
DRAIN TO WASTE
售樓處入口橋及溪剖面-B
SALES OFFICE ENTRY BRIDGE & STREAM
SECTION B
1:30.

详1C-2
1C平面大样 1:50
1750
1C-1剖面大样 1:50
1750
1C-2平面 1:15
1C-2立面 1:15

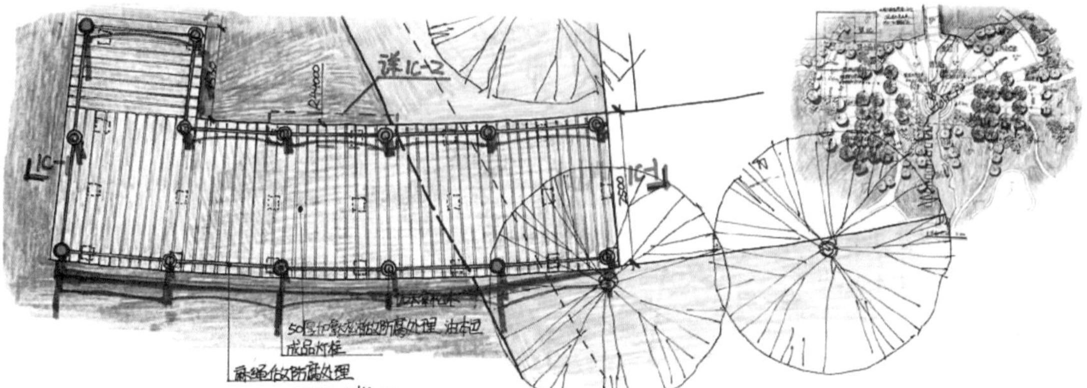

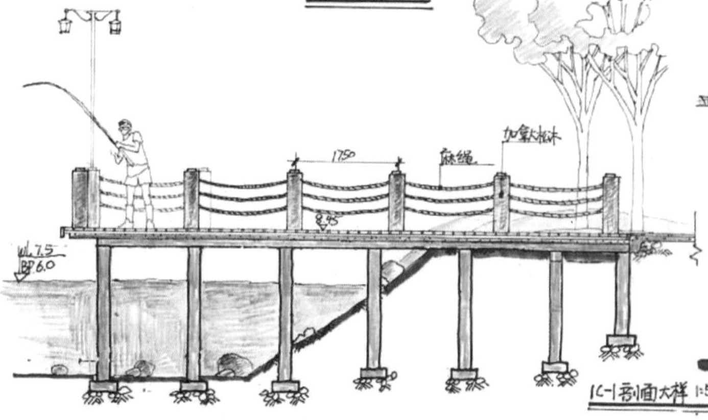

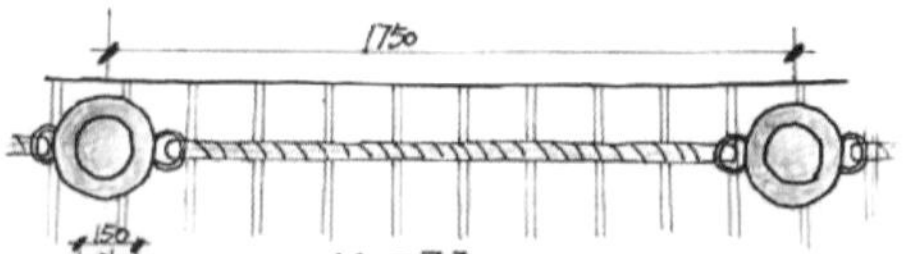

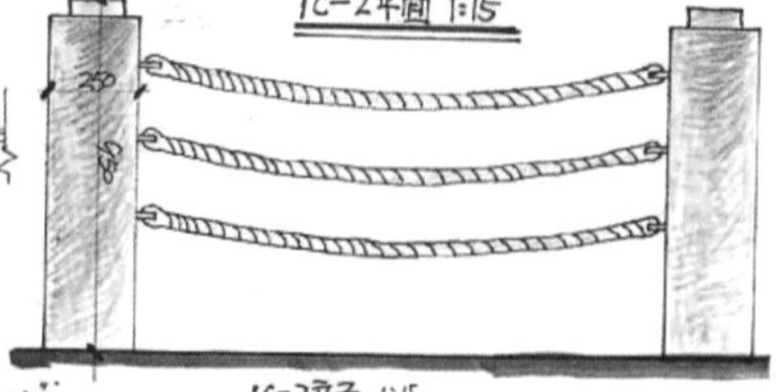

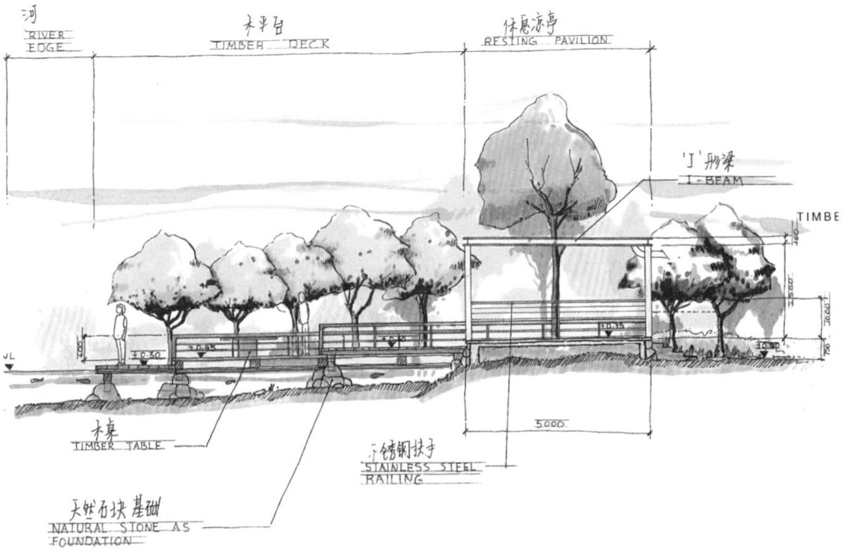

河
RIVER EDGE
木平台
TIMBER DECK
休憩凉亭
RESTING PAVILION
'J'形梁
I-BEAM
TIMBE
5000
木桌
TIMBER TABLE
不锈钢扶手
STAINLESS STEEL RAILING
天然石块基础
NATURAL STONE AS FOUNDATION

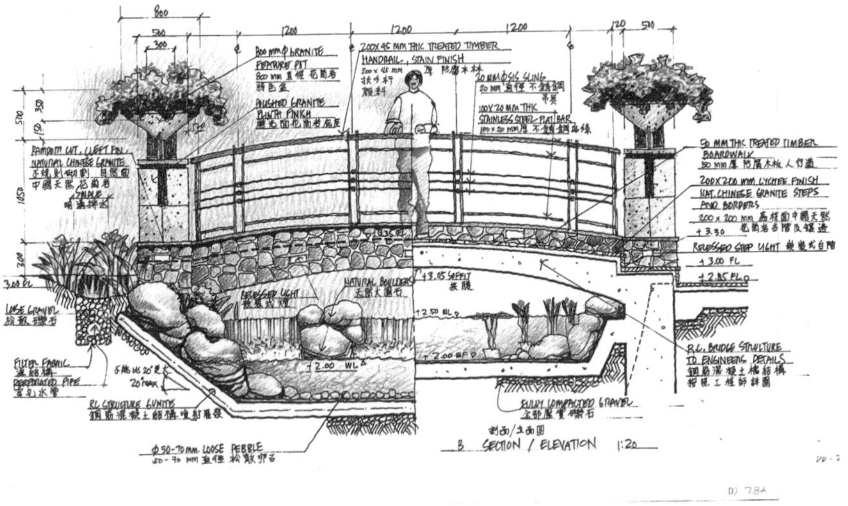

800
1200
1200
1200
200X45 MM THK TREATED TIMBER HANDRAIL, STAIN FINISH
50 MM THK TREATED TIMBER BOARDWALK
RECESSED STEP LIGHT
+3.00 FL
+2.85 FL
NATURAL BOULDERS
LOOSE GRAVEL
FILTER FABRIC
PERFORATED PIPE
RC STRUCTURE GUNITE
R.C. BRIDGE STRUCTURE TO ENGINEER'S DETAILS
ϕ50-70 MM LOOSE PEBBLE
剖面/立面图
B SECTION / ELEVATION 1:20

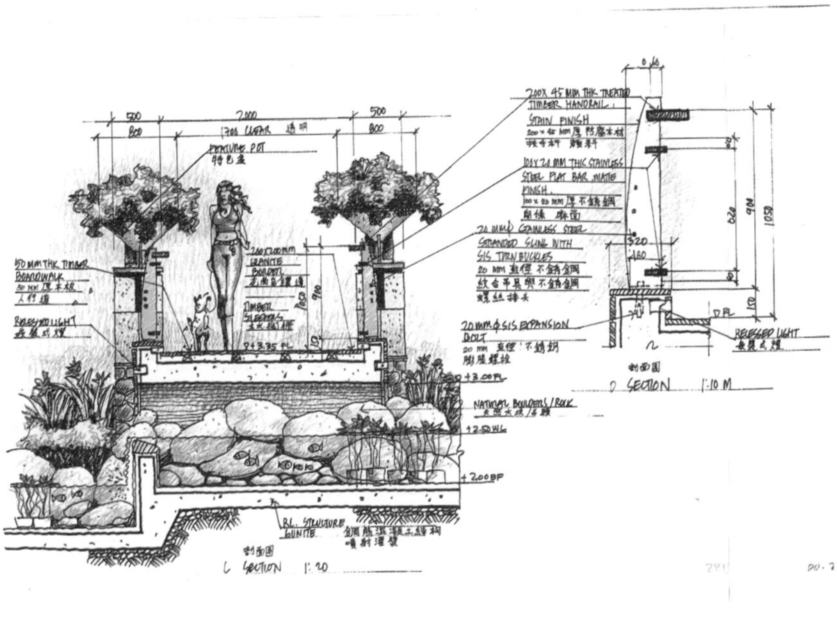

500
2000
500
800
800
FEATURE POT
特色盆
200X45 MM THK TREATED TIMBER HANDRAIL, STAIN FINISH
50 MM THK TIMBER BOARDWALK
RECESSED LIGHT
+3.00 FL
NATURAL BOULDERS / ROCK
+2.50 WL
+2.00 BF
RC STRUCTURE GUNITE
剖面图
C SECTION 1:20
剖面图
D SECTION 1:10 M
1050

桥

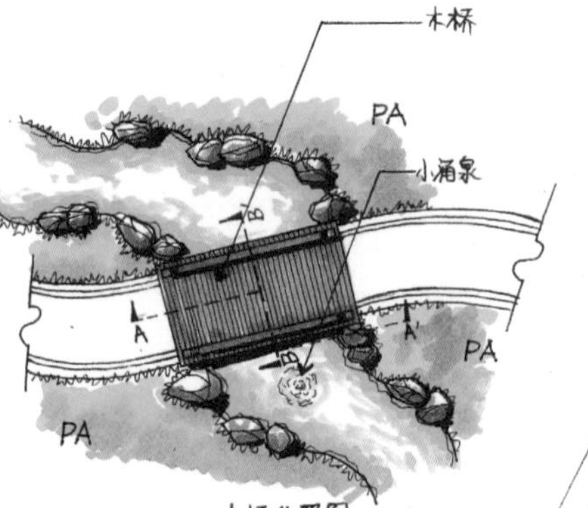

木桥
PA
小涌泉
Φ150MM木栏杆(菠萝格)
经防腐处理,上清漆
200MMX2500MM木板(菠萝格)
经防腐处理,上清漆
Φ20MM麻绳
4600
4400
120
2500
木桥位置图 scale 1:100
木桥平面图 scale 1:50
不锈钢套
Φ100MM木桩(菠萝格)
经防腐处理
钢筋混凝土结构
由专业工程师定
盆栽水生植物
A-A'剖面图 scale 1:50
B-B'剖面图 scale 1:50

景桥及特色喷水口详图
PLAN: BRIDGE
1:40 M.
景桥平面图
DETAIL: SPOUT
特色喷水口详图
DETAIL: RAILING
1:20 M.
扶手详图
1200
深圳·中航"阳光新苑"环境扩初设计
SZ002/DD19
L&A

特色景桥详图
CLAY TILE, COLOR: ANTIQUE BROWN
PART OF POND
LARGELY LAID 20-30MM RIVERSTONE COLOR: GREY
PLAN
景桥平面图
14000
75MMØ TREATED WOOD WEATHERED FIN.
85MMØ METAL BRACKET BRONZE FIN.
PRE-CAST CONC. BLOCK, ROUGH FINISH
NATURAL ROCK BOULDERS
ELEV/SECTION
景桥剖立面图
深圳·中航"阳光新苑"环境扩初设计
SZ002/DD26
L&A

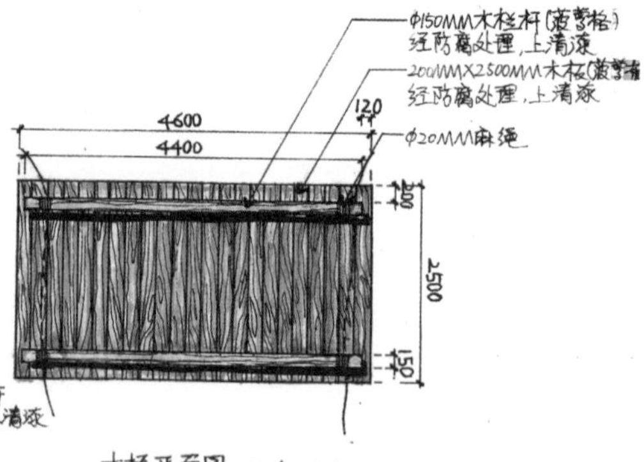

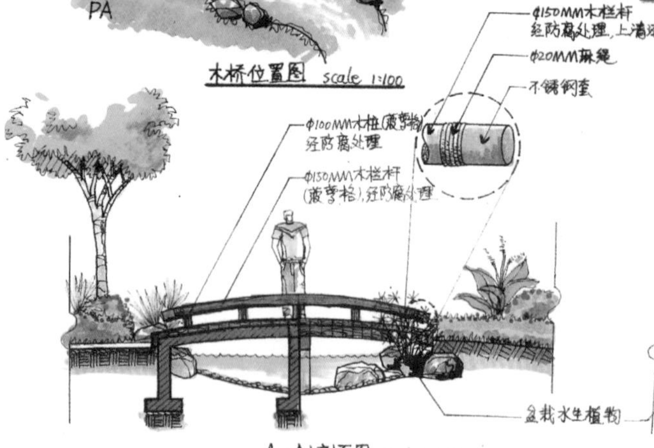

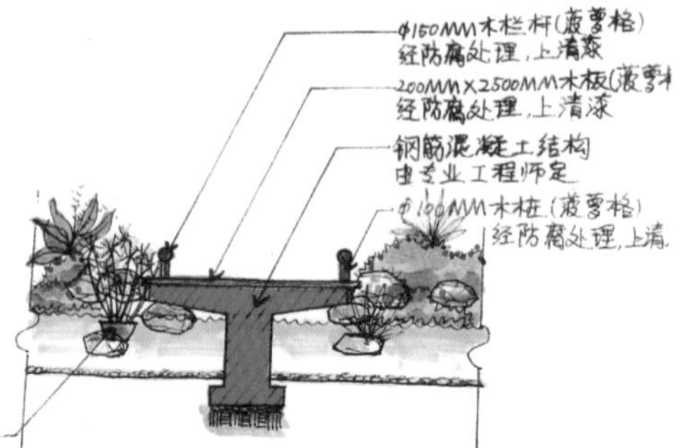

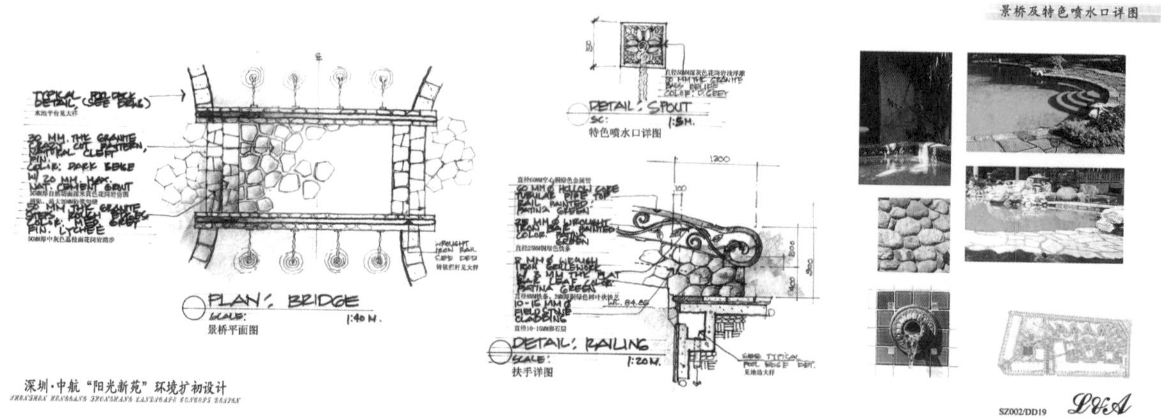

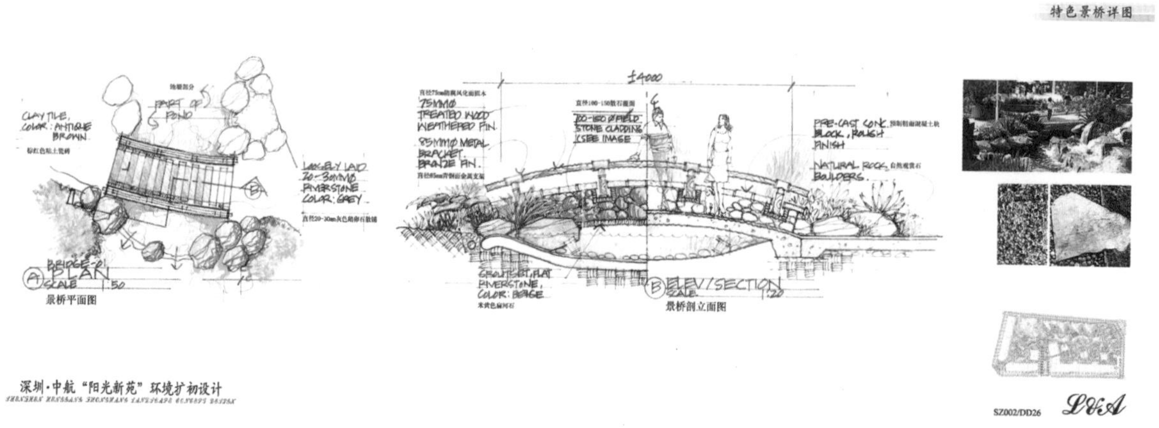

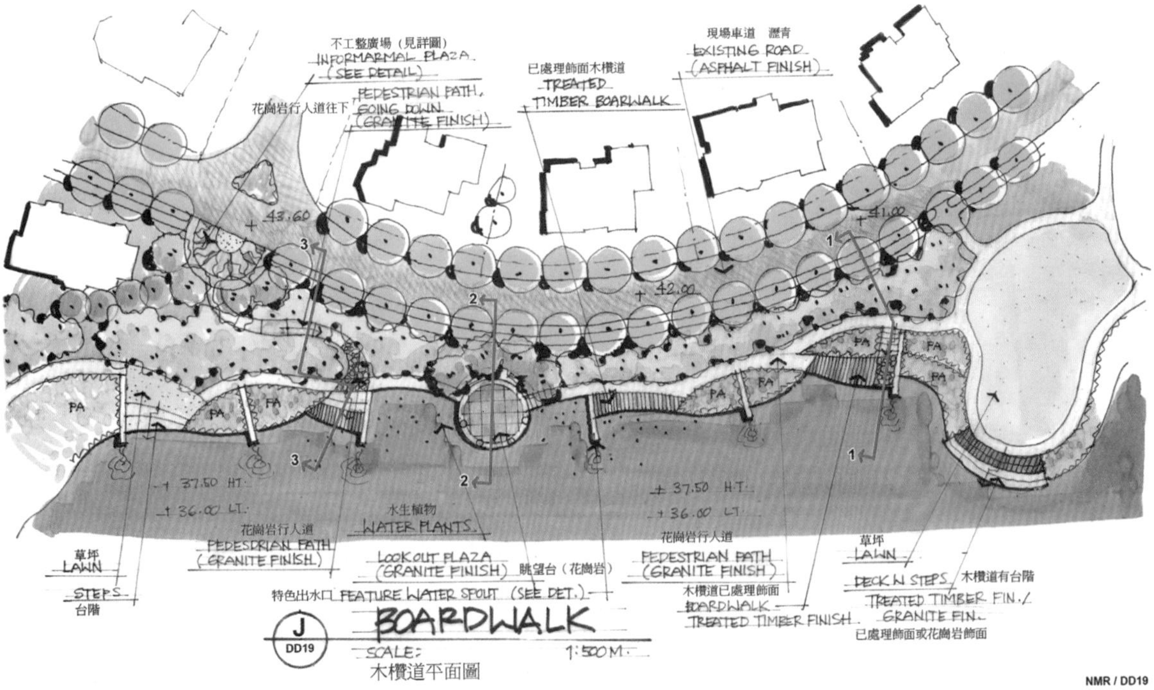
不工整廣場（見詳圖）
INFORMARMAL PLAZA (SEE DETAIL)
花崗岩行人道往下
PEDESTRIAN PATH, GOING DOWN (GRANITE FINISH)
已處理飾面木欖道
TREATED TIMBER BOARWALK
現場車道 瀝青
EXISTING ROAD (ASPHALT FINISH)
43.60
42.00
41.00
PA
± 37.50 HT.
± 36.00 LT.
水生植物
WATER PLANTS
花崗岩行人道
PEDESDRIAN PATH (GRANITE FINISH)
草坪
LAWN
STEPS
台階
LOOKOUT PLAZA (GRANITE FINISH) 眺望台（花崗岩）
特色出水口 FEATURE WATER SPOUT (SEE DET.)
PEDESTRIAN PATH (GRANITE FINISH)
木欖道已處理飾面
BOARDWALK
TREATED TIMBER FINISH
DECK N STEPS 木欖道有台階
TREATED TIMBER FIN./ GRANITE FIN.
已處理飾面或花崗岩飾面
J
DD19
BOARDWALK
SCALE: 1:500M
木欖道平面圖

NMR / DD19

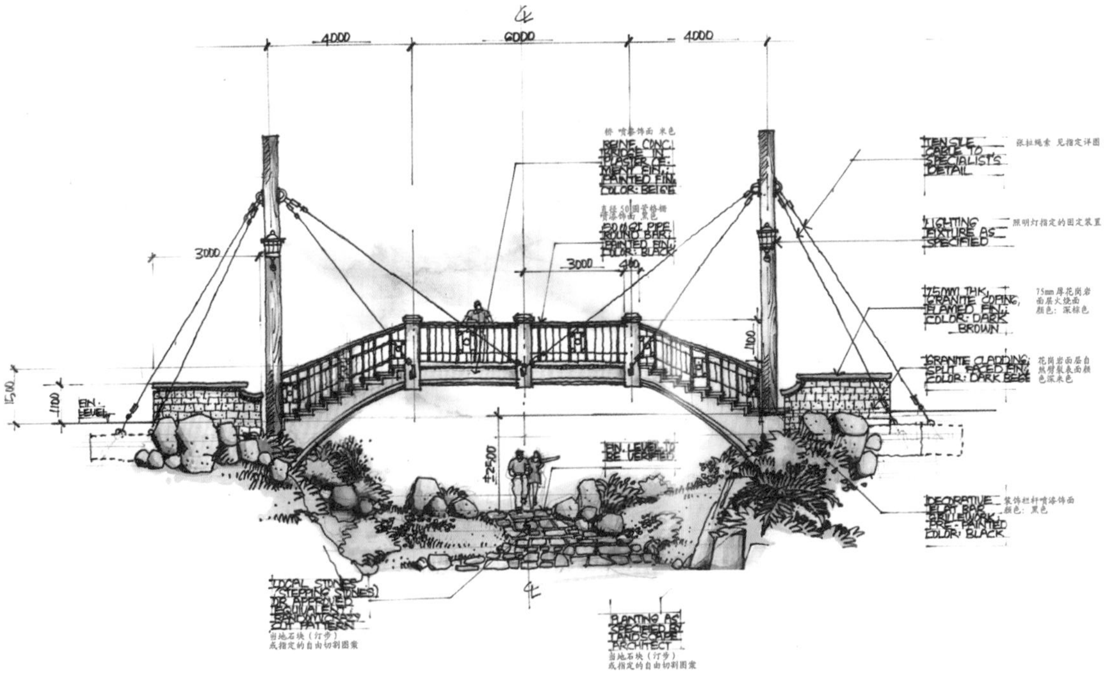
4000
6000
4000
3000
3000
1100
1500
1100
2500
桥 喷漆饰面 米色
REINF. CONC. BRIDGE IN PLASTER CEMENT FIN., PAINTED FIN. COLOR: BEIGE
PAINTED FIN. COLOR: BLACK
TENSILE CABLE TO SPECIALIST'S DETAIL
张拉绳索 见指定详图
LIGHTING FIXTURE AS SPECIFIED
照明灯指定的固定装置
75MM THK. GRANITE COPING, FLAMED FIN., COLOR: DARK BROWN
75mm厚花岗岩 面层火烧面 颜色：深棕色
GRANITE CLADDING, SPLIT FACED FIN., COLOR: DARK BEIGE
花岗岩面层自然劈裂表面颜色深米色
DECORATIVE FLAT BAR GRILLEWORK, PRE-PAINTED COLOR: BLACK
装饰栏杆喷漆饰面 颜色：黑色
FIN. LEVEL
FIN. LEVEL TO BE VERIFIED
当地石块（汀步）或指定的自由切割图案
PLANTING AS SPECIFIED BY LANDSCAPE ARCHITECT
当地石块（汀步）或指定的自由切割图案
A BRIDGE ELEV.
SCALE: 1:50M

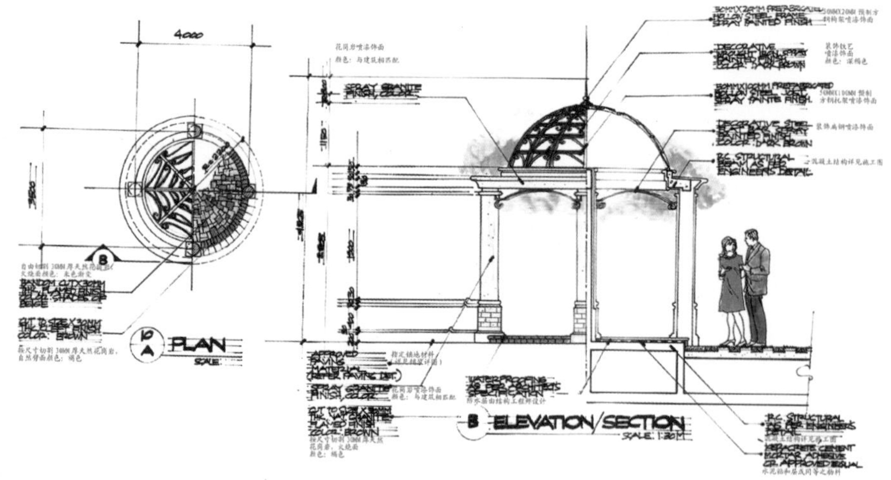

平面 / 屋顶平面

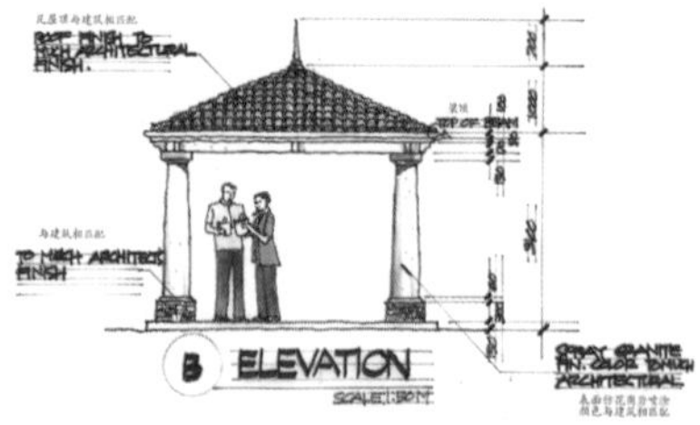

立面图

透视图

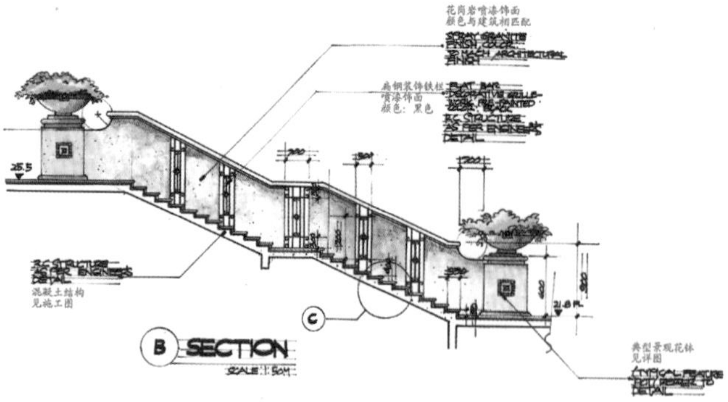

花钵示意性图片

台阶示意性图片

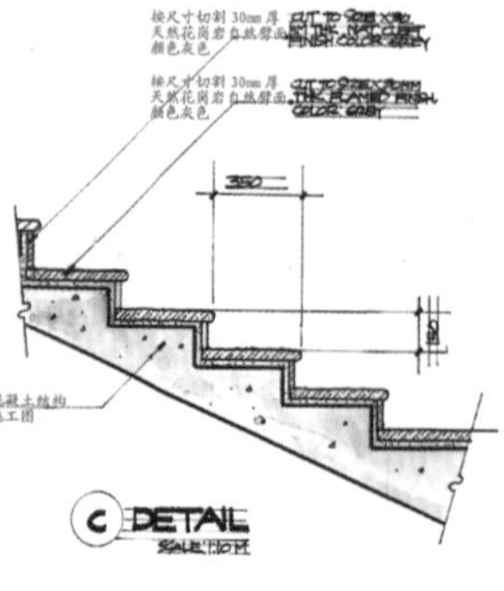